Anterior Cruciate Ligament Injuries Throughout the Life Span

Mustafa Karahan • Sachin R. Tapasvi
Annunziato Amendola
Can Doruk Basa
João Espregueira-Mendes
Jonathan D. Hughes
Editors

Anterior Cruciate Ligament Injuries Throughout the Life Span

Springer

ISAKOS
International Society of Arthroscopy, Knee Surgery and Orthopaedic Sports Medicine

Editors
Mustafa Karahan
Department of Orthopaedics
Acıbadem Mehmet Ali Aydinlar University
Istanbul, Türkiye

Annunziato Amendola
Department of Orthopaedic Surgery
Duke University Medical Center
Durham, NC, USA

João Espregueira-Mendes
Clínica Espregueira - FIFA Medical Centre of Excellence
Porto, Portugal

Sachin R. Tapasvi
Oyster & Pearl Hospital
Pune, Maharashtra, India

Can Doruk Basa
Tinaztepe University, Private Galen Hospital
Izmir, Türkiye

Jonathan D. Hughes
University of Pittsburgh/UPMC Freddie Fu Sports Medicine Center
Pittsburgh, PA, USA

ISBN 978-3-031-84500-0 ISBN 978-3-031-84501-7 (eBook)
https://doi.org/10.1007/978-3-031-84501-7

This Springer imprint is published by the registered company Springer Nature Switzerland AG
The registered company address is: Gewerbestrasse 11, 6330 Cham, Switzerland

Contents

Part I Basic Science

1 **Embryology and Histology of the Anterior Cruciate Ligament of the Knee in Relation to the Transition of Structure and Function Through Ages** 3
Nahum Rosenberg and Andrey A. Gritsyuk

2 **The Genetic Basis of ACL Injury** 9
Ehab M. Nazzal, Yunseo Linda Park, and Stephen J. Rabuck

3 **Hormones and ACL** 17
Vladan B. Stevanovic, Marko Bumbasirevic, Lazar Stijak, Marko Kadija, Zoran Bascarevic, and Miodrag Glisic

4 **Biomechanics of the Knee in Relation to ACL Rupture** 25
Nahum Rosenberg

5 **Natural History of ACL-Deficient Knee and Reconstructed Knee**. 31
Kuan-Hao Chen and Chih-Hwa Chen

Part II General Issues

6 **Physical and Psychosocial Characteristics of ACL Injury-Prone Groups** 43
Anna Nordenholm, Ramana Piussi, Jón Karlsson, and Eric Hamrin Senorski

7 **Frequency, Causality, and Accompanying Lesions of ACL Injuries in Regard to the Age Groups** 55
Can Doruk Basa and Elcil Kaya Bicer

8 **Graft Choices for Differing Age Groups, Physical Properties, and Sports**. 63
Patrick Djian and Marc Michnowski

9 **Pediatric ACL-Embryology and Development**. 81
Yuichi Hoshino and Yuta Nakanishi

10 **Pediatric ACL Tears: Natural History** 89
Paul Vincent Ryan

11 **Management of Pediatric ACL Injuries** 107
Melanio C. Acosta IV and Marc Anthony L. Castro

12 **Natural History and Prevention of ACL Injuries in Children** 123
Ariana Lott, Zachary Herman, and Bryson P. Lesniak

13 **Revision of Pediatric and Early Adolescent ACL Reconstruction After Re-injury** 133
Hazem A. Farouk, Ashraf Abdelkafy, and Hatem Galal Said

Part III Adult ACL Injuries

14 **Management of Adult ACL Injuries** 149
Philipp W. Winkler, Rafael Buerba Siller, Jan-Dierk Clausen, Adrian Deichsel, Benjamin B. Rothrauff, and Volker Musahl

15 **Age-Based Outcomes After Anterior Cruciate Ligament Reconstruction** 163
Richard Danilkowicz, Nathan L. Grimm, Jeffrey Liles, and Annunziato Amendola

16 **Management of Accompanying Pathologies with ACL Rupture in the Adult** 173
Volker Musahl, Abigail Boduch, Joseph D. Giusto, Ehab M. Nazzal, and Yoshiyuki Yahagi

17 **ACL Reconstruction and the Development of Early Arthrosis** 193
Richard Danilkowicz, Nathan L. Grimm, Jeffrey Liles, and Annunziato Amendola

18 **Risks and Complications of ACL Surgery** 201
Sachin R. Tapasvi, Amit Meena, and Anshu Shekhar

19 **Management of Revision ACL Reconstruction** 227
M. Süleyman Abul, Selim Ergün, and Mustafa Karahan

Part IV Female ACL Injuries

20 **Characteristics and Prevention of ACL Injuries in Females** 241
Yiğit Umur Cırdı, Ali Eray Günay, and Umut Akgün

Part V High Level Athletes ACL Injuries

21 **ACL Injuries in Female Athletes: The Ongoing Pandemic** 255
Rafael Serrano, Celeste Dix, Jose Maria Oliva Lozano, Sanjeev Malik, Nate Jones, and George Chiampas

22 **Management of ACL Injury for Different Sports Types** 265
Woo Chin Yee, Benedict Tan, and Yee Han Dave Lee

23 Return to Sports After ACL Reconstruction in High-Level Athletes 277
José Lito Mónico, Renato Andrade, Francesco Della Villa, Matthew Buckthorpe, Rogério Pereira, Fernando Fonseca, and João Espregueira-Mendes

24 Prevention of Anterior Cruciate Injuries in High-Level Athletes 299
Stephen E. Marcaccio, Joseph D. Giusto, and Jonathan D. Hughes

25 Risks and Complications Related to Anterior Cruciate Ligament Reconstruction in High-Level Athletes 307
Nicola Blucher and Brian M. Devitt

Part VI Young at Heart

26 Management of ACL Injuries in the Young at Heart (Elderly) Patients 327
Facundo Gigante, Manuel Perez Zabala, and Miguel Lapera

27 Management of ACL Tears in the Osteoarthritic Knee 345
R. D. Danilkowicz, Alberto Grassi, Ronald van Heerwaarden, and A. Amendola

Part VII Other Issues

28 Perioperative Concerns in ACL Reconstruction 353
Ignacio Cardona-Muñoz, Jesus I. Cardona-Medina, and Ariel De la Rosa

29 Bleeding and Clotting Control in ACL Reconstruction 359
Bertan Cengiz and Sinan Karaoglu

30 Prevention and Management of Infection in ACL Reconstruction 369
Anthony Truong and Christopher J. Vertullo

Part I

Basic Science

1 Embryology and Histology of the Anterior Cruciate Ligament of the Knee in Relation to the Transition of Structure and Function Through Ages

Nahum Rosenberg and Andrey A. Gritsyuk

1.1 Introduction

The anterior cruciate ligament (ACL) is essential for maintaining knee joint stability. Its structural disruption by degeneration is of significant concern. The embryological development and histological composition of the ACL are essential aspects of its function and vulnerability to injury. Understanding how this ligament forms during embryogenesis, its collagen-rich composition, and its hierarchical microstructure provides insights into its mechanical properties [1]. Moreover, knowledge of how the ACL changes with aging underscores the importance of injury prevention and rehabilitation strategies.

This overview presents the ACL embryonic origin and the subsequential factors contributing to ACL degeneration by age-related changes and following microtrauma. Understanding these factors can impact strategies to mitigate ACL degeneration and related injuries.

1.2 Embryology of ACL

The ACL, like other fibrous organs, is derived from mesenchymal origin [2]. Mesenchyme originates from the mesoderm as a loosely organized embryonic connective tissue containing mesenchymal stem cells (MSCs) [3] (Fig. 1.1).

During early embryogenesis, mesenchymal cells differentiate into connective tissues generating cells, including ligaments' fibroblasts. The ACL's development commences during early embryogenesis (in the tenth week of fetal development) when mesenchymal cells originate from the mesodermal layer of the limb bud. These undifferentiated cells subsequently migrate to the knee region, where they will give rise to the ACL [4].

During fetal development, various growth factors and signaling pathways, including transforming growth factor-beta (TGF-β), related to external mechanotransduction signaling, regulate the differentiation, proliferation, and alignment of fibroblasts [2]. This process is critical for establishing the ACL's unique structural characteristics.

1.2.1 Origin of Mesenchymal Cells

The ACL's mesenchymal cells originate from the paraxial and lateral plate mesoderm during the third to eighth weeks of gestation [2, 4].

N. Rosenberg (✉)
Specialists Center, National Insurance Institute, Haifa, Israel

A. A. Gritsyuk
Department of Traumatology, Orthopedics and Disaster Surgery, I.M. Sechenov First Moscow State Medical University, Moscow, Russia

M. Karahan et al. (eds.), *Anterior Cruciate Ligament Injuries Throughout the Life Span*,
https://doi.org/10.1007/978-3-031-84501-7_1

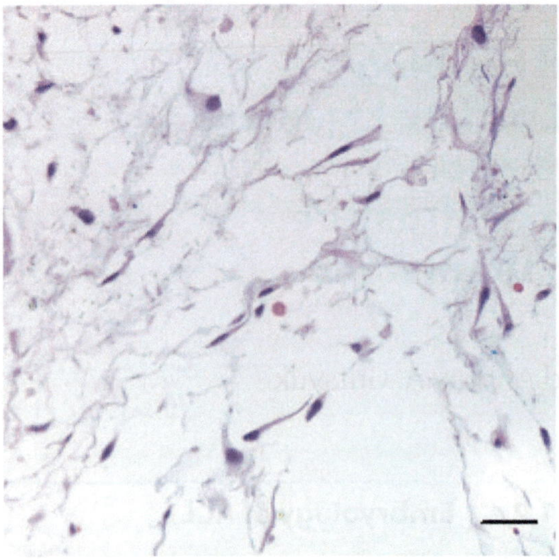

Fig. 1.1 Micrograph of mesenchymal tissue (Hematoxylin and Eosin stain, scale 50 micrometers)

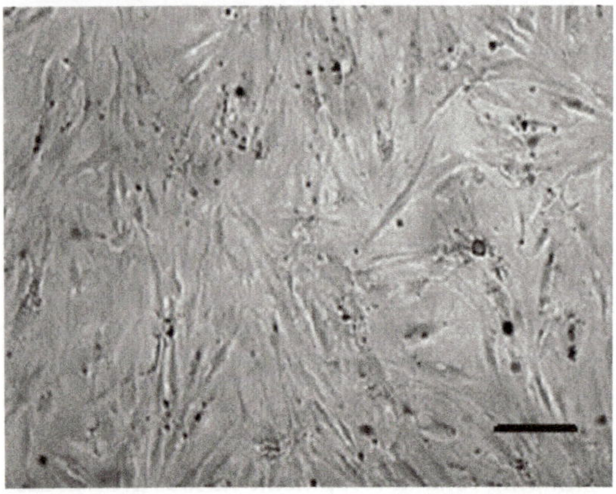

Fig. 1.2 Micrograph of mesenchymal stem cells (MSCs) in monolayer culture (scale 20 micrometers)

Mesenchymal cells have a spindle-shaped morphology with an elongated cell body. The cell membrane usually has multiple projections (filopodia and lamellipodia(, which are part of cell motility and interaction with the extracellular matrix (ECM) mechanisms [5]. These cells often exhibit a large, round, centrally located nuclei that contain one or more prominent nucleoli (Fig. 1.2). These cells migrate to the limb bud region and differentiate into fibroblasts, the primary cellular components of the ACL.

Complex molecular signaling pathways regulate the formation and differentiation of ACL precursor cells. Notably, the Wnt/β-catenin dependent and independent pathways have been identified as a critical player in limb, including ACL development, promoting fibroblast differentiation and collagen production [6]. The Wnt/β--catenin pathway is instrumental in directing mesenchymal cells toward ligament differentiation. This pathway orchestrates the expression of key transcription factors, such as Scleraxis (basic helix-loop-helix protein) and Mohawk, which are crucial for ligament development [7–10].

Growth factors like transforming growth factor-beta (TGF-β) also contribute to ACL formation. They stimulate cell proliferation, extracellular matrix synthesis, and tissue remodeling during embryogenesis [11].

1.2.2 Ligament Formation

The ACL begins to take shape around the ninth–tenth week of embryonic development. Initially, it is a loose and undifferentiated structure [4]. Then the fibroblasts start producing collagen, the primary structural protein of the ACL. Collagen fibers begin to organize, align, and interconnect, forming the mature ACL structure.

1.3 Histological Characteristics of the ACL

The ACL is a complex connective tissue structure with distinct histological features. Its composition and microstructure provide insights into its mechanical properties and function (Fig. 1.3).

The fetal ACL comprises two distinct bundles (anteromedial and posterolateral, separated by vascularized connective tissue) with high cellular content. This structure changes with age by reducing cellularity and subsequential decrease in ECM turnover and reduced ability for regeneration [12].

The ECM of ACL is characteristic of the general structure of ligaments and comprises primarily collagen, proteoglycans, and elastin [13].

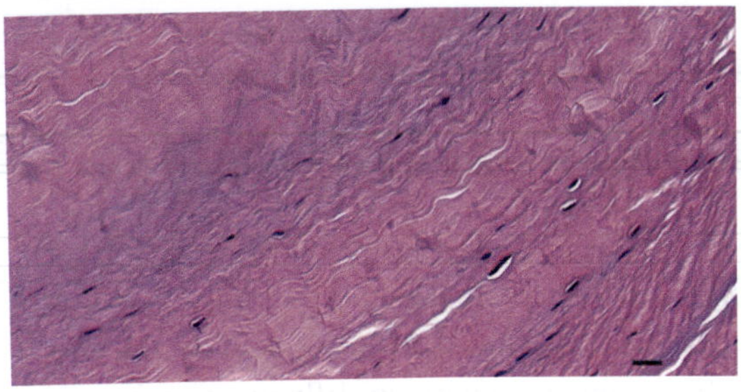

Fig. 1.3 A histological image of the normal anterior cruciate ligament (ACL): densely packed, parallel arrays of collagen fibers; interspersed among these collagen fibers are spindle-shaped ligamentocytes, aligned in the same direction as the collagen. Sparse vascular channels and nerve fibers are primarily located in the periphery, with the central portions of the ACL appearing relatively avascular and aneural (Hematoxylin and Eosin stain, scale 50 micrometers)

Collagen is the predominant ECM component of the ACL and is critical for its mechanical properties. Type I collagen is the most abundant collagen type within the ACL, comprising approximately 85–90% of its dry weight, with the addition of 10% of Type III collagen [13, 14]. Type I collagen is characterized by its high tensile strength, making it well suited for the ACL's load-bearing role. The collagen fibrils are hierarchically organized into fibrils, which bundle together to form fibers and, eventually, fascicles. This hierarchical structure allows the ACL to withstand high tensile forces. The crimp pattern of collagen fibrils is another key feature that allows the ACL to stretch under load before reaching its tensile limit [15].

Proteoglycans, which are also part of the ACL ECM, are molecules that are interspersed within the collagen matrix [16]. They consist of a core protein and long chains of glycosaminoglycans (GAGs). Proteoglycans, due to their electric charges, are responsible for retaining water within the ACL, contributing to its viscoelastic properties and lubrication.

Elastin is a fibrous protein that provides elasticity to the ACL. It is synthesized by ligamentocytes as tropoelastin, which then cross-link to form the mature elastin fiber network [17]. In the ACL, elastin makes up a relatively small fraction of the total dry weight compared to collagen. It is estimated to comprise around 1–2% of the dry weight [18]. Despite its lesser proportion, it contributes distinctive mechanical properties to the overall function of the ACL. Elastin in the ACL is primarily localized around individual collagen fibrils and within the endotendineum, the sheath-like structure that bundles collagen fibrils into fascicles [19]. This distribution suggests that elastin is responsible for enabling the ACL to return to its original shape after being subjected to mechanical strain at the elastic deformation range (a strain of ~2–4%) [19, 20], thereby contributing to the recovery and resilience of the ACL post motion [21].

The cellular component of ACL is comprised of ligamentocytes, which are fibroblasts embedded within the ECM and responsible for its maintenance by the synthesis, deposition, and remodeling of the ECM. The ligamentocytes are elongated and typically aligned along the direction of collagen fibers [22].

The ACL has a relatively poor blood supply, primarily from the middle genicular artery via the surrounding synovial tissue. The outer layer, or epiligament, exhibits the highest vascularization, while the core shows minimal vascularity [22]. This limited blood supply plays a significant role in the poor intrinsic healing potential of the ACL, and it makes the understanding of its histology critical for bioengineering solutions aiming to enhance ACL healing [23]. A fibrocartilaginous zone was identified five to ten millimeters proximal to the tibial ACL insertion in the anterior portion of the ACL; this part of the ACL is mostly avascular [24]. This limited vascularity reduces the ACL's ability to heal after injury and may partially explain its poor regenerative capacity.

The innervation of ACL includes sensory and sympathetic nerve fibers scattered throughout the ligament, predominantly located within the epiligament and surrounding adipose tissue. These nerve fibers are essential for proprioception and pain sensation [13].

Overall, the gross structure of the ACL exhibits a hierarchical organization of collagen fibers. At the macroscopic level, it appears as a cord-like structure. On a finer scale, the collagen fibers are organized into bundles or fascicles. These fascicles, in turn, consist of individual collagen fibrils, the smallest structural units of the ACL [22].

1.4 Aging and Degeneration

With age, the ACL undergoes histological changes. These changes can impact its mechanical properties and increase its susceptibility to injury. In older individuals above the age of 40 years, the ACL elastic modulus, ultimate tensile stress, and strain energy to failure are more than two times lower, indicating the intrinsic degeneration of the ACL [1].

Aging is associated with structural alterations in the ACL. Collagen fibers, its critical component, become disorganized and lose tensile strength over time. This results in changes in ligament elasticity and potentially increased vulnerability to injury [25] (Fig. 1.4).

Additionally, aging leads to decreased cellular activity within the ACL, impairing its ability to repair and regenerate. This effect might be attributed to the lower local content of MSCs. The diminished capacity for tissue remodeling contributes to the gradual degradation of the ACL [26].

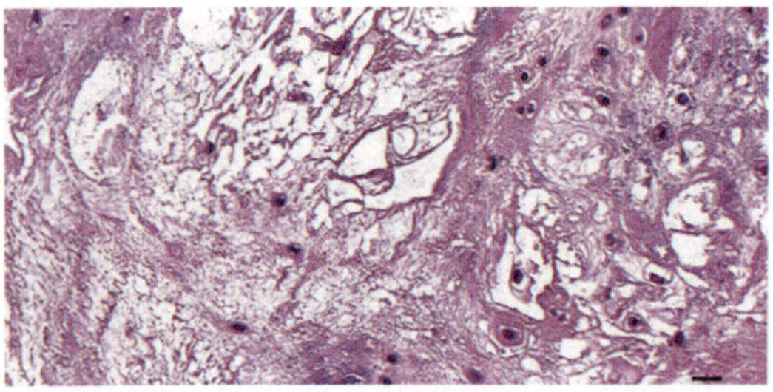

Fig. 1.4 A histological image of a degenerative anterior cruciate ligament (ACL). There is a noticeable disruption in the normally parallel and organized collagen fibers, which appear fragmented or disorganized. The ligamentocytes show signs of morphological changes, such as increased cellularity or a more rounded appearance, indicative of a cellular response to injury or degeneration. The extracellular matrix has a more disorganized structure due to an increased deposition of proteoglycans and changes in water content (Hematoxylin and Eosin stain, scale 50 micrometers)

1.5 Microtrauma-Induced Mechanical Overload

Repetitive microtrauma, such as that experienced by athletes in high-impact sports, can subject the ACL to mechanical overload. This chronic stress leads to microdamage accumulation and subsequential mechanical weakening over time [27]. Microtrauma triggers an inflammatory response within the ACL. Chronic inflammation can lead to the breakdown of collagen fibers and other extracellular matrix components, compromising ACL integrity [28].

Accordingly, understanding the impact of age-related changes and microtrauma on ACL degeneration is very important for assessing injury risk, especially in high-risk populations like older individuals and athletes. This knowledge can guide injury prevention programs.

1.6 Conclusion

ACL injuries are widespread in sports medicine and orthopedic practice; thus, exploring its histological composition aids in injury prevention and treatment. Therefore, tailoring treatment and rehabilitation strategies to address age-related and microtrauma-induced degeneration can improve outcomes [29].

Future clinical methods aimed at enhancing collagen synthesis and reducing inflammation via gene therapies are promising avenues for clinical intervention [30]. These methods are still mostly in the basic sciences stage but have promising therapeutic prospects.

The recognition of embryonic formation and further development and maintenance by the optimal cellular, humoral (related to body fluids), and biomechanical factors of the ACL can guide injury prevention strategies. ACL injuries are prevalent in athletes, and understanding its developmental vulnerabilities can inform preventive measures.

In tissue engineering, insights into ACL embryology can guide efforts to develop regenerative therapies for ACL injuries. By mimicking the embryonic environment with progenitor cells and signaling pathways, the foreseen clinical aim is to enhance ACL repair and regeneration [31].

Overall, ACL degeneration is a complex process influenced by age-related changes and microtrauma. Aging leads to structural and cellular alterations in the ACL, while microtrauma-induced degeneration involves mechanical overload and chronic inflammation. Recognizing these factors and their clinical implications is vital for injury prevention and developing effective treatment strategies. Further research into the cellular and molecular mechanisms governing ACL development and homeostasis may lead to innovative approaches for addressing ACL injuries and improving patient outcomes.

References

1. Noyes FR, Grood ES. The strength of the anterior cruciate ligament in humans and rhesus monkeys. J Bone Joint Surg. 1976;58(8):1074–82.
2. Benjamin M. The fascia of the limbs and back—a review. J Anat. 2002;201(4):455–67.
3. Langman J. Medical embryology. Baltimore: The Williams & Wilkins Company; 1981. p. 60, 123.
4. Mérida-Velasco JA, Sánchez-Montesinos I, Espín-Ferra J, Mérida-Velasco JR, Rodríguez-Vázquez JF, Jiménez-Collado J. Development of the human knee joint ligaments. Anat Rec. 1997;248(2):259–68. https://doi.org/10.1002/(SICI)1097-0185(199706)248:2<259::AID-AR13>3.0.CO;2-O. PMID: 9185992.
5. Discher DE, Mooney DJ, Zandstra PW. Growth factors, matrices, and forces combine and control stem cells. Science. 2009;324(5935):1673–7.
6. Geetha-Loganathan P, Nimmagadda S, Scaal M. Wnt signaling in limb organogenesis. Organogenesis. 2008;4(2):109–15. https://doi.org/10.4161/org.4.2.5857. PMID: 19279722; PMCID: PMC2634256.
7. Cserjesi P, Brown D, Ligon KL, Lyons GE, Copeland NG, Gilbert DJ, Jenkins NA, Olson EN. Scleraxis: a basic helix-loop-helix protein that prefigures skeletal formation during mouse embryogenesis. Development. 1995;121(4):1099–110. https://doi.org/10.1242/dev.121.4.1099. PMID 7743923.
8. Shukunami C, Takimoto A, Nishizaki Y, Yoshimoto Y, Tanaka S, Miura S, Watanabe H, Sakuma T, Yamamoto T, Kondoh G, Hiraki Y. Scleraxis is a transcriptional activator that regulates the expression of Tenomodulin, a marker of mature tenocytes and ligamentocytes. Sci Rep. 2018;8(1):3155. https://doi.org/10.1038/s41598-018-21194-3. PMID: 29453333; PMCID: PMC5816641.
9. Garcia-Lee V, Díaz-Hernandez ME, Chimal-Monroy J. Inhibition of WNT/β-catenin is necessary and sufficient to induce Scx expression in developing tendons of chicken limb. Int J Dev Biol. 2021;65(4-5-6):395–401. https://doi.org/10.1387/ijdb.200166jc. PMID: 32930359.
10. Cully M. Mohawk is downregulated in ACLs from knees with OA. Nat Rev Rheumatol. 2013;9:384. https://doi.org/10.1038/nrrheum.2013.88].
11. Massagué J. TGFβ signaling in context. Nat Rev Mol Cell Biol. 2012;13(10):616-630.
12. Ferretti M, Levicoff EA, Macpherson TA, Moreland MS, Cohen M, Fu FH. The fetal anterior cruciate ligament: an anatomic and histologic study. Arthroscopy. 2007;23(3):278–83. https://doi.org/10.1016/j.arthro.2006.11.006. PMID: 17349471.
13. Duthon VB, Barea C, Abrassart S, Fasel JH, Fritschy D, Ménétrey J. Anatomy of the anterior cruciate ligament. Knee Surg Sports Traumatol Arthrosc. 2006;14(3):204–13. https://doi.org/10.1007/s00167-005-0679-9. Epub 2005 Oct 19. PMID: 16235056.
14. Morales-Avalos R, Torres-González EM, Padilla-Medina JR, Monllau JC. ACL anatomy: is there still something to learn? Rev Esp Cir Ortop Traumatol. 2023;12:S1888-4415(23)00070-X. English, Spanish. https://doi.org/10.1016/j.recot.2023.02.005. Epub ahead of print. PMID: 36787832.
15. Woo SL, Hollis JM, Adams DJ, Lyon RM, Takai S. Tensile properties of the human femur-anterior cruciate ligament-tibia complex: the effects of specimen age and orientation. Am J Sports Med. 1991;19(3):217–25.
16. Kannus P. Structure of the tendon connective tissue. Scand J Med Sci Sports. 2000;10(6):312–20.
17. Kielty CM, Sherratt MJ, Shuttleworth CA. Elastic fibres. J Cell Sci. 2002;115(14):2817–28.
18. Amiel D, Frank C, Harwood F, Fronek J, Akeson W. Tendons and ligaments: a morphological and biochemical comparison. J Orthop Res. 1984;1(3):257–65.
19. Smith KD, Vaughan-Thomas A, Spiller DG, Innes JF, Clegg PD, Comerford EJ. The organisation of elastin and fibrillins 1 and 2 in the cruciate ligament complex. J Anat. 2011;218(6):600–7. https://doi.org/10.1111/j.1469-7580.2011.01374.x. Epub 2011 Apr 6. PMID: 21466551; PMCID: PMC3125893
20. Rosenberg N. Introductory chapter: Tendons – trauma, inflammation, degeneration, and treatment [Internet]. In: Tendons - trauma, inflammation, degeneration, and treatment. IntechOpen; 2023. Available from:. https://doi.org/10.5772/intechopen.110708.
21. Marieswaran M, Jain I, Garg B, Sharma V, Kalyanasundaram D. A review on biomechanics of anterior cruciate ligament and materials for reconstruction. Appl Bionics Biomech. 2018;2018:4657824. https://doi.org/10.1155/2018/4657824. PMID: 29861784; PMCID: PMC5971278.
22. Benjamin M, Ralphs JR. Fibrocartilage in tendons and ligaments--an adaptation to compressive load. J Anat. 1998;193(Pt 4):481–94. https://doi.org/10.1046/

j.1469-7580.1998.19340481.x. PMID: 10029181; PMCID: PMC1467873.
23. Dahners LE, Mullis BH. Effects of nonsteroidal anti-inflammatory drugs on bone formation and soft-tissue healing. J Am Acad Orthop Surg. 2000;8(3):190–200.
24. Petersen W, Tillmann B. Structure and vascularization of the cruciate ligaments of the human knee joint. Anat Embryol (Berl). 1999;200(3):325–34. https://doi.org/10.1007/s004290050283. PMID: 10463347.
25. Strocchi R, De Pasquale V, Facchini A, Raspanti M, Zaffagnini S, Marcacci M. Age-related changes in human anterior cruciate ligament (ACL) collagen fibrils. Ital J Anat Embryol. 1996;101(4):213–20. PMID: 9203869.
26. Cheng M, Johnson VM, Murray MM. Effects of age and platelet-rich plasma on ACL cell viability and collagen gene expression. J Orthop Res. 2012;30(1):79–85. https://doi.org/10.1002/jor.21496. Epub 2011 Jul 11. PMID: 21748791; PMCID: PMC3202019.
27. Woo SL, Abramowitch SD, Kilger R, Liang R. Biomechanics of knee ligaments: injury, healing, and repair. J Biomech. 2006;39(1):1–20.
28. Hasegawa A, Nakahara H, Kinoshita M, et al. Cellular and extracellular matrix changes in anterior cruciate ligaments during human knee aging and osteoarthritis. Arthritis Res Ther. 2013;15:R29. https://doi.org/10.1186/ar4165.
29. Hewett TE, Myer GD, Ford KR, et al. Biomechanical measures of neuromuscular control and valgus loading of the knee predict anterior cruciate ligament injury risk in female athletes: a prospective study. Am J Sports Med. 2006;34(2):197–206.
30. Amini M, Venkatesan JK, Liu W, Leroux A, Nguyen TN, Madry H, Migonney V, Cucchiarini M. Advanced gene therapy strategies for the repair of ACL injuries. Int J Mol Sci. 2022;23(22):14467. https://doi.org/10.3390/ijms232214467. PMID: 36430947; PMCID: PMC9695211.
31. Wang C, Hu Y, Zhang S, Ruan D, Huang Z, He P, Cai H, Heng BC, Chen X, Shen W. Application of stem cell therapy for ACL graft regeneration. Stem Cells Int. 2021;2021:6641818. https://doi.org/10.1155/2021/6641818. PMID: 34381504; PMCID: PMC8352687.

The Genetic Basis of ACL Injury

2

Ehab M. Nazzal, Yunseo Linda Park, and Stephen J. Rabuck

2.1 Introduction

There are a whole host of factors to consider when discussing the genetic basis of anterior cruciate ligament (ACL) injury. These range from factors related to biomechanics and biology, but also include sex-related factors, genetic markers, and heritability. Unraveling these specific genetic factors is crucial to understanding how they contribute to injury and provide the underpinnings of optimal prevention and treatment strategies.

2.2 Joint Hypermobility

One commonly discussed factor regarding ACL injury is joint hypermobility. Joint hypermobility is widely recognized as a significant risk factor for ACL injury. The American Physical Therapy Association's guidelines highlight familial predisposition as a risk factor for ACL injuries, indicating a genetic component to joint hypermobility and ligamentous injuries [1]. Generalized joint hypermobility refers to joint range of motion that exceeds normal limits, often due to alterations in collagen fibers that are essential for maintaining the structure of connective tissue. This increased hypermobility can result from heritable diseases such as Marfan and Ehlers-Danlos syndromes or may simply be a benign hypermobility finding. Variants in collagen genes and inflammatory pathway genes have also been associated with increased joint hypermobility [2–4]. Joint hypermobility is significantly more prevalent among females than males, with a study of college athletes showing 33.1% of females with a Beighton score ≥5, compared to 13.6% of males [5]. Young age has also been reported to be associated with joint hypermobility, with decreasing mobility seen with increased age [6]. Joint hypermobility can be evaluated using the Beighton scale, which can help assess risk of re-injury, guide individualized rehabilitation, and inform preventative strategies both before and after injury.

The association between joint hypermobility and risk of ACL injury is strongly supported by evidence. Studies have found that knee hyperextension and side-to-side differences in anterior tibial translation significantly increase the odds of ACL injury in young female athletes [7]. A 2015 study compared the non-injured knees of patients with non-contact ACL injuries to knees of a healthy cohort. Results showed that the ACL injury group had higher average anterior and internal rotational knee laxity, suggesting that increased knee laxity is a predisposing factor for ACL injuries [8]. Another study found that greater side-to-side differences in anterior knee laxity predicted medial and lateral meniscal injuries in patients with ACL injuries [9], indicating

E. M. Nazzal · Y. L. Park · S. J. Rabuck (✉)
Department of Orthopaedic Surgery, University of Pittsburgh Medical Center, Pittsburgh, PA, USA
e-mail: Rabucksj@upmc.edu

M. Karahan et al. (eds.), *Anterior Cruciate Ligament Injuries Throughout the Life Span*,
https://doi.org/10.1007/978-3-031-84501-7_2

that joint hyperlaxity can impact not only the ACL but also concomitant structures in the knee. It is hypothesized that this joint hyperlaxity amplifies the forces exerted on intra-articular knee ligaments, particularly during the tibiofemoral rotational and valgus moments typically associated with ACL injuries.

2.3 Sex Differences in ACL Injury

A discussion of the genetic basis of ACL injury would be incomplete without a summary of the differences examined between males and females. The risk of ACL injury among the male and female sex is perhaps the most often studied topic in the category of the genetics of ACL injury. These differences are mainly noted when comparing males and females within the same sport, where studies have demonstrated a two to eight times increased risk of ACL injury among the female population [10–13]. Many hypotheses have been proposed to explain this discrepancy, including increasing trends in female sports participation [14], hormonal differences, and the biomechanical and neuromuscular differences between males and females [10]. Though all of these are important to consider in the factor of sex as a genetic basis of ACL injury, yet the latter two are most relevant to the goals of this discussion.

2.3.1 Hormonal Variation

Joint hypermobility and ligamentous laxity are generally higher in females compared to males. While the reasons for this are multifactorial, one main contributing factor is the sex-specific hormonal differences. In particular, higher concentrations of serum estrogen and relaxin are present in females compared to males [15]. The consequence of higher estrogen levels in females compared to males is not entirely understood; however, one theory is that estrogen affects ligamentous metabolism, resulting in decreased stiffness and increased ligamentous laxity [16–18]. This has been demonstrated in an animal model study, where pubic ligaments were shown to have estrogen-dependent collagenase production [19], as well as an ACL tissue-culture model study that showed increasing estrogen concentrations resulted in decreased fibroblast and procollagen production [20]. This increased ligamentous laxity, while serving as a potential benefit in increasing flexibility and facilitating a more successful childbirth [18], may explain the higher risk of female ACL injury even among athletes participating in the same sports.

Hormonal differences have been investigated by evaluating ACL injury patterns within females during different phases of the menstrual cycle. Small-scale studies have looked at hormonal levels during specific times of the female menstrual cycle. An early study in 2002 investigated hormone levels of female athletes via saliva samples and found that a significantly greater number of ACL injuries occurred during the first 2 days of the female menstrual cycle [21]. Additionally, a 2013 article relating ACL injury to the menstrual cycle in female recreational skiers reported that injuries were 2.4 times more likely in the preovulatory phase (follicular and ovulatory) compared to the post-ovulatory phase (luteal), implicating higher estrogen levels as the main contributing factor [22]. This variation in injury risk with the menstrual cycle was supported in recent meta-analyses, which have demonstrated an effect of the menstrual cycle on anteroposterior knee laxity, especially during the preovulatory phase [23, 24]. While these findings are certainly interesting, they come from small studies and further research is necessary. However, this information highlights an important genetic factor in ACL injury risk and provides important diagnostic and counseling information for clinicians and female athletes (Fig. 2.1).

2.3.2 Biomechanics and Anatomic Differences

In addition to hormonal differences, a wide range of biomechanical and neuromuscular factors that are intrinsically different between males and

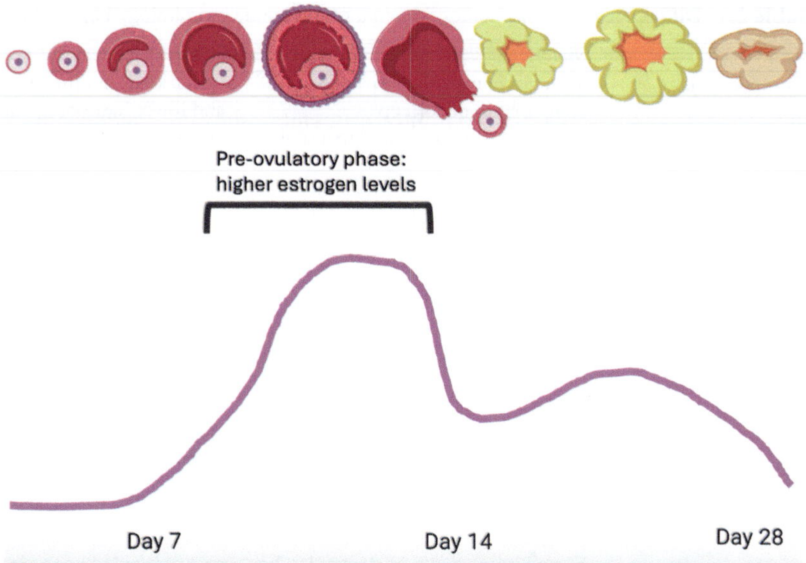

Fig. 2.1 Estrogen levels throughout a menstrual cycle. Elevated estrogen levels during the ovulatory phase of the menstrual cycle increase ACL injury risk by enhancing ligament laxity, altering neuromuscular control, and reducing ligament mechanical strength. These factors result in females being more susceptible to ACL injuries during this phase. ACL anterior cruciate ligament [22]

females may also contribute to the risk of ACL injury. Table 2.1 summarizes the differences in biomechanics, kinematics, and bone morphology.

Much attention has been placed on phases of jumping, landing, and side-to-side movement. Previous biomechanical research has shown that greater lower extremity valgus, mainly knee abduction during landing, is related to increased risk of ACL injury [25]. Additionally, increased knee valgus during cutting maneuvers by young females may contribute to risk of ACL injury [26]. Other knee-specific factors such as knee stiffness where females land with a low knee flexion angle and higher peak vertical ground reaction force can predispose them to ACL injury [27]. Increased anterior tibial translation in females [28] is likely related to several factors; however, it results in increased stress on the ACL and injury susceptibility.

Hip rotation and mechanics also may influence the risk of ACL injury with males demonstrating greater hip abduction angles than females during sport maneuvers [29], and the resultant increased external hip adduction moment coupled with the dynamic valgus moment of the knee especially in landing may explain the increased susceptibility to ACL injury in females [30–32].

Lastly, bone morphology, specifically at the intercondylar notch, has been implicated as a risk factor for ACL injuries. Decreased femoral intercondylar width has been associated with increased ACL injury, with females having more narrow intercondylar widths compared to their male counterparts [33]. This relationship between notch width and ACL injury has also been demonstrated in all-female comparisons, where one evaluation of a military cohort found that women with a notch width of less than 13 mm had a 16.8 times increased risk of ACL injury compared to those with wider notches [34]. The interplay of sex, intercondylar width, and ACL injury may be multifactorial, yet the prevailing theory is that smaller intercondylar notches may result in increased contact between the lateral notch and the ACL, resulting in increased strain on the ACL and thus greater injury risk [35].

2.3.3 Sex-Specific Differences in Neuromuscular Coordination

The differences in neuromuscular coordination between males and females, along with their implications for ACL injury, have been extensively studied and could warrant a separate

Table 2.1 Differences in biomechanics, kinematics, and bone morphology between males and females

	Male	Female
Biomechanics	Stronger hamstring activation compared to quadriceps Greater hip abductor and external rotator strength Less ligament laxity; greater stability Higher internal and valgus stiffness at various degrees of knee flexion More inter-limb asymmetries in muscle activity timing	Quadriceps dominant: higher quadriceps activation and lower hamstring activation Weaker hip abductors and external rotators, contributing to knee valgus Increased ligament laxity, particularly during menstruation Greater internal and valgus laxity from 0 to 50 degrees of flexion and greater anterior laxity at 50 degrees of flexion Commonly land with more extended knees and hips Fewer inter-limb differences in muscle activity patterns
Kinematics	Less knee valgus during dynamic movement Less internal rotation of tibia during movement More knee flexion during cutting and pivoting Flexed knees and hips during landing	Greater knee abduction angles and less knee adduction angles during high-impact activities Greater relative ACL elongation during athletic activities Increased internal rotation of tibia Less knee flexion during landing, commonly with more extended knees and hips; stiffer landing Upright trunk and hip posture; higher knee valgus
Bone morphology	Less valgus alignment Larger, more circular femoral notch Flatter posterior tibial slope Wider tibial plateau More vertical femoral shaft Patella alta less common	Greater Q-angle contributing to greater knee valgus and stress on ACL Narrower femoral notch; reduced space for ACL; increased risk of ACL impingement and stress Steeper posterior tibial slope, increasing anterior tibial translation and ACL strain Shallower medial tibial plateau, increasing contact stress on ACL Shorter tibial spines; less effective restraint against anterior tibial translation Lower tibiofemoral joint congruence; increased instability Patella alta more common

ACL anterior cruciate ligament, *Q-angle* quadriceps angle

chapter of their own. Generally, the main sex-specific differences are related to varied muscle coactivation patterns at the knee, specifically the quadriceps and hamstring. Females have a higher quadriceps-to-hamstring activation ratio compared to their male counterparts. This is important because preferential activation of the quadriceps in females increases the loading of the ACL with anterior shear forces, placing the ACL at risk for injury [36]. These findings have also been seen during landing patterns and side-cutting activities, in which biomechanical studies have shown both greater peak quadriceps activity and slower hamstring activation, resulting in increased joint compression forces, anterior tibial translation, and increased strain on the ACL [37–39]. As we will discuss in a later section, understanding these sex-specific differences is important, as addressing imbalances for female athletes is an important area of ACL injury prevention.

2.4 The Genomic Sequence of ACL Injury

While there has been extensive research on sex-specific differences as a genetic basis of ACL injury, the literature on specific genetic variants influencing ACL injury is much less understood. Perhaps the most implicated gene factor associated with risk of ACL injury is for collagen.

Collagen is a family of proteins with 28 different subtypes, comprising approximately 30% of total protein mass in mammals, providing structural support, and facilitating important mechanical function [40]. In the ACL, 85% of the collagen present is type I, and has been the focus of genetic research regarding ACL injury [41].

The first investigation in the role of genetic polymorphisms of type I collagen on ACL injury was in the early 2000s, with a description of the association of ACL injury from non-contact mechanisms with a specific chromosomal region of the *COL1A1* gene, irrespective of sex differences [42]. Subsequent studies showed an association with higher frequency of a specific *COL1A1* haplotype and reduced risk of ACL injury in professional soccer players [43], reaffirming the genetic basis of ACL injury on the molecular level. Apart from type I collagen, other forms of collagen have been implicated in the risk of ACL injury, including type III [44, 45], type V [44–47], type XII [44, 48], type XIV [47], and type XV collagen [47].

While other genetic polymorphisms have been investigated, including matrix metalloproteinases [42, 49, 50], growth differentiation factor 5 [51], elastin/fibrillin [52], and proteins involved in angiogenesis such as Vascular Endothelial Growth Factor Isoform-A (VEGFA) and Kinase Insertion Domain Receptor [53], these genetic associations are much less robust and should be considered with caution when determining the true genetic basis of ACL injury.

2.5 ACL Injury: A Heritable Trait

Lastly, heritability and familial differences have been investigated as evidence for the genetic basis of ACL injury. Research evaluating the role of familial differences in ACL injury dates back to the late 1990s from the University of Pittsburgh, as surgeons examining patients with bilateral ACL injuries found a significant difference in the incidence of family history of ACL injury, indicating a possible congenital risk [54]. Similarly, a 2016 study evaluated a Swedish ski team and found a family predisposition to ACL injury between team members and their parents [55].

While these studies started the discussion on the heritability of ACL injury, they relied mainly on reports of family history and are subject to various levels of bias. More recent heritability studies have attempted to further elucidate the role of familial predisposition to ACL injury. In 2020, a twin study over a 30-year period was performed within the Swedish registry, and a genetic contribution to ACL injury of just under 70% was found, demonstrating the possible contribution of familial clustering and ACL injury [56]. Another twin study performed among female twins in high-risk sports found similar findings, citing multiple shared biomechanical factors as the potential reason for increased risk of ACL injury [57]. More recently, in 2022, a systematic review with meta-analysis supported these findings, showing that a family history of ACL injury more than doubled the odds of sustaining a primary or subsequent ACL injury [58]. These studies show associations at best and not true causation. They do support the theory of heritability in ACL injury and are a starting point for further genomic evaluations to elucidate the genetic basis of ACL injury.

2.6 Future Directions and Implications

As the genetic basis of ACL injury evolves, further research will need to translate how genetics can change the management of ACL injury.

First, as many of the studies cited in this chapter have discussed, there is a powerful role of diagnosis in determining which athletes are at risk of ACL injury. Understanding the genetics at play has shown that female athletes, particularly those with poor neuromuscular control with landing and side-cutting mechanisms, should be cautioned about their risk of ACL injury. Additionally, while family history of ACL injury is not a causal factor in predicting future ACL injury, it is an important piece of information that should be included when counseling athletes and their family members. At this time, there is no indication for genome sequencing as a method of risk stratification for ACL injury, but as future investigations and prospective studies further delineate the role of these gene polymorphisms, they may play a role in diagnosis, treatment, and prevention.

Regarding treatment options, the genetic basis of ACL injury may become an important factor to consider when looking at graft options and evaluating possible barriers to a successful ACL reconstruction. Surgeons should consider the different muscle activation patterns between females and males when deciding whether the hamstring autograft is appropriate, or whether the quadriceps tendon or bone-patella tendon-bone autograft are favorable. Lastly, genomic sequencing and the role of collagen quality in ACL injury may inform surgeons on which patients to consider alternative graft options.

Lastly, the genetic basis of ACL injury has informed and will continue to educate athletes, surgeons, and physiotherapists alike on how to avoid ACL injury, as well as on rehabilitation after ACL reconstruction. Understanding the importance of muscle coactivation, core and hip rotational control, and proper landing mechanics that may differ between individuals based on their genetic makeup is invaluable, and helping athletes develop an optimized, strengthened kinetic chain may prevent ACL injury or subsequent injury after ACL reconstruction.

2.7 Conclusion

The genetic basis of ACL injury is a complicated topic that is multifactorial in nature. Our knowledge of the genetic factors contributing to an increased risk of ACL injury in athletes has grown and covers a wide breadth of topics from sex differences, biomechanics and neuromuscular activation, genetic markers, and heritability. While we may not be able to change the nonmodifiable components of genetics in regard to ACL injury, having an understanding of these factors allows for improved counseling of athletes, potential optimization of treatment strategies, and an enhanced rehabilitation system after ACL injury.

References

1. Logerstedt DS, Scalzitti D, Risberg MA, Engebretsen L, Webster KE, Feller J, et al. Knee stability and movement coordination impairments: knee ligament sprain revision 2017. J Orthop Sports Phys Ther. 2017;47(11):A1–A47.
2. Beckley S, Dey R, Stinton S, van der Merwe W, Branch T, September AV, et al. Investigating the association between COL1A1 and COL3A1 gene variants and knee joint laxity and ligament measurements. Clin Biomech (Bristol, Avon). 2022;100:105822.
3. Malecki K, Fabis-Strobin A, Salacinska K, Kwas K, Stelmach W, Beczkowski J, et al. Clinical significance of polymorphisms of genes encoding collagen (COL1A1, COL5A1) and their correlation with joint laxity and recurrent patellar dislocation in adolescents. Sci Rep. 2023;13(1):22300.
4. Suijkerbuijk MAM, Ponzetti M, Rahim M, Posthumus M, Hager CK, Stattin E, et al. Functional polymorphisms within the inflammatory pathway regulate expression of extracellular matrix components in a genetic risk dependent model for anterior cruciate ligament injuries. J Sci Med Sport. 2019;22(11):1219–25.
5. Decoster LC, Bernier JN, Lindsay RH, Vailas JC. Generalized joint hypermobility and its relationship to injury patterns among NCAA lacrosse players. J Athl Train. 1999;34(2):99–105.
6. Jansson A, Saartok T, Werner S, Renstrom P. General joint laxity in 1845 Swedish school children of different ages: age- and gender-specific distributions. Acta Paediatr. 2004;93(9):1202–6.
7. Myer GD, Ford KR, Paterno MV, Nick TG, Hewett TE. The effects of generalized joint laxity on risk of anterior cruciate ligament injury in young female athletes. Am J Sports Med. 2008;36(6):1073–80.
8. Mouton C, Theisen D, Meyer T, Agostinis H, Nuhrenborger C, Pape D, et al. Noninjured knees of patients with noncontact ACL injuries display higher average anterior and internal rotational knee laxity compared with healthy knees of a noninjured population. Am J Sports Med. 2015;43(8):1918–23.
9. Nakamae A, Sumen Y, Tsukisaka K, Deie M, Fujimoto E, Ishikawa M, et al. A larger side-to-side difference in anterior knee laxity increases the prevalence of medial and lateral meniscal injuries in patients with ACL injuries. Knee Surg Sports Traumatol Arthrosc. 2022;30(5):1560–7.
10. Devana SK, Solorzano C, Nwachukwu B, Jones KJ. Disparities in ACL reconstruction: the influence of gender and race on incidence, treatment, and outcomes. Curr Rev Musculoskelet Med. 2022;15(1):1–9.
11. Agel J, Rockwood T, Klossner D. Collegiate ACL injury rates across 15 sports: National Collegiate Athletic Association Injury Surveillance System Data Update (2004-2005 through 2012-2013). Clin J Sport Med. 2016;26(6):518–23.
12. Arendt E, Dick R. Knee injury patterns among men and women in collegiate basketball and soccer.

NCAA data and review of literature. Am J Sports Med. 1995;23(6):694–701.
13. Hewett TE, Ford KR, Myer GD. Anterior cruciate ligament injuries in female athletes: part 2, a meta-analysis of neuromuscular interventions aimed at injury prevention. Am J Sports Med. 2006;34(3):490–8.
14. Stracciolini A, Amar-Dolan L, Howell DR, Alex T, Berkner P, Sandstrom NJ, et al. Female sport participation effect on long-term health-related quality of life. Clin J Sport Med. 2020;30(6):526–32.
15. Herzberg SD, Motu'apuaka ML, Lambert W, Fu R, Brady J, Guise JM. The effect of menstrual cycle and contraceptives on ACL injuries and laxity: a systematic review and meta-analysis. Orthop J Sports Med. 2017;5(7):2325967117718781.
16. Legerlotz K, Nobis T. Insights in the effect of fluctuating female hormones on injury risk-challenge and chance. Front Physiol. 2022;13:827726.
17. Wojtys EM, Huston LJ, Boynton MD, Spindler KP, Lindenfeld TN. The effect of the menstrual cycle on anterior cruciate ligament injuries in women as determined by hormone levels. Am J Sports Med. 2002;30(2):182–8.
18. Chidi-Ogbolu N, Baar K. Effect of estrogen on musculoskeletal performance and injury risk. Front Physiol. 2018;9:1834.
19. Wahl LM, Blandau RJ, Page RC. Effect of hormones on collagen metabolism and collagenase activity in the pubic symphysis ligament of the Guinea pig. Endocrinology. 1977;100(2):571–9.
20. Liu SH, Al-Shaikh RA, Panossian V, Finerman GA, Lane JM. Estrogen affects the cellular metabolism of the anterior cruciate ligament. A potential explanation for female athletic injury. Am J Sports Med. 1997;25(5):704–9.
21. Slauterbeck JR, Fuzie SF, Smith MP, Clark RJ, Xu K, Starch DW, et al. The menstrual cycle, sex hormones, and anterior cruciate ligament injury. J Athl Train. 2002;37(3):275–8.
22. Lefevre N, Bohu Y, Klouche S, Lecocq J, Herman S. Anterior cruciate ligament tear during the menstrual cycle in female recreational skiers. Orthop Traumatol Surg Res. 2013;99(5):571–5.
23. Zazulak BT, Paterno M, Myer GD, Romani WA, Hewett TE. The effects of the menstrual cycle on anterior knee laxity: a systematic review. Sports Med. 2006;36(10):847–62.
24. Belanger L, Burt D, Callaghan J, Clifton S, Gleberzon BJ. Anterior cruciate ligament laxity related to the menstrual cycle: an updated systematic review of the literature. J Can Chiropr Assoc. 2013;57(1):76–86.
25. Hewett TE, Myer GD, Ford KR, Heidt RS, Colosimo AJ, McLean SG, et al. Biomechanical measures of neuromuscular control and Valgus loading of the knee predict anterior cruciate ligament injury risk in female athletes: a prospective study. Am J Sports Med. 2005;33(4):492–501.
26. McLean SG, Lipfert SW, van den Bogert AJ. Effect of gender and defensive opponent on the biomechanics of sidestep cutting. Med Sci Sports Exerc. 2004;36(6):1008–16.
27. Leppänen M, Pasanen K, Kujala UM, Vasankari T, Kannus P, Äyrämö S, et al. Stiff landings are associated with increased ACL injury risk in young female basketball and floorball players. Am J Sports Med. 2017;45(2):386–93.
28. Rosene JM, Fogarty TD. Anterior tibial translation in collegiate athletes with normal anterior cruciate ligament integrity. J Athl Train. 1999;34(2):93–8.
29. Leetun DT, Ireland ML, Willson JD, Ballantyne BT, Davis IM. Core stability measures as risk factors for lower extremity injury in athletes. Med Sci Sports Exerc. 2004;36(6):926–34.
30. Imwalle LE, Myer GD, Ford KR, Hewett TE. Relationship between hip and knee kinematics in athletic women during cutting maneuvers: a possible link to noncontact anterior cruciate ligament injury and prevention. J Strength Cond Res. 2009;23(8):2223–30.
31. Hewett TE, Ford KR, Myer GD, Wanstrath K, Scheper M. Gender differences in hip adduction motion and torque during a single-leg agility maneuver. J Orthop Res. 2006;24(3):416–21.
32. Pollard CD, Sigward SM, Powers CM. Gender differences in hip joint kinematics and kinetics during side-step cutting maneuver. Clin J Sport Med. 2007;17(1):38–42.
33. Fernández-Jaén T, López-Alcorocho JM, Rodriguez-Iñigo E, Castellán F, Hernández JC, Guillén-García P. The importance of the intercondylar notch in anterior cruciate ligament tears. Orthop J Sports Med. 2015;3(8):2325967115597882.
34. Shelbourne KD, Davis TJ, Klootwyk TE. The relationship between intercondylar notch width of the femur and the incidence of anterior cruciate ligament tears. A prospective study. Am J Sports Med. 1998;26(3):402–8.
35. Wilson R, Barhorst AA. Intercondylar notch impingement of the anterior cruciate ligament: a cadaveric in vitro study using robots. J Healthc Eng. 2018;2018:8698167.
36. LaBella CR, Hennrikus W, Hewett TE. Anterior cruciate ligament injuries: diagnosis, treatment, and prevention. Pediatrics. 2014;133(5):e1437–50.
37. Hewett TE, Myer GD, Ford KR. Anterior cruciate ligament injuries in female athletes:part 1, mechanisms and risk factors. Am J Sports Med. 2006;34(2):299–311.
38. Huston LJ, Wojtys EM. Neuromuscular performance characteristics in elite female athletes. Am J Sports Med. 1996;24(4):427–36.
39. Zazulak BT, Ponce PL, Straub SJ, Medvecky MJ, Avedisian L, Hewett TE. Gender comparison of hip muscle activity during single-leg landing. J Orthop Sports Phys Ther. 2005;35(5):292–9.
40. Ricard-Blum S. The collagen family. Cold Spring Harb Perspect Biol. 2011;3(1):a004978.

41. Frank CB. Ligament structure, physiology and function. J Musculoskelet Neuronal Interact. 2004;4(2):199–201.
42. Posthumus M, Collins M, van der Merwe L, O'Cuinneagain D, van der Merwe W, Ribbans WJ, et al. Matrix metalloproteinase genes on chromosome 11q22 and the risk of anterior cruciate ligament (ACL) rupture. Scand J Med Sci Sports. 2012;22(4):523–33.
43. Ficek K, Cieszczyk P, Kaczmarczyk M, Maciejewska-Karłowska A, Sawczuk M, Cholewinski J, et al. Gene variants within the COL1A1 gene are associated with reduced anterior cruciate ligament injury in professional soccer players. J Sci Med Sport. 2013;16(5):396–400.
44. O'Connell K, Knight H, Ficek K, Leonska-Duniec A, Maciejewska-Karlowska A, Sawczuk M, et al. Interactions between collagen gene variants and risk of anterior cruciate ligament rupture. Eur J Sport Sci. 2015;15(4):341–50.
45. Stępień-Słodkowska M, Ficek K, Maciejewska-Karłowska A, Sawczuk M, Ziętek P, Król P, et al. Overrepresentation of the COL3A1 AA genotype in Polish skiers with anterior cruciate ligament injury. Biol Sport. 2015;32(2):143–7.
46. Posthumus M, September AV, O'Cuinneagain D, van der Merwe W, Schwellnus MP, Collins M. The COL5A1 gene is associated with increased risk of anterior cruciate ligament ruptures in female participants. Am J Sports Med. 2009;37(11):2234–40.
47. Caso E, Maestro A, Sabiers CC, Godino M, Caracuel Z, Pons J, et al. Whole-exome sequencing analysis in twin sibling males with an anterior cruciate ligament rupture. Injury. 2016;47:S41–50.
48. Posthumus M, September AV, O'Cuinneagain D, van der Merwe W, Schwellnus MP, Collins M. The association between the COL12A1 gene and anterior cruciate ligament ruptures. Br J Sports Med. 2010;44(16):1160–5.
49. Lulińska E, Gibbon A, Kaczmarczyk M, Maciejewska-Skrendo A, Ficek K, Leońska-Duniec A, et al. Matrix metalloproteinase genes (MMP1, MMP10, MMP12) on chromosome 11q22 and the risk of non-contact anterior cruciate ligament ruptures. Genes (Basel). 2020;11(7):766.
50. Guo R, Aizezi A, Fan Y, Ji Z, Li W, Li Y, et al. Association between matrix metalloproteinase-3 gene polymorphisms and tendon-ligament injuries: evidence from a meta-analysis. BMC Sports Sci Med Rehabil. 2022;14(1):26.
51. Chen B, Li B, Qi YJ, Tie K, Chen LB. Association study between growth differentiation factor 5 polymorphism and non-contact anterior cruciate ligament rupture in Chinese Han population. Int J Clin Exp Med. 2015;8(12):22484–90.
52. Khoury LE, Posthumus M, Collins M, van der Merwe W, Handley C, Cook J, et al. ELN and FBN2 gene variants as risk factors for two sports-related musculoskeletal injuries. Int J Sports Med. 2015;36(4):333–7.
53. Rahim M, Gibbon A, Hobbs H, van der Merwe W, Posthumus M, Collins M, et al. The association of genes involved in the angiogenesis-associated signaling pathway with risk of anterior cruciate ligament rupture. J Orthop Res. 2014;32(12):1612–8.
54. Harner CD, Paulos LE, Greenwald AE, Rosenberg TD, Cooley VC. Detailed analysis of patients with bilateral anterior cruciate ligament injuries. Am J Sports Med. 1994;22(1):37–43.
55. Westin M, Reeds-Lundqvist S, Werner S. The correlation between anterior cruciate ligament injury in elite alpine skiers and their parents. Knee Surg Sports Traumatol Arthrosc. 2016;24(3):697–701.
56. Magnusson K, Turkiewicz A, Hughes V, Frobell R, Englund M. High genetic contribution to anterior cruciate ligament rupture: Heritability ~69. Br J Sports Med. 2020;bjsports-2020-102392. https://doi.org/10.1136/bjsports-202-102392. Online ahead of print.
57. Hewett TE, Lynch TR, Myer GD, Ford KR, Gwin RC, Heidt RS. Multiple risk factors related to familial predisposition to anterior cruciate ligament injury: fraternal twin sisters with anterior cruciate ligament ruptures. Br J Sports Med. 2010;44(12):848–55.
58. Hasani S, Feller JA, Webster KE. Familial predisposition to anterior cruciate ligament injury: a systematic review with meta-analysis. Sports Med. 2022;52(11):2657–68.

Hormones and ACL

3

Vladan B. Stevanovic, Marko Bumbasirevic, Lazar Stijak, Marko Kadija, Zoran Bascarevic, and Miodrag Glisic

3.1 Introduction

Sex differences in tendon and ligament injury risk are reported: the relative risk of tearing the anterior cruciate ligament (ACL) is two to six times greater in young female compared to male athletes even after socioeconomic, health, and lifestyle background variables and the level of sports participation have been taken into account [1]. Especially when circulating estrogen is peaking around the time of ovulation during the menstrual cycle, the risk of an ACL rupture seems to be enhanced [1]. These observations have led to a search for the underlying mechanism(s) involved in the discrepancy in the risk of injuries between men and women and between young and elderly women. A complex interaction involving several risk factors is probably in play (e.g., ligamentous laxity and size, limb alignment, notch dimensions and decreased neuromuscular control of knee, motion skill level, and muscular strength) [1, 2]. However, since estrogen receptors have been localized in tendons and ligaments, it has been suggested that at least part of this gender difference in risk might be explained by sex hormonal differences influencing tendon and ligament structure and biomechanical properties. Taken together, these observations indicate that sex differences in tendon and ligament biomechanical properties are significant in the period of life where the sex hormonal profile is markedly different between women and men. In line with this, the sex disparity in ACL injury risk is peaking in the teenage years where sex hormonal differences are the greatest [3].

In order to understand the effect of hormones on ACL injuries, it is necessary to consider the effect of testosterone, estrogen, and progesterone on connective tissue and muscles, in both men and women. Initial studies showed a greater influence of hormones on ACL injury in female subjects and thereby justified a greater frequency of injuries in women [1–3]. However, some more recent studies have shown that the concentration of hormones also affects the frequency of injury in men [4, 5]. The difference in hormonal status between men and women regarding sex hormones (testosterone, estrogen, and progesterone) has led many researchers to investigate the effect of these hormones on the ACL [6–11].

The hormonal milieu of an individual may affect the ACL and the knee joint in a multifold manner. Sex hormones act on the ACL directly, via their receptors located in the ligament.

V. B. Stevanovic (✉) · Z. Bascarevic · M. Glisic
Institute for Orthopaedics "Banjica", School of Medicine, University of Belgrade, Belgrade, Serbia

M. Bumbasirevic · M. Kadija
Clinic of Orthopaedic Surgery and Traumatology, Clinical Center of Belgrade, School of Medicine, University of Belgrade, Belgrade, Serbia

L. Stijak
Department of Anatomy, School of Medicine, University of Belgrade, Belgrade, Serbia

M. Karahan et al. (eds.), *Anterior Cruciate Ligament Injuries Throughout the Life Span*,
https://doi.org/10.1007/978-3-031-84501-7_3

However, the affinity of a hormone to overlap with the receptors of another hormone should not be overlooked. Thus, the function of the ACL may be affected by numerous products of the adrenal cortex, which show a partial affinity for testosterone receptors, as well as by certain estrogens (estrone, estriol), which are less present, but whose effect on estrogen receptors in the ACL cannot be ignored. On the other hand, the indirect effect of these hormones is reflected in their influence on muscular power, that is, on the muscles where testosterone and estrogen receptors also exist [12, 13]. In addition, testosterone affects early diaphyseal–epiphyseal fusion in the bones before the end of puberty, which, in turn, affects the physical constitution of an individual, that is, the person's height, the width of their pelvis, and the valgus knee angle. Finally, the effect of sex hormones on the psychological state of individuals should not be overlooked, as it often contributes to many injuries, including ACL injury. Future case-control studies with a larger number of subjects should consider the above-stated data.

The effect of hormones on ligament function is threefold. There is, first, an effect on the elasticity of the ligaments, but there is a difference in the type of sex hormones that increase or reduce the elasticity of the ligaments [14]. These hormones also alter the composition and cellular structure of ligaments rendering the ligament stronger or weaker. The third type of influence is exerted by the hormones through neuromuscular control affecting the function and movement of the limbs. This influence can be absolute or can vary during the menstrual cycle [15]. The effect of hormones on connective tissue in women is complicated by the different hormone concentrations during the menstrual cycle as well as different concentrations of the hormones excreted at different times of the day. Since the level of estrogens varies during the menstrual cycle of a woman, so does the risk of injury. The menstrual cycle correlates with the risk of non-contact ACL injuries in women. The mechanism itself has not yet been determined [16].

At the macroscopic level, women have a greater elasticity of the knee joint (larger movement amplitude) and a lesser resistance to tibial translation and rotation [17]. These differences do not only relate to the sagittal, but also to the frontal and transverse planes. On the other hand, women have a smaller ACL, which causes a lesser linear rigidity of the ACL, a lesser elongation, and a lesser absorption of energy, as compared to men [16]. Also, a greater elasticity of the joints and a reduction in muscle rigidity are among the factors contributing to the differences due to which women are more prone to ACL injury than men. Estrogen is probably not the only "culprit" responsible for these features of the ligaments; the ligaments are also influenced by progesterone, testosterone, and relaxin. To date, it remains unknown which hormone, and through which mechanism, acts on the physiological properties of the ACL, but it is known that hormones are a significant factor in the metabolism of muscles, collagen, and bones [16].

The ACL has been marked as an estrogen- and progesterone-sensitive tissue, due the fact that receptors for these hormones are in this ligament [9]. Also, the authors of some in vivo studies are of the opinion that progesterone and estrogen may lead to a decrease in the ability of the ACL to absorb force by lowering the level of collagen in it [14, 17]. Some in vitro studies have shown that 17-β estradiol decreases the synthesis of type I collagen, which affects the extension strength of the ACL [19]. As the concentration of these hormones is higher in women, it can be concluded that their influence on the ACL could be one of the reasons why women are more prone to injury [18–20].

3.2 Testosterone

The androgen most present in the body is testosterone, which, in addition to bringing about an increase in collagen in the male prostate gland [20], also leads to a greater regeneration capacity of the knee ligaments [21]. Androgen receptors are present in male tissues and in some tissues of women [22, 23]. In his in vitro study on a canine cell culture of the ACL, Ohno et al. [24] established that the potent form of testosterone, dihydrotestosterone, causes an increased expression of androgen receptors and an increased synthesis of collagen in the ACL fibrocytes, whereas cell division occurs 24–48 hours after treatment with this hormone.

Lovering and Romani [10] have concluded that the ACL is an androgen-sensitive tissue. The study has also found that, in addition to androgen receptors, the concentration of testosterone positively correlates with the firmness of the ACL during ovulation and that the ratio between 17-β estradiol and testosterone negatively correlates with the firmness of the ACL during ovulation.

The complexity of the influence of hormones on ACL injury has been established by Stijak et al. [25, 26]. According to their case-control study, which included 84 matched male and female athletes, testosterone had a different effect in men as compared to women. The concentration of testosterone in men with ACL rupture was higher (182 pg/mL) than the concentration of the testosterone in men without ACL rupture (149 pg/mL). On the other hand, women with ACL rupture had a lower concentration of testosterone than their matched control test subjects. As men with ACL rupture displayed mildly elevated testosterone values in their saliva, it can be hypothesized that in men the testosterone level had the effect of increasing muscle strength, thus creating a greater force, which, in turn, probably led to a greater strain of the knee joint, which could not keep up with the increase in muscular activity. In patients with a lower level of testosterone the muscular activity was not excessive, thus the knee joint suffered a lesser degree of force, which is why the injury that had occurred was clinically milder, without ACL rupture, that is, as can be seen from data obtained from the control group, the injuries were mostly meniscal distortions or tearing. Additionally, test subjects with ACL rupture had a greater generalized hyperelasticity than their matched pairs, which can be explained by the consequent extraglandular conversion of the elevated testosterone concentrations into 17-β estradiol, which leads to a greater elasticity of connective tissue. The results of the previous study speak in favor of the fact that male athletes taking exogenous testosterone (or its analogues) for improving performance, place themselves at a higher risk of ACL injury. Finally, Stijak et al. [26] hypothesized that an increased concentration of testosterone in patients with ACL rupture may influence the increase in muscle strength. However, testosterone is not alone in this activity; in fact, the hormones of the adrenal medulla (catecholamines) also play a significant role, which should be the subject of future study.

As opposed to men, where the concentration of testosterone was higher in the study group than in the control group, in women, the concentration ratio of hormones was reversed [25]. The subjects with intact ACL had significantly higher testosterone concentrations in the saliva (85 pg/mL) than the subjects with ACL rupture (63 pg/mL), which leads to the conclusion that a higher concentration of this hormone may act protectively. An intensified synthesis of collagen and a greater resilience of the ligaments are probably only a part of the proven influence of testosterone on connective tissue. The authors proposed that these influences are not the only ones, rather that the influence of testosterone on ACL rupture is exerted in several different ways and that further studies with a larger number of subjects should investigate the indirect influence of this hormone, which is exerted through muscle strength and neuromuscular control.

3.3 Estrogen and Progesterone

Estrogen is a hormone that is most present in women during the reproductive period, and, in addition to affecting secondary sex characteristics, a higher concentration of estrogen is also connected to the reduction of collagen synthesis and a decrease in fibroblast proliferation [27]. The most potent form of estrogen is 17-β estradiol, which is synthetized from testosterone, in women in the ovaries, and in men in the testicles and via extraglandular conversion of androgens [28].

ACL has been marked as estrogen- and progesterone-sensitive tissue since receptors for these hormones are in this ligament [28]. In their study of human ACL tissue culture, Yu et al. [28–30] established that the increase in 17-β estradiol concentration, even within the physiological range, led to the decrease in the proliferation of fibroblasts and in the synthesis of type I procollagen. As the concentration of these hormones is higher in women, it can be concluded that their influence on the ACL could be one of the reasons why women are more prone to injury.

Wentorf et al. [31] compared the dimensions and mechanical features of the ACL in ovariectomized and sham ovariectomized female monkeys. Even though the ACL of ovariectomized female monkeys was subjected, for two whole years, to far lower concentrations of estrogen than the sham ovariectomized female monkeys, neither metric nor differences in mechanical features of the ACL were found between the two groups. There was also no difference found regarding the patellar tendon.

Liu et al. [32] researched the distribution of estrogen receptors in the human ACL and established their existence in the nuclei of synoviocytes, fibroblasts, and the cells of ACL blood vessel walls. In their study on ovariectomized female rabbits, Slauterbeck et al. [28] established that an increase in the concentration of estrogen influences the decrease in the strength of the ACL.

The presence of generalized hyperelasticity should also be taken into consideration as one of the risk factors for ACL rupture. Seventeen-β estradiol, as well as other estrogens, acts on the increase of tissue elasticity and thereby on the increase of ligament elasticity. This effect manifests itself in the increase of generalized hyperelasticity, which is greater in men with a higher concentration of this hormone. Consequently, the concentration of estrogen in men with ACL rupture is, as expected, higher than in men without ACL rupture [26].

Other authors have highlighted a greater number of injuries during the follicular phase as compared to the luteal phase of the menstrual cycle [33–35]. Wojtys et al. [35] found a greater number of ACL ruptures during the ovulatory phase, and the phase of the menstrual cycle was determined on the basis of the metabolite concentrations of progesterone, estrogen, and luteinizing hormone in urine. A study performed on female skiers established a greater number of ACL injuries during the pre-ovulatory phase than during the post-ovulatory phase. In this study, the phase of the menstrual cycle was determined based on estradiol and progesterone concentrations in the serum [33].

It has been generally accepted, and many studies have also reported [33–35], that ACL injury in women is more frequent during the pre-ovulatory phase. As the concentrations of sex hormones are low in the pre-ovulatory phase, the question arises as to whether ACL injury can be attributed to the pre-ovulatory phase itself or to the low concentrations of sex hormones. In other words, are all women prone to a greater risk of ACL rupture during the pre-ovulatory phase, or does this apply only to women with a low level of sex hormones?

Stijak et al. [25] found a higher concentration of 17-β estradiol in patients with ACL rupture as compared to their matched test subjects. As one of the possible reasons for such a result, the authors state the conversion of extraglandular testosterone into 17-β estradiol, as test subjects with ACL rupture had higher testosterone concentrations as compared to their matched test subjects. This may indicate the existence of a direct correlation between these two hormones in the injured group, which was not established in the control group. However, even though consequential, the increase in 17-β estradiol concentration could lead to a greater laxity of ligaments. The authors stated that during the follicular phase, the concentrations of 17-β estradiol for both subgroups were almost identical. During ovulation and the luteal phase, within both subgroups, a significant rise in the concentration of 17-β estradiol was identified, with this rise being twice greater in the control group as compared to the examined group (1.6:0.75 pg/mL). The expected value of 17-β estradiol concentration during the luteal phase was 90% higher than during the follicular phase. Such results were, indeed, obtained in the female control subgroup (2.3:3.9 pg/mL). However, the female examined group demonstrated an increase of only around 30% of the value registered during the follicular phase (2.25:3.00 pg/mL). The authors believed the phase of the menstrual cycle is not crucial to ACL injury, rather it is the increase in hormone concentrations [25].

Most of the previous studies obtained information on the phase of the cycle based on hormone concentrations, which vary significantly. Thus, for instance, the concentration of progesterone in the saliva within the control group during the follicular phase was of a value close to the concentration of progesterone in the saliva within the examined group, but during the luteal phase [33–35]. It is indeed because of great individual differences during the follicular and ovulatory phases of the cycle that the phases can be misinterpreted and consequently lead to the wrong classification of the female subjects. Stijak et al. [25, 26] stated that the information regarding the day of the menstrual cycle at the time of sampling and the information on the duration of the menstrual cycle were more valid than the estimation of the day of the menstrual cycle based on the hormone concentration. Consequently, the phase of the cycle was eliminated as a potential risk factor and replaced by a more valid factor—the concentrations of 17-β estradiol and progesterone. Stijak et al. [25, 26] further stated that the risk linked to the pre-ovulatory phase is conditioned by the concentrations of progesterone and 17-β estradiol. They also stated that when female subjects (women naturally have lower concentrations of these hormones) reached the pre-ovulatory phase, their risk doubled, since this phase is characteristic of low hormone concentrations.

Progesterone can be found in the blood of men and women but its role in men is unclear [36]. In women, its main role is in the preparation of the endometrium for the implantation of the fertilized ovum, while in both sexes its secretion is intensified during stress and anxiety.

The influence of progesterone concentration on ACL rupture in men was not established in the study carried out by Stijak et al. [26]. Similar concentrations of this hormone in the saliva were found by Celec et al. [37] in a study on male subjects aged 20–21 years. Based on these results it could not be concluded that the progesterone concentration in the saliva affects ACL injury in men. Somewhat higher progesterone concentrations were registered in the saliva of older men (57–84 years of age) [38].

One case-control study [25] on test subjects with ACL rupture registered lower concentrations of progesterone in the saliva during both phases of the cycle, follicular (29:62 pg/mL) and luteal (75:129 pg/mL). In both of the subgroups, there was an identical rise in progesterone between the two phases; only the female test subjects with ACL rupture had a lower concentration of progesterone during both phases, which speaks in favor of the theory that the decreased concentration of progesterone is one of the risk factors for ACL injury in women.

Finally, we can conclude that monitoring the influence of sex hormones concentration levels on ACL injury is complicated and tiring work, which is not ruled by simple laws of summing up the individual effects of each hormone. Each person, male or female, is a different individual, and therefore the specific response of an organism to strain, stress, and physical and mental exhaustion varies from person to person. This most rapidly reflects on the hormone level immediately before injury, which is why sampling specimens for determining hormone concentration levels at a time when the individual is not under physical nor mental stress cannot adequately represent the state of the organism during competitive sport activities. The solution would be conducting long-term, prospective studies with no less than 50 subjects per study, as they could contribute to a more precise determining of the response mechanism. These studies should, at the beginning, monitor periodical hormone excretion (per day/seasonal; at rest/under stress; in a state of satisfaction/disappointment; etc.) for each individual, for the purpose of determining the situations when sudden changes in hormone concentration levels occur that could affect ACL injury in different ways. This would especially contribute to the understanding of the effect of hormone concentrations in women, in whom the phase (or to be more precise the day) of the menstrual cycle additionally complicates the interpreting of the results.

References

1. Oliva F, Piccirilli E, Berardi AC, et al. Hormones and tendinopathies: the current evidence. Br Med Bull. 2016;117(1):39–58.
2. Arendt EA. Musculoskeletal injuries of the knee: are females at greater risk? Minn Med. 2007;90:38–44.
3. Prodromos CC, Han Y, Rogowski J, et al. A meta-analysis of the incidence of anterior cruciate ligament tears as a function of gender, sport, and a knee injury-reduction regimen. Arthroscopy. 2007;23:1320–5.
4. Gwinn D, Wilcknes J, McDewitt E, Ross G, Tzu-Vheg K. The relative incidence of anterior cruciate ligament injury in man and woman at the United States Naval Academy. Am J Sports Med. 2000;28:98–102.
5. Uhorchak JM, Scoville CR, Williams GN, Arciero RA, Pierre P, Taylor DC. Risk factors associated with noncontact injury of the anterior cruciate ligament. A prospective four-year evaluation of 859 west point cadets. Am J Sports Med. 2009;31:831–42.
6. Arendt E, Agel J, Dick R. Anterior cruciate ligament injury patterns among collegiate men and woman. J Athl Train. 1999;34:86–92.
7. Adachi N, Nawata K, Maeta M, Kurozawa Y. Relationship of the menstrual cycle phase to anterior cruciate ligament injuries in teenaged female athletes. Arch Orthop Trauma Surg. 2008;128:473–8.
8. Bell DR, Blackburn JT, Norcorss MF, Ondrak KS, Hudson JD, Hackney AC, Padua DA. Estrogen and muscle stiffness have a negative relationship in females. Knee Surg Sports Traumatol Arthrosc. 2012;20:361–7.
9. Dragoo JL, Castillo TN, Braun HJ, Ridley BA, Kennedy AC, Golish SR. Prospective correlation between serum Relaxin concentration and anterior cruciate ligament tears among elite collegiate female athletes. Am J Sports Med. 2011;39:2175–80.
10. Lovering RM, Romani WA. Effect of testosterone on the female anterior cruciate ligament. Am J Physiol Regul, Interg Comp Physiol. 2005;289:R15–22.
11. Park SK, Stefanyshyn DJ, Loitz-Ramage B, Hart DA, Ronsky JL. Changing hormone levels during the menstrual cycle affect knee laxity and stiffness in healthy female subjects. Am J Sports Med. 2009;37:588–98.
12. Lemoine S, Granier P, Tiffoche C, Rannou-Bekono F, Thieulant ML, Delamarche P. Estrogen receptor alpha mRNK in human skeletal muscles. Med Sci Sports Exerc. 2003;35:439–43.
13. Sinha-Hikim I, Taylor WE, Gonzalez Cadavid NF, Zheng W, Bhasin S. Androgen receptor in human skeletal muscle and cultured muscle satellite cells: up-regulation by androgen treatment. J Clin Endocrinol Metab. 2004;89:5245–55.
14. Fung YCB. Biomechanics: mechanical properties of the living tissues. Berlin/Heidelberg: Springer – Verlag; 1981.
15. Jakob RP, Stäubli HU. The knee and the cruciate ligaments: anatomy, biomechanics, clinical aspect, reconstruction, complications, rehabilitation. Berlin/Heidelberg/New York/London/Paris/Tokyo/Hong Kong/Barcelona/Budapest: Springer – Verlag; 1990.
16. Renstrom P, Ljungquist A, Arendt E, Beynnon B, Fukubayashi T, Garrett W, et al. Non-contact ACL injuries in female athletes: an International Olympic Committee current concepts statement. Br J Sports Med. 2008;42:394–412.
17. Nguyen AD, Shultz SJ. Sex differences in lower extremity posture. J Orthop Sports Phys Ther. 2007;37:389–98.
18. Slauterbeck JR. The effect of hormone levels in female athletes on the frequency of the anterior cruciate ligament injuries. Paper presented at the National Athletic Trainers Association Annual Meeting, Los Angeles; 2000.
19. Arendt EA, Bershadsky B, Agel JA. Periodicity of non-contact ACL injuries during the menstrual cycle. J Gend Specif Med. 2001;5(2):16–26.
20. Asano K, Maruyama S, Usui T, Fujimoto N. Regulation of estrogen receptor alpha and beta expression by testosterone in the rat prostate gland. Endocr J. 2003;50:281–7.
21. Hamlet W, Liu S, Panossian V, Finerman G. Primary immunolocalisation of androgen target cells in the human anterior cruciate ligament. J Orthop Res. 1997;15:657–63.
22. Pagnotta A, Specchia N, Greco F. Androgen receptors in Dupuytren's contracture. J Orthop Res. 2002;20:163–8.
23. Davis S, Tran J. What are "normal" testosterone levels for woman? J Clin Endocrin Metab. 2001;86:1842–4.
24. Ohno H, Kowatari Y, Owaki M, Ohta J, Nakajima N, Yoshioka K, Mutoh K, Oyamada T. Effects of androgens on cultured cells derived from canine anerior cruciate ligament. Okajimas Folia Anat Jpn. 2012;89:35–8.
25. Stijak L, Kadija M, Djulejić V, Aksić M, Petronijević N, Marković B, Radonjić V, Bumbaširević M, Filipović B. The influence of sex hormones on anterior cruciate ligament rupture: female study. Knee Surg Sports Traumatol Arthrosc. 2015;23:2742–9.
26. Stijak L, Kadija M, Djulejić V, Aksić M, Petronijević N, Aleksić D, Radonjić V, Bumbaširević M, Filipović B. The influence of sex hormones on anterior cruciate ligament ruptures in males. Knee Surg Sports Traumatol Arthrosc. 2015;23:3578.
27. Liu S, Al-Shaikh RA, Panossian V, Finerman GA, Lane J. Estrogen affects the cellular metabolism of the anterior cruciate ligament. Am J Sports Med. 1997;25:704–9.
28. Slauterbeck J, Clevenger C, Lundberg W, Burchfield D. Estrogen levels alters the failure load of the rabbit anterior cruciate ligament. J Orthop Res. 1999;17:405–48.
29. Yu WD, Liu S, Hatch JD, Panossian V, Finerman GA. Effect of estrogen on cellular metabolism of the human anterior cruciate ligament. Clin Orthop Relat Res. 1999;366:229–38.
30. Yu WD, Panossian V, Hatch JD, Liu S, Finerman GA. Combined effect of estrogen and progesterone

on anterior cruciate ligament. Clin Orthop Relat Res. 2001;383:268–81.

31. Wentorf FA, Sudoh K, Moses C, Arendt EA, Carlson CS. The effect of estrogen on material and mechanical properties of the intra- and extra-articular knee structures. Am J Sports Med. 2006;34:1948–52.
32. Liu SH, Al-Shaikh RA, Panossian V, Yang RS, Nelson SD, Soleiman N, Ginerman GA, Lane JM. Primary immunolocalisation of estrogen and progesteron target cells in the human anterior cruciate ligament. J Orthop Res. 1996;14:526–33.
33. Beynnon BD, Johnson RJ, Braun S, Sargent M, Bernstein IM, Skelly JM, Vacek PM. The relationship between menstrual cycle phase and anterior cruciate ligament injury: a case control study of recreational alpine skiers. Am J Sports Med. 2006;34:757–64.
34. Slauterbeck JR, Fuzie SF, Smith MP, Clark RJ, Xu K, Starch DW, Hardy DM. The menstrual cycle, sex hormones, and anterior cruciate ligament injury. J Ath Train. 2002;37:275–8.
35. Wojtys EM, Huston L, Boynton MD, Spindler KP, Lindenfeld TN. The effect of the menstrual cycle on anterior cruciate ligament in women as determined by hormone levels. Am J Sports Med. 2002;30:182–8.
36. Andersen ML, Tufik S. Does male sexual behavior require progesterone? Brain Res Brain Res Rev. 2006;51:136–43.
37. Celec P, Ostatnikova D, Hodosy J, Skoknova M, Putz Z, Kudela M. Infradian rhythmic variations of salivary estradiol and progesterone in healthy man. Biol Rhythm Res. 2006;37:37–44.
38. Gavrilova N, Lindau ST. Salivary sex hormone measurement in a national, population-based study of older adults. J Gerontol B Psychol Sci Soc Sci. 2009;64B(S1):i94–i105.

Biomechanics of the Knee in Relation to ACL Rupture

4

Nahum Rosenberg

4.1 Introduction

The knee joint is a synovial hinge joint composed of the tibiofemoral and patellofemoral articulations. Biomechanically, the knee's integrity is maintained by passive structures, including ligaments and the joint capsule, and active structures, such as musculature [1–3]. Biomechanically designed for stability and mobility, it endures repetitive loads during daily and athletic movements with loads that might exceed 425% of body weight [3, 4]. The objectives of this chapter are to dissect the complex biomechanical interrelationships within the knee joint relevant to anterior cruciate ligament (ACL) function and dysfunction and to synthesize current research findings into a coherent narrative that may guide future scientific inquiry and clinical practice.

The ACL has the general biomechanical characteristics of a stabilizing ligament. The biomechanical properties of ligaments are typically characterized by their stress–strain behavior. Under tensile load, a typical ligament exhibits a nonlinear stress–strain relationship, which can be divided into three regions: the toe region, linear region, and failure region [5, 6] (Fig. 4.1). In the toe region, the crimped collagen fibers straighten, resulting in large deformations with relatively low stress levels. As the load increases, the ligament enters the linear region of elastic deformation, where the stress–strain relationship becomes more predictable and linear. This region is crucial for understanding the functional range of ligament behavior during normal joint motion. In the failure region, when structural irreversible destruction of the ligament occurs, the ligament experiences structural damage, leading to a loss of mechanical integrity and eventual rupture.

The viscoelastic nature of ligaments also plays a significant role in their biomechanical behavior. Unlike purely elastic materials, ligaments exhibit time-dependent properties, such as creep and stress relaxation [7]. Creep refers to the gradual elongation of the ligament under a constant load, while stress relaxation describes the decrease in stress under a constant strain. These properties are important for understanding how ligaments respond to sustained loads, as in the case of prolonged postures or repetitive movements.

The knee joint is a complex biomechanical structure vital for mobility and load bearing. From the point of view of kinematics, the knee joint is a modified hinge joint that allows for flexion, extension, and a small degree of rotation and lateral movement. The motion of the knee involves coordinated rolling and sliding of the tibial condyles on the femoral condyles, regulated by ligamentous structures and muscle contractions. The kinetics of the knee joint are dependent on the forces that are generated by

N. Rosenberg (✉)
Specialists Center, National Insurance Institute, Haifa, Israel

M. Karahan et al. (eds.), *Anterior Cruciate Ligament Injuries Throughout the Life Span*, https://doi.org/10.1007/978-3-031-84501-7_4

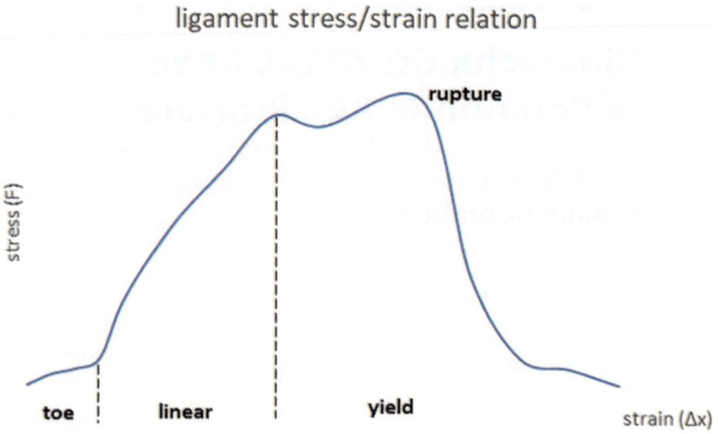

Fig. 4.1 A simplified schematic representation of the relation of force (F) per sectional area (stress) and structural deformation (strain) of the mechanically loaded ligament. The initial orientation of the ligament fibrils becomes more parallel with the force application (toe region), additional force brings the ligament into an elastic reversible phase (linear region), and when it exceeds a certain strain level, the ligament becomes irreversibly structurally damaged (yield or failure region) until the complete rupture. Δx change in length of the ligament

body weight, muscle activity, and external forces. These factors contribute to the complex loading conditions experienced by the knee joint during activities. Muscles surrounding the knee work with ligaments to control movement and stabilize the joint. The quadriceps and hamstrings play a significant role in modulating the forces applied to the ACL [8], which is one of the key components contributing to the knee's stability. The ACL is integral to maintaining the appropriate kinematics of the knee joint and is one of the most frequently injured ligaments. ACL injuries are a significant concern, particularly in sports medicine, due to their prevalence and the long recovery period [9]. Understanding the biomechanics of the knee in relation to ACL function and rupture is crucial for the development of preventive strategies, surgical techniques, and rehabilitation protocols.

The knee joint facilitates complex movements while bearing substantial loads, transferring forces to its stabilizers. ACL, as one of the main knee stabilizing structures, is subjected to applied forces in compression, distraction, and rotational planes, which are expressed in its intrinsic tension (Fig. 4.2). The ACL plays a critical role in maintaining knee stability by resisting excessive anterior translation, bending in the coronal plane, and rotation of the tibia relative to the femur [10].

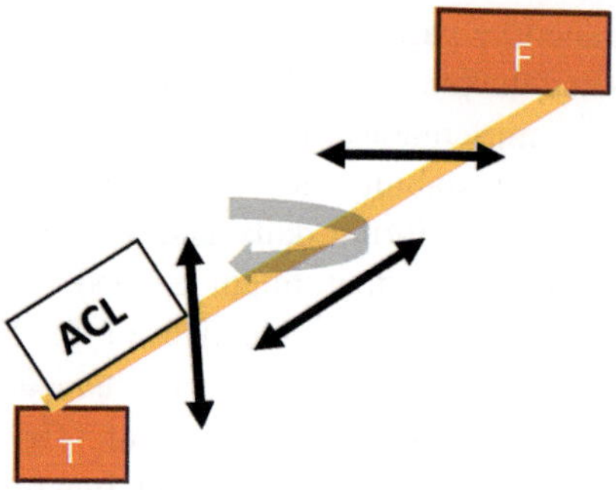

Fig. 4.2 Simplified schematic representation of forces acting on anterior cruciate ligament (ACL). The arrows present the vertical (distraction and compression), horizontal (translation), and torque forces, which are eventually expressed in the intrinsic tension force in the ACL. F femoral origin of ACL, T tibial insertion of ACL

Therefore, ACL injuries lead to knee instability, alterations in joint kinematics, and increased risk of subsequential knee joint disarrangement, including meniscal tears and osteoarthritis [11, 12]. The mechanism of ACL injury is multifactorial, encompassing dynamic interplays between anatomical, hormonal, and biomechanical factors. There is a well-documented discrepancy in ACL injury rates between men and women, with the latter having up to sixfold higher incidence of rupture. Several factors contribute to this disparity, including anatomical differences, hormonal influences, and biomechanical factors such as increased knee abduction moments and valgus alignment during dynamic tasks [13, 14].

4.2 The Role of the ACL in Knee Stability

The ACL, located centrally within the knee joint, is one of the key ligaments providing anteroposterior and rotatory stability. Its unique structure, composed of two main functional bundles, the anteromedial (AM) and posterolateral (PL) bundles, allows it to maintain joint integrity throughout the range of motion and serve distinct functions depending on the knee's position [15]. An additional intermediate bundle was also identified in cadaveric studies [15]. The ACL originates from the posteromedial aspect of the lateral femoral condyle, inserted into the tibial plateau. Biomechanically, the ACL resists anterior translation of the tibia relative to the femur and is a secondary restraint to tibial rotation and varus or valgus stress when the PL bundle significantly contributes to knee stability in full extension [16].

Complementing the ACL, other structures such as the posterior cruciate ligament (PCL), medial and lateral collateral ligaments, menisci, and surrounding musculature contribute to the complex stability of the knee.

The motion of the knee joint has six degrees of freedom that characterize tibiofemoral movements: flexion–extension, abduction–adduction, and internal–external rotation, and translations along three perpendicular axes [17].

The overall load in the knee during gait or athletic maneuvers is the result of shear and compressive forces causing moments and torques that act on the knee joint [18] (Fig. 4.2).

4.3 Mechanisms of ACL Injury

The ACL comprises a complex arrangement of collagen fibers that exhibit nonlinear, viscoelastic properties [19, 20] (Fig. 4.1). This means that its stiffness and strength vary depending on the rate and duration of the applied load [5–7]. The ligament's structural properties allow it to withstand substantial forces during dynamic activities. The ultimate tensile strength of the ACL has been reported to range between 1725 and 2350 N, with a strain rate of 20–25% at failure [21, 22].

During dynamic activities such as jumping, cutting, or pivoting, the knee joint undergoes a series of complex motions and loading patterns. The ground reaction forces, joint angles, muscle activations, and ligamentous constraints must all be coordinated to maintain stability. Disruptions in this coordination can result in non-physiologic loading of the ACL and lead to injury [23]. Studies utilizing video analysis and biomechanical modeling have shown that ACL injury typically occurs when the knee is near full extension (less than 40° of flexion), combined with valgus stress and internal tibial rotation [23, 24]. These positions place an excessive strain on the ACL, leading to its failure.

ACL injuries may be caused by contact or non-contact mechanisms [25]. The non-contact injuries occur following excessive axial and pivotal loadings while jumping or sudden change of movement directions. It was shown that non-contact mechanisms, such as landing from a jump with the knee extended and the tibia externally rotated, are among the most common causes of ACL injuries [26]. Contact injuries, direct or indirect, can also result in ACL ruptures when the knee is forced into positions that exceed the ligament's tensile capacity [25]. The direct injury occurs when a contact force is applied locally to the knee area, and the indirect mechanism follows force on other body areas and is transferred to the knee indirectly.

Non-contact injuries often involve deceleration or landing movements with a combination of knee valgus and tibial rotation stresses [27].

Overall, the effect of knee joint kinematics, including flexion angle, varus/valgus alignment, and tibial rotation, significantly influences ACL strain. ACL strain is typically minimized when the knee is in a neutral alignment and increases with hyperextension, valgus loading, and excessive internal or external tibial rotation. The muscles surrounding the knee, particularly the quadriceps and hamstrings, play a significant role in modulating the forces experienced by the ACL. The quadriceps are known to exert an anterior shear force on the tibia, which can increase ACL strain, especially when the knee is near full extension [18]. Conversely, the hamstrings apply a posterior shear force that can protect the ACL from excessive loading [28]. Therefore, neuromuscular control is essential for the coordinated action of muscles around a joint to stabilize it during dynamic movements. Poor neuromuscular control, especially of the quadriceps and hamstrings, can lead to dynamic valgus collapse of the knee, a common mechanism in non-contact ACL injuries [13].

As mentioned above, the risk factors contributing to ACL tears include intrinsic factors such as gender [12], anatomical variations [29], hormonal influences [12], genetic predisposition due to collagen variants imposing on the material properties of ACL [30], and neuromuscular control [29], and extrinsic factors like footwear, playing surfaces, and environmental conditions [29]. The anatomical intrinsic factors, which are related to the biomechanics of ACL failure, include the size and shape of the notch where the ACL resides. A narrower notch may predispose athletes to ACL tears due to mechanical impingement of the ligament, especially in young patients [31].

Footwear and surface interactions also play a role in ACL injury mechanics as part of the extrinsic factors. Traction between the shoe and the playing surface can influence the rotational forces transmitted to the knee, potentially increasing the risk of an ACL injury. For example, athletes playing on artificial turf may have a different risk profile compared to those on natural grass due to variations in surface friction and hardness [32]. Additionally, the type of activity or sport significantly affects the risk of ACL rupture. Sports that involve sudden changes in direction, jumping, or contact with other players have higher rates of ACL injuries [26].

4.4 Conclusion and Perspectives

The biomechanics of the knee joint in relation to ACL function and rupture is a multifaceted subject that integrates the anatomical, mechanical, and physiological aspects of the ligament and the surrounding structures. The biomechanical integrity of the knee joint is heavily dependent on the functionality of the ACL. The rupture of this ligament disrupts the delicate balance of forces within the knee, leading to altered joint kinematics and increased risk of secondary injuries and degenerative changes.

Advances in ACL reconstruction surgery aim to restore the knee's biomechanical function by replicating the native ACL's anatomy and tensioning behavior. Rehabilitation protocols leverage biomechanical principles to gradually reintroduce loads on the reconstructed ligament without compromising its integrity. Training program modifications should improve neuromuscular control and reduce the incidence of ACL injuries [33]. The future progress in the utilization of computerized identification and varying parameters, such as muscle forces, joint angles, and loading conditions (for instance using the Gait Analysis), should potentially identify injurious situations and accordingly update training programs for improved ACL tear prevention [34].

Accordingly, training methods focusing on neuromuscular control, proprioceptive exercises, and sports technique modification have shown potential in reducing ACL injury rates [33]. Ongoing research is essential for further elucidating the complexities of ACL injuries and for continually improving preventative and therapeutic strategies.

References

1. Benjaminse A, Gokeler A, van der Schans CP. Clinical diagnosis of an anterior cruciate ligament rupture: a meta-analysis. J Orthop Sports Phys Ther. 2006;36(5):267–88. https://doi.org/10.2519/jospt.2006.2011. PMID: 16715828.
2. Aagaard H, Verdonk R. Function of the normal meniscus and consequences of meniscal resection. Scand J Med Sci Sports. 1999;9(3):134–40. https://

doi.org/10.1111/j.1600-0838.1999.tb00443.x. PMID: 10380269.
3. Kutzner I, Heinlein B, Graichen F, Bender A, Rohlmann A, Halder A, Beier A, Bergmann G. Loading of the knee joint during activities of daily living measured in vivo in five subjects. J Biomech. 2010;43(11):2164–73. https://doi.org/10.1016/j.jbiomech.2010.03.046. PMID: 20537336.
4. Andriacchi TP, Mündermann A. The role of ambulatory mechanics in the initiation and progression of knee osteoarthritis. Curr Opin Rheumatol. 2006;18(5):514–8. https://doi.org/10.1097/01.bor.0000240365.16842.4e. PMID: 16896293.
5. Girgis FG, Marshall JL, Monajem A. The cruciate ligaments of the knee joint. Anatomical, functional and experimental analysis. Clin Orthop Relat Res. 1975;106:216–31. https://doi.org/10.1097/00003086-197501000-00033. PMID: 1126079.
6. Marieswaran M, Jain I, Garg B, Sharma V, Kalyanasundaram D, Kalyanasundaram D. A review on biomechanics of anterior cruciate ligament and materials for reconstruction. Appl Bionics Biomech. 2018;2018:2018.
7. Provenzano PP, Lakes RS. Interrelation of creep and relaxation: a modeling approach for ligaments. J Biomech Eng. 2002;124(6):714–7.
8. Aagaard P, Simonsen EB, Andersen J, Magnusson SP, Bojsen-møller F, Dyhre-Poulsen P. Antagonist muscle coactivation during isokinetic knee extension. Scand J Med Sci Sports. 2000;10:58–67.
9. Ardern CL, Webster KE, Taylor NF, Feller JA. Return to sport following anterior cruciate ligament reconstruction surgery: a systematic review and meta-analysis of the state of play. Br J Sports Med. 2016;50(12):776–82.
10. Markolf KL, Burchfield DM, Shapiro MM, Shepard MF, Finerman GA, Slauterbeck JL. Combined knee loading states that generate high anterior cruciate ligament forces. J Orthop Res. 1995;13(6):930–5. https://doi.org/10.1002/jor.1100130618. PMID: 8544031.
11. Monk AP, Davies LJ, Hopewell S, Harris K, Beard DJ, Price AJ. Surgical versus conservative interventions for treating anterior cruciate ligament injuries. Cochrane Database Syst Rev. 2016;4(4):CD011166. https://doi.org/10.1002/14651858.CD011166.pub2. PMID: 27039329; PMCID: PMC6464826.
12. Lohmander LS, Ostenberg A, Englund M, Roos H. High prevalence of knee osteoarthritis, pain, and functional limitations in female soccer players twelve years after anterior cruciate ligament injury. Arthritis Rheum. 2004;50(10):3145–52. https://doi.org/10.1002/art.20589. PMID: 15476248.
13. Hewett TE, Myer GD, Ford KR, et al. Biomechanical measures of neuromuscular control and valgus loading of the knee predict anterior cruciate ligament injury risk in female athletes: a prospective study. Am J Sports Med. 2005;33(4):492–501. https://doi.org/10.1177/0363546504269591.
14. Anderson AF, Dome DC, Gautam S, Awh MH, Rennirt GW. Correlation of anthropometric measurements, strength, anterior cruciate ligament size, and intercondylar notch characteristics to sex differences in anterior cruciate ligament tear rates. Am J Sports Med. 2001;29(1):58–66. https://doi.org/10.1177/03635465010290011501. PMID: 11206258.
15. Amis AA, Dawkins GP. Functional anatomy of the anterior cruciate ligament. Fibre bundle actions related to ligament replacements and injuries. J Bone Joint Surg Br. 1991;73(2):260–7.
16. Gabriel MT, Wong EK, Woo SL, Yagi M, Debski RE. Distribution of in situ forces in the anterior cruciate ligament in response to rotatory loads. J Orthop Res. 2004;22(1):85–9. https://doi.org/10.1016/S0736-0266(03)00133-5. PMID: 14656664.
17. Lloyd DG, Buchanan TS. Strategies of muscular support of varus and valgus isometric loads at the human knee. J Biomech. 2001;34(10):1257–67. https://doi.org/10.1016/s0021-9290(01)00095-1. PMID: 11522305.
18. Shin CS, Chaudhari AM, Andriacchi TP. The influence of deceleration forces on ACL strain during single-leg landing: a simulation study. J Biomech. 2007;40(5):1145–52. https://doi.org/10.1016/j.jbiomech.2006.05.004. Epub 2006 Jun 23. PMID: 16797556.
19. Woo SL, Hollis JM, Adams DJ, Lyon RM, Takai S. Tensile properties of the human femur-anterior cruciate ligament-tibia complex. The effects of specimen age and orientation. Am J Sports Med. 1991;19(3):217–25. https://doi.org/10.1177/036354659101900303. PMID: 1867330.
20. Kwan MK, Lin TH, Woo SL. On the viscoelastic properties of the anteromedial bundle of the anterior cruciate ligament. J Biomech. 1993;26(4–5):447–52. https://doi.org/10.1016/0021-9290(93)90008-3. PMID: 8478348.
21. Noyes FR, Grood ES. The strength of the anterior cruciate ligament in humans and Rhesus monkeys. J Bone Joint Surg Am. 1976;58(8):1074–82. PMID: 1002748.
22. Noyes FR, Butler DL, Grood ES, Zernicke RF, Hefzy MS. Biomechanical analysis of human ligament grafts used in knee-ligament repairs and reconstructions. J Bone Joint Surg Am. 1984;66(3):344–52. PMID: 6699049.
23. Besier TF, Lloyd DG, Cochrane JL, Ackland TR. External loading of the knee joint during running and cutting maneuvers. Med Sci Sports Exerc. 2001;33(7):1168–75. https://doi.org/10.1097/00005768-200107000-00014. PMID: 11445764.
24. Koga H, Nakamae A, Shima Y, Iwasa J, Myklebust G, Engebretsen L, Bahr R, Krosshaug T. Mechanisms for noncontact anterior cruciate ligament injuries: knee joint kinematics in 10 injury situations from female team handball and basketball. Am J Sports Med. 2010;38(11):2218–25. https://doi.org/10.1177/0363546510373570. Epub 2010 Jul 1. PMID: 20595545.

25. Takahashi S, Nagano Y, Ito W, Kido Y, Okuwaki T. A retrospective study of mechanisms of anterior cruciate ligament injuries in high school basketball, handball, judo, soccer, and volleyball. Medicine (Baltimore). 2019;98(26):e16030. https://doi.org/10.1097/MD.0000000000016030. PMID: 31261507; PMCID: PMC6616929.
26. Griffin LY, Agel J, Albohm MJ, Arendt EA, Dick RW, Garrett WE, Garrick JG, Hewett TE, Huston L, Ireland ML, Johnson RJ, Kibler WB, Lephart S, Lewis JL, Lindenfeld TN, Mandelbaum BR, Marchak P, Teitz CC, Wojtys EM. Noncontact anterior cruciate ligament injuries: risk factors and prevention strategies. J Am Acad Orthop Surg. 2000;8(3):141–50. https://doi.org/10.5435/00124635-200005000-00001. PMID: 10874221.
27. Boden BP, Dean GS, Feagin JA Jr, Garrett WE Jr. Mechanisms of anterior cruciate ligament injury. Orthopedics. 2000;23(6):573–8. https://doi.org/10.3928/0147-7447-20000601-15. PMID: 10875418.
28. More RC, Karras BT, Neiman R, Fritschy D, Woo SL, Daniel DM. Hamstrings--an anterior cruciate ligament protagonist. An in vitro study. Am J Sports Med. 1993;21(2):231–7. https://doi.org/10.1177/036354659302100212. PMID: 8465918.
29. Smith HC, Vacek P, Johnson RJ, Slauterbeck JR, Hashemi J, Shultz S, Beynnon BD. Risk factors for anterior cruciate ligament injury: a review of the literature - part 1: neuromuscular and anatomic risk. Sports Health. 2012;4(1):69–78. https://doi.org/10.1177/1941738111428281. PMID: 23016072; PMCID: PMC3435896.
30. Posthumus M, September AV, Keegan M, O'Cuinneagain D, Van der Merwe W, Schwellnus MP, Collins M. Genetic risk factors for anterior cruciate ligament ruptures: COL1A1 gene variant. Br J Sports Med. 2009;43(5):352–6. https://doi.org/10.1136/bjsm.2008.056150. Epub 2009 Feb 4. PMID: 19193663.
31. Domzalski M, Grzelak P, Gabos P. Risk factors for Anterior Cruciate Ligament injury in skeletally immature patients: analysis of intercondylar notch width using Magnetic Resonance Imaging. Int Orthop. 2010;34(5):703–7. https://doi.org/10.1007/s00264-010-0987-7. Epub 2010 Mar 24. PMID: 20333378; PMCID: PMC2903177
32. Dragoo JL, Braun HJ, Harris AH. The effect of playing surface on the incidence of ACL injuries in National Collegiate Athletic Association American Football. Knee. 2013;20(3):191–5. https://doi.org/10.1016/j.knee.2012.07.006. Epub 2012 Aug 21. PMID: 22920310.
33. Myer GD, Brent JL, Ford KR, Hewett TE. Real-time assessment and neuromuscular training feedback techniques to prevent ACL injury in female athletes. Strength Cond J. 2011;33(3):21–35. https://doi.org/10.1519/SSC.0b013e318213afa8. PMID: 21643474; PMCID: PMC3105897.
34. Di Stasi SL, Logerstedt D, Gardinier ES, Snyder-Mackler L. Gait patterns differ between ACL-reconstructed athletes who pass return-to-sport criteria and those who fail. Am J Sports Med. 2013;41(6):1310–8. https://doi.org/10.1177/0363546513482718. Epub 2013 Apr 5. PMID: 23562809; PMCID: PMC3732407.

Natural History of ACL-Deficient Knee and Reconstructed Knee

5

Kuan-Hao Chen and Chih-Hwa Chen

5.1 Introduction

The anterior cruciate ligament (ACL) is an important ligament that stabilizes the knee. It is frequently injured by sport accidents and trauma [1]. More and more ACL injuries occur every year, mostly during the high school and college years. The incidence of these injuries is slowly increasing, especially in females. This trend is likely due to their growing involvement in high school sports and other organized athletic activities. A recent study showed the annual incidence of ACL injuries was up to 0.068% per person-year in the general population, with the age-adjusted incidence higher in male compared to female [2]. However, among high school athletes, female players demonstrated a higher risk of ACL injuries compared to their male counterparts [3, 4]. For sex-comparable sports athletes, females pose two to three times higher risk per exposure [3–5]. Football poses the highest injury risk in male athletes, with the incidence of 0.8% per season, whereas soccer poses the highest injury risk in female athletes, with the incidence of 1.1% per season [3, 6]. In the Scandinavian population, alpine skiing was the leading cause for ACL injuries [7, 8]. Owing to participation in high-demand sports at an earlier age, and an expanded role of magnetic resonance imaging (MRI) for diagnosis, ACL injuries seem to be more frequently seen in pediatric patients, even in patients as young as 3 years old [9–12]. On the other hand, as average age and life expectancy rise, the demand for physical activity in the elderly population increases accordingly. Consequently, ACL injuries are becoming more frequent in the over-40-year-old population [13, 14].

While the clinical significance of ACL injuries rises, understanding the natural history of the ACL-deficient knee and reconstructed knee is of

K.-H. Chen
Department of Orthopedics, Shuang Ho Hospital, Taipei Medical University, New Taipei City, Taiwan

Graduate Institute of Biomedical Materials and Tissue Engineering, College of Biomedical Engineering, Taipei Medial University, Taipei, Taiwan

C.-H. Chen (✉)
Department of Orthopedics, Shuang Ho Hospital, Taipei Medical University, New Taipei City, Taiwan

Graduate Institute of Biomedical Materials and Tissue Engineering, College of Biomedical Engineering, Taipei Medial University, Taipei, Taiwan

Department of Orthopedics, School of Medicine, College of Medicine, Taipei Medical University, Taipei, Taiwan

School of Biomedical Engineering, College of Biomedical Engineering, Taipei Medical University, Taipei, Taiwan

Research Center of Biomedical Device, Taipei Medical University, Taipei, Taiwan

M. Karahan et al. (eds.), *Anterior Cruciate Ligament Injuries Throughout the Life Span*,
https://doi.org/10.1007/978-3-031-84501-7_5

profound importance. In this chapter, we will delineate the different characteristics of ACL injuries in each stage of life and the natural history of ACL injuries.

5.2 Characteristics and the Natural History of ACL Injuries in Preadolescent Children

ACL insufficiencies in this age may present with no trauma history. The nontraumatic ACL insufficiencies may be seen in children with generalized joint laxity, or those with congenital absence or attenuation of the ACL [15]. Historically, traumatic ACL injuries in the preadolescent population were thought to be uncommon. Studies in the 1970s–1980s reported the incidence of ACL rupture in skeletally immature patients suffering a knee injury being between 1.0 and 3.4% [9, 15, 16]. On the other hand, data from a US pediatric medical center between 2000 and 2009 showed that ACL injuries accounted for 6.3% of all sports injuries in the 5–12-year-old cohort [17]. A study analyzing the national database in the United States for ACL tears in pediatric patients between 2007 and 2011 showed a 4.5% increase in the diagnosis of ACL tears in the 5–9-year-old cohort [18]. Another report on Australian data on ACL reconstructions between 2003 and 2008 also showed a significant increase for males in the 5–14-year-old cohort [19]. Several reasons were related to this increase in frequency, including participation in high-intensity sports at younger ages, improvement in diagnostic resources, and increase in the demand for treatment of these injuries in the preadolescent children [20].

Differences in the presentation of such injuries and associated injuries between preadolescent children and adolescents were also reported. In general, partial tears of the ACL or anterior tibial spine avulsion fractures more commonly occurred in the preadolescent population compared with the adolescent population [20–22]. Though both the preadolescent population and the adolescent population shared the similar injury mechanisms, factors such as skeletal immaturity, weakness of the musculature surrounding the knee joint, and increased elasticity of the ligaments may have contributed to this greater incidence in ACL partial tears or bony avulsion injuries [23–25]. Concomitant meniscal and/or collateral ligamental injuries have been reported in preadolescents, with the incidence as high as 42–45% [22, 26]. Like in the adult population, medial meniscal injuries were the most seen associated injuries with ACL tears in the pediatric population [22, 27, 28]. The incidence of associated medial meniscal injuries in the preadolescents was positively correlated to patient's body weight and duration of time between injury and surgery [26, 29]. Regarding the effect of age and skeletal maturity on incidence of associated medial meniscal injuries in the pediatric population, results from studies varied [22, 26, 29].

Classically, ACL reconstruction for treating preadolescent ACL injuries was discouraged due to risk of physeal damage and premature closure of growth plate, leading to leg length discrepancies [11, 30]. However, the natural history of untreated ACL injuries was reported in an observational report that showed progressive instability of the knee, deterioration of function, inability to maintain preinjury activity level, and radiologic signs of early-onset osteoarthritis in a high proportion of these patients [11, 31] (Fig. 5.1). Chronic ACL deficiency may also have a substantial psychological effect on the affected individuals [32, 33]. In a recent systemic literature review, Chalmers et al. found ACL reconstructions were related to fewer subsequent meniscal injuries and significantly greater improvement in activity level, but no significant differences in function scores or development of osteoarthritis [34] (Fig. 5.2). Another systemic review on the natural history of ACL partial tears reported good functional

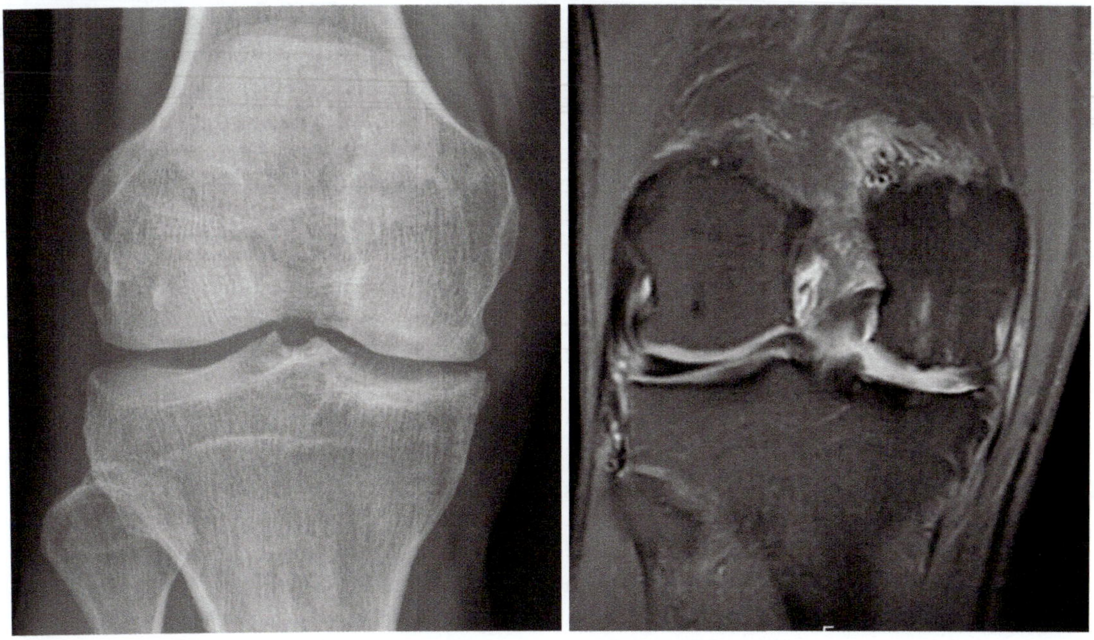

Fig. 5.1 Anterior–posterior radiograph and magnetic resonance imaging images of an untreated anterior cruciate ligament-injured knee in a 32-year-old male. The initial injury happened 20 years ago when he was 12 years old. The images showed Kellgren-Lawrence grade 3 medial-side osteoarthritis with marked meniscus tear, cartilage destruction, and osteophyte formation

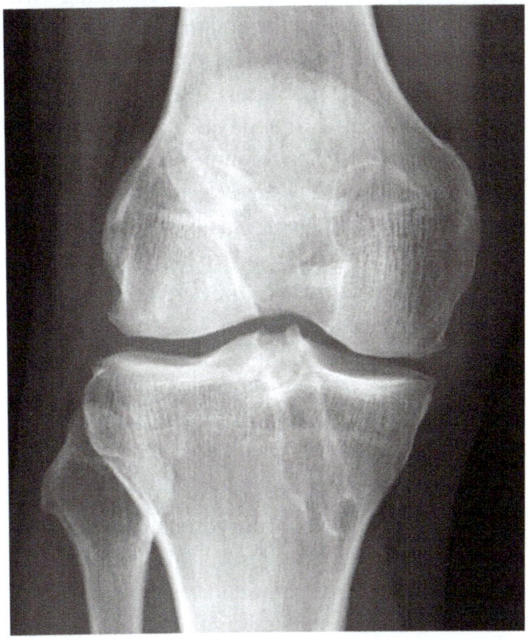

Fig. 5.2 Anterior–posterior radiograph of an anterior cruciate ligament-reconstructed knee in a 31-year-old female. The initial injury happened 18 years ago when she was 13 years old. The image showed Kellgren-Lawrence grade 2 medial-side osteoarthritis

outcomes at mid-term follow-up with half of the patients returning to their preinjury level of activity, but progressive laxity and high prevalence of knee pain were also observed [35]. Therefore, meticulous assessment of factors including bone age and maturity, comprehensive planning, and family consultation are all essential components to ensure a satisfactory treatment outcome for each patient.

In general, for a preadolescent patient with incomplete ACL rupture on MRI and those who have a normal to near-normal physical examination, nonsurgical treatment including activity modification, bracing, and physical therapy should be seriously considered [8, 36]. On the other hand, for preadolescent patients with repeated instability episodes and significant reduction in activity, physeal-sparing ACL reconstruction should be planned to improve knee function, avoid strict activity modification, and reduce additional intraarticular injuries [11, 36, 37].

5.3 Characteristics and the Natural History of Anterior Cruciate Ligament Injuries in Adolescents

Risk of ACL injuries in adolescents is significantly greater than preadolescents, and the incidence increased as the adolescent ages [10]. ACL injuries accounted for 10.6% of all sports injuries in 13–17-year-old cohort and were the leading cause of sports injuries for this population [17]. The highest incidence of ACL injuries was observed in males 17–18 years old and females 16–17 years old [4, 5, 10]. ACL reconstruction rates reported in a 20-year review in New York State was 8 times more in the 15–18-year-old cohort than the 11–14-year-old cohort [38]. In a population-based study for children and adolescents, ACL injuries from 1994 to 2013 revealed an annual increase of injury rate by 2.4 and 2.6% in males 15–16 years old and females 15–16 years old, respectively. In the same report, the injury rate in the 17–18-year-old population did not increase significantly [10].

A systemic review focusing on the risk and incidence of ACL injuries among high school athletes revealed that males experienced a greater number of ACL injuries than females. However, when considering the rate of injury per exposure, females exhibited a higher risk compared to males, with a relative risk of 1.57 (95% confidence interval [CI], 1.35–1.82) [6, 39]. In sex-comparable sports such as soccer, basketball, and baseball, females had a higher injury risk per exposure [6, 40, 41]. This vulnerability of female knees has been attributed to risk factors such as anatomical, environmental, neuromuscular, hormonal, and biomechanical differences between male and female [42–45].

Compared with preadolescents, ACL injuries in adolescents occur more in the substance of the ACL instead of tibial spine bony avulsions [20, 27, 28]. For associated meniscal or chondral injuries, body mass index (BMI) and duration of time between injury and surgery remained the most significant factors leading to increased injuries [22, 26, 29]. Age was not considered a significant factor for associated meniscal or chondral injuries [22, 26]. Dumont et al. reported more medial meniscal tears and more medial femoral condyle chondral injuries in patients >15 years old compared to patients <15 years old [29].

For managing adolescent ACL injuries, evaluation of knee stability, bone maturity, Tanner stage, and growth remaining are of profound importance. Several studies regarding synovial fluid changes after ACL injuries found in asymptomatic individuals with an ACL injury, the inflammatory enzymes and degradation products increased initially and returned to normal six months after the injury [46, 47]. Whereas for symptomatic patients, the inflammatory enzymes and degradation products were elevated regardless of the duration from injury [48]. Nonoperative treatment of chronic ACL injuries in symptomatic young competitive athletics was thus reported to be poor. Initially the patients may suffer difficulties with athletic activities and would eventually progress to difficulties with daily living [49, 50]. Even with rehabilitation, only one-third of patients improved, whereas the remaining two-thirds either remained the same or deteriorated over time [50].

Therefore, for patients with incomplete ACL injuries and near-normal physical examination, it may be viable to opt for nonsurgical treatment and delay any further surgical intervention until physeal closure may be feasible. However, this may lead to an increased risk of meniscal or chondral injuries when they return to sports activities [21, 29, 36]. The adolescent patients may need to decrease their activity level accordingly [8]. For adolescents with more than 2 years of growth remaining, all-epiphyseal or intraarticular-extraphyseal reconstruction should be considered. For those with less than 2 years of growth remaining, physeal-respecting transphyseal reconstruction or nonsurgical treatment until skeletal maturity for routine transphyseal reconstruction are the choices. For adolescents with chronic ACL injuries who reach skeletal maturity, routine transphyseal reconstruction as adults is feasible [36].

5.4 Characteristics and the Natural History of Anterior Cruciate Ligament Injuries in Young Adults

In adults, the incidence of ACL injuries is highest in early adulthood with males peaking in the 20–25-year-old age bracket and females in the 18–20-year-old age bracket. The incidence subsequently decreased with age [2, 39]. For early adulthood, the epidemiology is close to adolescents reaching maturity, with football, soccer, and lacrosse being the most common causes for ACL injuries. With advancing age, individuals tend to adopt a more sedentary lifestyle and engage in less physically demanding activities. This shift is mirrored in the reduced incidence of ACL injuries. Furthermore, the primary causes of injury transition to activities such as alpine skiing, soccer, and work-related incidents [51].

For young, active adults, nonoperative treatment for ACL injuries is poorly tolerated and often leads to recurrent instability and meniscal/chondral injuries [52]. In this group of patients, ACL reconstruction is often favored for future sport activities. However, Ardern et al. reported in a systemic review that, after ACL reconstruction, 82% of the patients returned to some level of sporting activity, 63% returned to the preinjury level, and only 44% returned to competitive sports [53]. Younger and male athletes were more likely to return to play after ACL reconstruction [54]. Factors favoring return to play included younger age, aggressive preoperative rehabilitation program, restoration of normal knee range of motion preoperatively especially full knee extension, and normal neuromuscular control [1, 55].

Osteoarthritis after ACL injury and subsequent ACL reconstruction is of great concern, especially in young and active group of patients. The incidence of osteoarthritis after ACL reconstruction was reported as high as 61% after 20 years of follow-up [1]. In a randomized control trial with 14-year follow-up, the prevalence of osteoarthritis increased by threefold in the ACL-reconstructed knee compared with the contralateral healthy knee [56] (Figs. 5.3 and 5.4).

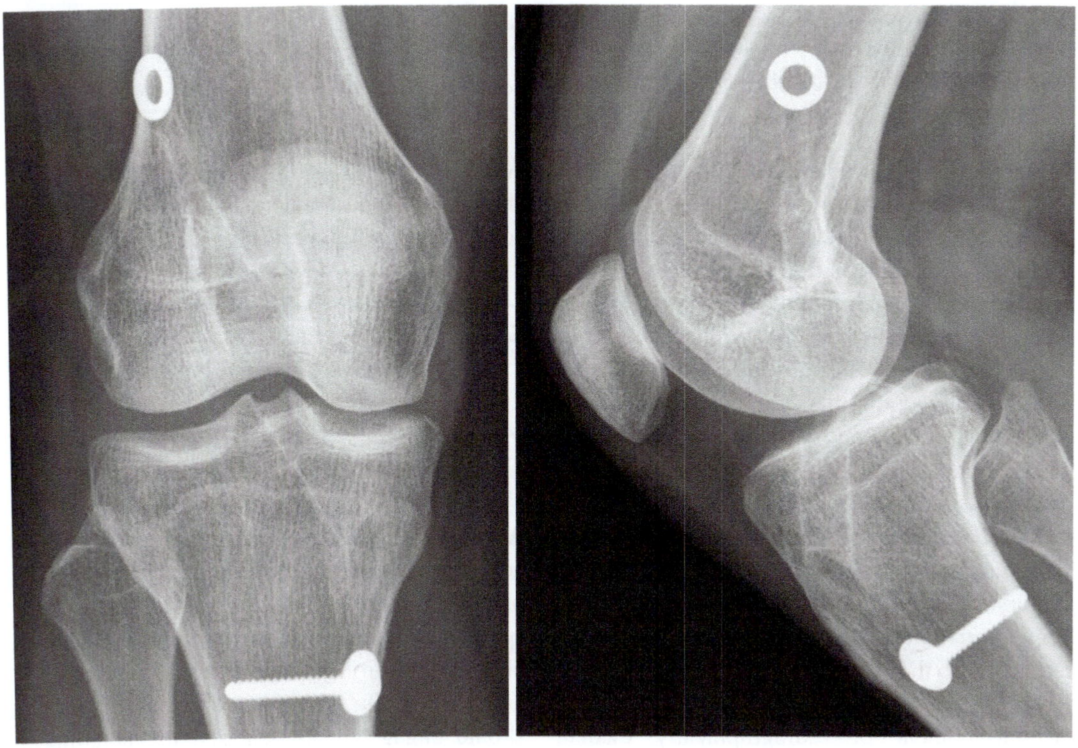

Fig. 5.3 Anterior–posterior and lateral radiographs of an anterior cruciate ligament-reconstructed knee in a 35-year-old male. The initial injury happened 15 years ago when he was 20 years old. The image showed no radiographic osteoarthritis

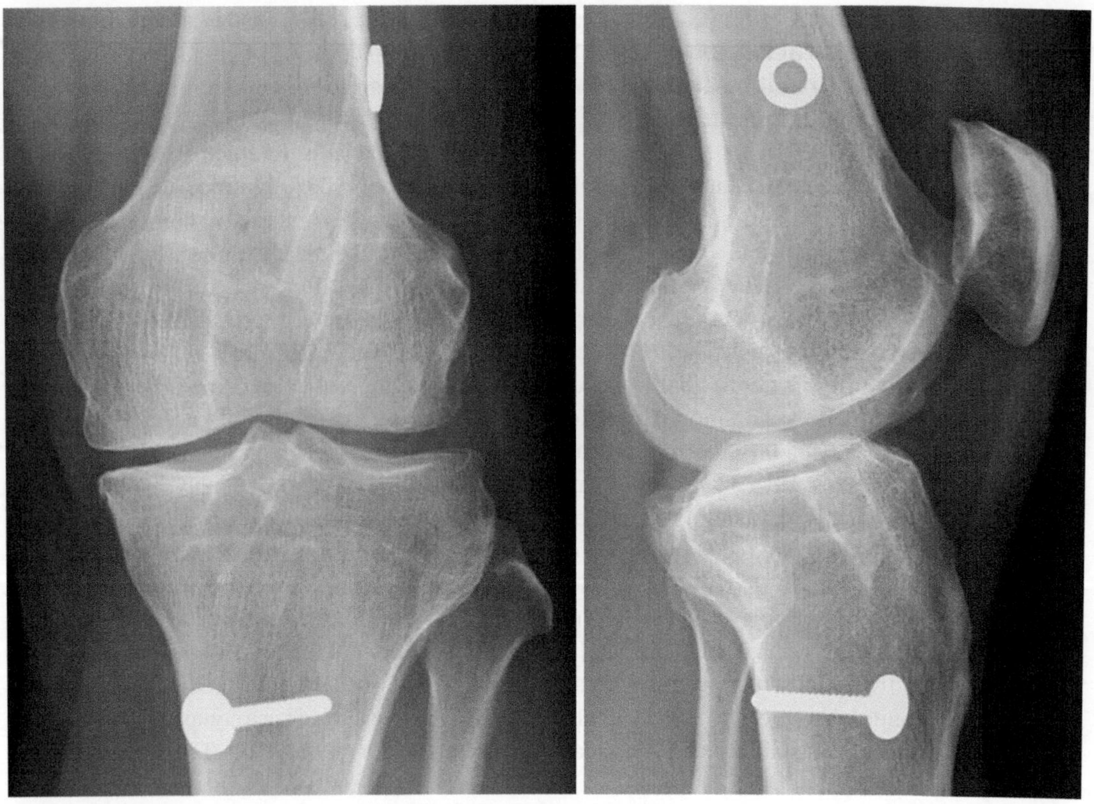

Fig. 5.4 Anterior–posterior and lateral radiographs of an anterior cruciate ligament-reconstructed knee in a 41-year-old male. The initial injury happened 14 years ago when he was 27 years old. The image showed Kellgren-Lawrence grade 3 medial-side osteoarthritis

Preoperative Tegner activity level, greater BMI, prior medial meniscectomy, and medial chondrosis were all related to development of osteoarthritis after ACL reconstruction [52, 56, 57]. The use of bone-patella tendon-bone autograft was also reported to increase the radiographic osteoarthritis risk [52, 58, 59]. The question remains whether ACL reconstruction alters the natural history of ACL injury. A systemic review of long-term outcomes comparing ACL reconstruction and nonoperative management of ACL injuries revealed that ACL reconstruction significantly decreased subsequent meniscal injuries and need for further surgery, and significantly improved patient's Tegner activity level. However, no significant differences in the Lysholm score, International Knee Documentation Committee (IKDC) score, or development of radiographically evident osteoarthritis were found [34]. Before managing ACL injuries in young patients, all these factors should be incorporated for a comprehensive solution.

5.5 Characteristics and the Natural History of Anterior Cruciate Ligament Injuries in Middle-Aged Adults and the Elderly

As the life expectancy lengthens with medical advancement, more middle-aged and elderly people remain physically active and involved in sports activities and competition. The Federal physical activity guidelines for Americans suggest that adults who are 65 years old or older should engage in physical activity including at least 150 minutes of moderate-intensity aerobic exercise per week, and muscle-strengthening activities of moderate to high intensity on two or more days a week. In a report in 2016 regarding the well-being of older Americans, it was found that an increasing proportion of Americans aged 65 years and above were adhering to the aforementioned guidelines. This indicates a positive trend toward improved physical fitness among older Americans. In 1998, about 6% of people aged 65 years and older met the guidelines, compared with 12% in 2014 [60]. In the same year, 37% of Americans met the guidelines for aerobic activity and 17% met the guidelines for muscle-strengthening activities. In general, age-related muscle atrophy begins at about 50 years of age [61]. Typically, it is expected that the maximum voluntary strength of an individual at the age of 60 years should retain at least 80–90% of their peak strength from their younger years [62]. However, the strength of the ACL tends to decline more rapidly than overall muscle strength. Research conducted by Woo et al. showed that the maximum load before failure of the ACL decreases significantly with age. The study found that the ACL's strength was 70% in middle-aged specimens (ages 40–50) and dropped to 30% in the older age group (ages 60–97), compared to the younger group (ages 22–35) [63].

As more physically fit middle-aged adults and elderly with preserved muscle strength but weakened ACL engage in physical activity, ACL ruptures may pose a substantial clinical problem in this age group. In a population-based study in New Zealand, the rate of ACL reconstruction in the 50–54-year-old population was approximately 10 per 100,000 person-years and decreased with age [39]. Another study focusing on the US population revealed that the annual incidence of ACL injuries in individuals over 46 years old was 20.4 per 100,000 person-years. Notably, the incidence significantly increased during the observation period from 1990 to 2010 [2].

Recent studies regarding patients older than 50 years showed encouraging results including clinical satisfaction, return to activity, and complication rate after ACL reconstruction [14, 64–66]. In this age group, the major mechanism of injury is skiing and other noncontact sports such as dancing and tennis, in contrast to the younger population, whose predominant causes of injury are contact sports such as soccer, basketball, and baseball [65, 66]. The rate of associated injuries to the meniscus or cartilage was as high as 90% [14, 65]. ACL reconstruction in patients older than 50 years resulted in high satisfaction rate of 88–96%, high rate of return to preinjury activity level of 86–88%, and a low complication rate of 6% [14]. All these results were comparable to the younger population [14, 64–67]. Despite the above encouraging results, Kinugasa et al. reported poorer synovial coverage and more partial tears of the ACL graft during second-look arthroscopy in the older age group (>50 years), compared to the younger patient groups [68]. No significant difference in clinical outcomes among age groups was found in the report [68]. Physicians should also be aware that the presence of preoperative osteoarthritis or limb malalignment may have a negative effect on the results of ACL reconstruction [14, 65]. Thus, careful patient selection for ACL reconstruction in this age group is the key to a successful surgical outcome.

5.6 Conclusion

ACL injury is not a condition exclusively encountered in young athletes. Earlier participation in competitive sports in the pediatric population and increased physical activities in the middle-aged and elderly population have expanded the need for treatment of ACL injuries to a wider population. If left untreated, the ACL injury may pose significant disability and limitations of physical activity throughout the life span and would cause associated intraarticular injuries later in life, ultimately leading to degenerative changes in the knee. Age, maturity, and physical demands of the patients all affect the epidemiology, characteristics, and treatment strategies of ACL injuries. Comprehensive evaluation and consultation for each patient are of great importance for tailoring proper treatments individually.

References

1. Ellman MB, Sherman SL, Forsythe B, LaPrade RF, Cole BJ, Bach BR. Return to play following anterior cruciate ligament reconstruction. J Am Acad Orthop Surg. 2015;23:283–96.
2. Sanders TL, Kremers HM, Bryan AJ, Larson DR, Dahm DL, Levy BA, Stuart MJ, Krych AJ. Incidence of anterior cruciate ligament tears and reconstruction: a 21-year population-based study. Am J Sports Med. 2016;44:1502–7.
3. Kaeding CC, Léger-St-Jean B, Magnussen RA. Epidemiology and diagnosis of anterior cruciate ligament injuries. Clin Sports Med. 2017;36:1–8.
4. Musahl V, Karlsson J. Anterior cruciate ligament tear. N Engl J Med. 2019;380:2341–8.
5. Ladenhauf HN, Graziano J, Marx RG. Anterior cruciate ligament prevention strategies: are they effective in young athletes - current concepts and review of literature. Curr Opin Pediatr. 2013;25:64–71.
6. Gornitzky AL, Lott A, Yellin JL, Fabricant PD, Lawrence JT, Ganley TJ. Sport-specific yearly risk and incidence of anterior cruciate ligament tears in high school athletes: a systematic review and meta-analysis. Am J Sports Med. 2016;44:2716–23.
7. Granan L-P, Forssblad M, Lind M, Engebretsen L. The Scandinavian ACL registries 2004–2007: baseline epidemiology. Acta Orthop. 2009;80:563–7.
8. Moksnes H, Engebretsen L, Eitzen I, Risberg MA. Functional outcomes following a non-operative treatment algorithm for anterior cruciate ligament injuries in skeletally immature children 12 years and younger. A prospective cohort with 2 years follow-up. Br J Sports Med. 2013;47:488.
9. Shaw L, Finch CF. Trends in pediatric and adolescent anterior cruciate ligament injuries in Victoria, Australia 2005–2015. Int J Environ Res Public Heal. 2017;14:599.
10. Beck NA, Lawrence JTR, Nordin JD, DeFor TA, Tompkins M. ACL tears in school-aged children and adolescents over 20 years. Pediatrics. 2017;139:e20161877.
11. McConkey MO, Bonasia DE, Amendola A. Pediatric anterior cruciate ligament reconstruction. Curr Rev Musculoskelet Med. 2011;4:37.
12. Waldrop JI, Broussard TS. Disruption of the anterior cruciate ligament in a three-year-old child. A case report. J Bone Joint Surg Am. 1984;66:1113–4.
13. Legnani C, Terzaghi C, Borgo E, Ventura A. Management of anterior cruciate ligament rupture in patients aged 40 years and older. J Orthop Traumatol. 2011;12:177–84.
14. Figueroa D, Figueroa F, Calvo R, Vaisman A, Espinoza G, Gili F. Anterior cruciate ligament reconstruction in patients over 50 years of age. Knee. 2014;21:1166–8.
15. DeLEE JC, Curtis R. Anterior cruciate ligament insufficiency in children. Clin Orthop Relat Res. 1983;172:112–8.
16. Clanton TO, DeLee JC, Sanders B, Neidre A. Knee ligament injuries in children. J Bone Joint Surg Am. 1979;61:1195–201.
17. Stracciolini A, Casciano R, Friedman HL, Meehan WP, Micheli LJ. Pediatric sports injuries: an age comparison of children versus adolescents. Am J Sports Med. 2013;41:1922–9.
18. Werner BC, Yang S, Looney AM, Gwathmey FW. Trends in pediatric and adolescent anterior cruciate ligament injury and reconstruction. J Pediatr Orthop. 2016;36:447–52.
19. Janssen KW, Orchard JW, Driscoll TR, van Mechelen W. High incidence and costs for anterior cruciate ligament reconstructions performed in Australia from 2003–2004 to 2007–2008: time for an anterior cruciate ligament register by Scandinavian model? Scand J Med Sci Sports. 2012;22:495–501.
20. Vaquero J, Vidal C, Cubillo A. Intra-articular traumatic disorders of the knee in children and adolescents. Clin Orthop Relat Res. 2005;432:97–106.
21. Mohtadi N, Grant J. Managing anterior cruciate ligament deficiency in the skeletally immature individual: a systematic review of the literature. Clin J Sport Med. 2006;16:457–64.
22. Prince JS, Laor T, Bean JA. MRI of anterior cruciate ligament injuries and associated findings in the pediatric knee: changes with skeletal maturation. AJR Am J Roentgenol. 2005;185:756–62.
23. Meyers MH, Mckeever FM. Fracture of the intercondylar Eminence of the Tibia. J Bone Joint Surg. 1970;52:1677–84.
24. Kocher MS, Mandiga R, Klingele K, Bley L, Micheli LJ. Anterior cruciate ligament injury versus Tibial spine fracture in the skeletally immature knee: a comparison of skeletal maturation and notch width index. J Pediatr Orthop. 2004;24:185.
25. Mayo MH, Mitchell JJ, Axibal DP, Chahla J, Palmer C, Vidal AF, Rhodes JT. Anterior cruciate ligament injury at the time of anterior Tibial spine fracture in young patients: an observational cohort study. J Pediatr Orthop. 2019;39:e668–73.
26. Vavken P, Tepolt FA, Kocher MS. Concurrent meniscal and chondral injuries in pediatric and adolescent patients undergoing ACL reconstruction. J Pediatr Orthop. 2018;38:105–9.
27. Wessel LM, Scholz S, Rüsch M. Characteristic pattern and management of intra-articular knee lesions in different pediatric age groups. J Pediatr Orthop. 2001;21:14.
28. Williams JS, Abate JA, Fadale PD, Tung GA. Meniscal and nonosseous ACL injuries in children and adolescents. Am J Knee Surg. 1996;9:22–6.
29. Dumont GD, Hogue GD, Padalecki JR, Okoro N, Wilson PL. Meniscal and chondral injuries associated with pediatric anterior cruciate ligament tears. Am J Sports Med. 2012;40:2128–33.
30. Larsen MW, Garrett WE, DeLee JC, Moorman CT. Surgical management of anterior cruciate liga-

ment injuries in patients with open Physes. J Am Acad Orthop Surg. 2006;14:736–44.
31. Aichroth PM, Patel DV, Zorrilla P. The natural history and treatment of rupture of the anterior cruciate ligament in children and adolescents. A prospective review. J Bone Joint Surg Br. 2002;84:38–41.
32. Filbay SR, Culvenor AG, Ackerman IN, Russell TG, Crossley KM. Quality of life in anterior cruciate ligament-deficient individuals: a systematic review and meta-analysis. Br J Sports Med. 2015;49:1033.
33. Filbay SR, Crossley KM, Ackerman IN. Activity preferences, lifestyle modifications and re-injury fears influence longer-term quality of life in people with knee symptoms following anterior cruciate ligament reconstruction: a qualitative study. J Physiother. 2016;62:103–10.
34. Chalmers PN, Mall NA, Moric M, Sherman SL, Paletta GP, Cole BJ, Bach BR. Does ACL reconstruction alter natural history & quest. J Bone Joint Surg. 2014;96:292–300.
35. Pujol N, Colombet P, Cucurulo T, et al. Natural history of partial anterior cruciate ligament tears: a systematic literature review. Orthop Traumatol Surg Res. 2012;98:S160–4.
36. DeFrancesco CJ, Storey EP, Shea KG, Kocher MS, Ganley TJ. Challenges in the management of anterior cruciate ligament ruptures in skeletally immature patients. J Am Acad Orthop Surg. 2018;26:e50–61.
37. Longo UG, Ciuffreda M, Casciaro C, Mannering N, Candela V, Salvatore G, Denaro V. Anterior cruciate ligament reconstruction in skeletally immature patients : a systematic review. Bone Joint J. 2016;99-B:1053–60.
38. Dodwell ER, LaMont LE, Green DW, Pan TJ, Marx RG, Lyman S. 20 years of pediatric anterior cruciate ligament reconstruction in New York State. Am J Sports Med. 2014;42:675–80.
39. Gianotti SM, Marshall SW, Hume PA, Bunt L. Incidence of anterior cruciate ligament injury and other knee ligament injuries: a national population-based study. J Sci Med Sport. 2009;12:622–7.
40. Joseph AM, Collins CL, Henke NM, Yard EE, Fields SK, Comstock RD. A multisport epidemiologic comparison of anterior cruciate ligament injuries in high school athletics. J Athl Train. 2013;48:810–7.
41. Shea KG, Grimm NL, Ewing CK, Aoki SK. Youth sports anterior cruciate ligament and knee injury epidemiology: who is getting injured? In what sports? when? Clin Sports Med. 2011;30:691–706.
42. Ingram JG, Fields SK, Yard EE, Comstock RD. Epidemiology of knee injuries among boys and girls in US high school athletics. Am J Sports Med. 2008;36:1116–22.
43. Gehring D, Melnyk M, Gollhofer A. Gender and fatigue have influence on knee joint control strategies during landing. Clin Biomech. 2009;24:82–7.
44. Benjaminse A, Otten E. ACL injury prevention, more effective with a different way of motor learning? Knee Surg Sports Traumatol Arthrosc. 2011;19:622–7.
45. Benjaminse A, Gokeler A, Dowling AV, Faigenbaum A, Ford KR, Hewett TE, Onate JA, Otten B, Myer GD. Optimization of the anterior cruciate ligament injury prevention paradigm: novel feedback techniques to enhance motor learning and reduce injury risk. J Orthop Sports Phys Ther. 2015;45:170–82.
46. Cameron M, Buchgraber A, Passler H, Vogt M, Thonar E, Fu F, Evans CH. The natural history of the anterior cruciate ligament-deficient knee: changes in synovial fluid cytokine and Keratan sulfate Concentrations. Am J Sports Med. 1997;25:751–4.
47. Dahlberg L, Fridén T, Roos H, Lark MW, Lohmander LS. A longitudinal study of cartilage matrix metabolism in patients with cruciate ligament rupture—synovial fluid concentrations of Aggrecan fragments, Stromelysin-1 and tissue inhibitor of Metalloproteinase-1. Rheumatology. 1994;33:1107–11.
48. Lohmander LS, Hoerrner LA, Dahlberg L, Roos H, Björnsson S, Lark MW. Stromelysin, tissue inhibitor of metalloproteinases and proteoglycan fragments in human knee joint fluid after injury. J Rheumatol. 1993;20:1362–8.
49. Noyes FR, Mooar PA, Matthews DS, Butler DL. The symptomatic anterior cruciate-deficient knee. Part I: the long-term functional disability in athletically active individuals. J Bone Joint Surg Am. 1983;65:154–62.
50. Noyes FR, Matthews DS, Mooar PA, Grood ES. The symptomatic anterior cruciate-deficient knee. Part II: the results of rehabilitation, activity modification, and counseling on functional disability. J Bone Joint Surg. 1983;65:163–74.
51. Zysk SP, Refior HJ. Operative or conservative treatment of the acutely torn anterior cruciate ligament in middle-aged patients. Arch Orthop Trauma Surg. 2000;120:59–64.
52. Murawski CD, van Eck CF, Irrgang JJ, Tashman S, Fu FH. Operative treatment of primary anterior cruciate ligament rupture in adults. J Bone Joint Surg. 2014;96:685–94.
53. Ardern CL, Webster KE, Taylor NF, Feller JA. Return to sport following anterior cruciate ligament reconstruction surgery: a systematic review and meta-analysis of the state of play. Br J Sports Med. 2011;45:596.
54. Brophy RH, Schmitz L, Wright RW, Dunn WR, Parker RD, Andrish JT, McCarty EC, Spindler KP. Return to play and future ACL injury risk after ACL reconstruction in soccer athletes from the Multicenter Orthopaedic Outcomes Network (MOON) Group. Am J Sports Med. 2012;40:2517–22.
55. Morris RC, Hulstyn MJ, Fleming BC, Owens BD, Fadale PD. Return to play following anterior cruciate ligament reconstruction. Clin Sports Med. 2016;35:655–68.
56. Barenius B, Ponzer S, Shalabi A, Bujak R, Norlén L, Eriksson K. Increased risk of osteoarthritis after anterior cruciate ligament reconstruction: a 14-year

follow-up study of a randomized controlled trial. Am J Sports Med. 2014;42:1049–57.
57. Group MK, Jones MH, Oak SR, et al. Predictors of radiographic osteoarthritis 2 to 3 years after anterior cruciate ligament reconstruction: data from the MOON on-site nested cohort. Orthop J Sports Med. 2019;7:2325967119867085.
58. Xie X, Liu X, Chen Z, Yu Y, Peng S, Li Q. A meta-analysis of bone–patellar tendon–bone autograft versus four-strand hamstring tendon autograft for anterior cruciate ligament reconstruction. Knee. 2015;22:100–10.
59. Roe J, Pinczewski LA, Russell VJ, Salmon LJ, Kawamata T, Chew M. A 7-year follow-up of Patellar Tendon and Hamstring Tendon grafts for arthroscopic anterior cruciate ligament reconstruction: differences and similarities. Am J Sports Med. 2005;33:1337–45.
60. Statistics FIF on A-R. Older Americans 2016: key indicators of Well-being. Federal Interagency Forum on aging-related statistics. Washington, DC: U.S. Government Printing Office; 2016.
61. Faulkner JA, Davis CS, Mendias CL, Brooks SV. The aging of elite male athletes: age-related changes in performance and skeletal muscle structure and function. Clin J Sport Med. 2008;18:501–7.
62. Menard D. The aging athlete. In: Harri M, CW, WDS, Micheli LJ, editors. . 2nd ed. Oxford: Oxford University Press; 1998. p. 787–814.
63. Woo SL-Y, Hollis JM, Adams DJ, Lyon RM, Takai S. Tensile properties of the human femur-anterior cruciate ligament-tibia complex: the effects of specimen age and orientation. Am J Sports Med. 1991;19:217–25.
64. Cinque ME, Chahla J, Moatshe G, DePhillipo NN, Kennedy NI, Godin JA, LaPrade RF. Outcomes and complication rates after primary anterior cruciate ligament reconstruction are similar in younger and older patients. Orthop J Sports Med. 2017;5:2325967117729659.
65. Dahm DL, Wulf CA, Dajani KA, Dobbs RE, Levy BA, Stuart MA. Reconstruction of the anterior cruciate ligament in patients over 50 years. J Bone Joint Surg Br. 2008;90-B:1446–50.
66. Blyth MJG, Gosal HS, Peake WM, Bartlett RJ. Anterior cruciate ligament reconstruction in patients over the age of 50 years: 2- to 8-year follow-up. Knee Surg Sports Traumatol Arthrosc. 2003;11:204–11.
67. Baker CL, Jones JC, Zhang J. Long-term outcomes after anterior cruciate ligament reconstruction in patients 60 years and older. Orthop J Sports Med. 2014;2:2325967114561737.
68. Kinugasa K, Mae T, Matsumoto N, Nakagawa S, Yoneda M, Shino K. Effect of patient age on morphology of anterior cruciate ligament grafts at second-look arthroscopy. Arthrosc: J Arthrosc Relat Surg. 2011;27:38–45.

Part II

General Issues

Physical and Psychosocial Characteristics of ACL Injury-Prone Groups

6

Anna Nordenholm, Ramana Piussi, Jón Karlsson, and Eric Hamrin Senorski

6.1 Introduction

The fact that several risk factors play a role in the risk of sustaining an injury to the anterior cruciate ligament (ACL) is inescapable. Factors thought to determine which individuals are predisposed to suffer an ACL injury have been frequently studied [1–11]. The risk of an ACL injury is multifactorial, with both intrinsic and extrinsic risk factors theorized to place some individuals at greater ACL injury risk compared with others [7]. Intrinsic risk factors are related to the individual's physical characteristics and are either modifiable, such as biomechanics and neuromuscular function, or non-modifiable, such as anatomy, sex, and age. Intrinsic risk factors also include psychological and psychosocial factors, which have not been studied as frequently as physical characteristics. Extrinsic risk factors are defined as environmental factors, that is, sports exposure, type of sport, equipment, climate, and social context. This chapter addresses risk factors for sustaining a primary ACL injury and attempts to summarize the characteristics that predispose an individual to injury. Moreover, the methodological quality of the research behind the conclusions is discussed.

A. Nordenholm
Sahlgrenska Sports Medicine Center, Gothenburg, Sweden

R. Piussi · E. Hamrin Senorski (✉)
Sahlgrenska Sports Medicine Center, Gothenburg, Sweden

Department of Rehabilitation and Health, Institute of Neuroscience and Physiology, Sahlgrenska Academy, University of Gothenburg, Gothenburg, Sweden
e-mail: eric.hamrin.senorski@gu.se

J. Karlsson
Sahlgrenska Sports Medicine Center, Gothenburg, Sweden

Department of Orthopaedics, Sahlgrenska University Hospital, Mölndal, Sweden

Department of Orthopaedics, Institute of Clinical Sciences, Sahlgrenska Academy, University of Gothenburg, Gothenburg, Sweden

6.2 Incidence of ACL Injury and the Influence of Sex, Age, and Exposure

The risk of sustaining an ACL injury in organized sport is estimated at 2.8% during a time frame of 1 year (incidence proportion) [5]. Accordingly, in a team of 36 athletes, one ACL injury every year is expected. The annual incidence of ACL reconstructions has been reported as 68.8 per 100,000 person-years in the general population when adjusted for age and sex [12]. When determining the ACL injury risk by sex, there are significant differences between females and males. For individuals involved in organized sports, females have a 1.5 times higher relative risk of sustaining

M. Karahan et al. (eds.), *Anterior Cruciate Ligament Injuries Throughout the Life Span*,
https://doi.org/10.1007/978-3-031-84501-7_6

an ACL injury compared with males during a specified time period of 1–25 years (3.5% vs. 2.0%) [5].

So, for females, there is one ACL injury in every 29 females compared with one in every 50 males. However, athletic exposure, that is, training time and competitive participation, must be taken into account when quantifying the risk of injury [13]. Studies of ACL injury risk in relation to athletic exposure report that females run a greater risk of ACL injury compared with males [14]. Females have a 1.7-fold increased relative ACL injury risk compared with males at a given athletic exposure, with an incidence rate ratio of 1.9 compared with 0.9 ACL injuries in 10,000 athletic exposures [5]. The sex disparity of ACL injury incidence rates is higher for amateur athletes compared with intermediate and elite athletes. Accordingly, female athletes at amateur level are especially at risk of ACL injury compared with their male counterparts [5].

In addition to disparities in incidence rates, there are sex-based differences during the life span in which females and males sustain their injury. The peak in annual ACL injury incidence per 100,000 person-years in females is observed between 14 and 18 years of age, compared with 19–25 years in males (Fig. 6.1) [12]. Accordingly, ACL injuries are most common in females during high-school years and in males during college or shortly after [12]. After 18 years of age, there is a steep decline in the ACL injury rate for females, which is likely a result of the reduction in competitive sports participation after high school. However, in males, the incidence decreases more slowly after around 25 years of age, which may indicate that male athletes continue to participate in organized sport to a greater extent at a higher age [12].

Prior to puberty, ACL injuries are rare and the injury rates do not differ between the sexes [3]. However, during puberty, both females and males demonstrate anatomic, neuromuscular, and hormonal differences that potentially influence the increase in ACL injury rates [3]. The hormonal milieu differs significantly between the sexes, due to an increase in sex hormones, estrogen in females and testosterone in males [15]. Despite the fact that the onset of puberty is associated with increased ligament stiffness, it has been suggested that the cyclic fluctuations in female hormones during the menstrual cycle affect the mechanical properties of the ligaments, resulting in increased knee laxity [16]. An association with an increased risk of ACL injury has not yet been established, however.

Throughout adolescence, young individuals undergo rapid increases in height and body mass, resulting in a higher center of mass and longer lever arms in the lower extremities [3]. In males, this adolescence growth spurt is also associated with increases in strength, power, and the ability to attenuate forces [3]. The same increase in physical capacity is not seen in adolescent females, which may partly explain the change in

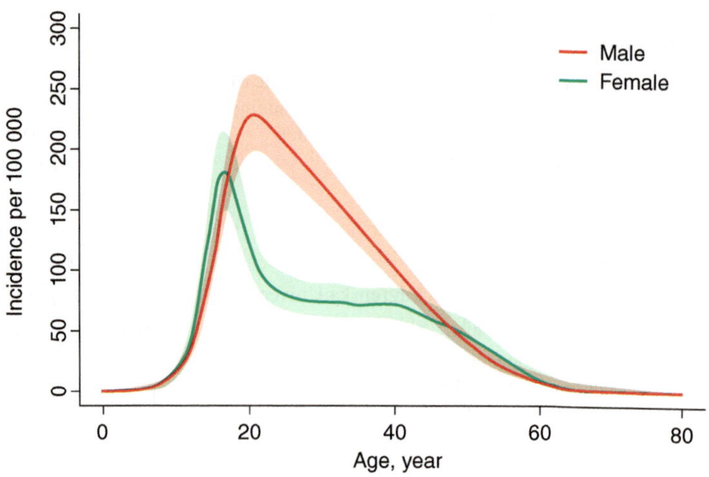

Fig. 6.1 Age-specific annual anterior cruciate ligament injury incidence per 100,000 person-years [12]

biomechanics found in females and their high incidence of ACL injuries at this age [3].

The ACL injury risk is strongly influenced by the type of exposure, that is, physical activity or sport. Sports that involve pivoting and cutting maneuvers, especially with sudden changes of speed, place the athlete at a greater risk of an ACL injury. When participating in the same high-risk sport, females have been reported to run a two- to eightfold greater risk of ACL injury compared with males [3]. Collision and contact sports have a high incidence rate, 1.88–2.10 in 10,000 athlete exposures for females and 0.87–1.12 for males [6]. The incidence rates for sports with no or limited athlete-to-athlete contact have been reported to have similar risks between the sexes, 0.36–0.71 for females and 0.21–0.29 for males in 10,000 athlete exposures [6]. However, non-contact sports involving fixed objects in combination with high-impact rotational landings, such as gymnastics and obstacle course races, expose the athletes, especially females, to a greater ACL injury risk, with an incidence rate of 4.80 in 10,000 athlete exposures in females compared with 1.75 in males [6].

In general, studies reporting on the incidence of ACL injury are observational cohort or case-control studies and have a moderate risk of bias [12, 17]. Most of studies reporting the incidence of ACL injury include large cohorts and the majority show an increased ACL injury risk in females compared with males. Based on this, there is moderate evidence that females run an increased risk of sustaining an ACL injury compared with males [6].

Fact Box 1: Statistical terms

Incidence proportion	The number of new injuries within a specified time period divided by the size of the population initially at risk
Incidence rate	The rate of incidence per exposure or per 100,000 person-years
Incidence rate ratio	The ratio of two rates of incidence
Odds ratio	The number of injuries divided by the number of non-injuries, not taking time or exposure into consideration
Relative risk/ risk ratio	The probability of injury in one group compared with another group

6.3 Neuromuscular and Anatomic Characteristics

To participate in sports can impose complex physical requirements on an athlete when movements such as pivoting, cutting, and landing are performed. Muscle strength and activation patterns in the lower extremities are important for overall function. Neuromuscular function is a term that is used in relation to injury and sports and its definition is an important starting point. Neuromuscular function has been defined as any of the aspects relating to the nervous system control of muscle activation and the factors contributing to task performance [18]. In the specific case of joint stability, neuromuscular control can be defined as the unconscious activation of dynamic restraints occurring before (and in response to) joint motion and loading, for the purpose of maintaining and restoring functional joint stability [18]. Sufficient attenuation of pivoting, cutting, and landing forces, as well as the ability to resist perturbations and external loads, is important when it comes to ACL injury prevention [3, 19].

Most ACL injuries occur during landing or lateral cutting movements and occur in non-contact mechanisms associated with high knee abduction forces and low knee flexion and lateral trunk motion over the involved leg [3]. During these movements, the ACL injury can happen as early as 40 milliseconds after the initial contact [20]. The neuromuscular function of the trunk and lower extremities differs between the sexes after the onset of puberty and these differences influence knee motion and the ability to manage high forces during sports movements [3]. After puberty, females exhibit measurable quadriceps dominance relative to the hamstrings, and present greater strength differences between the dominant and non-dominant leg [3]. There are also strength deficits in the hip muscles relative to body weight that can predispose females to ACL injury [7]. In addition, strength deficits in the hamstring muscles relative to the quadriceps have been associated with an increased ACL injury risk in females compared with matched male controls [7].

Females demonstrate different hip and knee joint muscular activation patterns, with greater knee abduction motions compared with males, which appear to predispose females to loads on the knee joint associated with the ACL injury mechanism [3]. Post-pubertal females exhibit less dynamic knee stability and reduced active stiffness in the knee joint compared with males [3]. There are also sex differences in trunk and hip neuromuscular control, where females exhibit greater ranges of hip and trunk motions during landings compared with males [3]. Greater ranges of hip and trunk motions, in combination with reduced active knee joint stiffness and increased joint laxity, probably increase the risk of ACL injury in females [3]. There is consistent evidence that a reduction in the ability to control trunk movement after rapid movements is strongly associated with ACL injury [3, 21]. However, there is robust evidence in support of the effectiveness of ACL injury prevention programs [22]. A meta-analysis of meta-analyses demonstrated a 50% reduction in the risk of both contact and non-contact ACL injuries in all athletes and a slightly higher reduction (67%) for non-contact injuries in female athletes [22].

Despite the fact that several neuromuscular injury risk factors have been identified, the role of the central nervous system (CNS) has not been thoroughly studied. There is, however, an increase in studies designed to determine the role of the CNS in managing movement patterns for the lower limbs. As the ACL is rich in mechanoreceptors, its rupture can lead to the cessation or differentiation of the afferent information from the affected joint to the CNS and can therefore result in a sensorimotor reorganization [23]. As a consequence, malfunctioning sensorimotor skills can affect thigh muscle strength, balance, and patterns of muscle activation [24]. The restoration of sensorimotor skills therefore plays a major role in the functional recovery of a patient with an ACL injury [24]. The idea that deficits in neurological function might be present prior to ACL injury and might therefore be a risk factor has been raised. Preliminary results from longitudinal studies show significantly less connectivity between the somatosensory cortex and the motor cortex, measured by functional magnetic resonance imaging, in collegiate athletes who go on to sustain an ACL injury [25, 26]. Similar results have also been reported in females classified as being at risk of suffering an ACL injury, defined by knee angles on the drop-jump test [27]. However, the current literature has major limitations in terms of small study samples and the fact that several areas in the brain have been analyzed, resulting in a high risk of type-1 error. In addition, the analyses have been performed at rest and do not account for the variability that can occur when an athlete is in motion. Despite limitations, the findings are novel. Considering the established relationship between neuromuscular function and ACL injury risk, the role of the CNS is an area of research that requires more attention.

Fact Box 2: Neuromuscular, anatomic, and psychosocial terms

Neuromuscular control	The unconscious activation of dynamic restraints occurring before (and in response to) joint motion and loading, for the purpose of maintaining and restoring functional joint stability [18]
Deficits in neuromuscular control	Deficits in muscle strength, power, coordination, or activation patterns that potentially increase injury risk due to increased joint loads [3]
Active knee joint stiffness	The elastic component of knee joint impedance depending on and able to be voluntarily controlled through muscle activation. Statistically calculated as the slope of the moment-angle curve [28]
Lateral femoral condyle ratio	Condylar depth relative to the total anteroposterior length of the condyle measured in the sagittal plane. Statistically calculated based on the intersection between the axis of the femoral shaft and the axis of the femoral condyle [1]
Notch width	The distance between the femoral condyles at the anterior outlet [1]
Notch width index	The ratio of notch width relative to the total condylar width [1]
Tibial slope	The posterior tilt of the tibial plateau relative to the longitudinal axis of the tibia [29]
Sports psychology	The cognitions, affects, and behaviors of sports participants and sports social culture as the social and cultural structures, climates, and processes influencing sports participants [30]

Among the anatomic risk factors for ACL injury, some pathologic morphologies of the knee joint have been associated with ACL injury [1]. There is evidence suggesting that femoral notch geometry/intercondylar shape is associated with ACL injury, where significant associations between an increased ACL injury risk and decreased notch width (distance between the femoral condyles at the anterior outlet), decreased notch width index (ratio of notch width to the total condylar width), as well as presence of notch stenosis, have been reported [1, 7, 11, 31]. These associations have been reported, regardless of age, and it has been suggested that females with a narrow notch width in particular run a great risk of ACL injury [7]. In addition, increased condylar depth (quantified as the lateral femoral condyle ratio), as well as an increased posterior tibial slope, has been correlated with increases in absolute laxity and side-to-side difference on a quantitative pivot-shift test (combined tibiofemoral internal rotation and anterior tibial translation) and, as a result, increased forces on the ACL (Fig. 6.2) [1]. These measurements may thus be indicative of

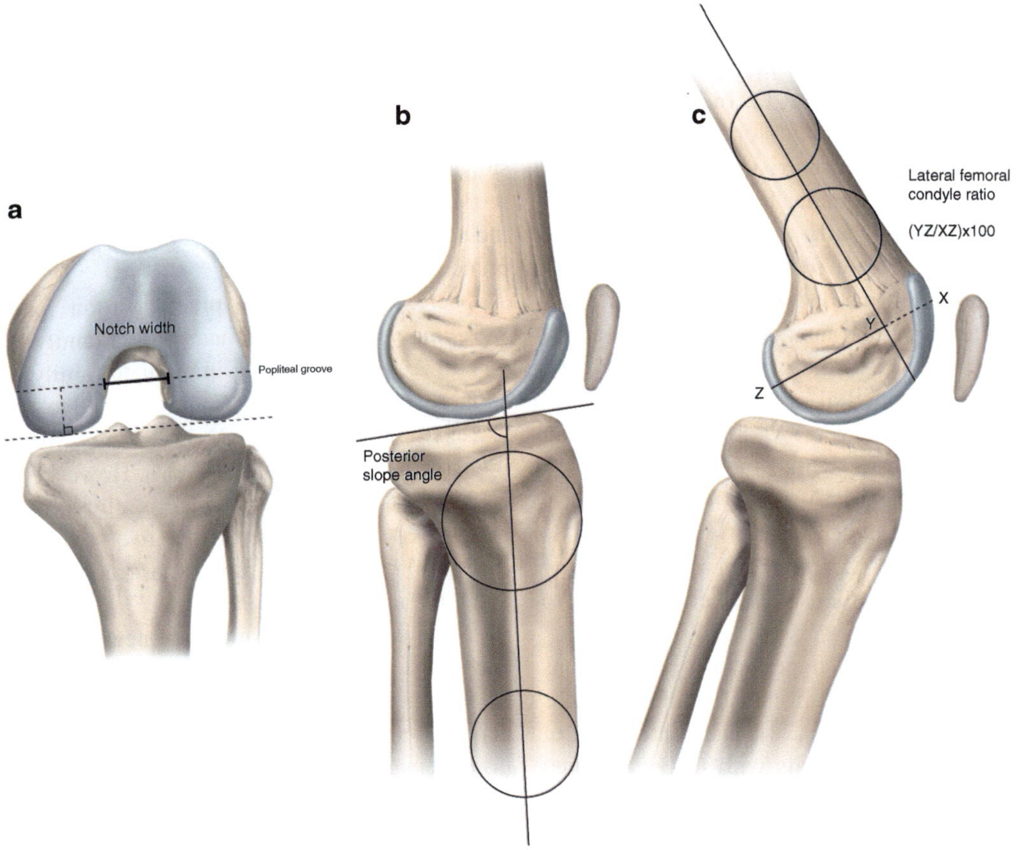

Fig. 6.2 Illustration of anatomic risk factors associated with anterior cruciate ligament injury. (**a**) Anterior view of the knee illustrating the measurement of the notch width, which is calculated as notch width relative to width ratio, which is calculated as notch width relative to total condylar width [1]. (**b**) Lateral view of the knee illustrating the posterior slope angle, which is calculated as the posterior tilt of the tibial plateau relative to the longitudinal axis of the tibia [29]. (**c**) Lateral view of the knee illustrating the lateral femoral condyle ratio. The long axis of the femoral shaft is determined by a line through the center of two circles centered on the femoral shaft and the axis of the femoral condyle is determined by a line between the most posterior and most anterior points of the lateral condyle. The ratio is calculated by dividing the distance from the intersection of these lines to the posterior end of the condyle (YZ) by the total anteroposterior length of the condyle (XZ) [1]. (Illustration created by Pontus Andersson at Pontus Andersson Art Production AB")

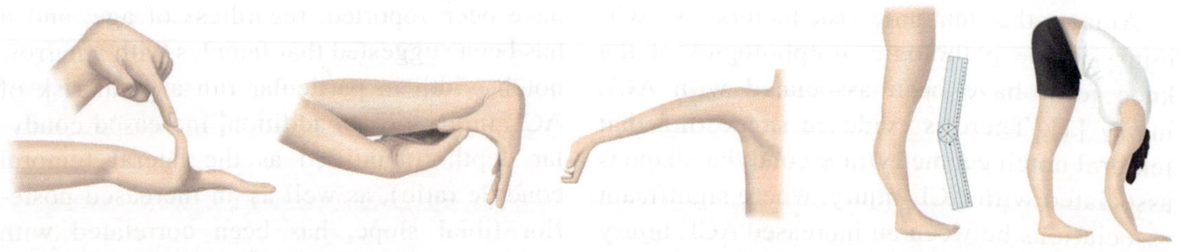

Fig. 6.3 The 9-point Beighton Score. The 9-point Beighton Score is the most commonly used measurement of generalized joint hypermobility (GJH), where a cut-off of ≥5 points is suggested to determine GJH [33]. One point is allocated for the ability to perform each of the mobility tests. One point is obtained for each thumb that touches the forearm when the passive apposition of the thumb to the flexor aspect of the forearm is performed. One point is obtained for each little finger that hyperextends beyond 90°, one point for each knee and elbow that hyperextend beyond 10°, and one point if the palms of the hands are able to rest flat on the floor when the forward flexion of the trunk with the knees extended is performed. (Illustration created by Pontus Andersson at Pontus Andersson Art Production AB")

increased rotatory knee laxity related to the ACL injury mechanism [1]. Increased medial and lateral tibial plateau slopes are also associated with an increased risk of ACL injury for males but not for females [29]. Moreover, poor tibiofemoral congruity and reduced cartilage and meniscal height increase ACL injury risk [1]. The medial meniscus is important to prevent anterior tibia translation, while the lateral meniscus provides rotatory knee stability [1]. Injuries to the medial or lateral meniscus significantly increase rotatory knee laxity during the pivot-shift test, which increases forces on the ACL and the risk of ACL injury [32]. Like many musculoskeletal injuries, individuals with a previous ACL injury run a higher risk of a new injury to the same knee (odds ratio [OR], 9.08) [7].

Generalized joint hypermobility (GJH) has been associated with primary unilateral ACL injury in several studies [33]. The relative risk of ACL injury in individuals with GJH compared with individuals without GJH is 2.8 for both sexes and 3.1 and 2.7, respectively, for males and females [33]. In individuals who have sustained an ACL injury, the presence of GJH is common (OR, 4.46) [33]. Consistent low-level evidence exists for GJH as a risk factor for primary unilateral ACL injury in males, but the evidence is conflicting with regard to females [33]. The 9-point Beighton Score is the most commonly used measurement of GJH, where a cut-off of ≥5 points is suggested to determine GJH—the higher the score, the higher the laxity. The Beighton Score includes five tests, of which four are performed bilaterally (Fig. 6.3).

There is a hereditary predisposition for an ACL injury, where individuals with a family history of ACL injury in the immediate family run an increased risk [2]. This finding has led to studies examining different DNA variants, mostly focusing on collagen types [4, 34]. However, the current results conflict and there is a high risk of bias [4, 7]. Future large prospective studies are therefore needed to draw robust conclusions about the influence of genetics.

When summarizing studies of neuromuscular and anatomic risk factors, there is a substantial heterogeneity of data, limiting the quality of the conclusions. Many studies have reported associations between knee joint morphology and ACL injury, but they do not provide any associated odds ratio. The interpretation of results is also complicated due to wide variations in inclusion criteria, measurement methodology, and nomenclature. Finally, there are strong limitations related to unreported information and inadequate control for confounding factors.

6.4 Psychosocial Characteristics

The psychological aspect of sports injuries is not as widely studied as the physical aspect, but it is attracting more attention. The vast majority of the available evidence on ACL injury and psychological factors is related to the recovery process [35]. Since the literature related to psychological risk factors for sport injuries is scarce, specific knowledge of/evidence relating to the ACL is limited.

Sports psychology is defined as "the cognitions, affects and behaviours of sport participants and sport social culture as the social and cultural structures, climates and processes influencing sport participants" [30]. In order to fully understand the potential impact of psychology on the ACL injury risk, the psychological burden associated with sport needs to be understood. In high-intensity sports, the athlete's training and competition are characterized by many factors, including the mental fortitude to push physical limits and a commitment to highly challenging goals [30]. Orthopedic injuries associated with macro-traumas should be seen in the light of a psychosocial sports culture, which is based on the disregard of pain and on performance success; the athlete has to do what it takes to succeed and, at times, even sacrifice health [30]. Both voluntary actions and social pressure may play a role in the relationship between sports participation and ACL injury risk [36].

The biopsychosocial influence on ACL injury has previously been studied by investigating the prediction and prevention/reduction of stress-related injuries [37], as well as identifying internal and external factors that characterize the ACL injury risk profile [38]. In ice hockey players, low vigor and fatigue, as measured by the Incredibly Short Profile of Mood States, have been reported to predict sports injuries [39]. Moreover, competence-based self-esteem and a reduced number of hours of sleep, together with changes in training volume, are risk factors for a new ACL injury in adolescent elite athletes [40].

The results of interviews with elite Swedish soccer players suggested that a combination of mental and physiological exhaustion might have predisposed players to an ACL injury [41]. The mental exhaustion may be due to overtraining (i.e., large chronic workout volumes), potential life stressors (such as family/sentimental situation, or difficult financial situation), and every player's personality (the ability of each individual to cope with stress). Mental exhaustion can then lead to reduced alertness and the capability to focus attention on the environment, placing a player at higher risk of ACL injury [41]. It has been suggested that reduced neurocognitive function, in terms of slower reaction time, speed in processing surrounding information, and visual and verbal memory, are risk factors for ACL injury [42, 43]. The results are based on 80 collegiate athletes who displayed lower neurocognitive function compared with a matched cohort in pre-season testing [42, 43]. Despite the large cohort and the strong methodological quality of the study, further research is needed to establish a potential relationship between neurocognitive function and ACL injury. One important factor when it comes to minimizing mental fatigue in elite athletes is the team's psychological climate, where social support given by teammates and the coach may reduce stress in cases of high levels of perceived stress [41].

Both psychological and psychosocial factors probably play a role in ACL injury risk. A sports injury is always followed by a psychological response and the present psychological status of an athlete greatly affects the recovery process of the individual [44]. The level of perceived stress in an athlete, related to participation, performance, and life outside sport (such as the death of a family member or a difficult economic

Table 6.1 Outcome measurements used to measure psychological factors related to ACL injury

Author, year	Psychological factor measured	Outcome measurement used
Swanik, 2007 [43]	Verbal memory, visual memory, processing speed, and reaction time	Computerized neurocognitive test battery (ImPACT)
Ivarsson, 2013 [46]	Trait anxiety	Swedish Universities Scales of Personality
Ivarsson, 2013 [46]	Life event stressors	The Life Events Survey for Collegiate Athletes (LESCA)
Ivarsson, 2013 [46]	Daily hassle	The Hassle and Uplift scale
Ivarsson, 2013 [46]	Coping strategies	The Brief COPE
Steffen, 2009 [47]	Task and ego goal orientation	Perception of Success Questionnaire (POSQ)
Steffen, 2009 [47]	Perceptions of the motivational climate in a team	Perceived Motivational Climate in Sport Questionnaire (PMCSQ)
Steffen, 2009 [47]	History of stressors	The Life Events Survey for Collegiate Athletes (LESCA)
Steffen, 2009 [47]	Somatic anxiety, worry, and concentration disruption	Norwegian Sports Anxiety Scale (SAS-n)
Steffen, 2009 [47]	Coping strategies	The Brief COPE
Von Rosen, 2017 [40]	Sleep	The Karolinska Sleep Questionnaire
Von Rosen, 2017 [40]	Perceived stress	The Perceived Stress Scale
Von Rosen, 2017 [40]	Competence-based self-esteem	The Competence-Based Self-Esteem Scale
Smith, 1997 [39]	Ability to be focused	The Positive States of Mind Inventory
Smith, 1997 [39]	Tension, depression, anger, vigor, fatigue, and confusion	The Incredibly Short Profile of Mood States

COPE Coping Orientation to Problem Experienced inventory; *ImPACT* Immediate Post-Concussion Assessment and Cognitive Testing

situation), has been linked to a higher injury risk [45]. A history of stressors, and the athlete's response to these stressors, has been associated with higher rates of sports injury [45]. Major life stresses have been associated with an increase in the risk of injury for an athlete [37]. In addition, both minor and major levels of stress, anxiety, and a mastery climate (supporting effort for individual or team development) have been associated with new injuries in soccer players across one season [46, 47]. On the other hand, perceived stress has not been reported to increase the risk of injury in adolescent elite athletes [40].

In general, studies analyzing the impact psychological factors can have on ACL injuries are longitudinal cohort studies with a moderate risk of bias and low generalizability. The findings are mostly based on small populations and there is a need for further studies on the subject. However, there is substantial evidence supporting the hypothesis that an increase in stress and perceived recovery appear to predict sports injury occurrence in sports [48] (Table 6.1).

6.5 Conclusion

There are several factors that can predispose individuals to a greater risk of ACL injury. One single risk factor alone is rarely able to explain why an injury has occurred. In most cases, an ACL injury is the result of a combination of several factors as well as coincidences. An ACL injury is more likely to occur during activities such as cutting, landing, and contact sports. Female athletes run a 1.7-fold greater risk of sustaining an ACL injury compared with male athletes at the same given exposure to risk activities. Moreover, females between 14 and 18 years of age and males between 19 and 25 years run a higher risk of sustaining an ACL injury. The risk is also increased for both sexes if the athlete is active at amateur level. Moreover, a higher level of perceived stress is associated with sports injury occurrence.

Deficits in strength relative to body weight in hip muscles or hamstring muscles relative to the quadriceps can increase the injury risk. Moreover, neuromuscular deficits such as a reduced ability to control trunk, hip, and knee abduction movements in the coronal plane during landings and cutting movements may expose the ACL to high forces and increase the associated injury risk. Neuromuscular deficits related to ACL injury risk are more common in female athletes compared with male athletes. Fortunately, they can be effectively targeted through appropriate injury-reduction training programs. Further, genetic and anatomic characteristics, such as increased femoral notch width, increased condylar depth, increased posterior tibial slope, cartilage and meniscal injuries, generalized joint hypermobility (for males), and previous ACL injuries in the immediate family may predispose an individual to injury. Our knowledge of whether psychosocial factors can predispose individuals to ACL injury is scarce, but preliminary evidence indicates that individuals exposed to a high level of perceived stress (such as the death of a family member) and fatigue might run an increased risk of injury. In addition, low competence-based self-esteem and an average of less than 8 hours of sleep during weekdays, together with quick changes in training volume, may increase the sports injury risk.

Taken together, individuals that present the following characteristics run an increased risk of sustaining an ACL injury: female sex; young age (14–25 years); being involved in high-risk sports at amateur level; high levels of perceived stress changes in movement patterns and/or strength deficits (Fig. 6.4). Most likely, the injury risk is also influenced by non-modifiable intrinsic factors related to the individual's genetic predisposition.

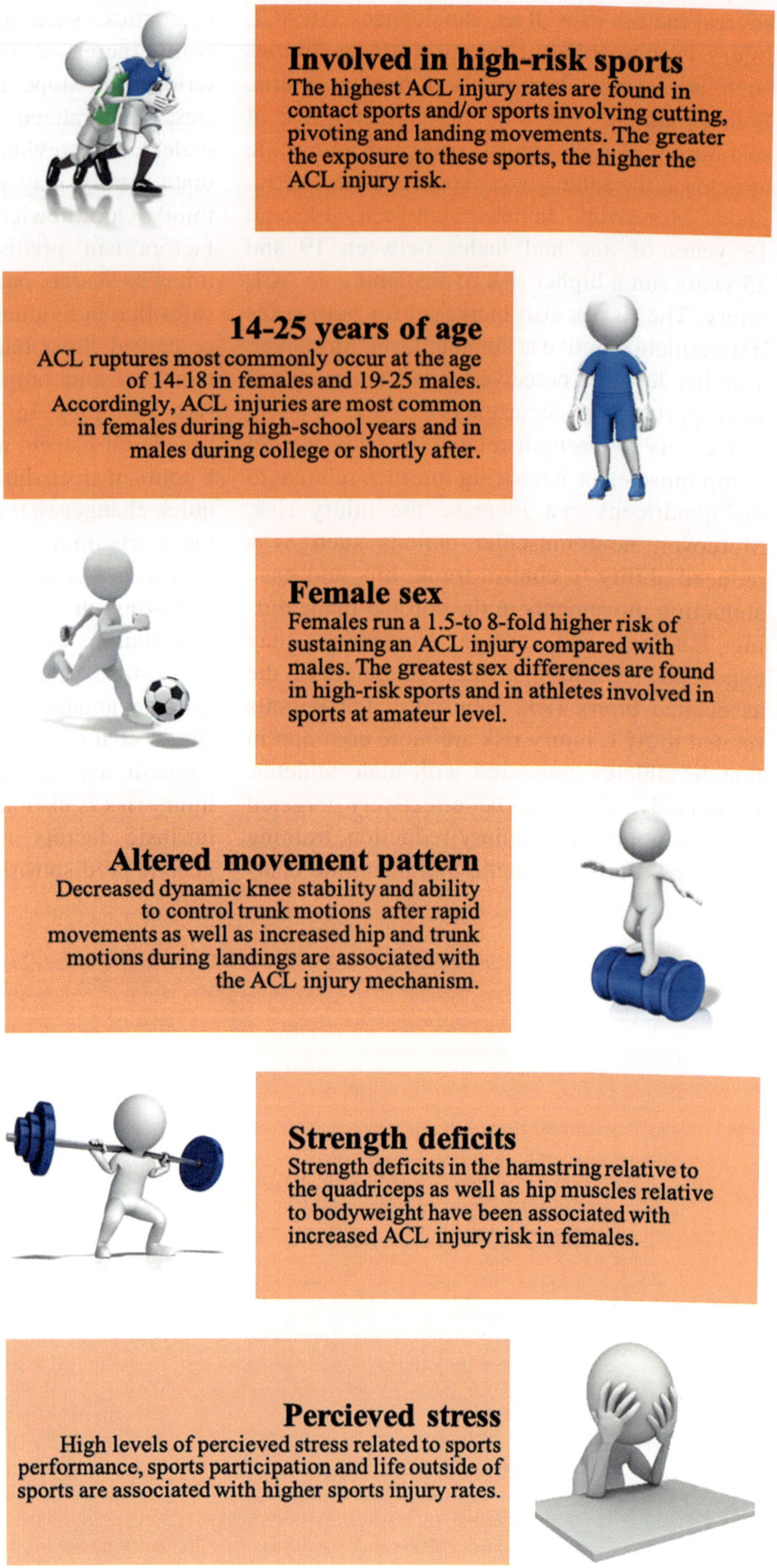

Fig. 6.4 Identifying the anterior cruciate ligament injury-prone individual—a summary of characteristics in individuals at increased risk

References

1. Bayer S, Meredith SJ, Wilson K, de Sa D, Pauyo T, Byrne K, et al. Knee morphological risk factors for anterior cruciate ligament injury: a systematic review. J Bone Joint Surg Am. 2020;102(8):703–18.
2. Flynn RK, Pedersen CL, Birmingham TB, Kirkley A, Jackowski D, Fowler PJ. The familial predisposition toward tearing the anterior cruciate ligament: a case control study. Am J Sports Med. 2005;33(1):23–8.
3. Hewett TE, Myer GD, Ford KR, Paterno MV, Quatman CE. Mechanisms, prediction, and prevention of ACL injuries: Cut risk with three sharpened and validated tools. J Orthop Res. 2016;34(11):1843–55.
4. Kaynak M, Nijman F, van Meurs J, Reijman M, Meuffels DE. Genetic variants and anterior cruciate ligament rupture: a systematic review. Sports Med (Auckland, NZ). 2017;47(8):1637–50.
5. Montalvo AM, Schneider DK, Yut L, Webster KE, Beynnon B, Kocher MS, et al. "What's my risk of sustaining an ACL injury while playing sports?" A systematic review with meta-analysis. Br J Sports Med. 2019;53(16):1003–12.
6. Montalvo AM, Schneider DK, Webster KE, Yut L, Galloway MT, Heidt RS Jr, et al. Anterior cruciate ligament injury risk in sport: a systematic review and meta-analysis of injury incidence by sex and sport classification. J Athl Train. 2019;54(5):472–82.
7. Pfeifer CE, Beattie PF, Sacko RS, Hand A. Risk factors associated with non-contact anterior cruciate ligament injury: a systematic review. Int J Sports Physi Ther. 2018;13(4):575–87.
8. Posthumus M, Collins M, September AV, Schwellnus MP. The intrinsic risk factors for ACL ruptures: an evidence-based review. Phys Sports Med. 2011;39(1):62–73.
9. Rafeeuddin R, Sharir R, Staes F, Dingenen B, George K, Robinson MA, et al. Mapping current research trends on neuromuscular risk factors of non-contact ACL injury. Phys Ther Sport. 2016;22:101–13.
10. Zeng C, Cheng L, Wei J, Gao SG, Yang TB, Luo W, et al. The influence of the tibial plateau slopes on injury of the anterior cruciate ligament: a meta-analysis. Knee Surg Sports Traumatol Arthrosc. 2014;22(1):53–65.
11. Zeng C, Gao SG, Wei J, Yang TB, Cheng L, Luo W, et al. The influence of the intercondylar notch dimensions on injury of the anterior cruciate ligament: a meta-analysis. Knee Surg Sports Traumatol Arthrosc. 2013;21(4):804–15.
12. Sanders TL, Maradit Kremers H, Bryan AJ, Larson DR, Dahm DL, Levy BA, et al. Incidence of anterior cruciate ligament tears and reconstruction: a 21-year population-based study. Am J Sports Med. 2016;44(6):1502–7.
13. Phillips LH. Sports injury incidence. Br J Sports Med. 2000;34(2):133–6.
14. Shi H, Jiang Y, Ren S, Hu X, Huang H, Ao Y. Sex differences in the knee orthopaedic injury patterns among recreational alpine skiers. BMC Sports Sci Med Rehabil. 2020;12(1):74.
15. Smith HC, Vacek P, Johnson RJ, Slauterbeck JR, Hashemi J, Shultz S, et al. Risk factors for anterior cruciate ligament injury: a review of the literature-part 2: hormonal, genetic, cognitive function, previous injury, and extrinsic risk factors. Sports Health. 2012;4(2):155–61.
16. Somerson JS, Isby IJ, Hagen MS, Kweon CY, Gee AO. The menstrual cycle may affect anterior knee laxity and the rate of anterior cruciate ligament rupture: a systematic review and meta-analysis. JBJS Rev. 2019;7(9):e2.
17. Bram JT, Magee LC, Mehta NN, Patel NM, Ganley TJ. Anterior cruciate ligament injury incidence in adolescent athletes: a systematic review and meta-analysis. Am J Sports Med. 2021;49(7):1962–72.
18. Riemann BL, Lephart SM. The sensorimotor system, part I: the physiologic basis of functional joint stability. J Athl Train. 2002;37(1):71–9.
19. Buckthorpe M. Optimising the late-stage rehabilitation and return-to-sport training and testing process after ACL reconstruction. Sports Med. 2019;49(7):1043–58.
20. Koga H, Nakamae A, Shima Y, Iwasa J, Myklebust G, Engebretsen L, et al. Mechanisms for noncontact anterior cruciate ligament injuries: knee joint kinematics in 10 injury situations from female team handball and basketball. Am J Sports Med. 2010;38(11):2218–25.
21. Della Villa F, Tosarelli F, Ferrari R, Grassi A, Ciampone L, Nanni G, et al. Systematic video analysis of anterior cruciate ligament injuries in professional male Rugby players: pattern, injury mechanism, and biomechanics in 57 consecutive cases. Orthop J Sports Med. 2021;9(11):23259671211048182.
22. Webster KE, Hewett TE. Meta-analysis of meta-analyses of anterior cruciate ligament injury reduction training programs. J Orthop Res. 2018;36(10):2696–708.
23. Kapreli E, Athanasopoulos S. The anterior cruciate ligament deficiency as a model of brain plasticity. Med Hypotheses. 2006;67(3):645–50.
24. Ingersoll CD, Grindstaff TL, Pietrosimone BG, Hart JM. Neuromuscular consequences of anterior cruciate ligament injury. Clin Sports Med. 2008;27(3):383–404. vii
25. Diekfuss JA, Grooms DR, Nissen KS, Schneider DK, Foss KDB, Thomas S, et al. Alterations in knee sensorimotor brain functional connectivity contributes to ACL injury in male high-school football players: a prospective neuroimaging analysis. Braz J Phys Ther. 2019;24:415.
26. Diekfuss JA, Grooms DR, Yuan W, Dudley J, Barber Foss KD, Thomas S, et al. Does brain functional connectivity contribute to musculoskeletal injury? A preliminary prospective analysis of a neural biomarker of ACL injury risk. J Sci Med Sport. 2019;22(2):169–74.

27. Bonnette S, Diekfuss JA, Grooms DR, Kiefer AW, Riley MA, Riehm C, et al. Electrocortical dynamics differentiate athletes exhibiting low- and high-ACL injury risk biomechanics. Psychophysiology. 2020;57(4):e13530.
28. Pfeifer S, Hardegger M, Vallery H, List R, Foresti M, Riener R, et al. Model-based estimation of active knee stiffness. IEEE Int Conf Rehabil Robot. 2011;2011:5975474.
29. Wang YL, Yang T, Zeng C, Wei J, Xie DX, Yang YH, et al. Association between Tibial Plateau slopes and anterior cruciate ligament injury: a meta-analysis. Arthroscopy. 2017;33(6):1248–59 e4.
30. Wiese-Bjornstal DM. Psychology and socioculture affect injury risk, response, and recovery in high-intensity athletes: a consensus statement. Scand J Med Sci Sports. 2010;20(Suppl 2):103–11.
31. Li H, Zeng C, Wang Y, Wei J, Yang T, Cui Y, et al. Association between magnetic resonance imaging-measured Intercondylar Notch dimensions and anterior cruciate ligament injury: a meta-analysis. Arthroscopy. 2018;34(3):889–900.
32. Rahnemai-Azar AA, Naendrup JH, Soni A, Olsen A, Zlotnicki J, Musahl V. Knee instability scores for ACL reconstruction. Curr Rev Musculoskelet Med. 2016;9(2):170–7.
33. Sundemo D, Hamrin Senorski E, Karlsson L, Horvath A, Juul-Kristensen B, Karlsson J, et al. Generalised joint hypermobility increases ACL injury risk and is associated with inferior outcome after ACL reconstruction: a systematic review. BMJ Open Sport Exerc Med. 2019;5(1):e000620.
34. John R, Dhillon MS, Sharma S, Prabhakar S, Bhandari M. Is there a genetic predisposition to anterior cruciate ligament tear? a systematic review. Am J Sports Med. 2016;44(12):3262–9.
35. te Wierike SC, van der Sluis A, van den Akker-Scheek I, Elferink-Gemser MT, Visscher C. Psychosocial factors influencing the recovery of athletes with anterior cruciate ligament injury: a systematic review. Scand J Med Sci Sports. 2013;23(5):527–40.
36. Murphy P, Waddington I. Are elite athletes exploited? Sport Soc. 2007;10(2):239–55.
37. Andersen MB, Williams JM. A model of stress and athletic injury: prediction and prevention. J Sport Exer Psychol. 1988;10(3):294.
38. Wiese-Bjornstal DM. Sport injury and college athlete health across the lifespan. J Issues Intercoll Athl. 2009;2(1):64–80.
39. Smith AM, Stuart MJ, Wiese-Bjornstal DM, Gunnon C. Predictors of injury in ice hockey players. A multivariate, multidisciplinary approach. Am J Sports Med. 1997;25(4):500–7.
40. von Rosen P, Frohm A, Kottorp A, Friden C, Heijne A. Multiple factors explain injury risk in adolescent elite athletes: applying a biopsychosocial perspective. Scand J Med Sci Sports. 2017;27(12):2059–69.
41. Ivarsson A, Johnson U, Karlsson J, Börjesson M, Hägglund M, Andersen MB, et al. Elite female footballers' stories of sociocultural factors, emotions, and behaviours prior to anterior cruciate ligament injury. Int J Sport Psychol. 2019;17(6):630–46.
42. Smith HC, Vacek P, Johnson RJ, Slauterbeck JR, Hashemi J, Shultz S, et al. Risk factors for anterior cruciate ligament injury: a review of the literature—part 2: hormonal, genetic, cognitive function, previous injury, and extrinsic risk factors. Sports Health. 2011;4(2):155–61.
43. Swanik CB, Covassin T, Stearne DJ, Schatz P. The relationship between neurocognitive function and noncontact anterior cruciate ligament injuries. Am J Sports Med. 2007;35(6):943–8.
44. Gennarelli SM, Brown SM, Mulcahey MK. Psychosocial interventions help facilitate recovery following musculoskeletal sports injuries: a systematic review. Phys Sportsmed. 2020;48:1–8.
45. Ivarsson A, Johnson U, Andersen MB, Tranaeus U, Stenling A, Lindwall M. Psychosocial factors and sport injuries: meta-analyses for prediction and prevention. Sports Med (Auckland, NZ). 2017;47(2):353–65.
46. Ivarsson A, Johnson U, Podlog L. Psychological predictors of injury occurrence: a prospective investigation of professional Swedish Soccer players. J Sport Rehabil. 2013;22:19–26.
47. Steffen K, Pensgaard AM, Bahr R. Self-reported psychological characteristics as risk factors for injuries in female youth football. Scand J Med Sci Sports. 2009;19(3):442–51.
48. Johnson U, Ivarsson A. Psychosocial factors and sport injuries: prediction, prevention and future research directions. Curr Opin Psychol. 2017;16:89–92.

Frequency, Causality, and Accompanying Lesions of ACL Injuries in Regard to the Age Groups

7

Can Doruk Basa and Elcil Kaya Bicer

7.1 Frequency

Anterior cruciate ligament (ACL) injury can occur at any age, often during sporting activities. The incidence may differ according to age, sex, activity level, and type of activity. Therefore, it is preferable to stratify the incidence of ACL injury according to these variables, in addition to the incidence of the general population, when reporting the frequency of ACL injury.

In the general population, the incidence of ACL injury ranges from 30 to 78 per 100,000 [1–4]. In a 21-year population-based study, Sanders et al. [3] showed that the incidence of ACL injury was higher in men than women (81.7 vs. 55.3 per 100,000), and that the annual incidence in the general population was 68.6 per 100,000. The incidence decreased with age in both sexes, despite being more pronounced in men. Notably, ACL injury occurred most frequently in men aged 19–25 years (241.0 per 100,000) and women aged 14–18 years (227.6 per 100,000). Although the incidence of ACL injury in the general population has tended to remain stable over time, the incidence in all age groups is increasing due to wider use of magnetic resonance imaging (MRI) and ACL reconstruction [3, 5]. Sporting activity is the most common etiology of ACL injury in all age groups. Regarding adolescents and young adults, that is, the groups most frequently affected by ACL injury, organized sports participation increased the risk by 4.0-fold for men and 8.5-fold for women. More than four to five sessions of physical activity per week is also a risk factor for ACL injury [2].

The incidence of ACL injury differs between professional athletes and the general population. The incidence of ACL injury is approximately 5348 per 100,000 person-years among professional female athletes [6]. This rate is significantly greater than that among the overall female population, when adjusted for age. In professional athlete populations, the incidence of ACL injury is usually measured as "injuries per 1000 hours of sports exposure." Among professional male athletes, the overall ACL injury rate varies from 0.066 to 0.076 per 1000 hours of sports exposure; rates during competition vary from 0.34 to 0.41 per 1000 hours, while rates during training vary from 0.017 to 0.04 per 1000 hours [7, 8]. These findings indicate that the incidence of ACL injury increases during competition [9].

Traditionally, pediatric ACL injuries were considered rare and tibial spine fractures were observed more frequently in children. The inci-

C. D. Basa (✉)
Tinaztepe University, Private Galen Hospital, İzmir, Turkey
e-mail: drcandorukbasa@gmail.com

E. Kaya Bicer
Faculty of Medicine, Department of Orthopaedics and Traumatology, Ege University, İzmir, Turkey

M. Karahan et al. (eds.), *Anterior Cruciate Ligament Injuries Throughout the Life Span*,
https://doi.org/10.1007/978-3-031-84501-7_7

dence of ACL injury increases in children who participate more in organized sports, including activities that require regular training throughout the year. In addition, the number of ACL injuries diagnosed and treated surgically in pediatric and adolescent patient populations is gradually increasing [10]. A nationwide study in Finland confirmed that the frequency of injury increased with age (i.e., from childhood to adolescence). In addition, the incidence of ACL injuries in pediatric patients has increased almost twofold in the past 15 years (17.7 per 100,000 to 31.5 per 100,000); the increase has been especially notable in girls aged 13–17 years [11]. In a study by Sanders et al. [3], the incidence of ACL injury was 2 per 100,000 in patients aged <13 years; this rate increased to 215 per 100,000 in patients aged 14–18 years, where this age group exhibited the highest incidence of injuries.

In a systematic review, biomechanical factors including increased peak knee adductor moments, knee valgus angles, hip and knee extension, and ground reaction forces were shown to be related to increased ACL loading in females younger than 13 years old, which should be encountered in the injury prevention strategies [5].

ACL injuries also differ by type among early childhood, adolescence, and adulthood life stages. Most ACL injuries in adults are midsubstance tears, whereas in children aged <11 years avulsion injury of tibial or femoral bone tends to be seen. The rate of ACL midsubstance tears is high in adolescents aged 11–14 years, while avulsion tears become less common once adolescents reach the age of 14–17 years [12].

The frequency of ACL injuries is higher in older individuals; with increased expectations regarding quality of life, participation in new sports activities in older age groups has also increased. There have been few studies regarding the incidence of ACL injury in older individuals. Sanders et al. [3] reported an incidence of ACL injury among adults aged ≥46 years of 20.4 per 100,000. A nationwide population-based study in Sweden reported a similar incidence of ACL injury among adults aged ≥50 years [1]. Although these incidence rates are lower than those reported for the general population, higher expectations regarding quality of life, higher rates of sports participation at an advanced age, and the development of new surgical techniques have led to an increased number of ACL reconstructions each year. Considering that some individuals with ACL injuries do not present to a hospital for treatment, and some admitted patients do not undergo ACL reconstruction, estimates of ACL injury incidence cannot be made only on the basis of the number of ACL reconstructions performed. Granan et al. [13] proposed that the incidence of ACL injury is 50–100% greater than the incidence of ACL reconstruction.

7.2 Etiology

ACL injuries can involve a contact or non-contact mechanism; analyses of ACL injuries in cause and prevention assessments tend to be stratified by mechanism. In the literature, 70–84% of all ACL injuries have a non-contact mechanism [14]. Video analysis, post-injury MRI studies, and in vivo studies have been performed to investigate non-contact injury mechanisms. Video analysis studies revealed that injured individuals tended to be in close contact with the opponent and experienced increased friction between the surface and shoe, such that they landed on the flat portion of their foot. Critical movements that contributed to the onset of injury included decreased ankle plantar flexion, low knee extension, and increased hip flexion [15]. The most likely scenarios in non-contact injuries include falling on the knee with full or nearly full extension, performing plant and cut maneuvers, as well as pivot movements on the knee with full or nearly full extension [16]. Bone bruises observed during post-injury MRI may be indicative of the severity and mechanism of injury. In patients with ACL injury, bone bruises are most common in the lateral femur and tibia. Bone bruises on the lateral side are considered more serious than those on the medial side [17].

ACL injury prevention programs have lowered injury rates. For example, a meta-analysis of studies examining ACL injury prevention programs for athletes revealed that the risk was decreased by 72% in female athletes aged <18 years, although in those aged ≥18 years the risk only decreased by 18% [18]. In a study of male footballers divided into two age groups (4–17 and 18–35 years), the ACL injury rate was nearly twofold higher in the younger group, although the sports exposure times were comparable [19]. Age <20 years has also been identified as a risk factor for ACL injury [20]; we suggest that age <18 years can be regarded as a risk factor. Notably, risk factors for ACL injury are multifactorial. All risk factors can be considered intrinsic or extrinsic. Factors such as weather, ground conditions (i.e., playing field conditions), and shoe type are regarded as environmental factors [21]. Personal risk factors include anatomical, hormonal, and biomechanical factors [22].

7.2.1 Anatomical Factors

Individuals with ACL injury often show distinct anatomical features. However, because anatomical features generally cannot be modified, they have minimal relevance in terms of prevention strategies.

7.2.1.1 Posterior Tibial Slope

The posterior tibial slope (PTS) is the angle between the long axis of the tibia and the inclination of the posterior tibial plateau. PTS may differ between the sexes and among ethnicities [23]. Although the ACL is important for preventing anterior translation of the tibia, the PTS also plays a role. Biomechanical studies have shown that PTS is effective in preventing anterior translation of the tibia and ensuring rotatory knee stability [24, 25]. In a meta-analysis of studies that investigated the relationship between the PTS and ACL injuries, increased lateral PTS and increased medial PTS both were related to an increased risk of ACL injury [26]. In another meta-analysis, Wang et al. [27] confirmed that increased medial and lateral PTS were both associated with an increased risk of ACL injury, albeit only in men.

The role of age in the relationship between the PTS and ACL injuries has been investigated. In a pediatric study, Vyas et al. [28] showed that the medial tibial slope was increased in children with versus without ACL injury. Furthermore, knee morphometry did not change with age, which indicated that morphometric characteristics are genetic. Although increased PTS appears to be a risk factor for ACL injury, PTS morphology does not change with age. Another study showed that increased PTS was associated with an increased risk of ACL injury, especially among individuals aged 20–49 years, and that increased PTS did not increase the risk of ACL injury among individuals aged ≥50 years [29].

7.2.1.2 Intercondylar Notch Width and ACL Size

In 1938, Palmer et al. demonstrated that decreased intercondylar notch width was an anatomical risk factor for the development of ACL injury [30]. However, this has proven controversial in the subsequent literature. Nevertheless, a meta-analysis of studies on the relationship between intercondylar notch dimensions and ACL injury demonstrated that individuals with ACL injuries had significantly narrower intercondylar notches compared to healthy controls [22, 31]. Intercondylar width measurement can also be instructive when investigating ACL volume. Notably, MRI studies showed a negative correlation between age and ACL volume [32], while no relationship was observed between age and the ratio of axial notch area to ACL volume; thus, this issue remains controversial.

7.2.1.3 Body Weight

Higher body weight and body mass index (BMI) are considered risk factors for ACL injury, presumably because of greater lower limb extension during landing. In a large-scale population study, Thein et al. [33] demonstrated that increased BMI was a risk factor for ACL injury in both sexes. Other studies have shown that the injury risk associated with higher BMI is further exacerbated in patients with increased lateral tibial slope [34, 35]. ACL injury can thus be prevented by weight control and neuromuscular training.

7.2.1.4 Generalized Joint Laxity

It has been observed that individuals with generalized joint laxity (GJL) are at higher risk of ACL injury. The relationship between GJL and ACL injury has been frequently discussed in the literature. However, GJL should be evaluated differently depending on the age group. Malfait et al. [36] stated that children, adolescents, and adults aged <50 years should be diagnosed with GJL based on different cutoff scores. A systematic review of studies on the relationship between ACL injury and GJL revealed that GJL increased the risk of injury by 4.4-fold, although the findings remain controversial [37]. GJL is considered a risk factor for ACL injury in populations including both men and women, as well as in those including men alone. Whether GJL is a risk factor for ACL injury in women is controversial, although hypermobility is more common in women and female sex is a risk factor for ACL injury [37].

7.2.2 Sex and Hormonal Factors

It is known that women sustain three- to sixfold more ACL injuries than men. This is presumably due to sex differences in biomechanics, dynamics, and anatomical features. An alternative theory is that hormonal fluctuations in women may further increase knee laxity during the menstrual cycle, where the laxity is already greater in women than men.

A meta-analysis revealed that the risk of ACL injury increased during the pre-ovulatory period [38], but this has since proven controversial. The relationship observed between hormonal fluctuation and risk of ACL injury implied that oral contraceptive use may prevent ACL injury. A large series of studies have shown that the use of oral contraceptives can reduce the risk of ACL injury by up to 20% [39, 40]. However, these studies were limited by their retrospective nature and inclusion of oral contraceptives, which are less frequently used by adolescents than adults. Prospective studies are needed to determine whether oral contraceptives can definitively prevent ACL injury [38].

In a systematic review regarding the role of serum relaxin levels on the risk of ACL injury, higher levels were found to be associated with increased risk in female athletes, which might be explained by the fact that relaxin binds to estrogen receptors in female ACLs and is associated with increased ligamentous laxity. Oral contraceptives may act on reducing serum relaxin levels. Monitoring relaxin levels may be encountered as a screening tool to determine the ACL injury risk [41].

7.2.3 Biomechanical Factors

During the aging process, the internal structure of the ACL deteriorates in a manner similar to that of other tissues in the body. Specifically, the cellular structure weakens, and degeneration, calcium deposition, and cystic deformation occur [42]. Animal experiments have revealed age-related changes in ACL mechanoreceptors [43]. In addition, knee joint position sense is decreased in elderly individuals [44]. Taken together, these findings suggest that ACL injuries in young and aged knees have different etiologies.

7.3 Accompanying Lesions

ACL injuries may be accompanied by some additional injuries, due to changes in knee kinematics during or after trauma. Especially in patients with chronic injury exhibiting rotatory knee instability, cartilage and meniscus injuries can occur, presumably due to changes in the tibiofemoral contact point in the medial compartment [45]. Some features of these accompanying lesions, and the age-related variations therein, are examined below.

7.3.1 Meniscus

Knee menisci essentially possess functions in load transmission and stability of the joint. Both medial and lateral menisci are secondary stabilizers of anterior tibial translation. In ACL-deficient knees, tibiofemoral contact surface alters and load on both menisci increases over time after the injury [46]. Subsequently, shear forces are increased, especially in the posterior medial meniscus, due to increased anterior tibial translation. Meniscal tears can also occur because ACL injury increases the secondary stabilizer effect of anterior translation of the medial meniscus [47].

A rotatory knee trauma causing an ACL injury may injure the meniscus simultaneously or may develop during repeated injuries. Delayed ACL reconstruction and recurrent knee rotatory traumas are known risk factors for meniscal tears in individuals with ACL injury [48]. Although increased time between injury and ACL reconstruction is often regarded as a risk factor for the development of meniscal tears, the reported cut-off times vary from six weeks to one year [49–51]. In a meta-analysis, Mehl et al. [52] investigated the prevalence of meniscus lesions in ACL-deficient knees. Medial meniscus tears were most commonly observed in patients with chronic ACL injury, whereas lateral meniscus tears were mostly encountered in the acute setting. Lateral meniscus tears are presumed to be more common during initial ACL injury because the lateral compartment of the knee is subluxated during this injury. The mobile nature of the lateral meniscus, and the reduced influence of anterior stabilization compared to the medial meniscus, may explain the low rate of lateral meniscus tears in patients with chronic ACL injury.

Ramp lesions are longitudinal tears at the meniscocapsular junction. An MRI study of knees with ACL injuries showed that prevalence of meniscal ramp lesions could be as high as 39.5%. It was reported that contact injury, presence of bone bruise at lateral femoral condyle, lateral tibial plateau, and also posteromedial tibial plateau as well as a Segond fracture might be suggestive for the presence of a meniscal ramp lesion [53, 54].

7.3.2 Cartilage

The incidence of chondral injuries accompanying acute ACL injuries has been reported to vary between 37% and 46% [55–57]. The older the patient and the longer the duration between injury and surgery, the greater the risk of developing chondral and meniscal lesions. Also, the mode of injury has been shown to influence the incidence of chondral injuries; non-contact injuries are associated with higher rates of chondral damage [57].

The locations of chondral damage in knees with ACL injury have been researched in MRI studies, which revealed that initially, lateral tibiofemoral compartment has been affected more commonly; during the follow-up period, these patients were shown to develop chondral injuries in other compartments of their injured knees [58].

In the setting of ACL injuries, the size of the bone bruise in the lateral compartment of the knee has prognostic importance in the development of chondral wear during follow-up. The larger bone bruises may have predictive roles [59].

7.3.3 Collateral Ligaments

The medial collateral ligament (MCL) and lateral collateral ligament have important effects on knee stability in the direction of varus–valgus. The MCL is also known to be important for rotatory knee stability. MCL injuries accompanying ACL injuries are not uncommon, with reported rates of up to 38% [60]. Treatment options for isolated collateral ligament injuries accompanying ACL injury include ACL reconstruction, nonoperative treatment, primary suture repair of the MCL, and MCL reconstruction [61]. Although good results have been reported following nonoperative treatment of MCL injuries accompanying ACL injury, the ACL graft load and revision rates are higher in patients undergoing nonoperative treatment [62, 63].

There have been few studies regarding the effect of age on the rate of MCL injury accompanying ACL injury. In one study, the mean age of patients who received nonoperative treatment for MCL injuries was 30.2 years, while the mean age of patients with MCL injuries who underwent suture repair was 32.6 years, and the mean age of patients with isolated ACL injuries was 27.8 years. Notably, patients with MCL injuries were significantly older [63].

7.3.4 Posterolateral Corner (PLC)

Posterolateral corner (PLC) injuries may accompany ACL injuries. An incidence of concurrent ACL and PLC injuries of up to 20% has been reported among imaging studies [64]. PLC injuries can often be missed if there is no clinical or radiological suspicion; this can lead to increased rotatory load on the reconstructed ACL, which may in turn increase the ACL graft failure rate.

7.4 Conclusion

The frequency, causality, and accompanying injuries vary according to age. Causality and frequency variability according to age are important for injury prevention strategies. Understanding the accompanying injuries and identifying them is important for complete treatment.

References

1. Nordenvall R, Bahmanyar S, Adami J, Stenros C, Wredmark T, Fellander-Tsai L. A population-based nationwide study of cruciate ligament injury in Sweden, 2001-2009: incidence, treatment, and sex differences. Am J Sports Med. 2012;40(8):1808–13.
2. Parkkari J, Pasanen K, Mattila VM, Kannus P, Rimpela A. The risk for a cruciate ligament injury of the knee in adolescents and young adults: a population-based cohort study of 46 500 people with a 9 year follow-up. Br J Sports Med. 2008;42(6):422–6.
3. Sanders TL, Maradit Kremers H, Bryan AJ, Larson DR, Dahm DL, Levy BA, et al. Incidence of anterior cruciate ligament tears and reconstruction: a 21-year population-based study. Am J Sports Med. 2016;44(6):1502–7.
4. Corry II, Webb J. Injuries of the sporting knee. Br J Sports Med. 2000;34(5):395.
5. Heering T, Rolley TL, Lander N, Fox A, Barnett LM, Duncan MJ. Identifying modifiable risk factors and screening strategies associated with anterior cruciate ligament injury risk in children aged 6 to 13 years: a systematic review. J Sports Sci. 2023;41(14):1337–62.
6. Pasanen K, Parkkari J, Kannus P, Rossi L, Palvanen M, Natri A, et al. Injury risk in female floorball: a prospective one-season follow-up. Scand J Med Sci Sports. 2008;18(1):49–54.
7. Rekik RN, Tabben M, Eirale C, Landreau P, Bouras R, Wilson MG, et al. ACL injury incidence, severity and patterns in professional male soccer players in a Middle Eastern league. BMJ Open Sport Exerc Med. 2018;4(1):e000461.
8. Walden M, Hagglund M, Magnusson H, Ekstrand J. ACL injuries in men's professional football: a 15-year prospective study on time trends and return-to-play rates reveals only 65% of players still play at the top level 3 years after ACL rupture. Br J Sports Med. 2016;50(12):744–50.
9. Chia L, De Oliveira SD, Whalan M, McKay MJ, Sullivan J, Fuller CW, et al. Non-contact anterior cruciate ligament injury epidemiology in team-ball sports: a systematic review with meta-analysis by sex, age, sport, participation level, and exposure type. Sports Med. 2022;52(10):2447–67.
10. Werner BC, Yang S, Looney AM, Gwathmey FW Jr. Trends in pediatric and adolescent anterior cruciate ligament injury and reconstruction. J Pediatr Orthop. 2016;36(5):447–52.
11. Weitz FK, Sillanpaa PJ, Mattila VM. The incidence of paediatric ACL injury is increasing in Finland. Knee Surg Sports Traumatol Arthrosc. 2020;28(2):363–8.
12. van der List JP, Mintz DN, DiFelice GS. The locations of anterior cruciate ligament tears in pediatric and adolescent patients: a magnetic resonance study. J Pediatr Orthop. 2019;39(9):441–8.
13. Granan LP, Bahr R, Steindal K, Furnes O, Engebretsen L. Development of a national cruciate ligament surgery registry: the Norwegian National Knee Ligament Registry. Am J Sports Med. 2008;36(2):308–15.
14. Alentorn-Geli E, Myer GD, Silvers HJ, Samitier G, Romero D, Lazaro-Haro C, et al. Prevention of non-contact anterior cruciate ligament injuries in soccer players. Part 1: mechanisms of injury and underlying risk factors. Knee Surg Sports Traumatol Arthrosc. 2009;17(7):705–29.
15. Carlson VR, Sheehan FT, Boden BP. Video analysis of Anterior Cruciate Ligament (ACL) injuries: a systematic review. JBJS Rev. 2016;4(11):e5.
16. Fauno P, Wulff JB. Mechanism of anterior cruciate ligament injuries in soccer. Int J Sports Med. 2006;27(1):75–9.
17. Shi H, Ding L, Jiang Y, Zhang H, Ren S, Hu X, et al. Bone Bruise distribution patterns after acute anterior cruciate ligament ruptures: implications for the injury mechanism. Orthop J Sports Med. 2020;8(4):2325967120911162.
18. Myer GD, Sugimoto D, Thomas S, Hewett TE. The influence of age on the effectiveness of neuromuscular training to reduce anterior cruciate ligament injury in female athletes: a meta-analysis. Am J Sports Med. 2013;41(1):203–15.
19. Schmikli SL, de Vries WR, Inklaar H, Backx FJ. Injury prevention target groups in soccer: injury

characteristics and incidence rates in male junior and senior players. J Sci Med Sport. 2011;14(3):199–203.
20. Leys T, Salmon L, Waller A, Linklater J, Pinczewski L. Clinical results and risk factors for reinjury 15 years after anterior cruciate ligament reconstruction: a prospective study of hamstring and patellar tendon grafts. Am J Sports Med. 2012;40(3):595–605.
21. Griffin LY, Agel J, Albohm MJ, Arendt EA, Dick RW, Garrett WE, et al. Noncontact anterior cruciate ligament injuries: risk factors and prevention strategies. J Am Acad Orthop Surg. 2000;8(3):141–50.
22. Zeng C, Gao SG, Wei J, Yang TB, Cheng L, Luo W, et al. The influence of the intercondylar notch dimensions on injury of the anterior cruciate ligament: a meta-analysis. Knee Surg Sports Traumatol Arthrosc. 2013;21(4):804–15.
23. Kacmaz IE, Topkaya Y, Basa CD, Zhamilov V, Er A, Reisoglu A, et al. Posterior tibial slope of the knee measured on X-rays in a Turkish population. Surg Radiol Anat. 2020;42(6):673–9.
24. Daniel DM, Stone ML, Dobson BE, Fithian DC, Rossman DJ, Kaufman KR. Fate of the ACL-injured patient. A prospective outcome study. Am J Sports Med. 1994;22(5):632–44.
25. Voos JE, Suero EM, Citak M, Petrigliano FP, Bosscher MR, Citak M, et al. Effect of tibial slope on the stability of the anterior cruciate ligament-deficient knee. Knee Surg Sports Traumatol Arthrosc. 2012;20(8):1626–31.
26. Zeng C, Cheng L, Wei J, Gao SG, Yang TB, Luo W, et al. The influence of the tibial plateau slopes on injury of the anterior cruciate ligament: a meta-analysis. Knee Surg Sports Traumatol Arthrosc. 2014;22(1):53–65.
27. Wang YL, Yang T, Zeng C, Wei J, Xie DX, Yang YH, et al. Association between Tibial Plateau slopes and anterior cruciate ligament injury: a meta-analysis. Arthroscopy. 2017;33(6):1248–59 e4.
28. Vyas S, van Eck CF, Vyas N, Fu FH, Otsuka NY. Increased medial tibial slope in teenage pediatric population with open physes and anterior cruciate ligament injuries. Knee Surg Sports Traumatol Arthrosc. 2011;19(3):372–7.
29. Sun YH, Chen LX, Jiao ZD, Wang L, Zhang RM, Fang J, et al. Age-related changes of posterior Tibial slope and its roles in anterior cruciate ligament injury. Int Surg. 2016;101(1–2):70–77.
30. Palmer I. On the injuries to the ligaments of the knee joint: a clinical study. 1938. Clin Orthop Relat Res. 2007;454:17–22. discussion 14
31. Li H, Zeng C, Wang Y, Wei J, Yang T, Cui Y, et al. Association between magnetic resonance imaging-measured intercondylar notch dimensions and anterior cruciate ligament injury: a meta-analysis. Arthroscopy. 2018;34(3):889–900.
32. Iriuchishima T, Ryu K, Fu FH. Evaluation of age-related differences in anterior cruciate ligament size. Knee Surg Sports Traumatol Arthrosc. 2019;27(1):223–9.
33. Thein R, Hershkovich O, Gordon B, Burstein G, Tenenbaum S, Derazne E, et al. The prevalence of cruciate ligament and Meniscus knee injury in young adults and associations with gender, body mass index, and height a large cross-sectional study. J Knee Surg. 2017;30(6):565–70.
34. Kizilgoz V, Sivrioglu AK, Aydin H, Ulusoy GR, Cetin T, Tuncer K. The combined effect of body mass index and Tibial slope angles on anterior cruciate ligament injury risk in male knees: a case-control study. Clin Med Insights Arthritis Musculoskelet Disord. 2019;12:1179544119867922.
35. Bojicic KM, Beaulieu ML, Imaizumi Krieger DY, Ashton-Miller JA, Wojtys EM. Association between lateral posterior Tibial slope, body mass index, and ACL injury risk. Orthop J Sports Med. 2017;5(2):2325967116688664.
36. Malfait F, Francomano C, Byers P, Belmont J, Berglund B, Black J, et al. The 2017 international classification of the Ehlers-Danlos syndromes. Am J Med Genet C Semin Med Genet. 2017;175(1):8–26.
37. Sundemo D, Hamrin Senorski E, Karlsson L, Horvath A, Juul-Kristensen B, Karlsson J, et al. Generalised joint hypermobility increases ACL injury risk and is associated with inferior outcome after ACL reconstruction: a systematic review. BMJ Open Sport Exerc Med. 2019;5(1):e000620.
38. Herzberg SD, Motu'apuaka ML, Lambert W, Fu R, Brady J, Guise JM. The effect of menstrual cycle and contraceptives on ACL injuries and laxity: a systematic review and meta-analysis. Orthop J Sports Med. 2017;5(7):2325967117718781.
39. Rahr-Wagner L, Thillemann TM, Mehnert F, Pedersen AB, Lind M. Is the use of oral contraceptives associated with operatively treated anterior cruciate ligament injury? A case-control study from the Danish Knee Ligament Reconstruction Registry. Am J Sports Med. 2014;42(12):2897–905.
40. Gray AM, Gugala Z, Baillargeon JG. Effects of oral contraceptive use on anterior cruciate ligament injury epidemiology. Med Sci Sports Exerc. 2016;48(4):648–54.
41. Berger GK, Rockov ZA, Byrne C, Trentacosta NE, Stone MA. The role of relaxin in anterior cruciate ligament injuries: a systematic review. Eur J Orthop Surg Traumatol. 2023;33(8):3319–26.
42. Hasegawa A, Otsuki S, Pauli C, Miyaki S, Patil S, Steklov N, et al. Anterior cruciate ligament changes in the human knee joint in aging and osteoarthritis. Arthritis Rheum. 2012;64(3):696–704.
43. Li B, Luo XD, Wen Y. Changes in mechanoreceptors in rabbits' anterior cruciate ligaments with age. J Mol Histol. 2019;50(3):229–37.
44. Relph N, Herrington L. The effects of knee direction, physical activity and age on knee joint position sense. Knee. 2016;23(3):393–8.
45. Li G, Moses JM, Papannagari R, Pathare NP, DeFrate LE, Gill TJ. Anterior cruciate ligament deficiency alters the in vivo motion of the tibiofemoral car-

tilage contact points in both the anteroposterior and mediolateral directions. J Bone Joint Surg Am. 2006;88(8):1826–34.
46. Atarod M, Frank CB, Shrive NG. Increased meniscal loading after anterior cruciate ligament transection in vivo: a longitudinal study in sheep. Knee. 2015;22(1):11–7.
47. Seon JK, Gadikota HR, Kozanek M, Oh LS, Gill TJ, Li G. The effect of anterior cruciate ligament reconstruction on kinematics of the knee with combined anterior cruciate ligament injury and subtotal medial meniscectomy: an in vitro robotic investigation. Arthroscopy. 2009;25(2):123–30.
48. Snoeker BA, Bakker EW, Kegel CA, Lucas C. Risk factors for meniscal tears: a systematic review including meta-analysis. J Orthop Sports Phys Ther. 2013;43(6):352–67.
49. Millett PJ, Willis AA, Warren RF. Associated injuries in pediatric and adolescent anterior cruciate ligament tears: does a delay in treatment increase the risk of meniscal tear? Arthroscopy. 2002;18(9):955–9.
50. Lawrence JT, Argawal N, Ganley TJ. Degeneration of the knee joint in skeletally immature patients with a diagnosis of an anterior cruciate ligament tear: is there harm in delay of treatment? Am J Sports Med. 2011;39(12):2582–7.
51. Dumont GD, Hogue GD, Padalecki JR, Okoro N, Wilson PL. Meniscal and chondral injuries associated with pediatric anterior cruciate ligament tears: relationship of treatment time and patient-specific factors. Am J Sports Med. 2012;40(9):2128–33.
52. Mehl J, Otto A, Baldino JB, Achtnich A, Akoto R, Imhoff AB, et al. The ACL-deficient knee and the prevalence of meniscus and cartilage lesions: a systematic review and meta-analysis (CRD42017076897). Arch Orthop Trauma Surg. 2019;139(6):819–41.
53. Bae BS, Yoo S, Lee SH. Ramp lesion in anterior cruciate ligament injury: a review of the anatomy, biomechanics, epidemiology, and diagnosis. Knee Surg Relat Res. 2023;35(1):23.
54. Cristiani R, van de Bunt F, Kvist J, Stalman A. High prevalence of meniscal ramp lesions in anterior cruciate ligament injuries. Knee Surg Sports Traumatol Arthrosc. 2023;31(1):316–24.
55. Spindler KP, Schils JP, Bergfeld JA, Andrish JT, Weiker GG, Anderson TE, et al. Prospective study of osseous, articular, and meniscal lesions in recent anterior cruciate ligament tears by magnetic resonance imaging and arthroscopy. Am J Sports Med. 1993;21(4):551–7.
56. Michalitsis S, Vlychou M, Malizos KN, Thriskos P, Hantes ME. Meniscal and articular cartilage lesions in the anterior cruciate ligament-deficient knee: correlation between time from injury and knee scores. Knee Surg Sports Traumatol Arthrosc. 2015;23(1):232–9.
57. Gupta R, Kapoor A, Soni A, Khatri S, Masih GD. Anterior cruciate ligament tear due to non-contact mode of injury associated with higher incidence of meniscal and chondral damage. J Clin Orthop Trauma. 2020;11(Suppl 3):S342–S5.
58. Davidson EJ, Figgie C, Nguyen J, Pedoia V, Majumdar S, Potter HG, et al. Chondral injury associated with ACL injury: assessing progressive chondral degeneration with morphologic and quantitative MRI techniques. Sports Health. 2023;16:19417381231205276.
59. Kia C, Cavanaugh Z, Gillis E, Dwyer C, Chadayammuri V, Muench LN, et al. Size of initial bone Bruise predicts future lateral chondral degeneration in ACL injuries: a radiographic analysis. Orthop J Sports Med. 2020;8(5):2325967120916834.
60. Zaffagnini S, Bonanzinga T, Marcheggiani Muccioli GM, Giordano G, Bruni D, Bignozzi S, et al. Does chronic medial collateral ligament laxity influence the outcome of anterior cruciate ligament reconstruction?: a prospective evaluation with a minimum three-year follow-up. J Bone Joint Surg Br. 2011;93(8):1060–4.
61. Liu X, Feng H, Zhang H, Hong L, Wang XS, Zhang J. Arthroscopic prevalence of ramp lesion in 868 patients with anterior cruciate ligament injury. Am J Sports Med. 2011;39(4):832–7.
62. Svantesson E, Hamrin Senorski E, Alentorn-Geli E, Westin O, Sundemo D, Grassi A, et al. Increased risk of ACL revision with non-surgical treatment of a concomitant medial collateral ligament injury: a study on 19,457 patients from the Swedish National Knee Ligament Registry. Knee Surg Sports Traumatol Arthrosc. 2019;27(8):2450–9.
63. Zhu J, Dong J, Marshall B, Linde MA, Smolinski P, Fu FH. Medial collateral ligament reconstruction is necessary to restore anterior stability with anterior cruciate and medial collateral ligament injury. Knee Surg Sports Traumatol Arthrosc. 2018;26(2):550–7.
64. Kinsella SD, Rider SM, Fury MS, Tepolt FA, Ecklund K, Kocher MS. Concomitant posterolateral corner injuries in skeletally immature patients with acute anterior cruciate ligament injuries. J Pediatr Orthop. 2020;40(6):271–6.

Graft Choices for Differing Age Groups, Physical Properties, and Sports

8

Patrick Djian and Marc Michnowski

8.1 Introduction

Anterior cruciate ligament (ACL) reconstruction is a common operation. The purpose of this surgery is to restore functional stability to the ACL deficient knee and avoid, over time, cartilage problems. The functional stability provided by the normal ACL is in both resisting anteroposterior translation and rotational subluxation. ACL injury is one of the most common problems in the orthopedic community. If patients suffer from recurrent knee instability, it is difficult for them to return to preinjury activity level. Therefore, ACL reconstruction has been developed to restore knee stability and allow the return to preinjury activity levels. ACL reconstruction can be performed using a variety of different surgical techniques as well as different graft materials. Over time, we have learned about the graft's mechanical strength and solved most of the fixation problems. However, there are two mains questions that are unsolved: the failure rate of ACL reconstruction has been reported to be between 4% and 30% [1–7]. Return to sports at the same level could take as long as 18–24 months [8, 9]. The development of osteoarthritis 10–20 years after ACL reconstruction is also an issue [10–12].

P. Djian (✉)
Cabinet Goethe, Paris, France

M. Michnowski
Paris, France

M. Karahan et al. (eds.), *Anterior Cruciate Ligament Injuries Throughout the Life Span*,
https://doi.org/10.1007/978-3-031-84501-7_8

Clinical outcome versus graft choice remains a controversial subject. There are still several intraoperative factors affecting clinical outcomes that remain widely debated. These include graft selection, tunnel position, graft fixation angle, initial graft tension, native tissue remnant preservation, and combined ligament injuries (anterolateral ligament (ALL), posterolateral corner of the knee (PLC), meniscus, and cartilage injuries at the time of surgery).

Surgeons have to consider several factors before choosing a graft for ACL reconstruction: These include harvest morbidity and complications, biomechanical properties of the graft, technique and fixation of the graft, iterative rupture, risk of future osteoarthritis, surgical time, and cost effectiveness.

Historically, autografts and especially bone-patellar tendon-bone (BPTB) autograft have been the gold standard [13]. The use of hamstring tendons (HT) as a quadruple graft has become more popular mostly due to the ease of harvesting and less residual anterior knee pain compared to BPTB [6, 14]. Other options include the use of the quadriceps tendon (QT) autograft (with or without bone plug), fascia lata autograft, allografts, and synthetics.

Several analyses and meta-analyses comparing the outcome of ACL reconstruction according to graft type have reported conflicting results [15, 16].

The purpose of this chapter is to discuss the current state of graft selection in the treatment of ACL injury-based on the literature.

8.2 Hamstring Tendons (HT) Autografts

8.2.1 One-Bundle Reconstruction (Fig. 8.1)

The first description of HT for ACL reconstruction was described by Alexander Edwards of Glasgow in 1926 [17]. The reconstruction was performed in the lab and was not anatomic and no clinical results were mentioned in the article. Then, anatomic ACL reconstruction with semitendinosus was proposed by Galeazzi [18] in 1934, with the tendon still distally attached to the pes anserinus. He reported three patients with a short follow-up. In 1938, Macey [19] in the United States described a similar technique to Galeazzi with the use of semitendinosus tendon with a two "incision technique." No clinical series was reported.

In 1950, Lindemann [20] developed the concept of "dynamic ACL reconstruction." He utilized a proximally based gracilis tendon, passed it through the posterolateral capsule into the inter-condylar notch, and passed it through a tibial tunnel. At an average follow-up of 2 years, six patients regained normal knee function and were able to return to work.

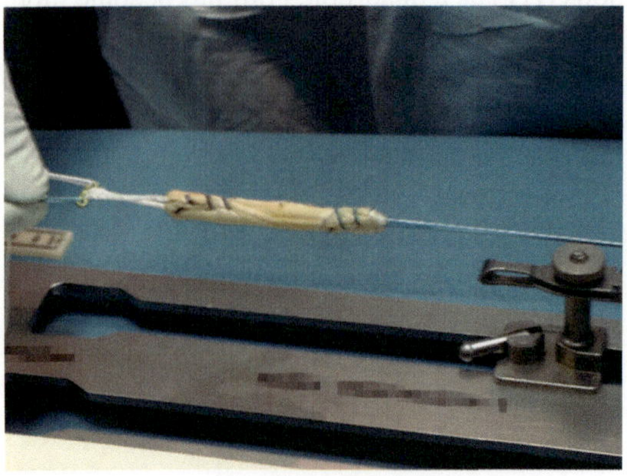

Fig. 8.1 One bundle hamstring preparation

In the United States, Cho [21] also adopted distally-attached semitendinosus as a graft source and in 1975 reported overall good results and no residual anterior drawer sign in five out of seven patients.

The first report of using the semitendinosus and the gracilis was made by Lipscomb [22] in 1982. The mechanical properties of this double-bundle graft have been criticized: Usually, the diameter of this simple graft is less than 8 mm. The principle, with a number of variations relating to the graft being free or attached at its distal end, be it single (2 strands) or double (4 strands) bundled, along with a multitude of proposed graft fixation techniques [23], would be adopted by all surgeons utilizing this graft. The graft could be left attached to the pes anserinus or detached. Marcacci [24] proposed using one of the strands of the graft to perform an anterolateral tenodesis. The results after ACL reconstruction using single bundle were not as good as believed: a positive "Pivot Shift" test of varying grades and proportions up to 25% could persist at clinical examination [25] after single-bundle reconstruction. This finding led surgeons to reconsider ACL anatomy in the beginning of 2000. The native ACL mainly consists of two distinct bundles: an anteromedial bundle (AMB) and a posterolateral bundle (PLB), named according to their tibial insertion sites [26]. Both bundles are critical to knee stability but functionally differently. The AMB is tight in flexion, whereas the PLB is tight in extension. The AMB controls anterior laxity, while the PLB controls both rotatory knee stability and anterior laxity. A biomechanical study found that the AMB is the primary restraint against anterior tibia translation, while the PLB secures the knee near full extension, particularly against rotatory loads [27].

Active flexion deficits in the four-strand Hamstring graft group were greater at the 8-month follow-up and 2-year follow-up but not at the 3-year follow-up [28–31].

Maletis [32] reported in a retrospective study of 9817 patients that when gender-specific analysis was performed, there was a 2.26 times (95% confidence interval [CI]: 1.15–4.44) increased risk of HT autograft revision in females compared with BPTB autograft. No increased risk was noted for males.

8.2.2 Partial Reconstructions

Arthroscopic double-bundle reconstruction has allowed us to progress by understanding the anatomy and also reflect on partial tears of the ACL. Suspected on clinical examination and revealed on magnetic resonance imaging (MRI), the diagnosis of a partial ACL tear has to be confirmed by arthroscopic evaluation. The percentage of partial ACL ruptures represents 10–15% of all ACL rupture [33, 34]. The techniques used to reconstruct the affected bundle are variable and all type of grafts can be used. The results of this type of reconstructions remain good with a significant improvement in anterior translation of the tibia relative to the preoperative measurement and a differential laxity measured at 1 mm [35, 36]. This partial reconstruction allows for improvement in the postoperative mechanical quality of the ACL, with a mechanically solid bundle protecting the graft and its fixation and allowing a more aggressive rehabilitation. Preservation of the vascularity at the level of the synovial envelope is required for graft healing [37]. Preservation of existing mechanoreceptors in the intact bundle improves the proprioceptive qualities of the knee; and, its ability to resume physical activity [38].

8.2.3 Double-Bundle Reconstruction (Fig. 8.2)

A recent meta-analysis [39] has revealed that double-bundle ACL reconstruction has some superior results in objective measurements of knee stability and protection against repeat ACL rupture or a new meniscal injury.

Recent studies that have compared anatomic single-bundle (SB) and double-bundle (DB) ACL reconstructions [16, 17] have shown that DB ACL reconstruction has better arthrometer differentials of 0.52 mm. Swantesson et al. [23] evaluated 22,460 patients issues from the Swedish National Register and found that DB ACL reconstruction was associated with a lower risk of revision surgery than SB ACL reconstruction.

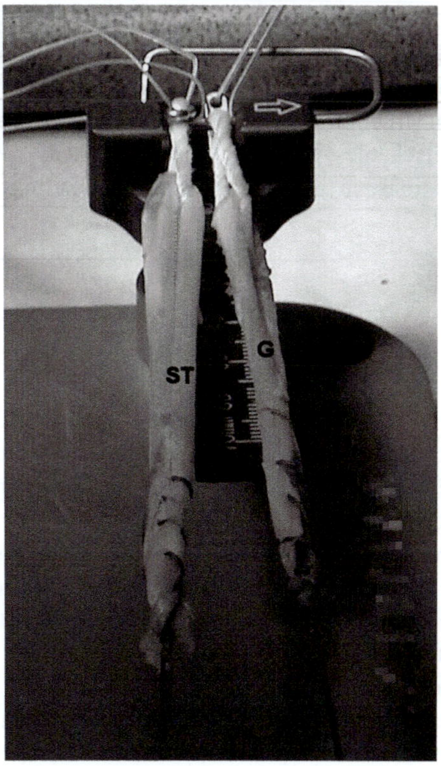

Fig. 8.2 Double-bundle hamstring preparation

8.2.4 Biomechanical Studies

Four-strand hamstring tendon autografts are stronger and stiffer than two-strand hamstring graft constructs or 10-mm wide patellar tendon autografts; graft preparation involves multiple-strand the harvested semitendinosus and gracilis tendons or quadrupling, if only semitendinosus is harvested, thereby conferring a maximum load to failure of approximately 4590 N [40]. The mechanical properties of all of these constructs are better than the native ACL [40–43]. The final graft ideally would measure at least 8 mm in diameter or greater, because it has been shown that failure rates increase with smaller diameter grafts [44]. Magnussen et al. [45] reported that use of hamstring autografts 8 mm in diameter or less in patients aged under 20 years was associated with higher revision rates. Recent biomechanical studies have revealed that DB ACL reconstruction restores more normal knee biomechanics than traditional SB ACL reconstructions, in response to rotatory loading [46, 47] [27].

8.2.5 Fixation of Hamstring Tendons

There are a variety of soft tissue fixation methods described in the literature, including interference screws, cortical suspensory fixation, staples, washers, and cross pins [48]. That means that there is no ideal fixation method for soft tissue grafts. The fixation methods can be divided in three categories: cortical, intra-tunnel, and at the aperture of the tunnel. No studies have been performed that report a statistically significant difference in clinical outcomes when comparing fixation methods [49, 50]. It has been reported that use of cortical suspensory on the femoral side causes tunnel widening mainly because of an increased distance between the location of fixation and the ACL femoral insertion site. Therefore, other femoral fixation devices such as cross-pin fixation have been developed to shorten the fixation distance. A recent meta-analysis [51] showed that there were no differences in clinical and functional outcomes between cortical suspensory fixation and cross-pin fixation. A review by Colvin et al. [52] compared intratunnel versus cortical fixation for the femoral side and found no difference in subjective and objective International Knee Documentation Committee (IKDC) scores; however, there was no statistical difference in failure with interference screw fixation.

The tibial fixation is usually regarded as the weakest point of the initial construct after soft tissue graft ACL reconstruction because of the inferior bone mineral density at the tibial metaphysis and the forces acting against the graft more parallel to the tibial tunnel [35, 53]. A biomechanical study published by Mayr [36] compared the interference screw and a cortical suspensory for tibial fixation under cyclic loading in a load-to-failure test. Grafts with tibial adjustable-length loop cortical suspensory fixation resulted in higher graft elongation during cyclic loading and showed higher ultimate failure loads in comparison with conventional graft preparation with tibial interference screw fixation at time zero. Balazs [54] reviewed 21 studies investigating hybrid fixation combining intratunnel aperture fixation and cortical suspensory fixation on the tibial side in soft tissue graft ACL reconstruction. The authors included both biomechanical and clinical studies. The authors found that the hybrid fixation of the tibial-side graft in ACL reconstruction showed stronger initial stability and less side-to-side laxity after healing, whereas patient-reported outcomes were not changed at 1–3 years follow-up. Such techniques, however, come with the risk of additional complications especially anterior tibial tenderness or skin irritation and added cost. In one clinical study [15], 10% of patients required a re-operation for hardware removal.

The use of HT has grown in popularity over the years because there is less anterior knee pain and less kneeling disability. Knee flexor strength deficits after HT harvest is a common complaint [55, 56]. The morbidity could be greater if gracilis tendon is harvested in addition to semitendinosus tendon [57–59]. As such, some surgeons recommended that gracilis tendon should be preserved if possible.

8.3 Bone-Patellar Tendon-Bone (BPTB) Autograft (Fig. 8.3)

BPTB had been the gold standard for ACL reconstruction for many years [60]. High strength, stiffness, consistency of the size of the graft, and faster graft incorporation with solid fixation using interference screws made this the choice of graft. Another advantage to using the BPTB autograft is its ease of harvest, requiring minimal dissection to access the patellar tendon. Recently in order to minimize the pain in relation to donor site, Pujol and Beaufils [61] have described a two-incision technique to avoid trauma to the branch of the saphenous nerve.

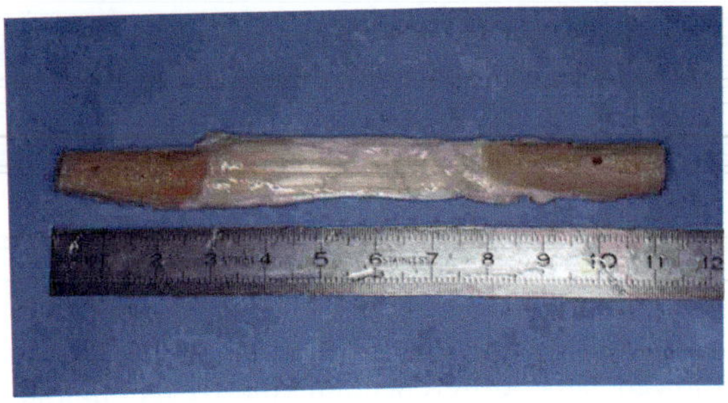

Fig. 8.3 BPTB

8.3.1 Biomechanical Studies

Woo et al. demonstrated that the maximum load to failure of the native ACL in younger patients (age 22–35 years) is 2160 N when it is tested in its anatomic orientation [46]. In clinical practice, 10 mm-wide BPTB autografts with ultimate loads to failure of 1880–2664 N are typically used. For BPTB graft, the stiffness is reported with values between 324 and 543 N/mm for grafts of 10 mm width [2, 27, 52, 62].

Recently, ACL anatomy was revisited by Śmigielski [63], who described the ACL as a ribbon like structure. A recent biomechanical study also showed that using BPTB graft, the ACL-reconstructed knee with a rectangular femoral tunnel more closely resembled the normal knee in biomechanical behavior than with a circular femoral tunnel [64].

However, due to donor site morbidity with BPTB autograft, there are a lot of complications such as anterior knee pain, kneeling pain, extensor strength deficit, patellar fracture, patellar tendon rupture, and postoperative osteoarthritis [7, 65, 66]. The incidence of patella fracture is rare (0.4–1.3%) and is thought to be due to deep and irregular saw cuts. Patellar tendon rupture is also rare (0.18–0.25%) and is thought to be due to poor centralization during harvest leading to the lateral or medial side being too thin [67]. Anterior knee pain and kneeling pain are fairly common with BPTB, with up to 52% of patients reporting anterior knee pain [15]. In this study, the severity of anterior knee pain was significantly higher in patients who received BPTB autografts. After 2 years of observation, there is no significant distinction between harvesting BPTB and HT [15].

8.3.2 Fixation

In ACL reconstruction, fixation methods can be categorized into compression, suspension, post, or hybrid fixation [10, 11]. Secure fixation is crucial for avoiding graft displacement and allowing for graft integration into the bone tunnel, no matter which implant is chosen for fixation [2].

The use of a screw in the tunnel to compress the graft against the cancellous tunnel wall (IFS) is a reproducible technique to achieve successful outcomes. This fixation can be done with metallic (often titanium), polyetheretherketone (PEEK), or bioabsorbable screws. The popularity of metallic screws has decreased because of the greater likelihood of clinical sequelae and the impact of projection artifacts on magnetic resonance imaging (MRI). Load to failure of bone block fixation with interference screws has been shown to be between 550 and 560 N, regardless of whether titanium screws or bio absorbable screws are used [35]. This is very close to the forces experienced by the reconstructed ligament in early rehabilitation, which can be as high as 450–500 N [68].

Suspension fixation of an ACL reconstructed graft is another commonly used technique that typically involves the use of an extracortical bone plug. Suspensory fixation can be achieved by both fixed-loop devices (FLDs) and adjustable-loop devices (ALDs) being commercially available. Multiple studies have found that FLDs have been associated with a higher load-to-failure [31–33].

Another way to fix an ACL reconstruction is to use a metal screw, with or without a washer, or a cross pin that acts as a stable, inflexible point of fixation in the bone separate from the tunnel aperture [69, 70]. Weiss et al. compared IFS fixation, screw-post and washer fixation, and screw-post and washer fixation with an additional IFS in an ACL reconstruction animal model. In their analysis, the hybrid fixation of a post and IFS yielded significantly higher final stiffness and higher yield load than the other fixation methods [70].

8.4 BPTB Versus HT: Comparative Studies

There are numerous articles in the literature with meta-analysis and analysis from National Registries. Several meta-analyses comparing HT and BPTB have been published. We will focus on mid and long-term follow-up.

Zhao et al. published a meta-analysis including only randomized controlled trials (RCTs) that compared BPTB and HT autografts for ACL reconstruction with a 5-year minimum follow-up. Fifteen RCTs, involving 1298 patients (610 patients in the BPTB group and 688 patients in the HT group), were included. In terms of clinical function, no significant difference was found in the objective IKDC score, Lysholm knee score, return to preinjury activity level, and Tegner activity level. There was no statistically significant difference in the Lachman test, pivot-shift test, and side-to-side difference. There was a significant difference in favor of the BPTB group in the domain of graft failure [71].

Hoge et al. compared the long-term outcome of ACL reconstruction performed using BPTB and HT autografts. Four RCTs with a total of 299 patients were included in the study. The mean follow-up ranged from 10.2 to 17 years (mean, 14.79 years). No significant differences in knee laxity or clinical outcome scores were demonstrated in any of the studies [72]. One study found that BPTB autografts were significantly more likely to have osteoarthritis identified by radiographic findings [72].

8.4.1 Graft Failure and Revision (BPTB vs. HT)

Traumatic reinjury to the ACL graft and/or contra-lateral ACL tears remain the most notable problems following ACL reconstruction. There is no consensus regarding the risk of ACL graft failure between BPTB and HT. In some studies with a high number of patients, the number of complete and partial traumatic reinjuries combined was significantly higher in the HT and double-bundle groups compared with the use of BPTB [15, 65].

However, no systematic reviews or meta-analyses of randomized or prospective studies showed any difference in the revision rate between HT and BPTB grafts [7, 73, 74], except one meta-analysis showing that HT graft failed at a higher rate than BPTB graft [6].

In all the studies, failure rates in both grafts were quite low and almost equivalent, and the small difference in failure rate should be presented and discussed with the patient.

8.4.2 Static Knee Laxity

8.4.2.1 Anterior Laxity

Anterior laxity can be assessed by the Lachman test which has been shown to be a noninvasive test with high specificity and high sensitivity [75]. Semiquantitative evaluation of the Lachman test can be done by using the KT 1000 arthrometer [76, 77], the Rolimeter [78] and the GNRB arthrometer [79]. The Lachman test is considered as a static evaluation of anterior knee laxity. This test involves only 1 degree of freedom of the knee.

Systematic reviews have shown no difference in manual or instrumented laxity between BPTB and HT autografts [28, 73] Poehling and Samuellson reported a systematic review of studies with less than 2 years' follow-up and studies with 2 years to less than 5 years and studies with 5 years or longer. The BPTB and HT groups did not differ in terms of manual or instrumented laxity. The meta-analysis did not reveal any differences in anterior laxity between the two grafts. Anderson et al. [80] published a summary of systematic and meta-analysis: they found no difference in knee laxity for BPTB and versus HT autografts.

Paterno et al. [81], in a systematic review, underlined the difference between male and females. Four of the six studies [7, 8, 10, 11] that examined sex differences in anteroposterior (AP) knee laxity between HT and BPTB grafts reported that females who had HT had significantly greater AP knee laxity than males who had an ACL reconstruction with HT.

8.4.2.2 Rotatory Knee Laxity

Rotatory knee laxity in ACL-injured knees is commonly represented by the pivot shift test. This test is considered as a dynamic evaluation of the knee. It tests the kinematics of the entire knee joint. It has been shown to be superior to anterior laxity tests such as Lachman when it comes to predicting patient-reported knee instability. It has shown to be correlated with poor subjective and objective outcome scores [82]; however, the pivot shift test is subjective with low interobserver reliability [79]. Kocher et al. examined a group of patients without concurrent meniscal or chondral injury who underwent primary ACL reconstruction, and they found that the pivot-shift examination, in contradistinction to instrumented knee laxity and Lachman examination, had significant associations with certain subjective symptoms and function after ACL reconstruction. Patients with IKDC grade B and C pivot-shift examinations had more symptoms of partial and full giving way and had more functional difficulty with cutting and twisting. Rotatory laxity assessment between the two grafts has not been found to differ in most meta-analyses [6, 73]. Xie et al. [7] in 2015, showed, in a meta-analysis, a difference in favor of the BPTB.

Desai [83] in a meta-analysis about anatomic single versus double-bundle ACL reconstruction found two studies with significant differences in manual pivot-shift test between single-bundle and double-bundle groups in favor of double-bundle ACL reconstruction [12, 40].

8.4.3 Muscle Strength

This is a common outcome in the literature. The examination is conducted at speeds ranging from 60, 90, 120, 240, and 300 degrees/second.

There are a various reports in the literature with testing of hamstring strength after harvesting semitendinosus or both gracilis and semitendinosus. In the reconstructed knee in which both semitendinosus and gracilis were used, hamstring strength was found to average 99% compared to the normal knee [22]. When the semitendinosus alone was used, there was no difference (102%) from the normal knee.

HT autografts induce a higher hamstring strength deficit compared with BPTB autografts [57–59]. With these findings, harvesting of HT should not be used in ballet dancers or Judoka.

Beynon et al. [84] reported a prospective randomized controlled: Twenty-eight patients underwent ACL reconstruction with BPTB, and twenty-eight patients were treated with a two-strand semitendinosus-gracilis autograft. They found no differences with Cybex testing regarding the extension strength at 1- and 3-year-follow-up. However, the patients with HT (semitendinosus and gracilis) graft had significant deficit in flexion strength at 240 degrees/second compared with those with BPTB reconstruction at 3-year follow-up.

Marder et al. [85] reported a study of 80 patients with two groups with BPTB and HT grafts. The average values for peak quadriceps torque at 60 degrees/second were 88% ± 17% for the BPTB grafts and 91% ± 19% for the HT (semitendinosus and gracilis) grafts, which was not statistically significant. The average values for hamstrings peak torque at 60 degrees/second were 91 ± 18% and 83 ± 16% for the BPTB and HT (semitendinosus and gracilis) grafts, respectively; this difference was statistically significant ($p = 0.025$).

8.4.4 Return to Preinjury Activity Level

The Tegner activity rating system was developed by Dr. Y. Tegner [86] as a measure of activity level.

In an Evidence-Based Medicine Series Systematic Review, Samuellson et al. [28] found that all but one [87] of the studies found no difference between the BPTB graft and the HT graft in terms of Tegner activities score [30, 84, 88–98].

In another meta-analysis, Xie et al. [7] showed a significant difference in an analysis of eight studies [29, 85, 87, 99–103]. Eight studies including a total of 507 patients (243 BPTB, 264 HT) reported data regarding the percentages of patients returning to preinjury activity level. The analysis showed that there was a significant difference between the two groups, and the odds ratio for return to preinjury activity level was 1.48 in favor of BPTB.

8.4.5 Patient-Reported Clinical Outcomes

International Knee Documentation Committee (IKDC) score [104] and Lysholm score [105] have been widely used to evaluate postoperative ACL reconstruction outcomes. Several meta-analysis and systematic review papers have shown that there are no differences in IKDC score and Lysholm between BPTB and HT autografts [7, 56, 73].

8.4.6 Osteoarthritis (OA)

Prevention of OA after ACL reconstruction is one of the major issues. However, despite surgery intended to stabilize the joint, individuals who undergo an ACL reconstruction have a three to five times greater risk of developing posttraumatic OA of the knee. OA has been reported in up to 40% of patients at long-term follow-up after ACL reconstruction [106].

Lecoq et al. [107] has shown that at 12 years of follow-up, neither graft technique was superior to the other in terms of the rate of OA. There are several factors associated with the risk of OA: meniscus status, cartilage injury and cartilage status during ACL reconstruction, age, and obesity. However, recent meta-analyses and systematic reviews revealed that BPTB graft resulted in an increased incidence of OA compared with HT grafts [7, 73, 108].

8.4.7 Pain

8.4.7.1 Anterior Knee Pain

Anterior knee pain is a common complication after ACL rupture or reconstruction. Seven studies did not report any difference in anterior knee pain in between BPTB and HT [29, 84, 86, 92, 97, 109, 110]. But others studies found less anterior knee pain with a significant difference in favor of HT [111] . It is interesting to note that the difference was largest between 3 and 8 months after the procedure. The prevalence of pain occurring within 2 years after ACL reconstruction was higher in the BPTB group ($p < 0.001$). However, there was no significant difference between the BPTB and HT groups with respect to the duration of the pain at 2 years [42].

8.4.7.2 Kneeling Pain

The most common test for assessing kneeling pain is the Knee-Walking test [62]. In the literature, we found three different situations: BPTB graft could give more pain than HT, BPTB graft could be equal to HT, or HT could give more pain than BPTB [112]. Xie [7] found in a meta-analysis that seven studies including a total of 548 patients (248 BPTB/300 HT) reported data regarding the occurrence of kneeling pain. The analysis showed the risk ratio for kneeling pain was 2.05 in favor of 4SHT.

8.4.8 Infection

Infection is the most serious complication after ACL reconstruction. Infection can result in ACL graft failure, hyaline cartilage loss, and knee stiffness. Several reviews have reported the incidence of septic arthritis after ACL reconstruction to range from 0.5% to 0.6% [113–115]. Surgical debridement under arthroscopy combined with antibiotic therapy was the recommended treatment.

Several recent studies have shown that vancomycin-soaked grafts are effective in preventing infection [116–121].

8.4.9 Ligamentization

A series of biologic processes that occur to the free tendon graft utilized to reconstruct the ACL is termed ligamentization. The time required for ligamentization after ACL reconstruction has been reported to range from 9 to 18 months under light microscopy [122], 13–30 months under electron microscopy [123], and within 1 year in a collagen cross-links analysis [124]. The current challenge is to find a better strategy to improve tendon–bone interface healing and promote maturation of the grafted tendon for faster recovery.

There are two sites which are different: the bone tunnels and the intra articular site.

Animal studies have analyzed the tendon remodeling in relation to the type of fixation [125–129]. Weiler [125] found that fixation with absorbable screw may alter the process of early remodeling. Singhatat et al. [127] made a comparison in between screw fixations and spiked washer (WasherLoc Arthrotek, Inc., Warsaw, Indiana). The strength and stiffness of the complex deteriorated after 4 weeks of implantation with the interference screw but were either maintained or improved with the use of the WasherLoc device. They postulated that aggressive rehabilitation after ACL reconstruction should only be allowed after ACL reconstruction with a fixation device that maintains strength and stiffness in the first few weeks of implantation [127].

Other animal studies have analyzed the intra-articular remodeling process of different graft types after ACL reconstruction. The autograft underwent necrosis, revascularization, and remodeling over time. Return of overall graft integrity and histological appearance was completed 6–12 months after ACL reconstruction without full restoration of the mechanical strength of the intact ACL [130].

8.4.10 Tunnel Enlargement

Most of the time, HT is associated with radiographic tunnel widening. Seven studies that assessed radiographic tunnel widening found that the HT graft caused greater tunnel widening than the BPTB [89, 91, 93, 100, 131].

Webster et al. clearly demonstrated tunnel enlargement at 4 months that did not significantly change at 1 and 2 years [125]. It appears that bone tunnel enlargement develops during the first few months after surgery, with little change thereafter [88, 95].

8.5 Quadriceps Tendon (QT) Autograft (Fig. 8.4)

With time, there is an increase in publications about the QT as a source of autograft tissue for use in ACL reconstruction.

In 2010, van Eck [132] reported in a study that QT was used in 2.5% of all ACL reconstructions. In 2014, Middelton [40] found that QT was used in 11% of all ACL reconstruction.

Kim [133] reported a study comparing the QT and BPTB with 48 patients with 24 months follow-up. There was no significant difference in between the two groups in terms of measurements of laxity, return to sports activities, and outcome scores (IKDC score). The mean anterior knee pain score was the same between the two groups. There was less kneeling pain with QT than BPTB.

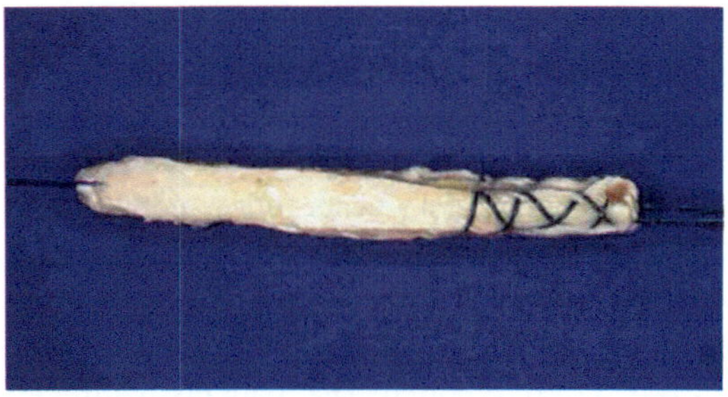

Fig. 8.4 Quadriceps tendon

Lund et al. [134], in a randomized controlled study, compared knee stability, kneeling pain, harvest site pain, sensitivity loss, and subjective clinical outcome after ACL reconstruction with either bone–patellar tendon–bone (BPTB) or quadriceps tendon–bone (QTB) autografts. The use of the QTB graft resulted in less kneeling pain, graft site pain, and sensitivity loss than seen with BPTB grafts. Slone et al. [135] performed a systematic review about the use of QT in the literature in which 14 studies were included. The authors compared QT to BPTB. Stability outcomes (Lachman, pivot-shift, and instrumented laxity testing), functional outcomes (International Knee Documentation Committee and Lysholm scores), overall patient satisfaction, range of motion, and complications were similar between QT and other graft options. Less donor-site morbidity was seen in patients who underwent QT ACL reconstructions. It is a safe, reproducible, and versatile graft that should be considered in future studies of ACL reconstruction.

Cavaignac et al. [136] found in a comparative study of 86 patients that the use of a QT in ACL reconstruction leads to equal or better functional outcomes than HT.

8.6 Iliotibial Band (ITB) Autograft

Stenbirk et al. [137] reported a study of 60 patients. The purpose was to assess whether the ITB autograft was a long-term reliable alternative to the BPTB autograft, using a prospective randomized controlled trial design. Similar ACL graft failures rates (16% for ITB and 13% for BPTB) were found. The ITB graft had an equal clinical outcome at 15 years follow-up. The ITB graft failure rate was comparable to the only other long-term follow-up case series [138] evaluating the ITB autograft and its failure rate.

Bak [139] reported 132 patients with ITB grafts at 4-year-follow-up. He noted 20% failure in women and suggested that ITB should not be recommended in women.

8.7 Allografts (Figs. 8.5 and 8.6)

Donor site morbidity has led to the use of allograft [140–142]. Lack of donor morbidity and shorter operative time are advantages of the use of allograft. But, allografts have a higher risk of revision compared with autografts [143, 144].

Mascarenhas et al. [145] reported a systematic review of overlapping meta analyses. Eight studies containing 15,819 patients were analyzed. Four analyses found no differences between allograft and autograft. Four analyses reported higher ACL graft ruptures in the allograft group.

Factors associated with high risk of ACL graft rupture include graft irradiation and bio cleanse graft processing [146].

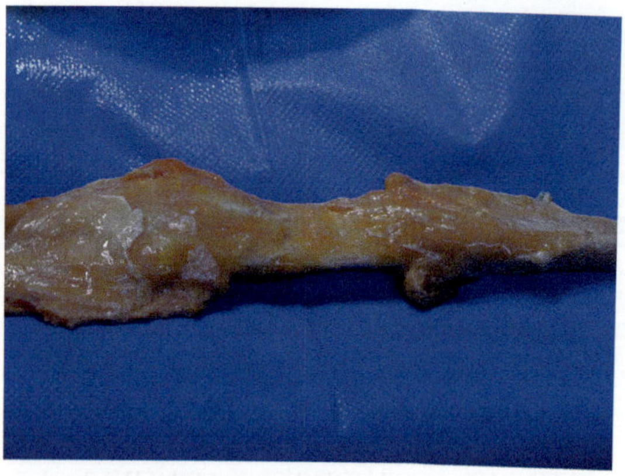

Fig. 8.5 Allograft anterior tibialis

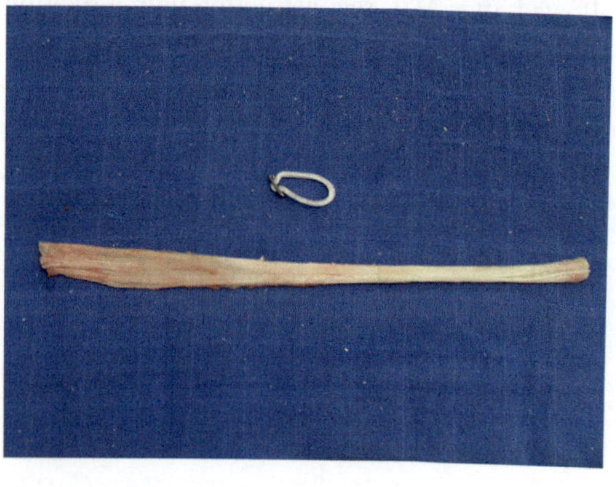

Fig. 8.6 Allograft extensor apparatus

Clinical results with nonirradiated and/or unprocessed allograft have been reported to be equal to those with autograft [145–151].

8.8 Synthetic Grafts

Synthetic grafts had gained in popularity because of easy availability, convenience, lack of disease transmission, and donor site morbidity and the potential for dramatically accelerated rehabilitation with return to sport significantly earlier than autograft and allograft.

First generations with Dacron [Stryker, Kalamazoo, MI], Goretex [W.L. Gore and Associates, Flagstaff, AZ], Klad [3M, St Paul, MN], Leeds Kieo [Xiros (formerly Neoligaments), Leeds, England] demonstrated high risks of complications such iterative rupture, infections, osteolysis, massive effusion, and OA [152–154].

Newer generations of synthetic grafts such as Ligament Advanced Reinforcement System (LARS) (Surgical Implants and Devices, Arc-sur-Tille, France) have been used with lower rates of failure.

In a systematic review, Batty et al. [155] reported an analysis of 85 papers including first generation and newer generation of synthetic devices. Preliminary results for newer-generation devices, specifically the LARS, showed lower reported rates of failure, revision, and sterile effusion/synovitis when compared with older devices. Revision rate was 2.6% for the LARS ligament.

Ventura et al. [152] reported a prospective study with 19 years follow-up.

A total of 51 patients were treated with artificial ACL reconstruction with polyethylene terephthalate (PET) synthetic ligaments between 1986 and 1990. A total of 7.5% were found to have ruptured their PET ligaments and 100% presented with degenerative OA at the final radiographic evaluation.

There are few studies [156–158] who have compared autografts and synthetics grafts with controversial results. There is a need for further prospective and randomized studies with a greater number of patients to justify the use of synthetic graft.

8.9 Conclusion

All different type of grafts can be used in this difficult surgery. There is definitively no best graft to use. There are some advantages and disadvantages for each graft.

For young age, the recommendation could be to use a BPTB graft. For women, there is a general theory that there is an increased anterior laxity after ACL reconstruction as compared with men and then BPTB graft could be recommended.

There is no significant difference in laxity between BPTB and HT. Evidence suggests that the BPTB graft is superior to HT in terms of ACL iterative rupture. There are no differences between BPTB and HT in clinical outcome, time to return to sports activities, or range of motion. During the first few months of surgery, the BPTB graft causes more anterior knee pain, but there is no noticeable difference after 2 years of follow-up compared to HT. With long follow-up, there is, perhaps, more OA with the use of BPTB. The use of HT can produce tunnel widening but there are no correlations between this event and clinical outcomes including knee laxity. The harvesting of semitendinosus and gracilis is associated with inferior knee flexion at high angles compared to semitendinosus grafts.

There are no differences in clinical outcome when SB and DB ACL reconstructions are compared.

Nowadays, QT is more frequently used for ACL reconstructions as the graft can be customized to reduce anterior knee pain.

The use of ITB can be an alternative to others graft. It is not recommended to use in women.

Allografts generally have slightly poorer results in terms of ACL graft rupture rates. However, they can be valuable in certain patient groups, such as revision patients who have had multiple graft harvests previously. The potential for donor site morbidity is eliminated, but they don't facilitate a faster return to sport.

Synthetic grafts are slowly regaining popularity with no donor site morbidity and the ability to perform multiligament reconstructions without compromising the BPTB or HT. Before recommending synthetic grafts, it's important to be aware of any prior complications and the need for further studies.

Evidence of hamstring weakness in flexion after harvesting of the semitendinosus/gracilis suggests preference for other grafts in sports such as gymnastics, wrestling and ballet dancer. Hamstring tears are common in football and sprinting sports, which may indicate why BPTB grafts are the most widely used in these activities.

Patellar tendinopathy is common in basketball and tennis leading some to suggest the avoidance of grafts involving the extensor mechanism in these activities.

At last, graft choice, therefore, needs to be chosen after an educated discussion with the patient regarding their expectations with regard to donor morbidity and speed of rehabilitation as well as the surgeon's personal experience.

References

1. Gifstad T, Foss OA, Engebretsen L, Lind M, Forssblad M, Albrektsen G, et al. Lower risk of revision with patellar tendon autografts compared with hamstring autografts: a registry study based on 45,998 primary ACL reconstructions in Scandinavia. Am J Sports Med. 2014;42(10):2319–28.
2. Runer A, Keeling L, Wagala N, Nugraha H, Özbek EA, Hughes JD, et al. Current trends in graft choice for anterior cruciate ligament reconstruction – part I: anatomy, biomechanics, graft incorporation and fixation. J Exp Orthop. 2023;10(1):37.
3. Fältström A, Kvist J, Hägglund M. High risk of new knee injuries in female soccer players after primary anterior cruciate ligament reconstruction at 5- to 10-year follow-up. Am J Sports Med. 2021;49(13):3479–87.
4. Persson A, Fjeldsgaard K, Gjertsen JE, Kjellsen AB, Engebretsen L, Hole RM, et al. Increased risk of revision with hamstring tendon grafts compared with patellar tendon grafts after anterior cruciate ligament reconstruction: a study of 12,643 patients from the Norwegian Cruciate Ligament Registry, 2004–2012. Am J Sports Med. 2014;42(2):285–91.
5. Rahr-Wagner L, Thillemann TM, Pedersen AB, Lind M. Comparison of hamstring tendon and patellar tendon grafts in anterior cruciate ligament reconstruction in a nationwide population-based cohort study: results from the Danish Registry of Knee Ligament Reconstruction. Am J Sports Med. 2014;42(2):278–84.
6. Samuelsen BT, Webster KE, Johnson NR, Hewett TE, Krych AJ. Hamstring autograft versus patellar tendon autograft for ACL reconstruction: is there a difference in graft failure rate? A meta-analysis of 47,613 patients. Clin Orthop. 2017;475(10):2459–68.
7. Xie X, Liu X, Chen Z, Yu Y, Peng S, Li Q. A meta-analysis of bone–patellar tendon–bone autograft versus four-strand hamstring tendon autograft for anterior cruciate ligament reconstruction. Knee. 2015;22(2):100–10.
8. Szymski D, Achenbach L, Weber J, Huber L, Memmel C, Kerschbaum M, et al. Reduced performance after return to competition in ACL injuries: an analysis on return to competition in the 'ACL registry in German Football'. Knee Surg Sports Traumatol Arthrosc. 2023;31(1):133–41.
9. Turk R, Shah S, Chilton M, Thomas TL, Anene C, Mousad A, et al. Return to sport after anterior cruciate ligament reconstruction requires evaluation of >2 functional tests, psychological readiness, quadriceps/hamstring strength, and time after surgery of 8 months. Arthrosc J Arthrosc Relat Surg. 2023;39(3):790–801.e6.
10. Luc B, Gribble PA, Pietrosimone BG. Osteoarthritis prevalence following anterior cruciate ligament reconstruction: a systematic review and numbers-needed-to-treat analysis. J Athl Train. 2014;49(6):806–19.
11. Wang LJ, Zeng N, Yan ZP, Li JT, Ni GX. Post-traumatic osteoarthritis following ACL injury. Arthritis Res Ther. 2020;22(1):57.
12. Friel NA, Chu CR. The role of ACL injury in the development of posttraumatic knee osteoarthritis. Clin Sports Med. 2013;32(1):1–12.
13. Dhammi IK, Rehan-Ul-Haq, Kumar S. Graft choices for anterior cruciate ligament reconstruction. Indian J Orthop. 2015;49(2):127–8.
14. Malige A, Baghdadi S, Hast MW, Schmidt EC, Shea KG, Ganley TJ. Biomechanical properties of common graft choices for anterior cruciate ligament reconstruction: A systematic review. Clin Biomech Bristol Avon. 2022;95:105636.
15. Rousseau R, Labruyere C, Kajetanek C, Deschamps O, Makridis KG, Djian P. Complications after anterior cruciate ligament reconstruction and their relation to the type of graft: a prospective study of 958 cases. Am J Sports Med. 2019;47(11):2543–9.
16. Grassi A, Nitri M, Moulton SG, Marcheggiani Muccioli GM, Bondi A, Romagnoli M, et al. Does the type of graft affect the outcome of revision anterior cruciate ligament reconstruction? A meta-analysis of 32 studies. Bone Joint J. 2017;99-B(6):714–23.
17. Edwards A. Operative repair of cruciate ligaments in severe trauma of knee. Br J Surg. 1926;13(51):432–8.
18. Galeazzi R. La ricostituzione dei ligamenti crociati del ginocchio. Atti e Memorie della. Soc Lomb Chir. 1934;13:302–17.
19. Macey H. A new operative procedure for the repair of ruptured cruciate ligaments of the knee joint. Surg Gynecol Obstet. 1939;69:108–9.

20. Lindemann K. Uber den plastischen Ersatz der Kreuzbänder durch gestielte Sehnenverpflanzungen [On the replacement of cruciate ligaments using pedunculated tendon transplants]. Z Orthop. 1950;79(2):316–34.
21. Cho K. Reconstruction of the anterior cruciate ligament by semitendinosus tenodesis. J Bone Joint Surg Am. 1975;57(5):608–12.
22. Brant Lipscomb A, Johnston RK, Snyder RB, Warburton MJ, Pressly GP. Evaluation of hamstring strength following use of semitendinosus and gracilis tendons to reconstruct the anterior cruciate ligament. Am J Sports Med. 1982;10(6):340–2.
23. Svantesson E, Sundemo D, Hamrin Senorski E, Alentorn-Geli E, Musahl V, Fu FH, et al. Double-bundle anterior cruciate ligament reconstruction is superior to single-bundle reconstruction in terms of revision frequency: a study of 22,460 patients from the Swedish National Knee Ligament Register. Knee Surg Sports Traumatol Arthrosc. 2017;25(12):3884–91.
24. Marcacci M, Zaffagnini S, Iacono F, Neri MP, Loreti I, Petitto A. Arthroscopic intra- and extra-articular anterior cruciate ligament reconstruction with gracilis and semitendinosus tendons. Knee Surg Sports Traumatol Arthrosc. 1998;6(2):68–75.
25. Chambat P, Vargas R, Fayard J, Lemaire B, Sonnery-Cottet B. Résultat des reconstructions du ligament croisé antérieur sous contrôle arthroscopique avec un recul supérieur à 15 ans. In: Chambat P, Neyret P (éd) Le genou et le sport du ligament à la prothèse. Sauramps Médical. 2008. p. 147–152.
26. Girgis FG, Marshall JL, Monajem A. The cruciate ligaments of the knee joint. Anatomical, functional and experimental analysis. Clin Orthop Relat Res. 1975;106:216–31.
27. Petersen W, Zantop T. Anatomy of the anterior cruciate ligament with regard to its two bundles. Clin Orthop Relat Res. 2007;454:35–47.
28. Samuelsson K, Andersson D, Karlsson J. Treatment of anterior cruciate ligament injuries with special reference to graft type and surgical technique: an assessment of randomized controlled trials. Arthrosc J Arthrosc Relat Surg. 2009;25(10):1139–74.
29. Shaieb MD, Kan DM, Chang SK, Marumoto JM, Richardson AB. A prospective randomized comparison of patellar tendon versus semitendinosus and gracilis tendon autografts for anterior cruciate ligament reconstruction. Am J Sports Med. 2002;30(2):214–20.
30. Ibrahim SAR, Al-Kussary IM, Al-Misfer ARK, Al-Mutairi HQ, Ghafar SA, Noor TAE. Clinical evaluation of arthroscopically assisted anterior cruciate ligament reconstruction: patellar tendon versus gracilis and semitendinosus autograft. Arthrosc J Arthrosc Relat Surg. 2005;21(4):412–7.
31. Anderson AF, Snyder RB, Lipscomb AB. Anterior cruciate ligament reconstruction. Am J Sports Med. 2001;29(3):8.
32. Maletis GB, Inacio MCS, Desmond JL, Funahashi TT. Reconstruction of the anterior cruciate ligament. Bone Joint J. 2013;95(5):6.
33. Siebold R, Fu F. Assessment and augmentation of symptomatic anteromedial or posterolateral bundle tears of the anterior cruciate ligament. Arthroscopy. 2008;24(11):1289–98.
34. Ochi M, Adachi N, Deie M, Kanaya A. Anterior cruciate ligament augmentation procedure with a 1-incision technique: anteromedial bundle or posterolateral bundle reconstruction. Arthroscopy. 2006;22(4):463–8.
35. Brand JC, Pienkowski D, Steenlage E, Hamilton D, Johnson DL, Caborn DNM. Interference screw fixation strength of a quadrupled hamstring tendon graft is directly related to bone mineral density and insertion torque. Am J Sports Med. 2000;28(5):705–10.
36. Mayr R, Heinrichs CH, Eichinger M, Coppola C, Schmoelz W, Attal R. Biomechanical comparison of 2 anterior cruciate ligament graft preparation techniques for tibial fixation: adjustable-length loop cortical button or interference screw. Am J Sports Med. 2015;43(6):1380–5.
37. Arnoczky S, Tarvin G, Marshall J. Anterior cruciate ligament replacement using patellar tendon. An evaluation of graft revascularization in the dog. J Bone Joint Surg Am. 1982;64(2):217–24.
38. Ochi M, Iwasa J, Uchio Y, Adachi N, Sumen Y. The regeneration of sensory neurones in the reconstruction of the anterior cruciate ligament. J Bone Joint Surg Am. 1999;81(5):902–6.
39. Tiamklang T, Sumanont S, Foocharoen T, Laopaiboon M. Double-bundle versus single-bundle reconstruction for anterior cruciate ligament rupture in adults. Cochrane Bone, Joint and Muscle Trauma Group, editor. Cochrane Database Syst Rev [Internet]. 2012 [cited 2020 Mar 23]. Available from: http://doi.wiley.com/10.1002/14651858.CD008413.pub2.
40. Middleton KK, Hamilton T, Irrgang JJ, Karlsson J, Harner CD, Fu FH. Anatomic anterior cruciate ligament (ACL) reconstruction: a global perspective. Part 1. Knee Surg Sports Traumatol Arthrosc. 2014;22(7):1467–82.
41. Wilson TW, Zafuta MP, Zobitz M. A biomechanical analysis of matched bone-patellar tendon-bone and double-looped semitendinosus and gracilis tendon grafts. Am J Sports Med. 1999;27(2):202–7.
42. Hammer D, Brown C, Steiner M, Hecker A, Hayes W. Hamstring tendon grafts for reconstruction of the anterior cruciate ligament: biomechanical evaluation of the use of multiple strands and tensioning techniques. J Bone Joint Surg Am. 1999;81(4):549–57.
43. Handl M, Drzik M, Cerulli G, Povýsil C, Chlpík J, Varga F, et al. Reconstruction of the anterior cruciate ligament: dynamic strain evaluation of the graft. Knee Surg Sports Traumatol Arthrosc. 2007;15(3):233–41.
44. Conte EJ, Hyatt AE, Gatt CJ, Dhawan A. Hamstring autograft size can be predicted and is a potential

risk factor for anterior cruciate ligament reconstruction failure. Arthrosc J Arthrosc Relat Surg. 2014;30(7):882–90.
45. Magnussen RA, Lawrence JTR, West RL, Toth AP, Taylor DC, Garrett WE. Graft size and patient age are predictors of early revision after anterior cruciate ligament reconstruction with hamstring autograft. Arthrosc J Arthrosc Relat Surg. 2012;28(4):526–31.
46. Woo SLY, Kanamori A, Zeminski J, Yagi M, Papageorgiou C, Fu FH. The effectiveness of reconstruction of the anterior cruciate ligament with hamstrings and patellar tendon: a cadaveric study comparing anterior tibial and rotational loads. J Bone Joint Surg Am. 2002;84(6):907–14.
47. Yagi M, Wong EK, Kanamori A, Debski RE, Fu FH, Woo SLY. Biomechanical analysis of an anatomic anterior cruciate ligament reconstruction. Am J Sports Med. 2002;30(5):660–6.
48. Shelton WR, Fagan BC. Autografts commonly used in anterior cruciate ligament reconstruction. Am Acad Orthop Surg. 2011;19(5):259–64.
49. Kousa P, Järvinen TLN, Vihavainen M, Kannus P, Järvinen M. The fixation strength of six hamstring tendon graft fixation devices in anterior cruciate ligament reconstruction: part II: tibial site. Am J Sports Med. 2003;31(2):182–8.
50. Kousa P, Järvinen TLN, Vihavainen M, Kannus P, Järvinen M. The fixation strength of six hamstring tendon graft fixation devices in anterior cruciate ligament reconstruction: part I: femoral site. Am J Sports Med. 2003;31(2):174–81.
51. Jiang H, Ma G, Li Q, Hu Y, Li J, Tang X. Cortical button versus cross-pin femoral fixation for hamstring anterior cruciate ligament reconstruction: a meta-analysis of randomized controlled trials. Am J Sports Med. 2018;46(9):2277–84.
52. Colvin A, Sharma C, Parides M, Glashow J. What is the best femoral fixation of hamstring autografts in anterior cruciate ligament reconstruction?: a meta-analysis. Clin Orthop Relat Res. 2011;469(4):1075–81.
53. Bartz RL, Mossoni K, Tyber J, Tokish J, Gall K, Siparsky PN. A biomechanical comparison of initial fixation strength of 3 different methods of anterior cruciate ligament soft tissue graft tibial fixation: resistance to monotonic and cyclic loading. Am J Sports Med. 2007;35(6):949–54.
54. Balazs GC, Brelin AM, Grimm PD, Dickens JF, Keblish DJ, Rue JPH. Hybrid tibia fixation of soft tissue grafts in anterior cruciate ligament reconstruction: a systematic review. Am J Sports Med. 2016;44(10):2724–32.
55. Gifstad T, Sole A, Strand T, Uppheim G, Grøntvedt T, Drogset JO. Long-term follow-up of patellar tendon grafts or hamstring tendon grafts in endoscopic ACL reconstructions. Knee Surg Sports Traumatol Arthrosc. 2013;21(3):576–83.
56. Spindler KP, Kuhn JE, Freedman KB, Matthews CE, Dittus RS, Harrell FE. Anterior cruciate ligament reconstruction autograft choice: bone-tendon-bone versus hamstring: does it really matter? A systematic review. Am J Sports Med. 2004;32(8):1986–95.
57. Sharma A, Flanigan DC, Randall K, Magnussen RA. Does gracilis preservation matter in anterior cruciate ligament reconstruction? A systematic review. Arthrosc J Arthrosc Relat Surg. 2016;32(6):1165–73.
58. Ardern C, Webster K. Knee flexor strength recovery following hamstring tendon harvest for anterior cruciate ligament reconstruction: a systematic review. Orthop Rev. 2009;1(2):12.
59. Yosmaoglu HB, Baltaci G, Ozer H, Atay A. Effects of additional gracilis tendon harvest on muscle torque, motor coordination, and knee laxity in ACL reconstruction. Knee Surg Sports Traumatol Arthrosc. 2011;19(8):1287–92.
60. Chechik O, Amar E, Khashan M, Lador R, Eyal G, Gold A. An international survey on anterior cruciate ligament reconstruction practices. Int Orthop. 2013;37(2):201–6.
61. Beaufils P, Gaudot F, Drain O, Boisrenoult P, Pujol N. Mini-invasive technique for bone patellar tendon bone harvesting: its superiority in reducing anterior knee pain following ACL reconstruction. Curr Rev Musculoskelet Med. 2011;4(2):45–51.
62. Vasconcelos W, Santos C, Ferracini A, Dejour D. Influence of anterior pain on results from anterior cruciate ligament reconstruction. Rev Bras Ortop. 2010;46:40–4.
63. Śmigielski R, Zdanowicz U, Drwięga M, Ciszek B, Ciszkowska-Łysoń B, Siebold R. Ribbon like appearance of the midsubstance fibres of the anterior cruciate ligament close to its femoral insertion site: a cadaveric study including 111 knees. Knee Surg Sports Traumatol Arthrosc. 2015;23(11):3143–50.
64. Suzuki T, Shino K, Otsubo H, Suzuki D, Mae T, Fujimiya M, et al. Biomechanical comparison between the rectangular-tunnel and the round-tunnel anterior cruciate ligament reconstruction procedures with a bone–patellar tendon–bone graft. Arthrosc J Arthrosc Relat Surg. 2014;30(10):1294–302.
65. Mohtadi NG, Chan DS, Dainty KN, Whelan DB. Patellar tendon versus hamstring tendon autograft for anterior cruciate ligament rupture in adults. Cochrane Bone, Joint and Muscle Trauma Group, editor. Cochrane Database Syst Rev [Internet]. 2011 [cited 2020 Jun 21]. Available from: http://doi.wiley.com/10.1002/14651858.CD005960.pub2.
66. Busam ML, Provencher MT, Bach BR. Complications of anterior cruciate ligament reconstruction with bone-patellar tendon-bone constructs: care and prevention. Am J Sports Med. 2008;36(2):379–94.
67. Nawabi DH, van der List JP, Williams A. Technical considerations for patellar tendon harvest. In: Nakamura N, Zaffagnini S, Marx RG, Musahl V, editors. Controversies in the Technical Aspects of ACL Reconstruction [Internet]. Berlin, Heidelberg: Springer Berlin Heidelberg; 2017 [cited 2020 Jul

17]. p. 103–20. Available from: http://link.springer.com/10.1007/978-3-662-52742-9_11.
68. Frank CB, Jackson DW. The science of reconstruction of the anterior cruciate ligament. J Bone Joint Surg Am. 1997;79(10):1556–76.
69. McDermott E, DeFoor MT, Blaber OK, Aman ZS, DePhillipo NN, Dekker TJ. Biomechanical comparison of anterior cruciate ligament reconstruction fixation methods and implications on clinical outcomes. Ann Joint [Internet]. 2023 [cited 2024 May 29];8. Available from: https://aoj.amegroups.org/article/view/7980.
70. Weiss FP, Possoli FADA, Costa IZ, Borges PC, Stieven Filho E, Kubrusly LF. Fixation of the anterior ligament graft at the tibial pole: biomechanical analysis of three methods. Rev Bras Ortop. 2019;54(6):697–702.
71. Zhao L, Lu M, Deng M, Xing J, He L, Wang C. Outcome of bone-patellar tendon-bone vs hamstring tendon autograft for anterior cruciate ligament reconstruction: a meta-analysis of randomized controlled trials with a 5-year minimum follow-up. Medicine (Baltimore). 2020;99(48):e23476.
72. Hoge CG, Matar RN, Khalil LS, Buchan JA, Johnson CM, Grawe BM. Outcomes following anterior cruciate ligament reconstruction with patellar tendon vs hamstring autografts: A systematic review of randomized controlled trials with a mean follow-up of 15 years. Arch Bone Joint Surg. 2022;10(4):311–9.
73. Poehling-Monaghan KL, Salem H, Ross KE, Secrist E, Ciccotti MC, Tjoumakaris F, et al. Long-term outcomes in anterior cruciate ligament reconstruction: a systematic review of patellar tendon versus hamstring autografts. Orthop J Sports Med. 2017;5(6):232596711770973.
74. Gabler CM, Jacobs CA, Howard JS, Mattacola CG, Johnson DL. Comparison of graft failure rate between autografts placed via an anatomic anterior cruciate ligament reconstruction technique: a systematic review, meta-analysis, and meta-regression. Am J Sports Med. 2016;44(4):1069–79.
75. Berruto M, Uboldi F, Gala L, Marelli B, Albisetti W. Is triaxial accelerometer reliable in the evaluation and grading of knee pivot-shift phenomenon? Knee Surg Sports Traumatol Arthrosc. 2013;21(4):981–5.
76. Daniel DM, Malcom LL, Losse G, Stone ML, Sachs R, Burks R. Instrumented measurement of anterior laxity of the knee. J Bone Joint Surg Am. 1985;67(5):720–6.
77. Boyer P, Djian P, Christel P, Paoletti X, Degeorges R. Reliability of the KT-1000 arthrometer (Medmetric) for measuring anterior knee laxity: comparison with Telos in 147 knees. Rev Chir Orthop Reparatrice Appar Mot. 2004;90(8):757–64.
78. Schuster AJ, McNicholas MJ, Wachtl SW, McGurty DW, Jakob RP. A new mechanical testing device for measuring anteroposterior knee laxity. Am J Sports Med. 2004;32(7):1731–5.
79. Robert H, Nouveau S, Gageot S, Gagnière B. A new knee arthrometer, the GNRB: experience in ACL complete and partial tears. Orthop Traumatol Surg Res OTSR. 2009;95(3):171–6.
80. Anderson MJ, Browning WM, Urband CE, Kluczynski MA, Bisson LJ. A systematic summary of systematic reviews on the topic of the anterior cruciate ligament. Orthop J Sports Med. 2016;4(3):232596711663407.
81. Paterno MV, Weed AM, Hewett TE. A between sex comparison of anterior-posterior knee laxity after anterior cruciate ligament reconstruction with patellar tendon or hamstrings autograft: a systematic review. Sports Med Auckl NZ. 2012;42(2):135–52.
82. Kocher MS, Steadman JR, Briggs KK, Sterett WI, Hawkins RJ. Relationships between objective assessment of ligament stability and subjective assessment of symptoms and function after anterior cruciate ligament reconstruction. Am J Sports Med. 2004;32(3):629–34.
83. Desai N, Björnsson H, Musahl V, Bhandari M, Petzold M, Fu FH, et al. Anatomic single- versus double-bundle ACL reconstruction: a meta-analysis. Knee Surg Sports Traumatol Arthrosc. 2014;22(5):1009–23.
84. Beynnon BD, Johnson RJ, Fleming BC, Kannus P, Kaplan M, Samani J, et al. Anterior cruciate ligament replacement: comparison of bone-patellar tendon-bone grafts with two-Strand hamstring grafts: a prospective, randomized study. J Bone Joint Surg Am. 2002;84(9):1503–13.
85. Marder RA, Raskind JR, Carroll M. Prospective evaluation of arthroscopically assisted anterior cruciate ligament reconstruction: patellar tendon versus semitendinosus and gracilis tendons. Am J Sports Med. 1991;19(5):478–84.
86. Tegner Y, Lysholm J. Rating systems in the evaluation of knee ligament injuries. Clin Orthop. 1985;198:43–9.
87. Maletis GB, Cameron SL, Tengan JJ, Burchette RJ. A prospective randomized study of anterior cruciate ligament reconstruction: a comparison of patellar tendon and quadruple-strand semitendinosus/gracilis tendons fixed with bioabsorbable interference screws. Am J Sports Med. 2007;35(3):384–94.
88. Lidén M, Ejerhed L, Sernert N, Laxdal G, Kartus J. Patellar tendon or semitendinosus tendon autografts for anterior cruciate ligament reconstruction: a prospective, randomized study with a 7-year follow-up. Am J Sports Med. 2007;35(5):740–8.
89. Zaffagnini S, Marcacci M, Lo Presti M, Giordano G, Iacono F, Neri MP. Prospective and randomized evaluation of ACL reconstruction with three techniques: a clinical and radiographic evaluation at 5 years follow-up. Knee Surg Sports Traumatol Arthrosc. 2006;14(11):1060–9.
90. Laxdal G, Kartus J, Hansson L, Heidvall M, Ejerhed L, Karlsson J. A prospective randomized comparison of bone-patellar tendon-bone and hamstring grafts for anterior cruciate ligament reconstruction. Arthrosc J Arthrosc Relat Surg. 2005;21(1):34–42.

91. Aglietti P, Giron F, Buzzi R, Biddau F, Sasso F. Anterior cruciate ligament reconstruction: bone-patellar tendon-bone compared with double semitendinosus and gracilis tendon grafts. A prospective, randomized clinical trial. J Bone Joint Surg Am. 2004;86(10):2143–55.
92. Ejerhed L, Kartus J, Sernert N, Köhler K, Karlsson J. Patellar tendon or semitendinosus tendon autografts for anterior cruciate ligament reconstruction? A prospective randomized study with a two-year follow-up. Am J Sports Med. 2003;31(1):19–25.
93. Harilainen A, Linko E, Sandelin J. Randomized prospective study of ACL reconstruction with interference screw fixation in patellar tendon autografts versus femoral metal plate suspension and tibial post fixation in hamstring tendon autografts: 5-year clinical and radiological follow-up results. Knee Surg Sports Traumatol Arthrosc. 2006;14(6):517–28.
94. Järvelä T. Double-bundle versus single-bundle anterior cruciate ligament reconstruction: a prospective, randomize clinical study. Knee Surg Sports Traumatol Arthrosc. 2007;15(5):500–7.
95. Sajovic M, Vengust V, Komadina R, Tavcar R, Skaza K. A prospective, randomized comparison of semitendinosus and gracilis tendon versus patellar tendon autografts for anterior cruciate ligament reconstruction: five-year follow-up. Am J Sports Med. 2006;34(12):1933–40.
96. Jansson KA, Linko E, Sandelin J, Harilainen A. A prospective randomized study of patellar versus hamstring tendon autografts for anterior cruciate ligament reconstruction. Am J Sports Med. 2003;31(1):12–8.
97. Beard DJ, Anderson JL, Davies S, Price AJ, Dodd CA. Hamstrings vs. patella tendon for anterior cruciate ligament reconstruction: a randomised controlled trial. Knee. 2001;8(1):45–50.
98. O'Neill DB. Arthroscopically assisted reconstruction of the anterior cruciate ligament. A prospective randomized analysis of three techniques. J Bone Joint Surg Am. 1996;78(6):803–13.
99. Sajovic M, Strahovnik A, Dernovsek MZ, Skaza K. Quality of life and clinical outcome comparison of semitendinosus and gracilis tendon versus patellar tendon autografts for anterior cruciate ligament reconstruction: an 11-year follow-up of a randomized controlled trial. Am J Sports Med. 2011;39(10):2161–9.
100. Feller JA, Webster KE. A randomized comparison of patellar tendon and hamstring tendon anterior cruciate ligament reconstruction. Am J Sports Med. 2003;31(4):564–73.
101. Svensson M, Sernert N, Ejerhed L, Karlsson J, Kartus JT. A prospective comparison of bone-patellar tendon-bone and hamstring grafts for anterior cruciate ligament reconstruction in female patients. Knee Surg Sports Traumatol Arthrosc. 2006;14(3):278–86.
102. Taylor DC, DeBerardino TM, Nelson BJ, Duffey M, Tenuta J, Stoneman PD, et al. Patellar tendon versus hamstring tendon autografts for anterior cruciate ligament reconstruction: a randomized controlled trial using similar femoral and tibial fixation methods. Am J Sports Med. 2009;37(10):1946–57.
103. Holm I, Oiestad BE, Risberg MA, Aune AK. No difference in knee function or prevalence of osteoarthritis after reconstruction of the anterior cruciate ligament with 4-strand hamstring autograft versus patellar tendon-bone autograft: a randomized study with 10-year follow-up. Am J Sports Med. 2010;38(3):448–54.
104. Irrgang JJ, Anderson AF, Boland AL, Harner CD, Kurosaka M, Neyret P, et al. Development and validation of the international knee documentation committee subjective knee form. Am J Sports Med. 2001;29(5):600–13.
105. Lysholm J, Gillquist J. Evaluation of knee ligament surgery results with special emphasis on use of a scoring scale. Am J Sports Med. 1982;10(3):150–4.
106. Øiestad BE, Engebretsen L, Storheim K, Risberg MA. Winner of the 2008 systematic review competition: knee osteoarthritis after anterior cruciate ligament injury. Am J Sports Med. 2009;37(7):1434–43.
107. Lecoq FA, Parienti JJ, Murison J, Ruiz N, Bouacida K, Besse J, et al. Graft choice and the incidence of osteoarthritis after anterior cruciate ligament reconstruction: a causal analysis from a cohort of 541 patients. Am J Sports Med. 2018;46(12):2842–50.
108. Magnussen RA, Carey JL, Spindler KP. Does autograft choice determine intermediate-term outcome of ACL reconstruction? Knee Surg Sports Traumatol Arthrosc. 2011;19(3):462–72.
109. Aune AK, Holm I, Risberg MA, Jensen HK, Steen H. Four-strand hamstring tendon autograft compared with patellar tendon-bone autograft for anterior cruciate ligament reconstruction. A randomized study with two-year follow-up. Am J Sports Med. 2001;29(6):722–8.
110. Eriksson K, Anderberg P, Hamberg P, Löfgren AC, Bredenberg M, Westman I, et al. A comparison of quadruple semitendinosus and patellar tendon grafts in reconstruction of the anterior cruciate ligament. J Bone Joint Surg Br. 2001;83(3):348–54.
111. Janani G, Suresh P, Prakash A, Parthiban J, Anand K, Arumugam S. Anterior knee pain in ACL reconstruction with BPTB graft – is it a myth? Comparative outcome analysis with hamstring graft in 1,250 patients. J Orthop. 2020;22:408–13.
112. Chantrelle M, Menu P, Crenn V, Grondin J, Daley P, Louguet B, et al. Consequences of anterior knee pain after anterior cruciate ligament reconstruction: a 2015–2020 cohort study. PLoS One. 2023;18(1):e0280146.
113. Wang C, Lee YHD, Siebold R. Recommendations for the management of septic arthritis after ACL reconstruction. Knee Surg Sports Traumatol Arthrosc. 2014;22(9):2136–44.

14. Kim SJ, Postigo R, Koo S, Kim JH. Infection after arthroscopic anterior cruciate ligament reconstruction. Orthopedics. 2014;37(7):477–84.
15. Saper M, Stephenson K, Heisey M. Arthroscopic irrigation and debridement in the treatment of septic arthritis after anterior cruciate ligament reconstruction. Arthrosc J Arthrosc Relat Surg. 2014;30(6):747–54.
116. Bohu Y, Klouche S, Sezer HB, Herman S, Grimaud O, Gerometta A, et al. Vancomycin-soaked autografts during ACL reconstruction reduce the risk of post-operative infection without affecting return to sport or knee function. Knee Surg Sports Traumatol Arthrosc. 2020;28:2578.
117. Offerhaus C, Balke M, Hente J, Gehling M, Blendl S, Höher J. Vancomycin pre-soaking of the graft reduces postoperative infection rate without increasing risk of graft failure and arthrofibrosis in ACL reconstruction. Knee Surg Sports Traumatol Arthrosc. 2019;27(9):3014–21.
118. Pérez-Prieto D, Torres-Claramunt R, Gelber PE, Shehata TMA, Pelfort X, Monllau JC. Autograft soaking in vancomycin reduces the risk of infection after anterior cruciate ligament reconstruction. Knee Surg Sports Traumatol Arthrosc. 2016;24(9):2724–8.
119. Phegan M, Grayson JE, Vertullo CJ. No infections in 1300 anterior cruciate ligament reconstructions with vancomycin pre-soaking of hamstring grafts. Knee Surg Sports Traumatol Arthrosc. 2016;24(9):2729–35.
120. Schuster P, Schlumberger M, Mayer P, Eichinger M, Geßlein M, Reddemann F, et al. Soaking of the graft in vancomycin dramatically reduces the incidence of postoperative septic arthritis after anterior cruciate ligament reconstruction. Knee Surg Sports Traumatol Arthrosc. 2020;28:2587.
121. Vertullo CJ, Quick M, Jones A, Grayson JE. A surgical technique using presoaked vancomycin hamstring grafts to decrease the risk of infection after anterior cruciate ligament reconstruction. Arthrosc J Arthrosc Relat Surg. 2012;28(3):337–42.
122. Dong S, Xie G, Zhang Y, Shen P, Huangfu X, Zhao J. Ligamentization of autogenous hamstring grafts after anterior cruciate ligament reconstruction: midterm versus long-term results. Am J Sports Med. 2015;43(8):1908–17.
123. Zaffagnini S, De Pasquale V, Marchesini Reggiani L, Russo A, Agati P, Bacchelli B, et al. Electron microscopy of the remodelling process in hamstring tendon used as ACL graft. Knee Surg Sports Traumatol Arthrosc. 2010;18(8):1052–8.
124. Marumo K, Saito M, Yamagishi T, Fujii K. The "ligamentization" process in human anterior cruciate ligament reconstruction with autogenous patellar and hamstring tendons: a biochemical study. Am J Sports Med. 2005;33(8):1166–73.
125. Weiler A, Peine R, Pashmineh-Azar A, Abel C, Südkamp NP, Hoffmann RFG. Tendon healing in a bone tunnel. Part I: biomechanical results after biodegradable interference fit fixation in a model of anterior cruciate ligament reconstruction in sheep. Arthrosc J Arthrosc Relat Surg. 2002;18(2):113–23.
126. Rodeo SA, Arnoczky SP, Torzilli PA, Hidaka C, Warren RF. Tendon-healing in a bone tunnel. A biomechanical and histological study in the dog. J Bone Joint Surg Am. 1993;75(12):1795–803.
127. Singhatat W, Lawhorn KW, Howell SM, Hull ML. How four weeks of implantation affect the strength and stiffness of a tendon graft in a bone tunnel: a study of two fixation devices in an extraarticular model in ovine. Am J Sports Med. 2002;30(4):506–13.
128. Voos JE, Musahl V, Maak TG, Wickiewicz TL, Pearle AD. Comparison of tunnel positions in single-bundle anterior cruciate ligament reconstructions using computer navigation. Knee Surg Sports Traumatol Arthrosc. 2010 Sep;18(9):1282–9.
129. Weiler A, Hoffmann RFG, Bail HJ, Rehm O, Südkamp NP. Tendon healing in a bone tunnel. Part II: histologic analysis after biodegradable interference fit fixation in a model of anterior cruciate ligament reconstruction in sheep. Arthrosc J Arthrosc Relat Surg. 2002;18(2):124–35.
130. Janssen RPA, van der Wijk J, Fiedler A, Schmidt T, Sala HAGM, Scheffler SU. Remodelling of human hamstring autografts after anterior cruciate ligament reconstruction. Knee Surg Sports Traumatol Arthrosc. 2011;19(8):1299–306.
131. Jansson KA, Harilainen A, Sandelin J, Karjalainen PT, Aronen HJ, Tallroth K. Bone tunnel enlargement after anterior cruciate ligament reconstruction with the hamstring autograft and endobutton fixation technique. A clinical, radiographic and magnetic resonance imaging study with 2 years follow-up. Knee Surg Sports Traumatol Arthrosc. 1999;7(5):290–5.
132. van Eck CF, Schreiber VM, Mejia HA, Samuelsson K, van Dijk CN, Karlsson J, et al. "Anatomic" anterior cruciate ligament reconstruction: a systematic review of surgical techniques and reporting of surgical data. Arthrosc J Arthrosc Relat Surg. 2010;26(9 Suppl):S2–12.
133. Kim SJ, Kumar P, Oh KS. Anterior cruciate ligament reconstruction: autogenous quadriceps tendon-bone compared with bone-patellar tendon-bone grafts at 2-year follow-up. Arthrosc J Arthrosc Relat Surg. 2009;25(2):137–44.
134. Lund B, Nielsen T, Faunø P, Christiansen SE, Lind M. Is quadriceps tendon a better graft choice than patellar tendon? A prospective randomized study. Arthrosc J Arthrosc Relat Surg. 2014;30(5):593–8.
135. Slone HS, Romine SE, Premkumar A, Xerogeanes JW. Quadriceps tendon autograft for anterior cruciate ligament reconstruction: a comprehensive review of current literature and systematic review of clinical results. Arthrosc J Arthrosc Relat Surg. 2015;31(3):541–54.
136. Cavaignac E, Coulin B, Tscholl P, Nik Mohd Fatmy N, Duthon V, Menetrey J. Is quadriceps tendon autograft a better choice than hamstring autograft for

anterior cruciate ligament reconstruction? A comparative study with a mean follow-up of 3.6 years. Am J Sports Med. 2017;45(6):1326–32.
137. Stensbirk F, Thorborg K, Konradsen L, Jørgensen U, Hölmich P. Iliotibial band autograft versus bone-patella-tendon-bone autograft, a possible alternative for ACL reconstruction: a 15-year prospective randomized controlled trial. Knee Surg Sports Traumatol Arthrosc. 2014;22(9):2094–101.
138. Yamaguchi S, Sasho T, Tsuchiya A, Wada Y, Moriya H. Long term results of anterior cruciate ligament reconstruction with iliotibial tract: 6-, 13-, and 24-year longitudinal follow-up. Knee Surg Sports Traumatol Arthrosc. 2006;14(11):1094–100.
139. Bak K, Jørgensen U, Ekstrand J, Scavenius M. Results of reconstruction of acute ruptures of the anterior cruciate ligament with an iliotibial band autograft. Knee Surg Sports Traumatol Arthrosc. 1999;7(2):111–7.
140. Noyes FR, Barber-Westin SD. Reconstruction of the anterior cruciate ligament with human allograft. Comparison of early and later results. J Bone Joint Surg Am. 1996;78(4):524–37.
141. Shino K, Inoue M, Horibe S, Nagano J, Ono K. Maturation of allograft tendons transplanted into the knee. An arthroscopic and histological study. J Bone Joint Surg Br. 1988;70(4):556–60.
142. Noyes FR, Barber SD, Mangine RE. Bone-patellar ligament-bone and fascia lata allografts for reconstruction of the anterior cruciate ligament. J Bone Joint Surg Am. 1990;72(8):1125–36.
143. Wasserstein D, Sheth U, Cabrera A, Spindler KP. A systematic review of failed anterior cruciate ligament reconstruction with autograft compared with allograft in young patients. Sports Health Multidiscip Approach. 2015;7(3):207–16.
144. Pallis M, Svoboda SJ, Cameron KL, Owens BD. Survival comparison of allograft and autograft anterior cruciate ligament reconstruction at the United States Military Academy. Am J Sports Med. 2012;40(6):1242–6.
145. Mascarenhas R, Erickson BJ, Sayegh ET, Verma NN, Cole BJ, Bush-Joseph C, et al. Is there a higher failure rate of allografts compared with autografts in anterior cruciate ligament reconstruction: a systematic review of overlapping meta-analyses. Arthrosc J Arthrosc Relat Surg. 2015;31(2):364–72.
146. Tejwani SG, Chen J, Funahashi TT, Love R, Maletis GB. Revision risk after allograft anterior cruciate ligament reconstruction: association with graft processing techniques, patient characteristics, and graft type. Am J Sports Med. 2015;43(11):2696–705.
147. Wei J, Bin YH, Bi QJ, Bao YT. A meta-analysis of anterior cruciate ligament reconstruction with autograft compared with nonirradiated allograft. Knee. 2015;22(5):372–9.
148. Sun K, Zhang J, Wang Y, Xia C, Zhang C, Yu T, et al. Arthroscopic reconstruction of the anterior cruciate ligament with hamstring tendon autograft and fresh-frozen allograft: a prospective, randomized controlled study. Am J Sports Med. 2011;39(7):1430–8.
149. Noh JH, Yi SR, Song SJ, Kim SW, Kim W. Comparison between hamstring autograft and free tendon Achilles allograft: minimum 2-year follow-up after anterior cruciate ligament reconstruction using EndoButton and Intrafix. Knee Surg Sports Traumatol Arthrosc. 2011;19(5):816–22.
150. Lawhorn KW, Howell SM, Traina SM, Gottlieb JE, Meade TD, Freedberg HI. The effect of graft tissue on anterior cruciate ligament outcomes: a multicenter, prospective, randomized controlled trial comparing autograft hamstrings with fresh-frozen anterior tibialis allograft. Arthrosc J Arthrosc Relat Surg. 2012;28(8):1079–86.
151. Mariscalco MW, Magnussen RA, Mehta D, Hewett TE, Flanigan DC, Kaeding CC. Autograft versus nonirradiated allograft tissue for anterior cruciate ligament reconstruction: a systematic review. Am J Sports Med. 2014;42(2):492–9.
152. Ventura A, Terzaghi C, Legnani C, Borgo E, Albisetti W. Synthetic grafts for anterior cruciate ligament rupture: 19-year outcome study. Knee. 2010;17(2):108–13.
153. Greis PE, Steadman JR. Revision of failed prosthetic anterior cruciate ligament reconstruction. Clin Orthop. 1996;325:78–90.
154. Paulos LE, Rosenberg TD, Grewe SR, Tearse DS, Beck CL. The GORE-TEX anterior cruciate ligament prosthesis: a long-term followup. Am J Sports Med. 1992;20:246–52.
155. Batty LM, Norsworthy CJ, Lash NJ, Wasiak J, Richmond AK, Feller JA. Synthetic devices for reconstructive surgery of the cruciate ligaments: a systematic review. Arthrosc J Arthrosc Relat Surg. 2015;31(5):957–68.
156. Tiefenboeck TM, Thurmaier E, Tiefenboeck MM, Ostermann RC, Joestl J, Winnisch M, et al. Clinical and functional outcome after anterior cruciate ligament reconstruction using the LARS™ system at a minimum follow-up of 10 years. Knee. 2015;22(6):565–8.
157. Tang LZ, Long ZX, Jiang Y, Zeng BF. Four-strand hamstring tendon autograft versus LARS artificial ligament for anterior cruciate ligament reconstruction. Int Orthop. 2010;34(1):45–9.
158. Pan X, Wen H, Wang L, Ge T. Bone–patellar tendon–bone autograft versus LARS artificial ligament for anterior cruciate ligament reconstruction. Eur J Orthop Surg Traumatol. 2013;23(7):819–23.

9 Pediatric ACL-Embryology and Development

Yuichi Hoshino and Yuta Nakanishi

9.1 Introduction

The anterior cruciate ligament (ACL) plays a pivotal role in knee joint function, undergoing intricate developmental processes from embryogenesis. These processes shape the structural and functional attributes of the ligament throughout life, underscoring its importance in preserving knee joint stability and function. In cases of ACL injury among pediatric patients, ACL reconstruction may be deemed appropriate, particularly for those engaged in sports activities. However, unlike adults, the proximity of the physis and tibial tuberosity to the bone tunnels needed for ACL graft fixation poses unique challenges in pediatric cases. To prevent the inadvertent violation of the physis and mitigate potential growth disturbances, it is imperative to possess an extensive comprehension of the developmental trajectory of the ACL within the knee joint. This chapter aims to thoroughly explore the anatomical growth of the ACL from embryonic stage through pediatric development, emphasizing its importance in mitigating growth-related complications.

Y. Hoshino (✉) · Y. Nakanishi
Department of Orthopaedic Surgery, Kobe University Graduate School of Medicine, Kobe, Japan

9.2 The Origin of the ACL in Embryonic Stage

Formation of the knee joint commences in the fourth week of gestation from vascular mesenchyme located between the femur and tibia, within the blastema of the femur and tibia. The ACL starts to form in the fetus as early as the eighth week of gestation [1–5]. Likewise, knee menisci might derive from a similar developmental pathway as the ACL, suggesting a close functional relationship between these structures. Around the eighth week after gestation, a discernible amount of fibers in the ACL begins to appear. By this stage, fibroblasts within the ligament have already aligned themselves along the axis of strain of the ACL [6]. At 9 weeks, the cruciate ligaments consist of numerous immature fibroblasts characterized by scanty cytoplasm and fusiform nuclei. In the following weeks, the primary change, aside from growth, is the heightened vascularity. During this period, fusiform fibroblasts express elevated levels of the angiogenic factor vascular endothelial growth factor (VEGF). Although VEGF expression is largely diminished after birth, it is strongly reactivated during the remodeling process of autologous tendon grafts used in ACL reconstruction [7].

An alternative theory for fetal ACL formation suggests a merging of ligamentous collagen fibers with periosteal fibers. Earlier embryologic investigations have shown that the fetal knee at

M. Karahan et al. (eds.), *Anterior Cruciate Ligament Injuries Throughout the Life Span*,
https://doi.org/10.1007/978-3-031-84501-7_9

9.5–10 weeks of gestational age closely resembles the adult knee in terms of all anatomical structures. The ACL in the adult population is recognized to comprise two bundles: the anteromedial and posterolateral bundles. These bundles start to become discernible by 16th week of gestation [2–5, 8]. During this developmental stage, the fetal ACL resembles its adult counterpart, but exhibits differences such as more parallel orientation of the bundles and broader femoral origins. Tena-Arregui and colleagues [9] performed arthroscopic observation in the knee joints of fetuses and found that the appearance of the ACL closely resembled that of the adult knee. However, they noted a difference in the femoral attachment of the ACL, which appeared more ribbon-like compared to the adult. Although two main bundles were already distinguishable, their orientation appeared more parallel compared to the adult ACL bundles [9]. Following the initial formation of the ligament, a period ensues wherein no significant organizational or compositional changes manifest throughout the remainder of fetal development [1, 2, 4, 10]. Behr et al. [11] focused on the development of the ACL in an anatomic study using fetal specimens and found that the femoral origin of the fetal ACL developed as a confluence of ligament fibers with periosteum at 20th week, vascular invasion into the epiphysis at 24th week, and establishment of a secure epiphyseal attachment by 36th week of gestation [11].

From a histological perspective, it is noteworthy that the fetal ACL demonstrates heightened cellularity and vascularity relative to its adult counterpart [8]. Throughout fetal development, the histological composition of the ACL undergoes significant alterations. Histogenesis investigations elucidate the development of collagen fibers and alignment of fibroblasts, shedding light on the ligament's evolving structural framework. These histological modifications enhance the ACL's strength and durability, laying the groundwork for its biomechanical functions in the future. The fetal ACL manifests two discernible bundles, segregated by a membranous septum, which closely resembles the structural configuration observed in the adult ligament. This architectural similarity underscores the developmental continuity and evolutionary consistency across different stages of ligament maturation [8]. At 36th week of gestation, there was interdigitation observed between the collagen fibers and the adjacent bone, accompanied by a transitional zone featuring fibrocartilage and mineralized fibrocartilage. These findings indicate the maturation process occurring at the femoral origin of the ACL in the fetal knee [11].

Currently, there remains a lack of consensus within the existing literature concerning the anatomical structures situated in the anterolateral aspect of the knee and the potential existence of a distinct ligament in this region. This ongoing debate persists, with particular emphasis on its relevance within the pediatric population, where anatomical considerations hold significant importance. As per the findings elucidated by Helito et al. [12] the anterolateral ligament (ALL) emerges during fetal development, showcasing anatomical and histological features that closely resemble those observed in the adult ALL. This exploration highlights the developmental continuum and shared morphological traits evident across different stages of ligament maturation [12]. Conversely, Sabzevari et al. [13] argued against the existence of a distinct ligamentous structure in the anterolateral aspect of the knee from a developmental perspective. They observed the presence of other anatomical structures such as the fibular collateral ligament (FCL), popliteus tendon, lateral capsule, patellar tendon, and medial collateral ligament (MCL). Furthermore, histological analysis of the anterolateral capsule (ALC) did not support the presence of any ligamentous structure, revealing instead a loose hypocellular connective tissue with less organized collagen fibers [13]. In a study conducted by Shea et al. [14], an attempt was made to characterize the ALL in patients with immature skeletons. However, the authors were only able to identify this structure in one out of the eight cases studied. These researchers proposed that the ALL might not be fully developed in the immature skeleton [14]. Moreover, Iseki et al. [15] embarked on a comprehensive investigation involving histological, immunohistochemical, and molecular analyses within the anterolateral capsule (ALC) of the knee. Their exhaustive research efforts culminated in the conclusion that despite meticulous examination, a discernible distinct ligament within the ALC remained elusive, as delineated in their referenced study [15].

9.3 The Postnatal Development of the ACL

As the ACL traverses from childhood through adolescence, the ligament adeptly adjusts to the evolving biomechanical demands, illustrating a dynamic interplay between developmental processes and functional requirements. The age at which the developmental processes of the ACL conclude remains a subject of ongoing controversy.

As individuals progress through life, their ACL undergoes shifts in its biomechanical characteristics, evolving to meet the changing demands placed upon it across different stages of development and activity levels. Murray et al. [16] conducted a comprehensive evaluation of the biomechanical outcomes associated with ACL healing in both skeletally immature and mature minipigs. Their investigation unveiled noteworthy findings, indicating that immature animals displayed markedly enhanced ligament healing capabilities compared to their mature counterparts. Moreover, they discerned that the structural integrity of the skeletally immature ligament surpassed that of mature animals by a significant margin. These findings shed valuable light on the differential healing processes and biomechanical attributes observed across varying developmental stages [16]. Woo et al. [17] examined the structural properties of the femur–ACL–tibia complex across different age groups: younger (22–35 years), middle-aged (40–50 years), and older (60–97 years) knees. They discovered that linear stiffness, ultimate load, and energy absorption significantly decreased with specimen age [17]. These findings align with the original observations of Noyes and Grood [18], who noted a decrease in linear stiffness and ultimate load in the ACL with advancing age [18]. Prior research has investigated age-dependent changes in ACL cell metabolism, uncovering higher rates of cell proliferation and migration in skeletally immature animals [19]. Moreover, skeletally immature animals demonstrated improved biomechanical responses to healing [16], possibly linked to a decrease in growth factor receptor numbers with age [20]. This intriguing relationship suggests a potential mechanism underlying the observed differences in healing outcomes across varying developmental stages, highlighting the complex interplay between age-related changes in growth factor signaling and biomechanical responses to injury. Wang et al. [21] investigated age-dependent alterations in the matrix and organization of ligament-to-bone insertion using a bovine model. They discovered structural and compositional changes at the insertion site that were age-dependent, with the skeletally immature group resembling articular cartilage, while the adult interface resembled fibrocartilaginous tissue. Collagen fiber orientation showed marked differences that intensified with age, and the extracellular matrix composition and cellularity were also found to vary with age [21]. It is conceivable that throughout the process of maturation, there is a gradual accrual of mechanical loading at the insertion site, and this phenomenon may vary with age, leading to adaptive changes in the organization and composition of the matrix at the interface. These age-dependent adaptations likely reflect the intricate interplay between mechanical stimuli and tissue remodeling processes, emphasizing the dynamic nature of structural modifications occurring at the ligament-bone interface during maturation.

Magnetic resonance imaging (MRI) studies were conducted to meticulously observe and document the in vivo developmental growth of the anterior cruciate ligament (ACL). This investigative approach allows for comprehensive insights into the intricate processes underlying ACL development across various stages of growth and maturation. In their study, Lima et al. [22] documented that the length of the ACL undergoes continuous growth until approximately 11–12 years of age. Moreover, their findings revealed a progressive verticalization not only of the ACL but also of the intercondylar notch roof across the range of ages analyzed [22]. This longitudinal assessment offers valuable insights into the dynamic developmental changes occurring within the ACL and its surrounding structures throughout childhood and adolescence. Meanwhile, Putur et al. [23] demonstrated morphological change of the ACL until the age of 18. In pediatric populations without any history of ACL injury, a minimum of two serial MRIs were taken, and comprehensive dimensional measurements of the ACL were performed. The study revealed three distinct phases of growth: individuals aged 1.5–5.75 years experienced an average growth of 2.25 mm per year; those aged 6–11.5 exhibited an average growth of 1.46 mm per year, with growth decelerating around age 11.75 and ceasing by 18.5 years. The growth model for ACL sagittal width demonstrated an increase of 0.45 mm per year from age 1.5 to 14.5, after which it reached a plateau. Meanwhile, coronal width exhibited a growth rate of 0.22 mm per year from age 1.5 to 18.75 [23]. Gender differences add complexity to the development of the ACL in the pediatric population. Initially, both males and females exhibit similar ACL morphology during early skeletal growth. However, notable sex disparities in ACL size and orientation emerge during later stages of skeletal growth and maturation, particularly during preadolescence [24].

9.4 Development of the ACL in Relation to Skeletal Changes

Over the course of the last two decades, there has been a threefold increase in the frequency of ACL reconstruction procedures carried out in the pediatric population [25]. It is important to note that there exist discernible differences in the clinical outcomes of ACL reconstruction procedures when comparing pediatric and adult populations. A systematic review and meta-analysis revealed a higher percentage of re-tears, with a 13% ipsilateral re-tear rate and a 14% contralateral ACL tear rate when pooling children and adolescents aged 6–19 years [26]. This may be attributed to the notable anatomical differences of the ACL concerning skeletal morphology between pediatric and adult populations.

The angle of the ACL has consistently shown an increase with age. According to Kim et al. [27] the ACL-tibial angles demonstrated a notable increase as age progressed during skeletal growth, eventually attaining levels comparable to those observed in adults following physeal fusion. On the other hand, the Blumensaat line-ACL angle remained constant after the age of 2 years [27]. The increase in the ACL angle with age could potentially influence the varying impact of posterior tibial slope on ACL tears across different age groups. An increase in the posterior tibial slope is commonly believed to heighten the risk of ACL rupture. In adult literature, the average slope of the proximal tibia is around 10°, with a variation of approximately 3° [28, 29]. Brandon et al. [30] reported that an increased posterior tibial slope was associated with ACL rupture in the adult population [30]. Similarly, teenaged pediatric subjects with open physes and ACL injuries exhibited a notably heightened medial tibial slope compared to subjects with open physes but without ACL injury [31]. Anchustegui et al. investigated the potential alterations in the posterior tibial slope throughout pediatric development, encompassing individuals aged 2–12 years. Their analysis revealed a lack of substantial variation in the posterior tibial slope across this age range, indicating a relatively stable anatomical feature during this critical period of growth and maturation [32]. However, a recent systematic review and meta-analysis presented a controversial revelation: an elevated posterior tibial slope was not identified as a significant risk factor for ACL tears in pediatric and adolescent patients, contrary to findings in the adult population [33].

As a potential risk factor for ACL injury, a narrow intercondylar notch has garnered attention. Domzalski et al. [34] compared MRI-measured notch width between pediatric groups with acute ACL injury and the normal population, discovering an association between a narrower intercondylar notch and ACL rupture risk [34]. This effect of a narrow intercondylar notch on the ACL might be consistent across pediatric and adult age groups.

MRI was employed to assess the tibial attachment of the ACL. As expected in the developmental process of the knee, the absolute positions of the ACL's anterior, central, and posterior tibial insertion points along the tibial anteroposterior direction demonstrated an increase with advancing age. However, when normalized as a percentage relative to the total length of the tibial anteroposterior width, the position of the ACL tibial attachment did not exhibit significant variation with age or sex [35]. This intriguing finding underscores the importance of considering anatomical nuances and developmental factors in understanding ACL morphology and its relationship with demographic variables.

The mechanisms leading to ACL tears in children closely mirror those observed in adults, where the ligament is subject to an intrasubstance lesion resulting from excessive tensile force [36]. However, in children, avulsion fractures are more prevalent occurrences. This propensity can be attributed to several factors, one of which is the relatively weaker strength of the incompletely ossified tibial plateau in comparison to the robustness of the ACL [37, 38]. Moreover, research indicates that ACL volume reaches adult levels before the onset of the growth spurt, implying a potential protective mechanism against ACL tears in pediatric knees during this crucial developmental phase [39]. This understanding highlights the importance of discerning the developmental timeline and maturation processes of ACL structures in pediatric populations. It emphasizes the need for tailored injury prevention strategies and clinical management approaches to effectively address the unique needs of young patients.

9.5 Considerations to Avoid Complications After the Pediatric ACL Reconstruction

The existence of open growth plates adds to the surgical complexity of ACL reconstruction in skeletally immature patients, and growth disturbance emerges as a notable clinical concern following pediatric ACL reconstruction [40, 41]. Although being less frequent nowadays, it can still occur with physeal-sparing, partial transphyseal, and transphyseal ACL reconstruction techniques [42]. When implementing the physeal-sparing technique, it is crucial to conduct a thorough morphological assessment.

The vertical height or depth of the tibial epiphysis has been suggested as a possible anatomical parameter to assist in determining the safe drilling of an all-epiphyseal tunnel [43–45]. Davis et al. [43] conducted an MRI study aiming to establish normal values for the height of the tibial epiphysis and the width of the lateral femoral condylar epiphysis in children and adolescents using MRI scans. The average height of the tibial epiphysis in both children and adolescents was found to be 15 mm. Interestingly, girls exhibited shorter tibial epiphyses compared to boys [43].

The distal femoral growth plate stands out as one of the largest and most rapidly growing physeal structures in the human body. Throughout skeletal development, it contributes to approximately 70% of the total length of the femur and 37% of the entire limb length. Collectively, it yields 0.9–1.0 cm of growth annually until reaching skeletal maturity [46, 47]. To minimize the risk of injury to the distal femoral growth plate during physeal-sparing ACL reconstruction, it is crucial to possess an in-depth and comprehensive understanding of the morphological characteristics of the physis in the distal femur. Ladenhauf et al. [48] conducted an MRI evaluation of physis morphology, revealing that while the physis appears linear in the anterior section of the femur, it displays a more concave shape in the posterior aspect of the medial and lateral condyles [48]. Therefore, it is essential for surgeons not to solely depend on lateral radiographs for the assessment of the anatomical characteristics of the distal femoral growth plate.

The complication rate following ACL reconstruction in the pediatric population exhibits considerable variability. While some earlier reviews have suggested a growth disturbance rate as low as 1%, subsequent studies have reported higher rates of postoperative deformity [40, 49, 50]. Prior systematic reviews regarding growth-related complications have often employed restricted date ranges for their literature searches, omitted case reports, small series, or studies reporting outcomes within heterogeneous cohorts, or solely encompassed studies reporting the presence of growth disturbance within their cohorts. Hence, meta-analyses have become increasingly common in recent times to compile existing data. Wong et al. [51] conducted a meta-analysis comprising 45 studies, which included 1321 patients and 1392 knees. They identified 58 cases of growth disturbances, with 16 necessitating corrective surgery (27.6%). Moreover, 3.7% exhibited angular deformities, primarily valgus, while 7.5% showed a limb length discrepancy of at least 1 cm [51]. Furthermore, Fury et al. [52] conducted the most extensive review, which comprised 3798 immature patients. They highlighted that previous literature has failed to sufficiently evaluate preoperative risks or acknowledge pre-existing deformities in immature patients undergoing ACL reconstruction. It was disclosed that the incidence of postoperative leg length discrepancy exceeding 10 mm and angular deformity was noted at 2.1% and 1.3%, respectively [52]. To accurately ascertain the complication rate following ACL reconstruction in the pediatric population, it is crucial to conduct ongoing patient monitoring until skeletal maturity is reached. This entails implementing radiographic surveillance and employing consistent, appropriate measurement methods to detect and track postoperative growth disturbances effectively.

9.6 Summary

Studies on the embryology and development of the ACL highlight distinct biomechanical properties and clinical outcomes between pediatric and adult populations. The development of the ACL commences at the eighth week of gestation, culminating in the formation of two distinct bundles by the 16th week of gestation. Subsequently, minimal structural changes occur until birth. Conversely, the existence of anatomical structures in the anterolateral aspect of the knee remains contentious during the prenatal period. Following birth, the ACL undergoes concurrent growth with skeletal development. Its length steadily increases until around 11–12 years of age, while the angle of the ACL becomes more vertical over time. Additionally, it is crucial to comprehend the development of the femur and tibia, including the locations of their growth plates, to prevent growth disturbances following ACL reconstruction. Understanding these characteristic features of the ACL and surrounding structures in skeletally immature populations is essential when determining treatment for ACL injuries in this age group.

References

1. Irarrázaval S, Albers M, Chao T, et al. Gross, arthroscopic, and radiographic anatomies of the anterior cruciate ligament: foundations for anterior cruciate ligament surgery. Clin Sports Med. 2017;36:9–23.
2. Gardner E, O'Rahilly R. The early development of the knee joint in staged human embryos. J Anat. 1968;102:289–99.
3. O'Rahilly R. The early prenatal development of the human knee joint. J Anat. 1951;85(2):166–70.
4. Haines RW. The early development of the femoro-tibial and tibio-fibular joints. J Anat. 1953;87(2):192–206.
5. Merida-Velasco JA, Sanchez-Montesinos I, Espin-Ferra J, et al. Development of the human knee joint ligaments. Anat Rec. 1997;248:259–68.
6. Petersen W, Tillmann B. Anatomy and function of the anterior cruciate ligament. Orthopade. 2002;31:710–8.
7. Petersen W, Unterhauser F, Pufe T, et al. The angiogenic peptide vascular endothelial growth factor (VEGF) is expressed during the remodeling of free tendon grafts in sheep. Arch Orthop Trauma Surg. 2003;123:168–74.

8. Ferretti M, Levicoff EA, Macpherson TA, et al. The fetal anterior cruciate ligament: an anatomic and histologic study. Arthroscopy. 2007;23:278–83.
9. Tena-Arregui J, Barrio-Asensio C, Viejo-Tirado F, et al. Arthroscopic study of the knee joint in fetuses. Arthroscopy. 2003;19:862–8.
10. Gray DJ, Gardner E. Prenatal development of the human knee and superior tibiofibular joints. Am J Anat. 1950;86:235–87.
11. Behr CT, Potter HG, Paletta GA Jr. The relationship of the femoral origin of the anterior cruciate ligament and the distal femoral physeal plate in the skeletally immature knee. An anatomic study. Am J Sports Med. 2001;29(6):781–7.
12. Helito CP, do Prado Torres JA, Bonadio MB, et al. Anterolateral ligament of the fetal knee: an anatomic and histological study. Am J Sports Med. 2017;45:91–6.
13. Sabzevari S, Rahnemai-Azar AA, Albers M, et al. Anatomic and histological investigation of the anterolateral capsular complex in the fetal knee. Am J Sports Med. 2017;45:1383–7.
14. Shea KG, Polousky JD, Jacobs JCJ, et al. The anterolateral ligament of the knee: an inconsistent finding in pediatric cadaveric specimens. J Pediatr Orthop. 2016;36:e51–4.
15. Iseki T, Rothrauff BB, Kihara S, et al. Paediatric knee anterolateral capsule does not contain a distinct ligament: analysis of histology, immunohistochemistry and gene expression. J ISAKOS. 2021;6(2):82–7.
16. Murray MM, Magarian EM, Harrison SL, et al. The effect of skeletal maturity on functional healing of the anterior cruciate ligament. J Bone Joint Surg Am. 2010;92:2039–49.
17. Woo SL, Hollis JM, Adams DJ, et al. Tensile properties of the human femuranterior cruciate ligament-tibia complex. The effects of specimen age and orientation. Am J Sports Med. 1991;19:217–25.
18. Noyes FR, Grood ES. The strength of the anterior cruciate ligament in humans and Rhesus monkeys. J Bone Joint Surg Am. 1976;58:1074–82.
19. Mastrangelo AN, Magarian EM, Palmer MP, et al. The effect of skeletal maturity on the regenerative function of intrinsic ACL cells. J Orthop Res. 2010;28:644–51.
20. Vavken P, Saad FA, Murray MM. Age dependence of expression of growth factor receptors in porcine ACL fibroblasts. J Orthop Res. 2010;28:1107–12.
21. Wang IE, Mitroo S, Chen FH, et al. Age-dependent changes in matrix composition and organization at the ligament-to-bone insertion. J Orthop Res. 2006;24:1745–55.
22. Lima FM, Debieux P, Astur DC, et al. The development of the anterior cruciate ligament in the paediatric population. Knee Surg Sports Traumatol Arthrosc. 2019;27:3354–63.
23. Putur DE, Slaven SE, Niu EL. ACL growth with age in pediatric patients: an MRI study. J Pediatr Orthop. 2020;40(8):438–47.
24. Hosseinzadeh S, Kiapour AM. Age-related changes in ACL morphology during skeletal growth and maturation are different between females and males. J Orthop Res. 2021;39(4):841–9.
25. Domzalski M, Karauda A, Grzegorzewski A, et al. Anterior cruciate ligament reconstruction using the transphyseal technique in prepubescent athletes: midterm, prospective evaluation of results. Arthroscopy. 2016;32(6):1141–6.
26. Kay J, Memon M, Marx RG, et al. Over 90% of children and adolescents return to sport after anterior cruciate ligament reconstruction: a systematic review and meta-analysis. Knee Surg Sports Traumatol Arthrosc. 2018;26:1019–36.
27. Kim HK, Laor T, Shire NJ, et al. Anterior and posterior cruciate ligaments at different patient ages: MR imaging findings. Radiology. 2008;247:826–35.
28. Genin P, Weill G, Julliard R. The tibial slope. Proposal for a measurement method. J Radiol. 1993;74:27–33.
29. Giffin JR, Vogrin TM, Zantop T, et al. Effects of increasing tibial slope on the biomechanics of the knee. Am J Sports Med. 2004;32:376–82.
30. Brandon ML, Haynes PT, Bonamo JR, et al. The association between posterior-inferior tibial slope and anterior cruciate ligament insufficiency. Arthroscopy. 2006;22:894–9.
31. Vyas S, van Eck CF, Vyas N, et al. Increased medial tibial slope in teenage pediatric population with open physes and anterior cruciate ligament injuries. Knee Surg Sports Traumatol Arthrosc. 2011;19:372–7.
32. Anchustegui N, Grimm NL, Milbrandt TA, et al. Medial and lateral posterior tibial slope in the skeletally immature: a cadaveric study. Orthop J Sports Med. 2022;10(4):23259671221088331. https://doi.org/10.1177/23259671221088331.
33. Farid AR, Pradhan P, Stearns SA, et al. Association between posterior tibial slope and ACL injury in pediatric patients: a systematic review and meta-analysis. Am J Sports Med. 2024;26:3635465231199649. https://doi.org/10.1177/03635465231199649. Online ahead of print.
34. Domzalski M, Grzelak P, Gabos P. Risk factors for anterior cruciate ligament injury in skeletally immature patients: analysis of intercondylar notch width using magnetic resonance imaging. Int Orthop. 2010;34:703–7.
35. Swami VG, Mabee M, Hui C, et al. MRI anatomy of the tibial ACL attachment and proximal epiphysis in a large population of skeletally immature knees: reference parameters for planning anatomic physeal-sparing ACL reconstruction. Am J Sports Med. 2014;42:1644–51.
36. Boden BP, Sheehan FT, Torg JS, et al. Noncontact anterior cruciate ligament injuries: mechanisms and risk factors. J Am Acad Orthop Surg. 2010;18:520–7.
37. Reynders P, Reynders K, Broos P. Pediatric and adolescent tibial eminence fractures: arthroscopic cannulated screw fixation. J Trauma. 2002;53:49–54.
38. Wiley JJ, Baxter MP. Tibial spine fractures in children. Clin Orthop Relat Res. 1990;255:54–60.
39. Tuca M, Hayter C, Potter H, et al. Anterior cruciate ligament and intercondylar notch growth plateaus

prior to cessation of longitudinal growth: an MRI observational study. Knee Surg Sports Traumatol Arthrosc. 2016;24:780–7.
40. Collins MJ, Arns TA, Leroux T, et al. Growth abnormalities following anterior cruciate ligament reconstruction in the skeletally immature patient: a systematic review. Arthroscopy. 2016;32:1714–23.
41. Kocher MS, Saxon HS, Hovis WD, et al. Management and complications of anterior cruciate ligament injuries in skeletally immature patients: survey of the Herodicus Society and the ACL Study Group. J Pediatr Orthop. 2002;22:452–7.
42. Peterson DC, Ayeni OR. Pediatric anterior cruciate ligament reconstruction outcomes. Curr Rev Musculoskelet Med. 2016;9:339–47.
43. Davis DL, Chen L, Young ST. Evaluation of epiphyses in the skeletally immature knee using magnetic resonance imaging: a pilot study to analyze parameters for anterior cruciate ligament reconstruction. Am J Sports Med. 2013;41:1579–85.
44. Lawrence JT, Bowers AL, Belding J, et al. All-epiphyseal anterior cruciate ligament reconstruction in skeletally immature patients. Clin Orthop Relat Res. 2010;468:1971–7.
45. Parikh SN. Transepiphyseal replacement of the anterior cruciate ligament in skeletally immature patients: a preliminary report. J Bone Joint Surg Am. 2011;93:e12.
46. White J, Stubbins S. Growth arrest for equalizing leg lengths. JAMA. 1944;126:1146–9.
47. Dimeglio A. Growth in pediatric orthopaedics. In Weinstein SL, editor. Morrissy R. Lovell and Winter's pediatric orthopedics. Philadelphia: Lippincott, Williams & Wilkins; 2006. p. 58.
48. Ladenhauf HN, Jones KJ, Potter HG, et al. Understanding the undulating pattern of the distal femoral growth plate: implications for surgical procedures involving the pediatric knee: a descriptive MRI study. Knee. 2020;27:315–23.
49. Faunø P, Rømer L, Nielsen T, et al. The risk of transphyseal drilling in skeletally immature patients with anterior cruciate ligament injury. Orthop J Sports Med. 2016;4(9):2325967116664685.
50. Pierce TP, Issa K, Festa A, et al. Pediatric anterior cruciate ligament reconstruction: a systematic review of transphyseal versus physeal-sparing techniques. Am J Sports Med. 2017;45:488–94.
51. Wong SE, Feeley BT, Pandya NK. Complications after pediatric ACL reconstruction: a meta-analysis. J Pediatr Orthop. 2019;39:e566–71.
52. Fury MS, Paschos NK, Fabricant PD, et al. Assessment of skeletal maturity and postoperative growth disturbance after anterior cruciate ligament reconstruction in skeletally immature patients: a systematic review. Am J Sports Med. 2022;50:1430–41.

10 Pediatric ACL Tears: Natural History

Paul Vincent Ryan

10.1 Summary

Anterior cruciate ligament (ACL) injuries in skeletally immature patients were previously considered rare, and the pediatric equivalent identified as the tibial spine avulsion (Figs. 10.1 and 10.2). These injuries have since been delineated, and, although related in many aspects, have been recognised as separate clinical entities. The number of ACL injuries and surgeries in children continues to rise, partly due to changes in sports attendance and training, and partly due to improved clinical and radiological diagnostic measures. Before embarking on the management of pediatric patients with ACL injuries, it is important to understand the natural history of the untreated, or non-operatively managed injury—the likely clinical outcome with regard to return to activities and sporting level, as well as the risks of further instability and progressive injury to the already injured knee, especially to chondral and meniscal elements. This then allows comparative assessment with operative interventions, and their outcomes, considering potential complications of surgery such as physeal damage and resultant limb deformity.

The attending physician along with the patient and their family is then empowered to proceed with a jointly agreed management plan.

P. V. Ryan (✉)
Department of Orthopaedic Surgery, University of Kwa-Zulu Natal, Inkosi Albert Luthuli Central Hospital, Durban, South Africa
e-mail: paulrya@ialch.co.za

M. Karahan et al. (eds.), *Anterior Cruciate Ligament Injuries Throughout the Life Span*,
https://doi.org/10.1007/978-3-031-84501-7_10

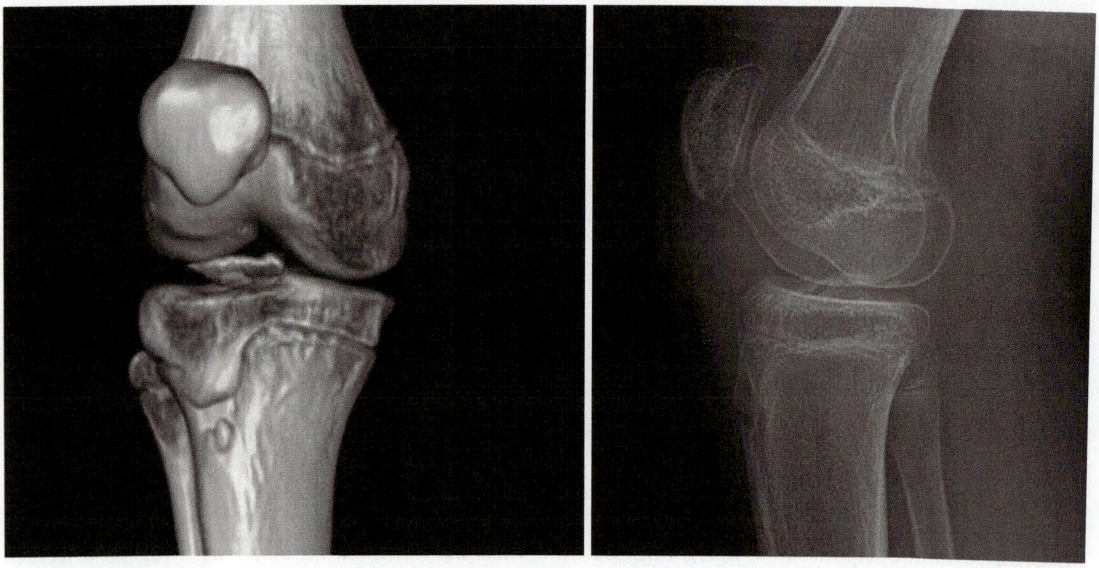

Figs. 10.1 and 10.2 Tibial spine avulsion injury pre-op and 3 months post-trans-osseous suture repair

10.2 Epidemiology

Previously considered rare entities, anterior cruciate ligament (ACL) injuries in the pediatric population have increased in incidence as reported by numerous authors. When this group is further subdivided according to age, notable differences in incidence and the relative increase in incidence over time can be seen between groups. Although there is no exact uniformity in the description of these age subdivisions, generally they are seen as: 1. Youngest group—prepubescent, aged around 8 years, or 0–12 years. 2. Mid-aged group—pubescent, aged 13–15 years; 3. Older group—adolescent around the age of 16–17 years or within 2 years of skeletal maturity. Grouping according to age is important as injury patterns/associations, management options and complications of management may differ between the groups.

Although the incidence or risk of mid-substance ACL injury increases as one gets older, the greatest increase in incidence has been noted by many to be within the middle aged, or pubescent group, and specifically in females.

Examining the exact incidence in the Finnish population under the age of 18 years, over a 17-year period, Weitz et al. [1] reported an overall incidence of ACL injury of 23.3 per 100,000 person-years. The highest incidence was seen in 17-year olds with 113.5 injuries per 100,000 person-years, but the fastest growing group was females aged 13–15 years, who they reported had increased by 143% over the study time-period.

Beck et al. [2] in a retrospective epidemiological study using 20 years of insurance data concerning patients between the ages of 6 and 18 years with ACL injuries showed an increase in approximately 2.3% per year. In their study, there were statistically significant differences in incidence between male and female patients in each age subgroup, and whilst the peak incidence in girls occurred around the age of 16 years, it occurred 1 year later in boys, at the age of 17 years. In all age groups except 17–18 years, the incidence of injury in female patients was higher than those of males. They postulate that the male predominance in the slightly older group is due to an increase participation in high energy contact sports, a drop off in the number of females participating in sports with age, as well as possible drop-out of female patients who may already have had an ACL injury. Interestingly, over their study period, all age groups of female patients showed significant increases in ACL injury, whilst this increased incidence was only seen in male patients aged 15–16 years.

Comstock et al. [3], when evaluating sports injuries in high school level athletes in the United

States in the 2017–2018 school year, reported the knee to be the third most common injury site (after head/face and ankle), making up 14% of all reported injuries. Within the subgroup of knee injuries, ACL injuries were 18.6% overall (17.3% in males and 20.9% in females). In a previous 2005–2011 report on knee injuries in school athletes [4], the authors identified the groups/sports associated most with ACL injuries were football players and female soccer players with injury rates of 1.17 injuries per 100,000 exposures. In gender comparable sports (soccer, volleyball etc.), girls were much more likely to suffer ACL injury than boys (relative risk 2.38).

The increased incidence of injuries and ACL reconstructive surgery in young female patients has been confirmed by numerous authors. In a review of ACL reconstruction in a Canadian population, the authors reported majority of the female patients undergoing ACL reconstruction surgery were aged below 20 years [5]. Dodwell et al. [6] reported on the rate of ACL reconstruction in the state of New York, demonstrating an almost threefold increase in rates of surgeries over a 20-year period, which they considered an indirect measure of increasing ACL injuries. In this cohort, differing from other series, the incidence of ACL reconstruction surgery was higher in male patients—the authors postulated that although female athletes had an increased risk of injury per exposure, male athletes may spend more time in at-risk activities.

Analysing the incidence of knee injuries, and specifically ACL injuries relative to other injuries in a group of pediatric and adolescent soccer players, Shea et al. [7] reviewed insurance claims and demonstrated a relative increase in knee and ACL injuries compared to other injuries with increasing age. They confirmed the relatively higher risk of female athletes sustaining an ACL injury compared to their male counterparts.

In a recent systematic review and meta-analysis of ACL injury risk in various sporting activities, Montalvo et al. [8] identified certain activities (gymnastics and obstacle course races) outside of the group of traditional high risk contact sports. They grouped these activities under the term 'fixed-object high-impact rotational landing (HIRL)'. In both contact sports, as well as the HIRL groups, girls had a higher incidence of injury, whilst in the HIRL group, the relative risk of injury was 5.51 ($p < 0.001$) when compared to their male counterparts.

Postulated reasons for the increased incidence of ACL injuries in the pediatric population include increased participation in sports, increase in intensity and period of training, focus on single sports, year-round training, and increased tibial slope. Postulated risks/associations specific to female athletes include degree of ligamentous/joint laxity, cross-sectional size of ACL, changes in bony anatomy –notch size and volume, reduced muscle mass and less dynamic stability, genetic and hormonal predisposition [9, 10].

10.3 Understanding the Natural History: Changing Perceptions on Management

With the growing number of publications on the natural history of untreated/non-operatively managed ACL injuries, a noticeable effect is the change in management perceptions over time. This is likely due in part to recognition of the deleterious effect of repeated instability, and progressive meniscal and chondral degeneration and in part to improvements in surgical technique, including physeal sparing/respecting procedures.

Kocher et al. [11] in 2002 surveyed two groups of surgeons (the Herodicus society and the ACL Study Group) in order to extract expert opinion on decision making/practice patterns for pediatric ACL injury. For an 8-year-old child, 58% would embark on initial conservative management, 16% would opt for surgical intervention, whilst 26% would proceed with surgery only once skeletally mature. For injury in a 13-year-old child, 51% would treat non-operatively, 34% would intervene surgically early on, whilst 14% would wait till skeletal maturity.

In 2018, 71 surgeons—members of the Pediatric Research in Sports Medicine (PRiSM) Society—were surveyed with a questionnaire based partly on that of Kocher's previous work, in order to see changes in management approach [12]. In the younger patient (aged 8 years) with an acute ACL injury, the number of surgeons that would treat non-surgically with bracing and activity modification had dropped down to only 7%, whilst a few (3%) would wait till skeletally maturity, and the majority (90%) would opt for early surgical intervention. In the case of an older patient with 2 years of growth left and an acute ACL injury, the shift towards surgery was even more pronounced, with 98% advising early surgical treatment, and only 1.5% of the surveyed opting to manage non-operatively—0% would delay until skeletal maturity.

In a paper published in 2019, Accadbled et al. [13] reported on practice patterns of surgeons belonging to the European Pediatric Orthopedic Society (EPOS) and the Pediatric Orthopedic Society of North America (POSNA). Participants were questioned about management choices for 8- and 13-year-old patients with ACL injuries. Sixty per cent favoured early ACL reconstruction in the prepubescent group, whilst this number grew to 79% in the older pubescent group. Of note in the study was the differences in management approaches between members of the two societies—this was most evident in the management of the younger (8 years old) patient, where 65% of the European surgeons favoured initial non-operative management, and ACL reconstruction when symptomatic; compared to 69% of American surgeons favoring earlyACL reconstruction (within the first 3 months following injury).

Changing perceptions and predisposition towards early surgical intervention, coupled with an increasing incidence of injury, has led to greater number of surgeries. In an analysis of ACL reconstruction rates from an Australian database, Zbrojkiewicz et al. [14] reported the fastest growing group to be aged 5–14 years, with annual increases in 7.7% for boys and 8.8% for girls.

10.4 Natural History of Untreated/Delayed Treatment of ACL Injuries

10.4.1 Natural History of Partial ACL Tears

There are clinical and radiological challenges in identifying partial tears of the ACL in children. The presence of a haemarthrosis after acute knee injury should alert the attending physician to the high possibility of intraarticular lesions. Meniscal and/or ACL tears are frequently encountered, with osteochondral fractures less commonly associated. Stanitski et al. [15] reported a 47% incidence of ACL tears, and 47% meniscal tears in patients under 12 years with acute knee injury and haemarthrosis, and in the age group 13–18 years, a 65% incidence of ACL tears.

Magnetic resonance imaging (MRI) sensitivity for ACL injuries of 71% has been reported, but smaller/partial tears may be missed on initial scans [16]. Children have a degree of ligamentous laxity, and the clinical effect of partial tears may only result in a mild increase in laxity. That noted, the natural history of partial tears and the progression to complete tears requiring surgery is worth expanding upon.

In their series of 32 arthroscopically confirmed partial ACL injuries, Noyes et al. [17] reported on eight boys aged 16 years or younger at the time of injury. Of these, only two partial tears progressed to full ACL tears—both were noted to have estimated tear size of 75% of the native ligament. The other remaining six patients had 25% or less of the ligament injured, and despite one patient having two further re-injuries, none progressed to full ACL tears requiring ACL reconstruction. They highlight the factors for predicting tear progression to be tear extent, an increase in initial translation and re-injury with giving way.

In their review of 32 ACL injuries in pediatric population (mean follow-up of 8.2 years), Kannus et al. [18] reported on 25 partial (grade 2 ACL injuries. Eighty-four per cent of these patients were able to return to sport at the same level, and strength loss in the injured limb was noted to be minimal. Whilst there were 36% mild re-injuries, none required surgical intervention, and no patients developed secondary arthrosis.

In contrast, Kocher et al. [19] reported 31% (14 of 45) patients requiring ACL reconstruction for the progression of partial to full thickness ACL tears. In this study, confirmation of partial ACL tears was made arthroscopically, and tear characteristics were delineated. In their analysis, tears >50%, predominantly in the posterolateral area, in older patients and those with grade B pivot shift were associated with subsequent ACL econstruction. Interestingly, in the remaining 31 atients, even though they had not required ACL econstruction, those with tears >50%, and predominantly in the posterolateral area, had significantly lower knee satisfaction and activities of laily living (ADL) scores.

The intrinsic healing capacity of the ACL has been noted by a few authors [20, 21], and it would appear then that small tears without significant clinical instability or demonstratable laxity have a reasonable chance of healing without progression to full tears (barring a significant re-injury).

10.4.2 Natural History of Non-operatively Managed ACL Injuries

Because of the inherent risks of ACL reconstruction surgery in the skeletally immature, non-operative management has been considered—either as definitive, or as interim treatment until skeletal maturity.

Non-operative management in the form of casting or bracing, physiotherapy and activity modification has been advocated by treating physicians with varying results. Table 10.1 outlines some of the published series on non-operative management.

In a 1989 study by Kannus et al. [18], a heterogeneous group of isolated and combined knee ligament injuries was presented with follow-up at a mean of over 8 years. Although the three patients with complete isolated ACL tear are not separated from the rest in the analysis, what is evident is that grade 3 injuries, whether isolated or combined, result in chronic severe instability, diminished return to sports, reduction in strength, reduced knee scores and a high rate of secondary osteoarthritis as seen on radiographs.

Angel et al. [22] in the same year reviewed a group of 27 patients with arthroscopically confirmed ACL tears, using a guided written questionnaire and telephone evaluation. Again, a heterogeneous group comprised 18 patients with partial ACL tears and 9 with complete ACL tears. In the complete ACL tears, one patient had surgical repair of a proximal avulsion. In the remaining eight patients with complete ACL tears, only one was able to return to sports at their previous level, whilst in the patients with partial ACL tears, 50% returned to sport at their previous level. A 50% return to sport appears poor, but it must be noted that five of these patients (38%) had associated pathology (meniscal tears and other), and the authors did not describe the amount and functionality of the remaining intact ACL seen at the time of initial arthroscopy. At the time of publication, seven patients had undergone or required ACL reconstruction, and a further four had surgery for internal derangement of the knee.

In 1995, Mizuta et al. [23] published their series of 18 non-operatively managed ACL injuries. Asides from the three patients that had repaired meniscal tears, who were initially put into long leg cast, all others underwent rehabilitation, and advised to wear an ACL brace during sports. Poor results were seen in the majority: only one patient returned to pre-injury level of sports; 17 patients reported instability with giving way; 6 went on to ACL reconstruction, 5 of which had new meniscal lesions seen at surgery; all 18 complained of some degree of pain, and on final review, almost two thirds demonstrated radiographic arthrosis.

Although not specifically evaluating non-operative management as an end-point, in the study by Aronowitz [24] all 21 adolescents identified with mid-substance ACL injuries failed treatment with bracing, activity modification and rehab and required surgery.

A number of studies include patients with early surgical intervention along with non-operatively managed patients (Table 10.2). There is occasional selection bias inherent in the study methodologies. For example, patients with concomitant meniscal injuries may be offered early surgical intervention, whilst isolated ACL injury may be treated non-operatively. Patients may also request early surgery in the belief that this will enable them to return to high level sporting activity. That noted, evaluation of the non-operative groups in these studies gives further insight into the natural history of the ACL deficient knee.

In 1988, McCarrol et al. [25] published their experience with children under the age of 14 years at the time of injury. Of the 16 patients treated without surgical stabilisation, 9 gave up

Table 10.1 Studies of non-operative management of pediatric ACL injuries

Author	Year	Patient no	Age	Follow-up	Clinical score	Return to sport	Laxity/ instability	Assoc initial injuries	Secondary injuries	Secondary surgery required	Secondary arthrosis	Other reported outcome
Kannus et al. [18]	1988	25 partial (grade 2)	14.8 ± 2.4	8.2 years (±2.3)	Lysholm 91 ± 10 Marshall 44 ± 4	84% same level	No gross instability		36% mild re-injury—cons Mx	Nil	Nil	Strength loss minimal in partial injuries, noticeable in grade 3
		7 complete (grade 3)			Lysholm 74 ± 17 Marshall 30 ± 7	71% reduced activity level	Severe or gross instability in all			2/7 ACLR	4/7 definitive evidence	
Angel et al. [22]	1989	27 (18 partial, 9 complete)	14.3 (8–18)	4.3 years (2.2–12.3)		11 pre-injury level 7 lower level 9 no return	15 reported giving way		3 meniscal tears, 1 loose body found at scope	4 scopes (IDK), 4 ALCR (3 pending ACLR)		Pain in 12, swelling in 10, locking in 4
Mizuta et al. [23]	1995	18	12.8 (10–15)	4.3	Mod Lysholm 64.3	1 same level 13 lower 4 no return	17 reported giving way	Meniscal tears in 13 (3 repairs)	5/6 patients had new meniscal lesions at ACLR	6 ACLR	7 grade 0 3 grade 1 6 grade 2 2 grade 3	Pain in 18, swelling in 11, locking in 10

Table 10.2 Studies including non-operative management and Sx stabilisation/mixed management

Author	Year	Patient/knee numbers Non-op ACLR	Age	Follow-up	Clinical score	Return to sport	Laxity/instability	Associated initial injuries	Secondary injuries	Secondary surgery required	Secondary arthrosis	Other reported outcome
McCarrol et al. [25]	1988	16 non-op	13	27 months		7	All unstable	4 med, 2 lat menisc				KT-1000 1 mm
		24 Sx		26 months		22 still competing	4 single instability episodes	12 med, 6 lat menisc, 1 OC lesion	1 meniscal tear in EA stabilisation	1ACLR + MM repair		
Graf et al. [29]	1992	8 non-op	14.5			Unsuccessful	All unstable	2 debrided lateral meniscal tears	7/8 new meniscal lesions	7/8 ACLR		
		4 Sx				IA ACLR able to (2)	IA ACLR stable (2) Progressive instability in other stabilisations (2)	4 repaired menisci	2 new meniscal tears	2 ACLR (conversion from extraarticular stab procedure)		
Pressman [28]	1997	13 non-op 6 repair 23 ACLR	14.4	5.3 years (2–13)	Zarins and Rowe better in ACLR		Lachman and pivot shifts better in ALCR group					KT-1000 Lysholm
Janarv et al. [27]	1996	7 15	11.9 13.3		Lysholm Tegner		2.4 (0–5) 1.5 (−2–4)	3 med, 3 lat meniscal partial resections		68% of non-op manged required ACLR at final review		

(continued)

Table 10.2 (continued)

Author	Year	Patient/ knee numbers Non-op ACLR	Age	Follow-up	Clinical score	Return to sport	Laxity/ instability	Associated initial injuries	Secondary injuries	Secondary surgery required	Secondary arthrosis	Other reported outcome
Aichroth et al. [29]	2002	23 (33 initially)	12.5	72 months	Lysholm 52.4 Tegner 4.2	'Very few'	Initial: all 23 grade 2 or 3 Lachman	14 meniscal tears 3 OC fractures		4 med, 7 lateral meniscectomy	10/23	Marked improvement in pivot and Lachman
		47	13	49 months	IKDC A or B in 77%		Initial: all grade 2 or 3	17 meniscal tears	3 re-ruptures	3 re-ruptures treated		
Arbes et al. [30]	2007	8 non-op 4 ACLR 3 ACLR at maturity		5.4 years	IKDC	4/8 pre-inj level 4/4 1/3	7.8 (2–10) 4.8 mm (4–7) 4.3 (2–7)				85% overall	Rolimeter laxity KOOS QOL best in ACLR group
Streich et al. [31]	2010	12 non-op	11	70 months (41–85)	IKDC 87 Lysholm 84		4.3 ± 2.9 mm	Nil	4 med 2 lat meniscal	7/12 ACLR		LLD, angular deformity, ROM—no difference between groups
		16 ACLR	11		IKDC 95 Lysholm 93		1.8 ± 2.4 mm	12 meniscal 4 osteochondral				

sports due to instability, and the 7 who were able to continue sport reported instability episodes despite bracing.

Graf et al. [26] evaluated a group of 12 ACL injured patients with arthrography and subsequent arthroscopy. Four patients had repairable meniscal injuries, and at the time of surgery underwent an accompanying ACL reconstruction. The remaining eight patients did not have ACL reconstruction and were treated with rehabilitation and bracing. All the latter patients complained of instability and inability to return to sports, and seven of the eight developed new meniscal tears. At the end of the study period, 11 of the 12 had required ACL reconstruction. This cross-over is commonly noted in similar comparative studies.

In 1996, Janarv et al. [27] reported on a series of 24 children under the age of 15 years at the time of injury. Five demanded immediate ACL reconstruction and the remaining participated in a three-month rehab programme. At the time of review, 68% of those who attempted non-operative management had undergone ACL reconstruction– some at skeletal maturity, others earlier due to instability, meniscal lesions or the inability to attain a desired activity level. Of the patients who remained non-operative at the conclusion of the study, seven were included for comparison with an operative group of 15 patients. Clinical laxity and muscle performance were not noticeably worse, and there were no new meniscal lesions in the non-operative group. However, it must be noted that by definition this non-operative group were patients who have not had a clinical indication for a change to operative management and their knees have been stable enough to cope with non-operative treatment.

In 1997, Pressman et al. [28] presented a retrospective review which included 40 patients (42 knees) divided into three groups: non-operative management, ACL repair and ACL reconstruction. Non-operative knees showed statistically significantly inferior results with regard to Zarins and Rowe score, as well as knee stability (Lachmans and pivot shift tests) compared to those with ACL reconstruction. They concluded that non-operative management cannot be supported over ACL reconstruction, and that delaying surgery until skeletal maturity should also not be advocated.

In 2002, Aichroth et al. [29] published their experience with the management of ACL injuries in children and adolescents. In the initial study period, patients were treated non-operatively with physio/rehabilitation, bracing and activity modification. They noted a significant number of these patients developed instability and poor function, with 10 of 33 requiring ACL reconstruction for progressive instability, and a further 7 requiring surgery for meniscal lesions. The majority of 23 patients who did not have ACL reconstruction were unable to continue with sports and the knee function progressively deteriorated with severe knee instability. Due to the poor results, they then changed management protocol to early ACL reconstruction. In the ACL reconstruction group, they reported satisfactory results (International Knee Documentation Committee Subjective Knee Form (IKDC) score grade A or B) in 36 of 47 (77%) patients with only 3 re-ruptures. Knee stability was also markedly improved in the ACL reconstruction group.

Arbes et al. [30] in 2007 reported on their experience with ACL tears to evaluate the best management protocol. Of the six groups of management they describe, early ACL reconstruction resulted in the best clinical outcomes and measured knee stability. Of the patients managed non-operatively, half complained of instability episodes and were unable to return to sports at their pre-injury level.

As noted above, patients may be pre-selected for early ACL reconstruction if they have concomitant intraarticular injuries that require early surgical treatment. Recently, Streich et al. [31] compared early ACL reconstruction (associated meniscal or chondral injuries) with a group of isolated ACL injuries managed non-operatively. At review, patients with early ACL reconstruction performed better with regard to laxity evaluation as well as clinical knee scores. Over half (58%) of non-operatively managed knees had required ACL reconstruction, and at the time of ACL reconstruction, 6 of 7 required additional surgery in the way of partial meniscal resection.

Meta-analyses by Vavken et al. [32], Ramski et al. [33], and Dunn et al. [34] confirmed the increased incidence of laxity/instability, meniscal lesions and inferior knee outcome scores following non-operative management when compared to early ACL reconstruction.

10.5 Delaying Surgery Until Skeletal Maturity

In order to avoid the potential complications of ACL reconstruction surgery (physeal damage/arrest and growth disturbance) in the pediatric population, a number of clinicians have opted to delay surgery until skeletal maturity (Table 10.3).

According to Moksnes et al. [35], the Norwegian treatment algorithm is conservative with delay of surgery until skeletal maturity. Children undergo rehabilitation and are only operated on in the event of repeated instability episodes or displaced meniscal lesions. In 2008, Moksnes et al. [35] reported reasonable functional results and an acceptable rate of secondary meniscal tears in a group of patients under the age of 12 years. Patients with acute ACL injuries undertook a period of structured rehabilitation, were encouraged to return to activities and advised to wear a protective brace when participating in sports where there was risk of pivoting. Of the 26 patients (27 knees) included at review, six patients (7 knees) had undergone ACL reconstruction (instability or symptomatic meniscal tear). Besides the three patients with meniscal tears noted at ACL reconstruction, there were two others who had arthroscopic partial meniscal resections, without concomitant ACL reconstruction. The authors reported an overall incidence of 9.5% meniscal injuries in the non-operative group. In the 20 children with unreconstructed knees, 13 (65%) reported being able to return to their pre-injury activity level.

In 2013 [36], Moksnes et al. published on a group of ACL injuries which they followed prospectively (46 patients all under 12 years old at the time of injury). In this cohort, despite 40% having meniscal pathology on initial MRI scan, none underwent any form of early surgery. They reported reasonable functional results in terms of strength, symmetrical knee function, hop test, Knee injury and Osteoarthritis Outcome Score (KOOS) scores; however, at 2 years only 50% reported a level 1 activity as their main activity (pre-injury involvement in level 1 activity was 85%), and ten patients required surgery for persistent instability, secondary meniscal tears or unacceptable limitation of activity. They suggested that in patients under the age of 12 years, non-operative management should be considered, understanding that the patients' main activity level will likely drop one level. Should the patient wish to return to level 1 sports (pivoting sports such as soccer), then they suggested ACL reconstruction near skeletal maturity (Table 10.4).

Activity restriction is emphasised by most authors when prescribing non-operative management. If a patient is adamant regarding return to high-risk sporting activity, they should first be assessed using the criteria set out by Fitzgerald et al. [37] as to their suitability for such a rehabilitation programme. The criteria include a hop tests score, ADL score, global knee function score and the presence of one or less further instability episode prior to assessment. Thereafter, the patient should embark on a structured phased rehabilitation programme including secondary injury prevention measures in the way of functional and neuromuscular training [36, 38].

Table 10.3 Studies for delaying surgery till skeletal maturity

Author	Year	Patient no	Age	Time to Sx	Follow-up	Clinical score	Return to sport	Laxity/ instability	Associated initial injuries	Secondary injuries	Secondary surgery required	Secondary arthrosis	Other reported outcome
Moksnes [35]	2008	26 (27 knees)	10.1 (5.3–12.7)		3.9 (1.9–9)	Lysholm, KOOS, IKDC, VAS, activity level	65% to pre-injury level	KT 1000 5.5 mm (±2.4)	2 partial meniscectomies		ACLR group: 3 meniscal surgeries @ ACLR (7 knees). 2 further meniscal resections		
Moksnes [36]	2013	46	11 ± 1.5		3.2 (±1.1)	KOOS, IDKS	50% level 1 sport at 2 years (85% at pre-injury scoring)	7 persistent instability	18 (40%) meniscal tears—not operated on	4 meniscal tears	10 ACLR		Single leg hop test, muscle strength grading

Table 10.4 Studies comparing early surgery versus delayed till skeletal maturity

Author	Year	Patient no (E = early L = late)	Age	Time to Sx	Follow-up	Clinical score	Return to sport	Laxity/ instability	Associated injuries	Secondary injuries	Other reported outcome
Woods and O'Connor [39]	2004	Early group 1 = 24 2 = 58 3 = 21 4 = 13		<1 weeks 2–6 weeks 6–26 >26					Meniscal 1 = 44% 2 = 42% 3 = 43% 4 = 69% ($p = 0.052$)		Additional surgeries 1 = 33% 2 = 24% 3 = 24% 4 = 62%
		L = 13	13.8	70.3 weeks					46% meniscal		Physes group = 31%
Henry et al. [40]	2009	E = 29	11.5 ± 2.2	13.5 months	27	IKDC (subjective) 94.6 (72–100)	89.7% same or superior level	KT differential 1.93 mm ±1.2	12 meniscal tears seen at Sx		Valgus growth deformity in 1—spontaneous resolution
		L = 27	13.3 ± 2	30 months	27	82.4 (23–100)	65.3% same or superior level	1.76 mm ±2	1 meniscal lesion on MRI (only 5 MRI scans)	19 meniscal lesions seen at surgery	

10.6 Comparing Early Surgery Versus Delaying till Skeletal Maturity

Woods and O'Çonnor [39] reported results with a management plan of delaying surgery until skeletal maturity, but with strict activity restriction. They reported no statistically significant increase in rates of meniscal lesions or additional injuries when compared to a control group of skeletally mature adolescents. The control group was sub-divided according to time delay to surgery. In the control subgroup delayed the longest (>6 months) and without strict activity restriction, there was a higher incidence of meniscal lesions at 69%, but due to small study numbers, this did not reach statistical significance ($p = 0.052$). They stressed the need for strict activity restriction and brace wearing if deciding to delay reconstruction until skeletal maturity.

Henry et al. [40] in 2009 reported on two groups of patients—one group treated in a pediatric hospital orthopaedic unit with early ACL reconstruction, the other in an adult hospital orthopaedic unit with delayed surgery at skeletal maturity. Early post injury (within 3 months) MRI evaluation revealed no significant difference in incidence of meniscal lesions, but at surgery, in the group treated with delayed ACL reconstruction, they had developed significantly more meniscal tears (19 lesions in 27 knees versus 12 lesions in the 29 knees treated with early surgery). The increased number of meniscal lesions was to the medial side, and there was no significant increase in the rate of lateral meniscal lesions. Although the knee scores and radiological analysis were similar between the groups, the mean follow-up was only 27 months, and possibly too early for secondary degeneration to be clinically or radiologically evident.

10.7 Effect of Surgical Delay and Recurrent Injury

In an attempt to analyse the risk of further harm in delaying surgery, Lawrence et al. [41] evaluated a group of 70 ACL reconstructions, 29 (41%) of which were performed more than 12 weeks post injury. They reported that delay in treatment was independently associated (odds ratio 4.1) with medial meniscal tears, and when the delay was more than 12 weeks, was further associated with increased severity/irreparability of medial meniscal lesions, as well as higher grade lateral and patella–trochlear chondral lesions. Similarly, Millet et al. [42] reported increased incidence of medial meniscal lesions when comparing acute ACL reconstruction and delayed ACL reconstruction more than 6 weeks post injury (11 vs. 36% incidence), whilst the incidence of lateral meniscal lesions was not statistically significantly different between groups.

Dumont et al. [43] divided 370 ACL reconstruction patients into an early (<150 days) and late group. They too reported a statistically significant increase in medial meniscal lesions in the delayed group (37.8 vs. 53.5%, $p = 0.14$), and no increase in lateral lesions. Predictive factors for increased rates for medial meniscal lesions were increased age and weight. The presence of chondral lesions was associated with meniscal lesions in the same compartment.

With increased waiting times, not only the number, but also the severity of meniscal tears may increase. Newman et al. [44] showed a significant correlation with advancing time and the presence of an irreparable meniscal tear. In their study, both attempted return to sports prior to ACL reconstruction, as well as obesity, were predictive in the adolescent group for the need for further procedures at the time or ACL reconstruction, whilst in the younger group, delay more than 3 months correlated with more severe chondral lesions. Raad et al. [45] found similarly that attempted non-operative management, body mass index (BMI) and time to surgery were related to meniscal lesions.

Contrasting this somewhat, Funahashi et al. [46] in a retrospective review of patients identified from a large integrated health care system reported no significant increase in meniscal or cartilage lesions (or combined lesions) with increased waiting time of 1 or even 2 years. They did report, however, a significant association between number of clinical encounters (>5 visits to healthcare provider for new pain/swelling) and combined cartilage and meniscal injury suggesting that more symptomatic/unstable knees were prone to secondary injuries.

Knee instability and its association with meniscal lesions (both medial and lateral) were well described by Anderson et al. [47]. In their study of 135 ACL reconstructions divided into acute, subacute and chronic, they reported a significant association between instability episodes, delay to surgery greater than 3 months and an attempt to return to sports prior to ACL reconstruction with significantly worse lesions.

10.8 Failure of Non-operative Management: Conversion to ACL Reconstruction After Embarking on Non-operative Management

As can be noted from many of the above studies, cross-over of management with attempted non-operative or delayed operative management is frequent. This cross-over may be due to repeated knee instability episodes, development of new symptomatic meniscal or cartilage lesions or the inability to return to previous activity level. The incidence of cross-over ranges: 68% Janarv et al. [27], 23% Moksnes [35], 32.5% Moksnes [48], 70.9% Raad et al. [45], 87.5% Graf et al. [26] and 58% Streich et al. [31].

10.9 Return to Sporting Activity

Few studies report significant percentages of patients returning to their previous level of sporting activity following non-operative management. Moksnes et al. [35] reported 65% were able to return to same level and 35% returned to a lower level. However, a multitude of studies (Tables 10.1 and 10.2) report repeat episodes of knee instability and further knee injury with an unrestricted return to sports. Bracing and activity restriction is advocated by most physicians during non-operative management, or whilst awaiting physeal closure and delayed surgery.

Although the ability to return to sports following ACL reconstruction had been questioned by those advocating non-operative management, there is a growing body of evidence supporting ACL reconstruction. McCarrol et al. [25] reported 55 of 60 patients were able to return to same level of sports following ACL reconstruction with only three re-ruptures. Dekker et al. [49] reported 91% return to sports at a 'high' rate, but with an alarming incidence of second ACL injury (19% re-rupture, 13% contralateral ACL injuries). In this manuscript, an early return to sport was predictive of a second ACL injury, and a slower return to sports was thus advocated. They postulated that besides muscular strength and neuromuscular training, adequate time for biological recovery was needed.

Arnowitz et al. [24] reported 16 of 19 (84%) patients returned to same sport after ACL reconstruction, and a meta-analysis by Kay et al. [50] revealed over 80% returning to competitive sport. The trend favouring early ACL reconstruction over delayed surgery or non-operative management was confirmed in meta-analyses by Dunn et al. [34] and Ramski et al. [33].

10.10 Associated Injuries at the Time of Injury: MRI

Injury to the knee severe enough to cause ACL rupture can cause a wide spectrum of associated injuries. The attending physician should actively seek out associated pathology in the injured joint—pre-operatively with a comprehensive clinical and radiological evaluation and intra-operatively with a thorough arthroscopic assessment (Fig. 10.3). Although not all lesions require surgical attention, the type and severity of associated injuries can help prognosticate clinical outcomes. Whereas arthrography and diagnostic arthroscopy were the modalities of past years, improvements in magnetic resonance scanning—both in image acquisition and interpretation/understanding—have resulted in it being the investigation of choice for assessing both the soft tissues and osseous structures.

Laremko et al. [51] published a pictorial review of MRI findings/associated injuries in pediatric patients with surgically confirmed ACL ruptures. The study highlights important points when considering the radiological assessment: The normal ACL may appear quite thin or attenuated. A thickened and oedematous appearing tendon (Figs. 10.4 and 10.5) is likely recently torn. Posteromedial capsular strains or incomplete/small tears may not be seen at surgery if interval healing has taken place between the time of the scan and that of operation. Injuries to the posterolateral corner are uncommon, but when they do occur, often involve the popliteomeniscal fascicles and the adjacent lateral meniscus. Lateral tibial plateau impression fractures may be confused with residual physeal tissue, but surrounding oedema suggests injury. In this study, it was noted that when medial collateral injuries were present, they were more commonly injured proximally at the femoral attachment, which is different from adult injury patterns.

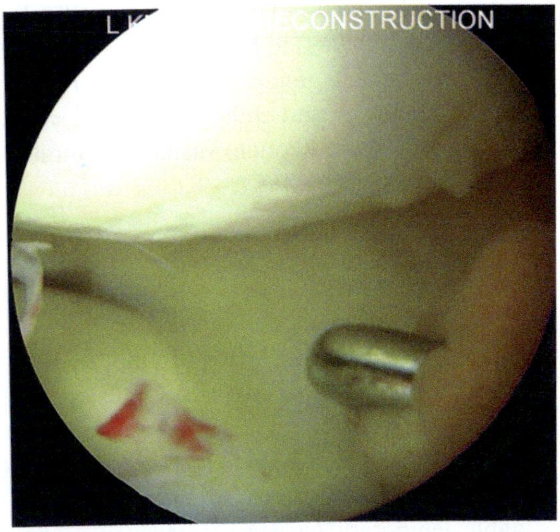

Fig. 10.3 Lateral femoral chondral lesion, 14-year-old female patient 4 weeks post anterior cruciate ligament injury

Improvements in MRI techniques assist in the assessment of the injured knee. Oeppen et al. [52] reviewed MRI scans of 82 patients with open physes. They highlighted the high incidence of chondral injuries after knee trauma but conceded that less severe chondral injuries (grade 1 and 2) may be underreported. The presence of subchondral oedema should alert the attending physician as this is frequently seen with chondral injuries.

Meniscal lesions may be frequently found on initial scans. In a prospective MRI study on

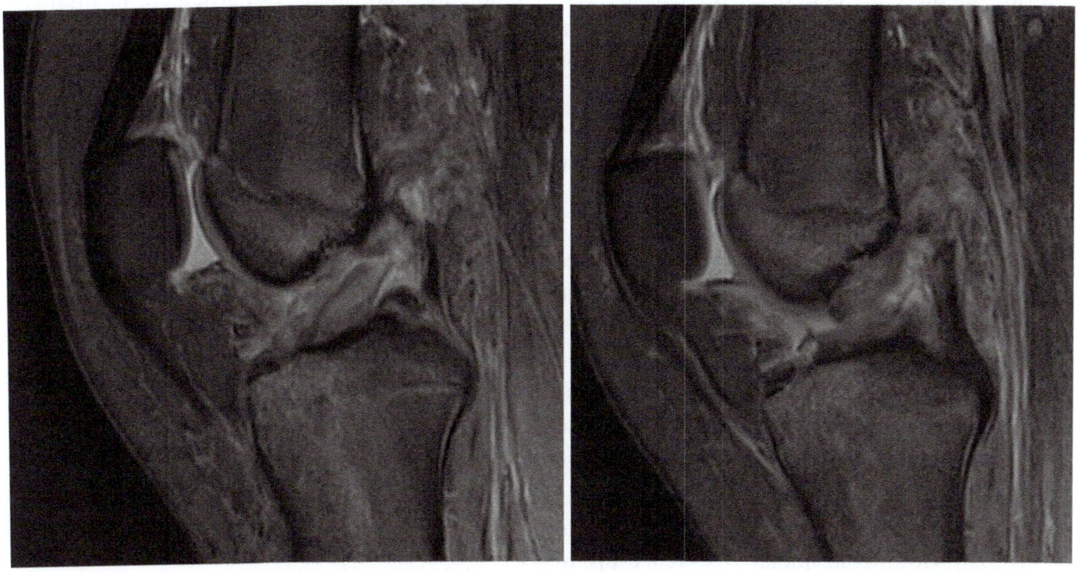

Figs. 10.4 and 10.5 Thickened oedematous anterior cruciate ligament—complete tear at surgery. Associated posterior cruciate ligament tear

non-operatively managed ACL injuries, Moksnes et al. [48] noted 46.3% meniscal lesions on initial scans.

10.11 Associated Long-Term Injuries: Risk of Contralateral ACL Injury, Risk of Osteoarthritis

Patients and their parents should be adequately counselled regarding the long-term outcomes of non-operatively managed ACL injuries. Besides repeat knee instability episodes, progressive meniscal and chondral damage, and the inability to return to same level sports, non-operative patients carry with them the long-term risk of developing osteoarthritis, as well as a risk of contralateral ACL rupture.

Aichroth et al. [29] reviewed a group of 23 patients managed non-operatively. At a mean of 72 months, with patients nearing the end of their second decade, almost half (10 of 23) demonstrated radiographic features of secondary joint degeneration, with a 'substantial Fairbank sign' in all, joint space narrowing in and large osteophytes in 4.

Kannus et al. [18] reported no cases of osteoarthritis in their 25 patients with partial ACL injuries, but 4 of the 7 with complete ACL injuries showed definitive evidence of degeneration at review (average 8.2 ± 2.3 years).

In the study by Mizuta et al. [23], at a mean of 54 months, 11 of 18 (61%) patients demonstrated radiographic osteoarthritis.

Although ACL reconstruction appears to improve knee stability and reduce the incidence of further meniscal/cartilage damage [53], the long-term protective effect on the development of degenerative joint disease is less clear. Mansson et al. [54] reviewed a group of adolescents 10–20 years post delayed ACL reconstruction and reported significantly more osteoarthritic changes in the operative knees.

In the long-term follow-up series by Pressman et al. [28] (follow-up mean 5.3 years up to 13 years), 4 of the 40 (10%) children had sustained a contralateral ACL injury that required ACL reconstruction. This approximates the 13% incidence of contralateral ACL injury reported by Dekker et al. [49].

To identify patients at risk of a contralateral ACL injury after ACL reconstruction, Patel et al. [55] published a prognostic study having analysed 498 children post ACL reconstruction. The overall incidence of contralateral ACL tear was 7% at a mean of 2.7 (±2.1) years. They reported female gender and young age (specifically younger than 15 years) to be independent associated factors. Female gender increased the risk by an odds ratio of 3.5 and children under 15 years were 3.1 times more likely to sustain a contralateral ACL injury.

10.12 Summary

Complete intrasubstance tears of the ACL in skeletally immature patients are infrequently encountered, but the incidence is increasing. Non-operative management in the way of physiotherapy, bracing and activity modification may be a viable temporary measure in the very young child, or if awaiting skeletal maturity; however, patients and their parents should be warned about the risk of further injury to the knee—especially if attempts are made to participate in high-risk sporting activities.

The natural history of the untreated, or non-operatively managed knee, tends towards the inability to return to sporting activities, repeated episodes of knee instability, further internal derangement of the knee, osteoarthritis and inferior clinical outcomes.

References

1. Weitz FK, Sillanpaa PJ, Mattila VM. The incidence of paediatric ACL injury is increasing in Finland. Knee Surg Sports Traumatol Arthrosc. 2020;28(2):363–8.
2. Beck NA, Lawrence JTR, Nordin JD, DeFor TA, Tompkins M. ACL tears in school-aged children and adolescents over 20 years. Pediatrics. 2017;139(3):e20161877.
3. Comstock RD, Pierpoint LA, Arakkal BS, Bihl JH. Summary report, national high school sports-

related injury surveillance study, 2017–2018 school year online PDF report.
4. Swenson DM, Collins CL, Best TM, Flanigan DC, Fields SK, Comstock RD. Epidemiology of knee injuries among U.S. high school athletes, 2005/2006–2010/2011. Med Sci Sports Exerc. 2013;45(3):462–9.
5. Zhang Y, McCammon J, Martin RK, Prior HJ, Leiter J, MacDonald PB. Epidemiological trends of anterior cruciate ligament reconstruction in a Canadian Province. Clin J Sport Med. 2018;30:e207–13.
6. Dodwell ER, Lamont LE, Green DW, Pan TJ, Marx RG, Lyman S. 20 years of pediatric anterior cruciate ligament reconstruction in New York state. Am J Sports Med. 2014;42(3):675–80.
7. Shea KG, Pfeiffer R, Wang JH, Curtin M, Apel PJ. Anterior cruciate ligament injury in pediatric and adolescent soccer players: an analysis of insurance data. J Pediatr Orthop. 2004;24(6):623–8.
8. Montalvo AM, Schneider DK, Webster KE, Yut L, Galloway MT, Heidt RS Jr, et al. Anterior cruciate ligament injury risk in sport: a systematic review and meta-analysis of injury incidence by sex and sport classification. J Athl Train. 2019;54(5):472–82.
9. Swami VG, Mabee M, Hui C, Jaremko JL. Three-dimensional intercondylar notch volumes in a skeletally immature pediatric population: a magnetic resonance imaging-based anatomic comparison of knees with torn and intact anterior cruciate ligaments. Arthroscopy. 2013;29(12):1954–62.
10. O'Malley MP, Milewski MD, Solomito MJ, Erwteman AS, Nissen CW. The association of tibial slope and anterior cruciate ligament rupture in skeletally immature patients. Arthroscopy. 2015;31(1):77–82.
11. Kocher MS, Saxon HS, Hovis WD, Hawkins RJ. Management and complications of anterior cruciate ligament injuries in skeletally immature patients: survey of the Herodicus Society and the ACL Study Group. J Pediatr Orthop. 2002;22(4):452–7.
12. Popkin CA, Wright ML, Pennock AT, Vogel LA, Padaki A, Redler LH, et al. Trends in management and complications of anterior cruciate ligament injuries in pediatric patients: a survey of the PRiSM Society. J Pediatr Orthop. 2018;38(2):e61–e5.
13. Accadbled F, Gracia G, Laumonerie P, Thevenin-Lemoine C, Heyworth BE, Kocher MS. Paediatric anterior cruciate ligament tears: management and growth disturbances. A survey of EPOS and POSNA membership. J Child Orthop. 2019;13(5):522–8.
14. Zbrojkiewicz D, Vertullo C, Grayson JE. Increasing rates of anterior cruciate ligament reconstruction in young Australians, 2000–2015. Med J Aust. 2018;208(8):354–8.
15. Stanitski CL, Harvell JC, Fu F. Observations on acute knee hemarthrosis in children and adolescents. J Pediatr Orthop. 1993;13(4):506–10.
16. Umans H, Wimpfheimer O, Haramati N, Applbaum YH, Adler M, Bosco J. Diagnosis of partial tears of the anterior cruciate ligament of the knee: value of MR imaging. AJR Am J Roentgenol. 1995;165(4):893–7.
17. Noyes FR, Mooar LA, Moorman CT 3rd, McGinniss GH. Partial tears of the anterior cruciate ligament. Progression to complete ligament deficiency. J Bone Joint Surg Br. 1989;71(5):825–33.
18. Kannus P, Jarvinen M. Knee ligament injuries in adolescents. Eight year follow-up of conservative management. J Bone Joint Surg Br. 1988;70(5):772–6.
19. Kocher MS, Micheli LJ, Zurakowski D, Luke A. Partial tears of the anterior cruciate ligament in children and adolescents. Am J Sports Med. 2002;30(5):697–703.
20. Steadman JR, Cameron-Donaldson ML, Briggs KK, Rodkey WG. A minimally invasive technique ("healing response") to treat proximal ACL injuries in skeletally immature athletes. J Knee Surg. 2006;19(1):8–13.
21. Nguyen DT, Ramwadhdoebe TH, van der Hart CP, Blankevoort L, Tak PP, van Dijk CN. Intrinsic healing response of the human anterior cruciate ligament: an histological study of reattached ACL remnants. J Orthop Res. 2014;32(2):296–301.
22. Angel KR, Hall DJ. Anterior cruciate ligament injury in children and adolescents. Arthroscopy. 1989;5(3):197–200.
23. Mizuta H, Kubota K, Shiraishi M, Otsuka Y, Nagamoto N, Takagi K. The conservative treatment of complete tears of the anterior cruciate ligament in skeletally immature patients. J Bone Joint Surg Br. 1995;77(6):890–4.
24. Aronowitz ER, Ganley TJ, Goode JR, Gregg JR, Meyer JS. Anterior cruciate ligament reconstruction in adolescents with open physes. Am J Sports Med. 2000;28(2):168–75.
25. McCarroll JR, Rettig AC, Shelbourne KD. Anterior cruciate ligament injuries in the young athlete with open physes. Am J Sports Med. 1988;16(1):44–7.
26. Graf BK, Lange RH, Fujisaki CK, Landry GL, Saluja RK. Anterior cruciate ligament tears in skeletally immature patients: meniscal pathology at presentation and after attempted conservative treatment. Arthroscopy. 1992;8(2):229–33.
27. Janarv PM, Nystrom A, Werner S, Hirsch G. Anterior cruciate ligament injuries in skeletally immature patients. J Pediatr Orthop. 1996;16(5):673–7.
28. Pressman AE, Letts RM, Jarvis JG. Anterior cruciate ligament tears in children: an analysis of operative versus nonoperative treatment. J Pediatr Orthop. 1997;17(4):505–11.
29. Aichroth PM, Patel DV, Zorrilla P. The natural history and treatment of rupture of the anterior cruciate ligament in children and adolescents. A prospective review. J Bone Joint Surg Br. 2002;84(1):38–41.
30. Arbes S, Resinger C, Vecsei V, Nau T. The functional outcome of total tears of the anterior cruciate ligament (ACL) in the skeletally immature patient. Int Orthop. 2007;31(4):471–5.
31. Streich NA, Barie A, Gotterbarm T, Keil M, Schmitt H. Transphyseal reconstruction of the anterior cruciate ligament in prepubescent athletes. Knee Surg Sports Traumatol Arthrosc. 2010;18(11):1481–6.

32. Vavken P, Murray MM. Treating anterior cruciate ligament tears in skeletally immature patients. Arthroscopy. 2011;27(5):704–16.
33. Ramski DE, Kanj WW, Franklin CC, Baldwin KD, Ganley TJ. Anterior cruciate ligament tears in children and adolescents: a meta-analysis of nonoperative versus operative treatment. Am J Sports Med. 2014;42(11):2769–76.
34. Dunn KL, Lam KC, Valovich McLeod TC. Early operative versus delayed or nonoperative treatment of anterior cruciate ligament injuries in Pediatric patients. J Athl Train. 2016;51(5):425–7.
35. Moksnes H, Engebretsen L, Risberg MA. Performance-based functional outcome for children 12 years or younger following anterior cruciate ligament injury: a two to nine-year follow-up study. Knee Surg Sports Traumatol Arthrosc. 2008;16(3):214–23.
36. Moksnes H, Engebretsen L, Eitzen I, Risberg MA. Functional outcomes following a non-operative treatment algorithm for anterior cruciate ligament injuries in skeletally immature children 12 years and younger. A prospective cohort with 2 years follow-up. Br J Sports Med. 2013;47(8):488–94.
37. Fitzgerald GK, Axe MJ, Snyder-Mackler L. Proposed practice guidelines for nonoperative anterior cruciate ligament rehabilitation of physically active individuals. J Orthop Sports Phys Ther. 2000;30(4):194–203.
38. Moksnes H, Engebretsen L, Risberg MA. Management of anterior cruciate ligament injuries in skeletally immature individuals. J Orthop Sports Phys Ther. 2012;42(3):172–83.
39. Woods GW, O'Connor DP. Delayed anterior cruciate ligament reconstruction in adolescents with open physes. Am J Sports Med. 2004;32(1):201–10.
40. Henry J, Chotel F, Chouteau J, Fessy MH, Berard J, Moyen B. Rupture of the anterior cruciate ligament in children: early reconstruction with open physes or delayed reconstruction to skeletal maturity? Knee Surg Sports Traumatol Arthrosc. 2009;17(7):748–55.
41. Lawrence JT, Argawal N, Ganley TJ. Degeneration of the knee joint in skeletally immature patients with a diagnosis of an anterior cruciate ligament tear: is there harm in delay of treatment? Am J Sports Med. 2011;39(12):2582–7.
42. Millett PJ, Willis AA, Warren RF. Associated injuries in pediatric and adolescent anterior cruciate ligament tears: does a delay in treatment increase the risk of meniscal tear? Arthroscopy. 2002;18(9):955–9.
43. Dumont GD, Hogue GD, Padalecki JR, Okoro N, Wilson PL. Meniscal and chondral injuries associated with pediatric anterior cruciate ligament tears: relationship of treatment time and patient-specific factors. Am J Sports Med. 2012;40(9):2128–33.
44. Newman JT, Carry PM, Terhune EB, Spruiell MD, Heare A, Mayo M, et al. Factors predictive of concomitant injuries among children and adolescents undergoing anterior cruciate ligament surgery. Am J Sports Med. 2015;43(2):282–8.
45. Raad M, Thevenin Lemoine C, Berard E, Laumonerie P, Sales de Gauzy J, Accadbled F. Delayed reconstruction and high BMI z score increase the risk of meniscal tear in paediatric and adolescent anterior cruciate ligament injury. Knee Surg Sports Traumatol Arthrosc. 2019;27(3):905–11.
46. Funahashi KM, Moksnes H, Maletis GB, Csintalan RP, Inacio MC, Funahashi TT. Anterior cruciate ligament injuries in adolescents with open physis: effect of recurrent injury and surgical delay on meniscal and cartilage injuries. Am J Sports Med. 2014;42(5):1068–73.
47. Anderson AF, Anderson CN. Correlation of meniscal and articular cartilage injuries in children and adolescents with timing of anterior cruciate ligament reconstruction. Am J Sports Med. 2015;43(2):275–81.
48. Moksnes H, Engebretsen L, Risberg MA. Prevalence and incidence of new meniscus and cartilage injuries after a nonoperative treatment algorithm for ACL tears in skeletally immature children: a prospective MRI study. Am J Sports Med. 2013;41(8):1771–9.
49. Dekker TJ, Godin JA, Dale KM, Garrett WE, Taylor DC, Riboh JC. Return to sport after pediatric anterior cruciate ligament reconstruction and its effect on subsequent anterior cruciate ligament injury. J Bone Joint Surg Am. 2017;99(11):897–904.
50. Kay J, Memon M, Marx RG, Peterson D, Simunovic N, Ayeni OR. Over 90% of children and adolescents return to sport after anterior cruciate ligament reconstruction: a systematic review and meta-analysis. Knee Surg Sports Traumatol Arthrosc. 2018;26(4):1019–36.
51. Jaremko JL, Guenther ZD, Jans LB, Macmahon PJ. Spectrum of injuries associated with paediatric ACL tears: an MRI pictorial review. Insights Imaging. 2013;4(3):273–85.
52. Oeppen RS, Connolly SA, Bencardino JT, Jaramillo D. Acute injury of the articular cartilage and subchondral bone: a common but unrecognized lesion in the immature knee. AJR Am J Roentgenol. 2004;182(1):111–7.
53. Kay J, Memon M, Shah A, Yen YM, Samuelsson K, Peterson D, et al. Earlier anterior cruciate ligament reconstruction is associated with a decreased risk of medial meniscal and articular cartilage damage in children and adolescents: a systematic review and meta-analysis. Knee Surg Sports Traumatol Arthrosc. 2018;26(12):3738–53.
54. Mansson O, Sernert N, Rostgard-Christensen L, Kartus J. Long-term clinical and radiographic results after delayed anterior cruciate ligament reconstruction in adolescents. Am J Sports Med. 2015;43(1):138–45.
55. Patel NM, Bram JT, Talathi NS, DeFrancesco CJ, Lawrence JTR, Ganley TJ. Which children are at risk for contralateral anterior cruciate ligament injury after ipsilateral reconstruction? J Pediatr Orthop. 2020;40(4):162–7.

Management of Pediatric ACL Injuries

11

Melanio C. Acosta IV and Marc Anthony L. Castro

11.1 Introduction

Treatment for pediatric anterior cruciate ligament (ACL) injuries may be more challenging when compared to adult ACL injury management. ACL injuries in the pediatric population are not so clear-cut, as injuries may fall anywhere on a spectrum including tibial avulsions, partial ACL tears, and complete tears. Pediatric ACL injuries are typically managed based on skeletal age of the patient and their possible remaining growth [1, 2]. With concern for physeal injury and iatrogenic growth disturbances, there is an ongoing debate on the most appropriate treatment to pediatric ACL injuries [3–8]. Using conventional adult ACL reconstructive techniques, the physis of skeletally immature patients is at risk during drilling of graft tunnels [9–11]. As a result non-operative treatment or delaying ACL reconstruction until skeletal maturity has been an option to avoid growth disturbances due to physeal injury. [8, 12–16]. Many different surgical techniques have been explored in pursuit to improve safety and avoid these complications in these skeletally immature patients.

M. C. Acosta IV
Makati Medical Center, Manila, Philippines

O.A.S.I. Bioresearch Foundation Gobbi N.P.O., Milan, Italy
e-mail: melanio.acosta@oasiortopedia.it

M. A. L. Castro (✉)
Makati Medical Center, Manila, Philippines

Philippine Orthopedic Institute, Inc., Manila, Philippines
e-mail: marcanthony.castro@makatimed.net.ph

11.2 Decisions for Non-surgical and Surgical Management

In 2018, the International Olympic Committee published a consensus statement addressing fundamental clinical questions regarding pediatric ACL injuries [17]. They outline 3 common goals for treatment of children with ACL injury. These include the following: (1) the restoration of a well-functioning, stable knee that allows the patient a healthy and active lifestyle, (2) the reduction of the impact of existing or possible risk for meniscal or chondral pathology, degenerative joint changes, and need for surgery in the future, and (3) the need to minimize risk of growth arrest and subsequent femur and tibial deformity.

Based on these 3 goals, they advocated two treatment options that can help pediatric patients with ACL injuries regardless if they have associated knee injuries or not. The treatment options emphasized were a non-surgical route with high-quality rehabilitation alone and ACL reconstruction with high-quality rehabilitation [17].

For children undergoing a non-surgical route with high-quality rehabilitation alone or those who underwent ACL reconstruction undergoing

M. Karahan et al. (eds.), *Anterior Cruciate Ligament Injuries Throughout the Life Span*,
https://doi.org/10.1007/978-3-031-84501-7_11

postoperative high-quality rehabilitation, in terms of high-quality rehabilitation, children cannot simply be viewed as "little adults." The exercises and goals of rehabilitation must be individualized and modified with less expectations that they are able to understand, follow, and perform unsupervised exercises. With this, there is an increased need for clinically supervised rehabilitation for children [17–21]. Patients who are very young with open physis should have a rehabilitation program with less emphasis on development of muscular strength and hypertrophy and more of a focus on the dynamic neuromuscular control and proprioception of multiple joints [17, 22]. When pediatric patients are older, more mature, and during puberty, there is an increase in androgenic hormones. With this increase in hormones, rehabilitation programs can more closely look like adult programs with more focus on heavier, external strength training [17, 23].

Children with ACL injury can be put into a four-phase rehabilitation program with those who underwent ACL reconstruction needing an additional prehabilitation phase. There should be an emphasized importance of patients meeting specific functional goals prior to moving from one phase to the next and return activities should be dependent on the patient's ability to achieve the return to sport criteria [17, 24, 25]. The recommended functional tests and return to sport criteria for children with ACL injury can be seen in Summary Box 11.1 [17].

In patients undergoing high-quality rehabilitation alone, some clinicians advocate the use of a protective brace during physical activities. Bracing may also be used during the prehabilitation phase in patients who eventually will undergo ACL reconstruction [17, 26]. However, studies have shown that in patients with complete tears, activity modification and bracing may not be effective and may lead to persistent instability of the knee and the development or progression of meniscal tears and cartilage damage [8, 10, 13, 18–21, 27–38].

In pediatric patients with ACL injuries, it is important to discuss with the patient and their family the appropriate management options with

Summary Box 11.1 Six fundamental clinical questions and relevant consensus statement topic(s)

For patients who choose ACL reconstruction
Prehabilitation
Full active extension and at least 120° active knee flexion
Little to no effusion
Ability to hold terminal knee extension during single leg standing (Fig. 11.2)
For adolescents: 90% limb symmetry on muscle strength tests
For patients who choose ACL reconstruction OR non-surgical treatment
Phase 1 to phase 2
Full active knee extension and 120° active knee flexion
Little to no effusion
Ability to hold terminal knee extension during single leg standing
Phase 2 to phase 3
Full knee range of motion
80% limb symmetry on single leg hop tests, with adequate landing strategies
Ability to jog for 10 min with good form and no subsequent effusion
For adolescents: 80% limb symmetry on muscle strength tests
Phase 3 to phase 4: sport participation (return to sport criteria), and continued injury prevention
Single-leg hop tests >90% of the contralateral limb (with adequate strategy and movement quality)
Performed gradual increase in sport-specific training without pain and effusion
Confident in knee function
Knowledge of high injury-risk knee positioning, and ability to maintain low-risk knee positioning in advanced sport-specific actions
Mentally ready to return to sport
For adolescents: 90% limb symmetry on muscle strength tests
Muscle strength testing should be performed using isokinetic dynamometry or handheld dynamometry/1 repetition maximum. The type of test and experience of the tester are highly likely to influence the results. If using handheld dynamometry/l repetition maximum, consider increasing the limb symmetry criterion cut-off by 10% (i.e. 90% limb symmetry becomes 100% limb symmetry). Clinicians who do not have access to appropriate strength assessment equipment should consider referring the patient elsewhere for strength evaluation

ACL Anterior cruciate ligament

respect to the current skeletal maturity status, any associated intra-articular injuries, and their functional goals. Children with ACL injuries with repairable associated knee injuries (e.g., repairable meniscal lesions or osteochondral defects), children with recurrent, symptomatic instability of the knee after completing high-quality rehabilitation, and children who experience unacceptable restrictions in their activities (i.e., inability to accept or follow activity modifications that can avoid knee instability) delaying surgery until skeletal maturity may be indicated for ACL reconstruction [17, 33].

11.3 Surgical Management

There are many different techniques investigated for ACL reconstruction in the pediatric population. These techniques can be broadly categorized based on the bone tunnel locations in relation to the physis—(1) transphyseal techniques where tunnels are drilled across the physis, (2) transepiphyseal techniques where tunnels are drilled through the epiphysis, sparing the physis, (3) hybrid techniques where typically the femoral physis is spared and transphyseal tunnels are used in the tibia, and (4) extraphyseal techniques where ACL reconstruction is done without bone tunnels [39].

11.4 Transphyseal ACL Reconstruction

Transphyseal ACL reconstruction (Fig. 11.1) in children is similar to the typical ACL reconstruction seen in adults. This approach is usually employed in adolescents near skeletal maturity, especially if the physes are closed.

Advantages of transphyseal ACL reconstruction include placement of the grafts in more anatomic locations and since these techniques are similar to the typical ACL reconstruction done in adults, surgeons are more familiar with these techniques, potentially resulting in decreased operative time and complications. The main disadvantage to transphyseal ACL reconstruction can be the potential for damage to the physes causing potential growth arrest or disturbance [40, 41].

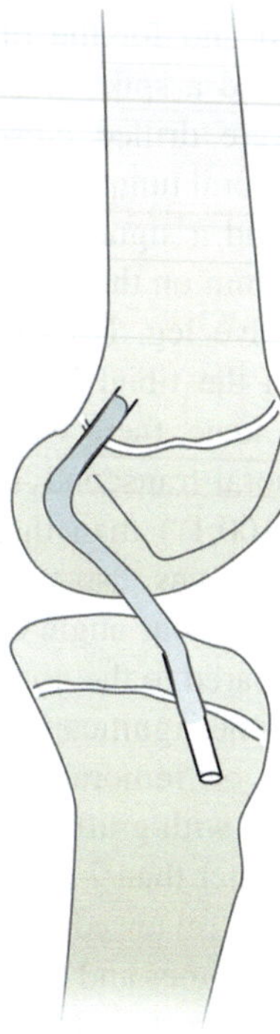

Fig. 11.1 Transphyseal ACL reconstruction. Tunnels are drilled through the physeal growth plate. (Diagram by A. Salvador). ACL anterior cruciate ligament

With regard to femoral tunnel drilling, it is suggested to avoid tunnels more eccentrically located across the distal femoral physis. This may cause the tunnels to exit closer to the perichondral ring leading to drilling across a larger area of the physis potentially causing more damage [42–45]. As a result, many surgeons advocate for the utilization of a transtibial drilling technique to produce more vertical oriented femoral tunnels resulting in lower rates of growth disturbance [43, 46–53].

In 2016, Faunø et al. [54] used a femoral and tibial transphyseal ACL reconstruction using 4-stranded hamstring autografts in 33 patients. All patients had open physes in the distal femur and proximal tibia at the time of surgery and were all skeletally mature at the time of final follow-up. For the femoral fixation, suspensory button fixation was used and for the tibial fixation, a bicortical screw and a spike washer were used. Tibial tunnels were drilled close to the tibial tuberosity and femoral tunnels were drilled transtibially. They found a significant mean femoral shortening of 3.5 mm on the operative compared to the non-operative leg. No significant difference was seen in the tibial length. In terms of angulation at the knee, there was a significantly more valgus femoral transcondylar angle on the operative knee (100.1°) than the non-operative knee (98.5°). There was also significantly more varus tibial transcondylar angle on the operative knee (87.2°) compared to the non-operative knee (88.2°). There was no significant difference in leg length inequality or femoral-tibial axis when comparing patients with graft sizes 7 mm or less and those with greater than 7 mm.

Calvo et al. [46] reviewed the 10-year, long-term functional outcomes and complications seen in skeletally immature patients who underwent arthroscopic transphyseal intra-articular ACL reconstruction with hamstring autograft. The average preoperative Lysholm and subjective International Knee Documentations Committee (IKDC) scores were 55 (range, 35–71) and 40 (range, 17–61), respectively. At final follow-up, average Lysholm and subjective IKDC scores were 94 (range, 55–100) and 92 (range, 44–100), respectively, showing a significant improvement ($p < 0.001$). They also noted that all patients were able to return to sport. However, only 88.89% of patients were able to return to a level equal or above preinjury level. On the patients who were able to undergo KT-1000 arthrometry, there was an average difference of 2.6 mm (range, −2.7 to 7.0 mm) when comparing the operative knee to the contralateral knee. Most patients, 77.7%, had less than 5 mm difference compared to the contralateral side, with no significant difference in functional scores in patients who had <5 mm difference compared to those who had >5 mm difference. During the follow-up period, 14.8% of patients underwent revision surgery due to either a traumatic rupture of the ACL graft during sport or the development of instability without any clear traumatic event, while 7.4% of patients developed chondral medial femoral condyle lesions and 3.7% of patients developed a lateral meniscus tear. No significant leg length difference was seen with an average of +1.6 mm (range −4 to 8 mm). There was also no significant valgus angulation when comparing the operative side to the contralateral side, 3.62°(range, 3.4–4.3°) and 4.05°(range, 3.6–4.2°), respectively.

11.5 All-Epiphyseal ACL Reconstruction

To avoid potential for growth disturbances causing leg length inequality and femoral-tibial axis deviations, all-epiphyseal (sometimes referred as transepiphyseal or intraepiphyseal) ACL reconstruction techniques were developed. All-epiphyseal ACL reconstruction (Fig. 11.2) may be more technically challenging, using bone tunnels that remain entirely within the proximal tibial and distal femoral epiphysis, without injuring critical structures of the knee such as the physis, anterolateral ligament, lateral collateral ligament, popliteus tendon, and articular cartilage [55, 56].

The goal for all-epiphyseal tunnels is to achieve graft orientation that is anatomic and similar to the native and intact ACL while restoring proper biomechanics of the knee joint [57–62]. In femoral transepiphyseal tunnels, the surgeon drills the tunnel in the epiphysis in a more horizontal course compared to what is seen in conventional transphyseal techniques, never crossing the physis. When viewed from a lateral view, the femoral tunnel should be located in the posterior quarter between the anterior and posterior condyles, opening at the posterior lateral condylar wall [63]. When drilling tibial transepiphyseal tunnels, the tunnels should be placed at the ACL footprint just medial to the free edge of the lateral meniscus and as well as at the tibial eminence seen on the coronal plane and just posterior to the Blumensaat's line on the sagittal plane [64].

There are many different surgical techniques for all-epiphyseal ACL reconstruction. First

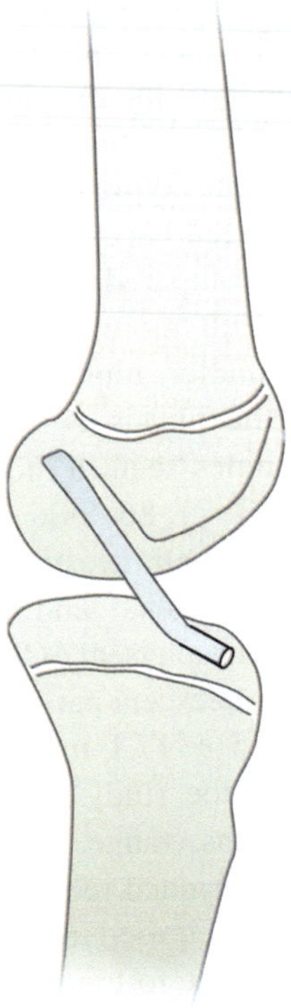

Fig. 11.2 All-epiphyseal ACL reconstruction. Tunnels are positioned entirely within the proximal tibial and distal femoral epiphysis. (Diagram by A. Salvador). ACL anterior cruciate ligament

described by Anderson et al. [64] in 2003, outside-in bone tunnels that remain within the epiphysis of the femur and tibia are made using two separate incisions. The quadruple hamstring autograft is secured at the femur using suspensory button fixation, while the tibial end of the graft is secured over a metaphyseal post.

The classic Anderson all-epiphyseal technique was then modified by Pennock et al. [65] in which a suspensory ArmorLink (OrthoPediatrics, Warsaw, IN) device was used for the tibial fixation instead of the aforementioned post. The purpose of this modification was to eliminate the risk of growth disturbance secondary to tethering of the metaphyseal-located tibial post described in the classic technique. On the femoral side a ShieldLoc (OrthoPediatrics, Warsaw, IN) ring and interference screw system replaced the original Endobutton (Smith & Nephew, Andover, MA) device.

Another modification to the classic Anderson all-epiphyseal technique was done by Lykissas et al. [66] in which a split tibial tunnel was used, ensuring that the bone tunnels, graft, and graft fixation were all exclusively inside the epiphysis. Using this technique there was an increased safety margin for keeping the tunnel within the tibial epiphysis by decreasing the tibial tunnel size to 4.5–5.55 mm by splitting the tibial graft in half, compared to the original 7–8 mm tunnels, thus, circumventing any damage of the growth plate. This technique minimized the risk of growth disturbance by avoiding drilling through or tethering over the physis.

Lawrence et al. [62] also presented an all-epiphyseal technique using Retroscrews (Arthrex, Naples, FL) for tibial graft fixation. This technique refrained from violating the physes during tunnel and graft placement along with graft fixation. Femoral tunnels were drilled parallel to the epiphysis and fixed with an interference screw, while the tibial tunnel was drilled with a RetroDrill (Arthrex, Naples, FL), placing it in a more anatomic location on the ACL insertion, and the tibial graft was fixed with a RetroScrew. This technique, however, required the use of an intraoperative computed tomography scan.

Due to concerns about injuring the physis, articular surface, or dilatation of typical bone tunnels when using interference screws, especially in the tibial epiphysis which typically has a minimum tunnel size requirement of at least 20 mm, McCarthy et al. [67] developed an "all-inside" all-epiphyseal technique using bone sockets instead of bone tunnels and buttons instead of tenodesis screws. The "all-inside" technique utilized guidance of an image intensifier and the use of retrograde drilling in the production of the femoral and tibial sockets. The hamstring graft was then passed retrograde into the femoral and tibial tunnels and fixed with suspensory buttons.

With the development of many various all-epiphyseal techniques, the height of the tibial epiphysis, limited safe tibial tunnel placement, and epiphyseal height in relation to the size of the tunnel drilled all act as limiting factors [68–70].

Gupta et al. [71] reviewed 17 studies [62, 65, 67, 72–85] encompassing 545 skeletally immature patients who underwent all-epiphyseal ACL reconstruction. They presented that all-epiphyseal ACL reconstruction techniques were able to restore good knee function, had excellent return to sport rates, and low complication rates. Four studies with a total of 96 patients measured postoperative subjective IKDC scores and reported a mean score of 90. Four studies with a total of 96 patients reported that all patients reported a normal Lachman test postoperatively, while five studies showed 77.8% of the total patients had a negative pivot shift test postoperatively. Eleven studies reported return to sport, in which 93.2% returned to sport in 8–22 months postoperatively, however, only 77.9% returned to a level equal or above pre-injury level. The overall complication rate seen was 9.8%. The most common complication was ACL graft ruptures seen in 49.5% of the reported complications. All other complications were as follows—26% were unspecified, 15.1% were limb overgrowths, 5.7% were limb length discrepancies, 3.8% were arthrofibrosis, 1.9% were knee stiffness, and 1.9% were soft tissue infection. To minimize growth-related complications, they concluded that all-epiphyseal ACL reconstruction should be mostly considered in patients with at least 2 years of skeletal growth remaining, while transphyseal and extra-physeal reconstruction considered for patients with less than 2 years of growth remaining or those who are skeletally mature.

11.6 Hybrid Techniques

As all-epiphyseal ACL reconstruction techniques are typically used for very young patients with open physes, hybrid techniques (sometimes referred to as partial transphyseal techniques), such as those that employ a combination of transepiphyseal femoral tunnels and transphyseal tibial tunnels, can be used in patients closer to skeletal maturity [55, 63, 86, 87].

These hybrid techniques exploit the lower risk of growth disturbance, leg length inequality, and angular deformity seen in transphyseal tibial tunnel techniques due to the smaller contribution to growth of the proximal tibial physis compared to the distal femur, where growth disturbances are more typically due to damage from tunnels located eccentrically in the distal femoral physis compared to centrally located tibial tunnels. With more central and vertical tibial tunnels, smaller, more circular holes are made through the physis compared to larger, more elliptical holes seen in more oblique tunnels [2, 51, 54, 63, 71, 86–89].

Chambers et al. [90] investigated growth disturbances, complications, and functional outcomes of partial transphyseal ACL reconstruction in pediatric and adolescent patients. With a mean follow up of 31.5 ± 17.1 months, 24 patients were included in the study. Within a mean of 29.2 ± 17.3 months (range, 1.5–49.5 months), 29.2% of patients needed reoperation. A rate of 8.3% of patients required reoperation due to ACL graft failure due to traumatic injuries during sport. As well, 4.2% of patients required hemiepiphysiodesis of bilateral medial distal femurs to correct a genu valgum deformity that was present prior to the primary ACL reconstruction. Functionally, the mean pediatric IKDC scores were 94.8 ± 5.3 (range, 84.7–100.0) for patients who returned surveys and 91.7% of patients were able to return back to unrestricted athletics at an average of 11.3 ± 3.3 months (range, 9–23 months). The true postoperative growth disturbance rate of 16.7% was seen with 12.5% having limb shortening and another 12.5% having a significant lateral translation of mechanical axis deviation. Among the population studied, no patients who had remaining growth less than 2 years had growth disturbances. There was an overall mean limb length difference of −3.0 ± 6.1 mm, with 12.5% of patients having more than 1 cm shortening on the operative side, in which femoral shortening was the major cause for this discrepancy. There were also no patients who had significant changes in medial lateral distal femoral angle or medial proximal tibial angles.

11.7 Extraphyseal Techniques

Physeal-sparing, extraphyseal techniques (Fig. 11.3) commonly using combined intra- and extra-articular ACL reconstruction has also been used in ACL-deficient patients. These techniques are typically used in skeletally immature patients with open physes, still have significant remaining growth, or smaller knees secondary to congenital anomalies or previous injury making the epiphysis exceptionally small for all-epiphyseal ACL reconstruction. Through these techniques, the risk for physeal damage and growth disturbance can be minimized by avoiding tunnels through or near the physis and use of intraosseous fixation techniques [14, 91–93]. These techniques were initially proposed and developed by MacIntosh and Darby [94] modified by Micheli et al. [14] and later Kocher et al. [91, 92].

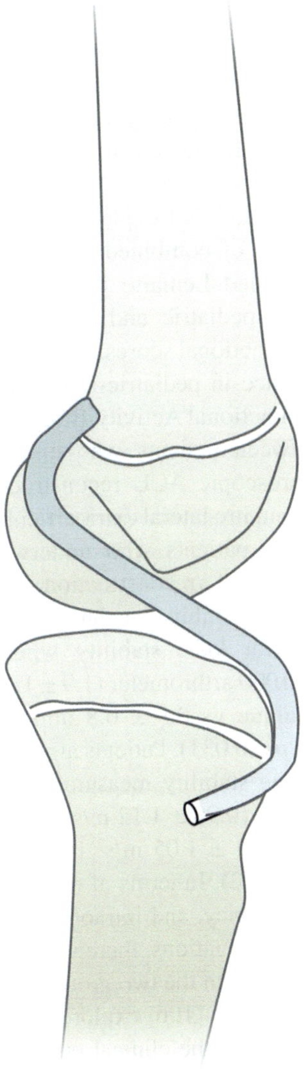

Fig. 11.3 Extraphyseal ACL reconstruction. Graft is placed around the physeal growth plate avoiding tunnels through or near the physis. (Diagram by A. Salvador). ACL anterior cruciate ligament

A physeal-sparing, combined intra- and extra-articular ACL reconstruction using an autologous iliotibial band was developed and recently described by Feroe et al. [95]. Using standard supine position with the operative knee hanging off the table, the iliotibial band (ITB) harvest is started with an incision along the lateral joint line extending proximally over the iliotibial band. After proper exposure of the ITB, the anterior and posterior borders of the iliotibial band are incised 2 cm proximal to the Gerdy's tubercle to secure a graft size of 1 cm wide. The ITB graft is then harvested using a tendon stripper with goals of obtaining a graft of at least 15 cm. The proximal end of the graft is then tubularized to aid in graft passage. The distal end of the graft is released from the lateral patellar retinaculum and joint capsule while remaining attached to the Gerdy's tubercle distally. The tibia is then prepared by making a 3–4 cm incision in the area of the pes anserinus insertion. With the knee in extension, standard anterolateral and anteromedial portals are created. Using a clamp through the anteromedial portal, the free end of the ITB graft is fed through the over-the-top position and pulled out of the anterolateral incision. The graft is then pulled through the intercondylar notch and out the anteromedial portal. A small trench is made through the tibial incision in the anteromedial aspect of the proximal tibial epiphysis. This trench is located inferior to the intermeniscal ligament and helps with graft passage for a more posterior placement. The graft is then pulled through the tibiofemoral joint, under the intermeniscal ligament in the trench created, and out the tibial incision. The graft is then fixed with the knee in 90° flexion with the foot in neutral position to the intermuscular septum and posterolateral femoral condyle periosteum. The periosteum is then longitudinally incised distal to the proximal tibial physis with the knee in extension. Periosteal flaps are created and the graft is sutured to these flaps under proper tension. A post can also be used to assist in the fixation of the graft, if necessary.

In comparing different ACL reconstruction techniques and the incidence of growth disturbances in pediatric and adolescent populations, Patil et al. [90, 96] evaluated 78 studies, and 2693 cases of pediatric and adolescent ACL reconstruction were identified in their systematic review. The growth disturbance rate was 2.6%, where 75.9% were males with a mean age of 12.1 years. Tibial transphyseal tunnels had a 2.7% growth disturbance rate, tibial intraepiphyseal tunnels had a 2.6% rate, and tibial extraphyseal had no growth disturbances. For femoral tunnels, femoral transphyseal had a growth disturbance rate of 2.9%, femoral intraepiphyseal tunnels had a rate of 3.3%, and femoral extraphyseal tunnels had a rate of 1.1%. Femoral intraepiphyseal tunnels had the highest rate of growth disturbances among all individual techniques, while a combination of tibial transphyseal and femoral intraepiphyseal tunnels had the highest rates of growth disturbance. Extraphyseal tunnels for both the tibia and femur showed the lowest rates for growth disturbance. There were 166 ACL graft ruptures from 2120 cases reported, giving an overall risk of 7.8%. Transphyseal graft tunnel techniques had the highest risk for ACL graft rupture with an ACL graft rupture rate of 8.7% in tibia transphyseal techniques and 9.3% in femoral transphyseal techniques. Extraphyseal graft tunnel techniques had the lowest risk for ACL graft rupture with tibial extraphyseal techniques having an ACL graft rupture rate of 5.1% and 6.0% for femoral extraphyseal techniques. Angular deformities were found as a result of asymmetric physeal growth disturbances, while reduced anterior tibial slope, varus, and valgus deformities were most likely due to physeal arrests in the anterior proximal tibia, medial proximal tibia, and lateral distal femur, respectively.

11.8 Combined Anterior Cruciate Ligament Reconstruction and Lateral Extra-Articular Tenodesis

With high levels of ACL graft rupture in patients who continue to play sports after undergoing ACL reconstruction, there has been increasing research on lateral extra-articular and anterolateral ligament augmentation techniques to help stabilize and protect ACL reconstructions [97–102]. By augmenting ACL reconstructions with lateral extra-articular tenodesis and anterolateral ligament procedures, there is increased protection against increased or residual anterolateral instability by decreasing tension on the graft [103, 104].

Perelli et al. [105] explored the outcomes of the technique of combined ACL reconstruction with a modified Lemaire lateral extra-articular tenodesis in pediatric and adolescent patients. Regarding functional scores, there was no significant difference in pediatric-IKDC and Pediatric Pediatric Functional Activity Brief Scale (FABS) scores between patients who underwent combined arthroscopic ACL reconstruction with a modified Lemaire lateral extra-articular tenodesis compared to patients who underwent isolated arthroscopic ACL reconstruction. Patients who underwent the combined technique had improved anteroposterior knee stability when measured with a KT-1000 arthrometer (1.9 ± 1.1 mm in isolated technique vs 0.8 ± 0.8 mm in combined technique; $p = 0.031$). Patients also had improved rotatory knee stability measured using a KiRA accelerometer (0.98 ± 1.12 m/s^2 in isolated technique vs −0.59 ± 1.05 m/s^2 in combined technique; $p = 0.012$). In terms of return to sport, leg length discrepancy, and intraoperative and postoperative complications, there was no significant difference between the two groups.

Carrozzo et al. [106] explored studies by the SANTI group on the clinical outcomes of combined ACL reconstruction and lateral extra-articular tenodesis procedures in skeletally immature patients. After reviewing 5 studies comprising 381 pediatric and adolescent patients, they noted a 4.65% overall ACL graft failure rate among those patients who underwent the combined technique. Return to play was 95.11% with a mean Lysholm score of 94.51 and a Pediatric IKDC score of 93.39. Only 1.9% had coronal plane deformities and only 0.8% was seen to have leg length discrepancies.

11.9 ACL Repair

Open ACL repair was popular and the gold standard treatment for ACL tears in the 1970s and 1980s [107–112] stemming from the work of Robson [113], Palmer [114, 115], and O'Donoghue [116, 117]. However, after further studies during that time showing mid-term deterioration of outcomes with high reinjury rates along with improved results of ACL reconstruction in comparison to repair, ACL reconstruction was adopted to be the new gold standard in the 1990s [118–128].

With improved knowledge on ACL anatomy, healing and improvement on arthroscopic techniques, there has been an increasing interest in arthroscopic ACL repair in the pediatric and adolescent population. The advantages of arthroscopic ACL repair over reconstruction are the circumvention of the need for autograft harvesting and its possible complications, the maintenance of the native ACL tissue preserving its innate proprioceptive function, the avoidance of the need to create graft tunnels, its less invasive nature resulting in a reduction of soft tissue trauma, and the maintenance of the knees bony anatomy allowing easier revisions [129–134].

Acute injuries of the proximal ACL with good, viable ligament remnants are best indicated for ACL repair. Proximal ACL tears are most ideal due its increased healing potential secondary to increased vascularity [135, 136]. There are generally two types of repairs used in pediatric populations. Direct repairs use suspensory methods or suture anchors typically for proximal or distal ACL injuries or osseous avulsions from the proximal femur [137–140], while augmented repairs employ the aid of a suture tape brace, an intra-ligament stabilization device, or absorbable collagen matrix [141–147].

Dabis et al. [142] concluded that proximal ACL tears managed with direct ACL repair was safe, effective, and had excellent short-term outcomes. With their repairs, there was no physeal injuries that led to any lower extremity deformity or leg length discrepancy. There were also increased median Lysholm, Tegner, and Knee injury and Osteoarthritis Outcome Score (KOOS) scores at 1 year follow-up with scores of 91.5, 7 (range, 5–9), and 93.7, respectively, and further improvement of scores at 2 year follow-up with scores of 95, 7 (range, 6–10), and 96.5, respectively. There was also no significant difference in rotatory knee stability and anterior translation compared to the uninjured side with a Lachman side-to-side difference of 1.2 mm and pivot shift side-to-side difference of 0.3 m/s^2. No ACL graft ruptures or ACL revisions were reported.

Knapik et al. [148] compared the results of skeletally immature patients who underwent ACL reconstruction with those who underwent arthroscopic ACL repair. The study included 14 different articles constituting 443 patients—164 of which underwent all-epiphyseal ACL reconstruction, 252 underwent extraphyseal ACL reconstruction, and 27 underwent ACL repair. The mean final follow-up time was 52.1 ± 16.1 months (range, 24–78 months). They showed that at final follow-up there was no significant difference in the mean Lysholm scores among those treated with all-epiphyseal ACL reconstruction (94.8 ± 2.7), extraphyseal ACL reconstruction (95.6 ± 3.1), or ACL repair techniques (96.8 ± 4.5), as well as mean IKDC scores 93.5 ± 3.0, 90.1 ± 3.7, 92.1 ± 1.8, respectively. Return to sport or activity was significantly higher in those who underwent extraphyseal ACL reconstruction (97%) compared to those who had all-epiphyseal ACL reconstruction (87%) or ACL repair (80%). There was also a significant increase in reported leg length discrepancy (LLD) in those who underwent all-epiphyseal ACL reconstruction (12%) compared to those who underwent extraphyseal ACL reconstruction (0.8%), while ACL repair had no reports of LLD. Limb overgrowth was seen in 85% of patients following all-epiphyseal ACL reconstruction, while 100% of patients who underwent extraphyseal ACL reconstruction had limb overgrowth. Varus and Valgus malalignment were not significantly different between all-epiphyseal and extraphyseal ACL reconstructions, both at 0.6%. ACL repair resulted in no angular deformities being reported. The most common complication was ACL graft rupture which occurred in 7.9% of all-epiphyseal ACL reconstruction patients, 3.6% of extraphyseal ACL reconstructions patients, and 41% of ACL repair patients; however, there was no statistically significant correlation.

11.10 Authors Preferred Surgical Technique

ACL reconstruction in the skeletally immature can be a very demanding procedure and a high level of surgical skill is required to achieve good to excellent results. Our favored autograft is the hamstring tendon, quadruple folded. This is a pure soft tissue graft which will eliminate physeal complication issues in terms of harvesting (genu recurvatum) and use of bone blocks in tunnels (growth arrest). We drill our tunnels to a maximum size of 8 mm using a transphyseal technique on the tibial side and a physeal sparring retrograde drilling with a Flipcutter (Arthrex, Naples, FL) fluoroscopically guided. Kercher et al. demonstrated a complication rate of less than 3% when drill size was kept to 8 mm or less [149]. Fixation of our grafts are done using extracortical suspensory fixation methods. We absolutely avoid intraosseous fixation techniques, as the ACL study group has shown a high incidence of growth disturbances with the use of intraosseous fixation.

11.11 Conclusion

Pediatric ACL injuries differ from adult injuries. The most appropriate type of treatment should be decided after factors such as the patient's age, athletic expectations and level, and additional injuries are discussed with the family. All treatments performed in the pediatric age each have their own pros and cons.

References

1. Anderson CN, Anderson AF. Pediatric ACL: evaluation and management. Curr Orthop Pract. 2014;25(4):312–20.
2. Milewski MD, Beck NA, Lawrence JT, Ganley TJ. Anterior cruciate ligament reconstruction in the young athlete: a treatment algorithm for the skeletally immature. Clin J Sport Med. 2011;30(4):801–10.
3. Kocher MS, Saxon HS, Hovis WD, Hawkins RJ. Management and complications of anterior cruciate ligament injuries in skeletally immature patients: survey of the Herodicus Society and the ACL Study Group. J Pediatr Orthop. 2002;22:452–7.
4. Koman JD, Sanders JO. Valgus deformity after reconstruction of the anterior cruciate ligament in a skeletally immature patient. A case report. J Bone Joint Surg Am. 1999;81:711–5.
5. Mohtadi N, Grant J. Managing anterior cruciate ligament deficiency in the skeletally immature individual: a systematic review of the literature. Clin J Sport Med. 2006;16:457–64. https://doi.org/10.1097/01.jsm.0000248844.39498.1f.
6. Woods GW, O'Connor DP. Delayed anterior cruciate ligament reconstruction in adolescents with open physes. Am J Sports Med. 2004;32:201–10. https://doi.org/10.1177/0363546503258868.
7. DerSimonian R, Laird N. Meta-analysis in clinical trials. Control Clin Trials. 1986;7(3):177–88.
8. Moksnes H, Engebretsen L, Eitzen I, Risberg MA. Functional outcomes following a non-operative treatment algorithm for anterior cruciate ligament injuries in skeletally immature children 12 years and younger: a prospective cohort with 2 years follow-up. Br J Sports Med. 2013;47(8):488–94.
9. Janarv PM, Wikstrom B, Hirsch G. The influence of transphyseal drilling and tendon grafting on bone growth: an experimental study in the rabbit. J Pediatr Orthop. 1998;18(2):149–54.
10. Kocher MS, Micheli LJ, Zurakowski D, Luke A. Partial tears of the anterior cruciate ligament in children and adolescents. Am J Sports Med. 2002;30(5):697–703.
11. McConkey MO, Bonasia DE, Amendola A. Pediatric anterior cruciate ligament reconstruction. Curr Rev Musculoskelet Med. 2011;4(2):37–44.
12. Andrews M, Noyes FR, Barber-Westin SD. Anterior cruciate ligament allograft reconstruction in the skeletally immature athlete. Am J Sports Med. 1994;22(1):48–54.
13. McCarroll JR, Shelbourne KD, Porter DA, Rettig AC, Murray S. Patellar tendon graft reconstruction for midsubstance anterior cruciate ligament rupture

in junior high school athletes: an algorithm for management. Am J Sports Med. 1994;22(4):478–84.
14. Micheli LJ, Rask B, Gerberg L. Anterior cruciate ligament reconstruction in patients who are prepubescent. Clin Orthop Relat Res. 1999;364:40–7.
15. Moksnes H, Engebretsen L, Risberg MA. Performance-based functional outcome for children 12 years or younger following anterior cruciate ligament injury: a two to nine-year follow-up study. Knee Surg Sports Traumatol Arthrosc. 2008;16(3):214–23.
16. Moksnes H, Engebretsen L, Risberg MA. Prevalence and incidence of new meniscus and cartilage injuries after a nonoperative treatment algorithm for ACL tears in skeletally immature children: a prospective MRI study. Am J Sports Med. 2013;41(8):1771–9.
17. Ardern CL, Ekås G, Grindem H, Moksnes H, Anderson A, Chotel F, Cohen M, Forssblad M, Ganley TJ, Feller JA, Karlsson J, Kocher MS, LaPrade RF, McNamee M, Mandelbaum B, Micheli L, Mohtadi N, Reider B, Roe J, Seil R, et al. 2018 International Olympic Committee consensus statement on prevention, diagnosis and management of paediatric anterior cruciate ligament (ACL) injuries. Knee Surg Sports Traumatol Arthrosc. 2018;26(4):989–1010. https://doi.org/10.1007/s00167-018-4865-y.
18. Graf BK, Lange RH, Fujisaki CK, Landry GL, Saluja RK. Anterior cruciate ligament tears in skeletally immature patients: meniscal pathology at presentation and after attempted conservative treatment. Arthroscopy. 1992;8:229–33.
19. Aichroth PM, Patel DV, Zorrilla P. The natural history and treatment of rupture of the anterior cruciate ligament in children and adolescents: a prospective review. J Bone Joint Surg Br. 2002;84(1):38–41.
20. Millett PJ, Willis AA, Warren RF. Associated injuries in pediatric and adolescent anterior cruciate ligament tears: does a delay in treatment increase the risk of meniscal tear? Arthroscopy. 2002;18(9):955–9.
21. Henry J, Chotel F, Chouteau J, Fessy MH, Bérard J, Moyen B. Rupture of the anterior cruciate ligament in children: early reconstruction with open physes or delayed reconstruction to skeletal maturity? Knee Surg Sports Traumatol Arthrosc. 2009;17(7):748–55.
22. Trivedi V, Mishra P, Verma D. Pediatric ACL injuries: a review of current concepts. Open Orthop J. 2017;11:378–88. https://doi.org/10.2174/1874325001711010378.
23. Bergeron MF, Mountjoy M, Armstrong N, Chia M, Côté J, Emery CA, et al. International Olympic Committee consensus statement on youth athletic development. Br J Sports Med. 2015;49:843–51.
24. van Melick N, van Cingel RE, Brooijmans F, Neeter C, van Tienen T, Hullegie W, et al. Evidence-based clinical practice update: practice guidelines for anterior cruciate ligament rehabilitation based on a systematic review and multidisciplinary consensus. Br J Sports Med. 2016;50:1506–15.
25. Ardern CL, Glasgow P, Schneiders A, Witvrouw E, Clarsen B, Cools A, et al. 2016 Consensus statement on return to sport from the First World Congress in sports physical therapy, Bern. Br J Sports Med. 2016;50:853–64.
26. Moksnes H, Engebretsen L, Seil R. The ESSKA paediatric anterior cruciate ligament monitoring initiative. Knee Surg Sports Traumatol Arthrosc. 2016;24:680–7.
27. Arendt EA, Grossfeld SL. Patient selection for ACL reconstruction. Sports Med Arthrosc Rev. 1996;4:328–35.
28. Dunn KL, Lam KC, Valovich McLeod TC. Early operative versus delayed or nonoperative treatment of anterior cruciate ligament injuries in pediatric patients. J Athl Train. 2016;51(5):425–7. https://doi.org/10.4085/1062-6050.51.5.11.
29. Kannus P, Järvinen M. Knee ligament injuries in adolescents. J Bone Joint Surg Br. 1988;70:772–6.
30. Ramski DE, Kanj WW, Franklin CC, Baldwin KD, Ganley TJ. Anterior cruciate ligament tears in children and adolescents: a meta-analysis of nonoperative versus operative treatment. Am J Sports Med. 2014;42(11):2769–76. https://doi.org/10.1177/0363546513510889.
31. McCarroll JR, Rettig AC, Shelbourne KD. Anterior cruciate ligament injuries in the young athlete with open physes. Am J Sports Med. 1988;16(1):44–7.
32. Streich NA, Barié A, Gotterbarm T, Keil M, Schmitt H. Transphyseal reconstruction of the anterior cruciate ligament in prepubescent athletes. Knee Surg Sports Traumatol Arthrosc. 2010;18(11):1481–6.
33. Beynnon BD, Johnson RJ, Abate JA, Fleming BC, Nichols CE. Treatment of anterior cruciate ligament injuries, part I. Am J Sports Med. 2005;33(10):1579–602. https://doi.org/10.1177/0363546505279913.
34. Madelaine A, Fournier G, Sappey-Marinier E, et al. French Arthroscopic Society. Conservative management of anterior cruciate ligament injury in paediatric population: about 53 patients. Orthop Traumatol Surg Res. 2018;104(8):S169–73.
35. Mizuta H, Kubota K, Shiraishi M, Otsuka Y, Nagamoto N, Takagi K. The conservative treatment of complete tears of the anterior cruciate ligament in skeletally immature patients. J Bone Joint Surg Br. 1995;77(6):890–4.
36. Pressman AE, Letts RM, Jarvis JG. Anterior cruciate ligament tears in children: an analysis of operative versus nonoperative treatment. J Pediatr Orthop. 1997;17(4):505–11.
37. Janarv PM, Nyström A, Werner S, Hirsch G. Anterior cruciate ligament injuries in skeletally immature patients. J Pediatr Orthop. 1996;16(5):673–7.
38. Angel KR, Hall DJ. Anterior cruciate ligament injury in children and adolescents. Arthroscopy. 1989;5(3):197–200.
39. Lykissas MG, Gkiatas I, Kontakis G. Surgical considerations and treatment algorithm for ACL tear. In: Parikh S, editor. The pediatric anterior cruci-

ate ligament. Cham: Springer; 2018. https://doi.org/10.1007/978-3-319-64771-5_10.
40. Pierce TP, Issa K, Festa A, Scillia AJ, McInerney VK. Pediatric anterior cruciate ligament reconstruction: a systematic review of transphyseal versus physeal-sparing techniques. Am J Sports Med. 2017;45(2):488–94. https://doi.org/10.1177/0363546516638079.
41. Hudgens JL, Dahm DL. Treatment of anterior cruciate ligament injury in skeletally immature patients. Int J Pediatr. 2012;2012:932702. https://doi.org/10.1155/2012/932702.
42. Shea KG, Belzer J, Apel PJ, Nilsson K, Grimm NL, Pfeiffer RP. Volumetric injury of the physis during single-bundle anterior cruciate ligament reconstruction in children: a 3-dimensional study using magnetic resonance imaging. Arthrosc J Arthrosc Relat Surg. 2009;25(12):1415–22. https://doi.org/10.1016/j.arthro.2009.06.023.
43. Cruz AI Jr, Lakomkin N, Fabricant PD, Lawrence JT. Transphyseal ACL reconstruction in skeletally immature patients: does independent femoral tunnel drilling place the physis at greater risk compared with transtibial drilling? Orthop J Sports Med. 2016;4(6):2325967116650432. https://doi.org/10.1177/2325967116650432.
44. Ford LT, Key JA. A study of experimental trauma to the distal femoral epiphysis in rabbits. J Bone Joint Surg Am Vol. 1956;38-A(1):84–92.
45. Peterson HA. Partial growth plate arrest and its treatment. J Pediatr Orthop. 1984;4(2):246–58. https://doi.org/10.1097/01241398-198403000-00015.
46. Calvo R, Figueroa D, Gili F, Vaisman A, Mocoçain P, Espinosa M, León A, Arellano S. Transphyseal anterior cruciate ligament reconstruction in patients with open physes: 10-year follow-up study. Am J Sports Med. 2015;43(2):289–94. https://doi.org/10.1177/0363546514557939.
47. Redler LH, Brafman RT, Trentacosta N, Ahmad CS. Anterior cruciate ligament reconstruction in skeletally immature patients with transphyseal tunnels. Arthrosc J Arthrosc Relat Surg. 2012;28(11):1710–7. https://doi.org/10.1016/j.arthro.2012.04.145.
48. Cohen M, Ferretti M, Quarteiro M, Marcondes FB, de Hollanda JP, Amaro JT, Abdalla RJ. Transphyseal anterior cruciate ligament reconstruction in patients with open physes. Arthrosc J Arthrosc Relat Surg. 2009;25(8):831–8. https://doi.org/10.1016/j.arthro.2009.01.015.
49. Kohl S, Stutz C, Decker S, Ziebarth K, Slongo T, Ahmad SS, Kohlhof H, Eggli S, Zumstein M, Evangelopoulos DS. Mid-term results of transphyseal anterior cruciate ligament reconstruction in children and adolescents. Knee. 2014;21(1):80–5. https://doi.org/10.1016/j.knee.2013.07.004.
50. Frosch KH, Stengel D, Brodhun T, Stietencron I, Holsten D, Jung C, Reister D, Voigt C, Niemeyer P, Maier M, Hertel P, Jagodzinski M, Lill H. Outcomes and risks of operative treatment of rupture of the anterior cruciate ligament in children and adolescents. Arthrosc J Arthrosc Relat Surg. 2010;26(11):1539–50. https://doi.org/10.1016/j.arthro.2010.04.077.
51. Kocher MS, Smith JT, Zoric BJ, Lee B, Micheli LJ. Transphyseal anterior cruciate ligament reconstruction in skeletally immature pubescent adolescents. J Bone Joint Surg Am Vol. 2007;89(12):2632–9. https://doi.org/10.2106/JBJS.F.01560.
52. Hui C, Roe J, Ferguson D, Waller A, Salmon L, Pinczewski L. Outcome of anatomic transphyseal anterior cruciate ligament reconstruction in Tanner stage 1 and 2 patients with open physes. Am J Sports Med. 2012;40(5):1093–8. https://doi.org/10.1177/0363546512438508.
53. Kim SJ, Shim DW, Park KW. Functional outcome of transphyseal reconstruction of the anterior cruciate ligament in skeletally immature patients. Knee Surg Relat Res. 2012;24(3):173–9. https://doi.org/10.5792/ksrr.2012.24.3.173.
54. Faunø P, Rømer L, Nielsen T, Lind M. The risk of transphyseal drilling in skeletally immature patients with anterior cruciate ligament injury. Orthop J Sports Med. 2016;4(9):2325967116664685. https://doi.org/10.1177/2325967116664685.
55. Gausden EB, et al. Surgical options for anterior cruciate ligament reconstruction in the young child. Curr Opin Pediatr. 2015;27(1):82–91.
56. Shea KG, Cannamela PC, Fabricant PD, Terhune EB, Polousky JD, Milewski MD, Ganley TJ, Anderson AF. Lateral radiographic landmarks for ACL and LCL footprint origins during all-epiphyseal femoral drilling in skeletally immature knees. J Bone Joint Surg. 2016;99:506.
57. Kennedy A, Coughlin DG, Metzger MF, Tang R, Pearle AD, Lotz JC, et al. Biomechanical evaluation of pediatric anterior cruciate ligament reconstruction techniques. Am J Sports Med. 2011;39(5):964–71.
58. McCarthy MM, Tucker S, Nguyen JT, Green DW, Imhauser CW, Cordasco FA. Contact stress and kinematic analysis of all-epiphyseal and over-the-top pediatric reconstruction techniques for the anterior cruciate ligament. Am J Sports Med. 2013;41(6):1330–9.
59. Stonestreet MJ, Jones KC, Kirkpatrick MS, Shah KS, Frampton CE, Morscher MA, et al. All-epiphyseal ACL reconstruction improves tibiofemoral contact: an in vitro study. J Pediatr Orthop. 2012;32(1):15–20.
60. Fabricant PD, Jones KJ, Delos D, Cordasco FA, Marx RG, Pearle AD, et al. Reconstruction of the anterior cruciate ligament in the skeletally immature athlete: a review of current concepts: AAOS exhibit selection. J Bone Joint Surg Am. 2013;95(5):e28.
61. Anderson AF. Transepiphyseal replacement of the anterior cruciate ligament using quadruple hamstring grafts in skeletally immature patients. J Bone Joint Surg Am. 2004;86-A(Suppl. 1(Pt 2)):201–9.
62. Lawrence JT, Bowers AL, Belding J, Cody SR, Ganley TJ. All-epiphyseal anterior cruciate ligament

reconstruction in skeletally immature patients. Clin Orthop Relat Res. 2010;468(7):1971–7.
63. Zbojniewicz AM, Meyers AB, Wall EJ. Postoperative imaging of anterior cruciate ligament reconstruction techniques across the spectrum of skeletal maturity. Skeletal Radiol. 2016;45(4):517–30.
64. Anderson AF. Transepiphyseal replacement of the anterior cruciate ligament in skeletally immature patients. A preliminary report. J Bone Joint Surg Am. 2003;85-A(7):1255–63. http://www.ncbi.nlm.nih.gov/pubmed/12851350.
65. Pennock A, Chambers H, Turk R, Parvanta K, Dennis M, Edmonds E. Use of a modified all-epiphyseal technique for anterior cruciate ligament reconstruction in the skeletally immature patient. Orthop J Sports Med. 2018;6:232596711878176.
66. Lykissas MG, Nathan ST, Wall EJ. All-epiphyseal anterior cruciate ligament reconstruction in skeletally immature patients: a surgical technique using a split tibial tunnel. Arthrosc Tech. 2012;1(1):e133–9. https://doi.org/10.1016/j.eats.2012.05.005.
67. McCarthy MM, Graziano J, Green DW, Cordasco FA. All-epiphyseal, all-inside anterior cruciate ligament reconstruction technique for skeletally immature patients. Arthrosc Tech. 2012;1(2):e231–9. https://doi.org/10.1016/j.eats.2012.08.005.
68. Davis DL, Almardawi R, Mitchell JW. Analysis of the tibial epiphysis in the skeletally immature knee using magnetic resonance imaging: an update of anatomic parameters pertinent to physeal-sparing anterior cruciate ligament reconstruction. Orthop J Sports Med. 2016;4(6):2325967116655313. https://doi.org/10.1177/2325967116655313.
69. Parikh SN. Transepiphyseal replacement of the anterior cruciate ligament in skeletally immature patients: a preliminary report. J Bone Joint Surg Am. 2011;93(4):e12. Author reply e12. http://www.ncbi.nlm.nih.gov/pubmed/21325577.
70. Swami VG, Mabee M, Hui C, Jaremko JL. MRI anatomy of the tibial ACL attachment and proximal epiphysis in a large population of skeletally immature knees: reference parameters for planning anatomic physeal-sparing ACL reconstruction. Am J Sports Med. 2014;42(7):1644–51. https://doi.org/10.1177/0363546514530293.
71. Gupta A, Tejpal T, Shanmugaraj A, Horner NS, Gohal C, Khan M. All-epiphyseal anterior cruciate ligament reconstruction produces good functional outcomes and low complication rates in pediatric patients: a systematic review. Knee Surg Sports Traumatol Arthrosc. 2020;28(8):2444–52. https://doi.org/10.1007/s00167-020-06085-3.
72. Akinleye SD, Sewick A, Wells L. All-epiphyseal ACL reconstruction: a three-year follow-up. Int J Sports Phys Ther. 2013;8(3):300.
73. Greenberg EM, Albaugh J, Ganley TJ, Lawrence JT. Rehabilitation considerations for all epiphyseal ACL reconstruction. Int J Sports Phys Ther. 2012;7(2):185.
74. Cordasco FA, Black SR, Price M, Wixted C, Heller M, Asaro LA, et al. Return to sport and reoperation rates in patients under the age of 20 after primary anterior cruciate ligament reconstruction: risk profile comparing 3 patient groups predicated upon skeletal age. Am J Sports Med. 2019;47(3):628–39.
75. Cordasco FA, Mayer SW, Green DW. All-inside, all-epiphyseal anterior cruciate ligament reconstruction in skeletally immature athletes: return to sport, incidence of second surgery, and 2-year clinical outcomes. Am J Sports Med. 2017;45(4):856–63.
76. Cruz A, Fabricant P, McGraw M, Rozell J, Ganley T, Wells L. All-epiphyseal ACL reconstruction in children. J Pediatr Orthop. 2017;37:204–9.
77. Roberti di Sarsina T, Macchiarola L, Signorelli C, Grassi A, Raggi F, Marcheggiani Muccioli G, et al. Anterior cruciate ligament reconstruction with an all-epiphyseal "over-the-top" technique is safe and shows low rate of failure in skeletally immature athletes. Knee Surg Sports Traumatol Arthrosc. 2018;27:498–506.
78. Patel N, DeFrancesco C, Talathi N, Bram J, Ganley T. All-epiphyseal anterior cruciate ligament reconstruction does not increase the risk of complications compared with pediatric transphyseal reconstruction. J Am Acad Orthop Surg. 2019;27:e752–7.
79. Ho B, Edmonds E, Chambers H, Bastrom T, Pennock A. Risk factors for early ACL reconstruction failure in pediatric and adolescent patients. J Pediatr Orthop. 2018;38(7):388–92.
80. Tuca M, Greditzer H, Gausden E, Uppstrom T, Potter H, Cordasco F, et al. Sequential MRI study of graft integrity and signal following pediatric all-epiphyseal ACL reconstruction: does the "sharp turn" at the socket of the distal femoral aperture matter? Orthop J Sports Med. 2017;5:2325967117S00431. https://doi.org/10.1177/2325967117S00431.
81. Graziano J, Chiaia T, de Mille P, Nawabi DH, Green DW, Cordasco FA. Return to sport for skeletally immature athletes after ACL reconstruction: preventing a second injury using a quality of movement assessment and quantitative measures to address modifiable risk factors. Orthop J Sports Med. 2017;5(4):2325967117700599.
82. Nawabi DH, Jones KJ, Lurie B, Potter HG, Green DW, Cordasco FA. Physeal-specific MRI analysis of growth plate disturbance following all-inside anterior cruciate ligament reconstruction in skeletally immature patients: does a physeal-sparing technique offer any advantage? Orthop J Sports Med. 2013;1:2325967113S00025. https://doi.org/10.1177/2325967113S00025.
83. Koch PP, Fucentese SF, Blatter SC. Complications after epiphyseal reconstruction of the anterior cruciate ligament in prepubescent children. Knee Surg Sports Traumatol Arthrosc. 2013;24(9):2736–40.
84. Lanzetti R, Pace V, Ciompi A, Perugia D, Spoliti M, Falez F, et al. Over the top anterior cruciate ligament

reconstruction in patients with open physes: a long-term follow-up study. Int Orthop. 2020;44:771–8.
85. Wall EJ, Ghattas PJ, Eismann EA, Myer GD, Carr P. Outcomes and complications after all-epiphyseal anterior cruciate ligament reconstruction in skeletally immature patients. Orthop J Sports Med. 2017;5(3):2325967117693604.
86. Falciglia F, et al. Anterior cruciate ligament reconstruction in adolescents (Tanner stages 2 and 3). Knee Surg Sports Traumatol Arthrosc. 2016;24(3):807–14.
87. Guzzanti V, Falciglia F, Stanitski CL. Preoperative evaluation and anterior cruciate ligament reconstruction technique for skeletally immature patients in Tanner stages 2 and 3. Am J Sports Med. 2003;31(6):941–8.
88. Aronowitz ER, Ganley TJ, Goode JR, Gregg JR, Meyer JS. Anterior cruciate ligament reconstruction in adolescents with open physes. Am J Sports Med. 2000;28:168–75.
89. Sankar WN, Carrigan RB, Gregg JR, Ganley TJ. Anterior cruciate ligament reconstruction in adolescents: a survivorship analysis. Am J Orthop (Belle Mead NJ). 2008;37:47–9.
90. Chambers CC, Monroe EJ, Allen CR, Pandya NK. Partial transphyseal anterior cruciate ligament reconstruction: clinical, functional, and radiographic outcomes. Am J Sports Med. 2019;47(6):1353–60. https://doi.org/10.1177/0363546519836423.
91. Kocher MS, Garg S, Micheli LJ. Physeal sparing reconstruction of the anterior cruciate ligament in skeletally immature prepubescent children and adolescents. Surg Tech J Bone Joint Surg Am Vol. 2006;88(Suppl 1 Pt 2):283–93. https://doi.org/10.2106/JBJS.F.00441.
92. Kocher MS, Heyworth BE, Fabricant PD, Tepolt FA, Micheli LJ. Outcomes of physeal-sparing ACL reconstruction with iliotibial band autograft in skeletally immature prepubescent children. J Bone Joint Surg Am Vol. 2018;100(13):1087–94. https://doi.org/10.2106/JBJS.17.01327.
93. Willimon SC, Jones CR, Herzog MM, May KH, Leake MJ, Busch MT. Micheli anterior cruciate ligament reconstruction in skeletally immature youths: a retrospective case series with a mean 3-year follow-up. Am J Sports Med. 2015;43(12):2974–81. https://doi.org/10.1177/0363546515608477.
94. MacIntosh DL, Darby TA. Lateral substitution reconstruction. In proceedings and reports of universities, colleges, councils and associations. J Bone Joint Surg. 1976;58B:142.
95. Feroe AG, Hassan MM, Kocher MS. Physeal-sparing anterior cruciate ligament reconstruction with iliotibial band autograft in the skeletally immature knee. Arthrosc Tech. 2022;11(9):e1597–603. https://doi.org/10.1016/j.eats.2022.05.005.
96. Patil V, Rajan P, Hayter E, Bartlett J, Symons S. Growth disturbances following paediatric anterior cruciate ligament reconstruction: a systematic review. Cureus. 2023;15(6):e40455. https://doi.org/10.7759/cureus.40455.
97. Weber AE, Zuke W, Mayer EN, et al. Lateral augmentation procedures in anterior cruciate ligament reconstruction: anatomic, biomechanical, imaging, and clinical evidence. Am J Sports Med. 2019;47:740–52.
98. Geeslin AG, Moatshe G, Chahla J, et al. Anterolateral knee extra-articular stabilizers: A robotic study comparing anterolateral ligament reconstruction and modified Lemaire lateral extra-articular tenodesis. Am J Sports Med. 2018;46:607–16.
99. Jesani S, Getgood A. Modified Lemaire lateral extraarticular tenodesis augmentation of anterior cruciate ligament reconstruction. JBJS Essent Surg Tech. 2019;9:e41.
100. Sonnery-Cottet B, Vieira TD, Ouanezar H. Anterolateral ligament of the knee: diagnosis, indications, technique, outcomes. Arthroscopy. 2019;35:302–3.
101. Sonnery-Cottet B, Daggett M, Fayard JM, et al. Anterolateral Ligament Expert Group consensus paper on the management of internal rotation and instability of the anterior cruciate ligament-deficient knee. J Orthop Traumatol. 2017;18:91–106.
102. Slette EL, Mikula JD, Schon JM, et al. Biomechanical results of lateral extra-articular tenodesis procedures of the knee: a systematic review. Arthroscopy. 2016;32:25922611.
103. Caterine S, Litchfield R, Johnson M, Chronik B, Getgood A. A cadaveric study of the anterolateral ligament: re-introducing the lateral capsular ligament. Knee Surg Sports Traumatol Arthrosc. 2015;23:3186–95.
104. Kennedy MI, Claes S, Fuso FAF, et al. The anterolateral ligament: an anatomic, radiographic, and biomechanical analysis. Am J Sports Med. 2015;43:1606–15.
105. Perelli S, Costa GG, Terron VM, Formagnana M, Bait C, Espregueira-Mendes J, Monllau JC. Combined anterior cruciate ligament reconstruction and modified Lemaire lateral extra-articular tenodesis better restores knee stability and reduces failure rates than isolated anterior cruciate ligament reconstruction in skeletally immature patients. Am J Sports Med. 2022;50(14):3778–85. https://doi.org/10.1177/03635465221128926.
106. Carrozzo A, Monaco E, Saithna A, Annibaldi A, Guy S, Ferreira A, Vieira TD, Ferretti A, Sonnery-Cottet B. Clinical outcomes of combined anterior cruciate ligament reconstruction and lateral extra-articular tenodesis procedures in skeletally immature patients: a systematic review from the SANTI Study Group. J Pediatr Orthop. 2023;43(1):24–30. https://doi.org/10.1097/BPO.0000000000002236.
107. Feagin JA, Abbott HG, Rokous JR. The isolated tear of the anterior cruciate ligament. J Bone Joint Surg Am. 1972:1340–1.

108. Weaver JK, Derkash RS, Freeman JR, Kirk RE, Oden RR, Matyas J. Primary knee ligament repair revisited. Clin Orthop Relat Res. 1985;199:185–91.
109. Odensten M, Lysholm J, Gillquist J. Suture of fresh ruptures of the anterior cruciate ligament. A 5-year follow-up. Acta Orthop Scand. 1984;55:270–2.
110. Warren RF. Primary repair of the anterior cruciate ligament. Clin Orthop Relat Res. 1983;172:65–70.
111. England RL. Repair of the ligaments about the knee. Orthop Clin North Am. 1976;7:195–204.
112. Marshall JL, Warren RF, Wickiewicz TL, Reider B. The anterior cruciate ligament: a technique of repair and reconstruction. Clin Orthop Relat Res. 1979:97–106.
113. Robson AW. VI. Ruptured crucial ligaments and their repair by operation. Ann Surg. 1903;37:716–8.
114. Palmer I. On the injuries to the ligaments of the knee joint. Acta Chir Scand. 1938:53.
115. Palmer I. On the injuries to the ligaments of the knee joint: a clinical study. 1938. Clin Orthop Relat Res. 2007;454:17–22. discussion 14.
116. O'Donoghue DH. An analysis of end results of surgical treatment of major injuries to the ligaments of the knee. J Bone Joint Surg Am. 1955;37-A:1–13. passim.
117. O'Donoghue DH. Surgical treatment of fresh injuries to the major ligaments of the knee. J Bone Joint Surg Am. 1950;32A:721–38.
118. van der List JP, DiFelice GS. Primary repair of the anterior cruciate ligament: a paradigm shift. Surgeon. 2017;15(3):161–8. https://doi.org/10.1016/j.surge.2016.09.006.
119. Feagin JA Jr, Curl WW. Isolated tear of the anterior cruciate ligament: 5-year follow-up study. Am J Sports Med. 1976;4:95–100.
120. Sandberg R, Balkfors B, Nilsson B, Westlin N. Operative versus non-operative treatment of recent injuries to the ligaments of the knee. A prospective randomized study. J Bone Joint Surg Am. 1987;69:1120–6.
121. Straub T, Hunter RE. Acute anterior cruciate ligament repair. Clin Orthop Relat Res. 1988;227:238–50.
122. Odensten M, Hamberg P, Nordin M, Lysholm J, Gillquist J. Surgical or conservative treatment of the acutely torn anterior cruciate ligament. A randomized study with short-term follow-up observations. Clin Orthop Relat Res. 1985;198:87–93.
123. Grontvedt T, Engebretsen L. Comparison between two techniques for surgical repair of the acutely torn anterior cruciate ligament. A prospective, randomized follow-up study of 48 patients. Scand J Med Sci Sports. 1995;5:358–63.
124. Grontvedt T, Engebretsen L, Benum P, Fasting O, Molster A, Strand T. A prospective, randomized study of three operations for acute rupture of the anterior cruciate ligament. Five-year follow-up of one hundred and thirty-one patients. J Bone Joint Surg Am Vol. 1996;78:159–68.
125. Kaplan N, Wickiewicz TL, Warren RF. Primary surgical treatment of anterior cruciate ligament ruptures. A long-term follow-up study. Am J Sports Med. 1990;18:354–8.
126. Engebretsen L, Benum P, Fasting O, Molster A, Strand T. A prospective, randomized study of three surgical techniques for treatment of acute ruptures of the anterior cruciate ligament. Am J Sports Med. 1990;18:585–90.
127. Sherman MF, Lieber L, Bonamo JR, Podesta L, Reiter I. The long-term followup of primary anterior cruciate ligament repair. Defining a rationale for augmentation. Am J Sports Med. 1991;19:243–55.
128. Andersson C, Odensten M, Good L, Gillquist J. Surgical or nonsurgical treatment of acute rupture of the anterior cruciate ligament. A randomized study with long-term follow-up. J Bone Joint Surg Am Vol. 1989;71:965–74.
129. Andersson C, Odensten M, Gillquist J. Knee function after surgical or nonsurgical treatment of acute rupture of the anterior cruciate ligament: a randomized study with a long-term follow-up period. Clin Orthop Relat Res. 1991;264:255–63.
130. Griffith TB, Allen BJ, Levy BA, Stuart MJ, Dahm DL. Outcomes of repeat revision anterior cruciate ligament reconstruction. Am J Sports Med. 2013;41:1296i301.
131. Dhillon MS, Bali K, Prabhakar S. Differences among mechanoreceptors in healthy and injured anterior cruciate ligaments and their clinical importance. Muscles Ligaments Tendons J. 2012;2:38–43.
132. Aga C, Wilson KJ, Johansen S, Dornan G, La Prade RF, Engebretsen L. Tunnel widening in single- versus double bundle anterior cruciate ligament reconstructed knees. Knee Surg Sports Traumatol Arthrosc. 2016;25:1316. https://doi.org/10.1007/s00167-016-4204-0.
133. Barrett DS. Proprioception and function after anterior cruciate reconstruction. J Bone Joint Surg Br. 1991;73:833–7.
134. Busam ML, Provencher MT, Bach BR Jr. Complications of anterior cruciate ligament reconstruction with bone-patellar tendon-bone constructs: care and prevention. Am J Sports Med. 2008;36:379–94.
135. Toy BJ, Yeasting RA, Morse DE, McCann P. Arterial supply to the human anterior cruciate ligament. J Athl Train. 1995;30(2):149–52.
136. Nguyen DT, Ramwadhdoebe TH, van der Hart CP, Blankevoort L, Tak PP, van Dijk CN. Intrinsic healing response of the human anterior cruciate ligament: an histological study of reattached ACL remnants. J Orthop Res. 2014;32(2):296–301. https://doi.org/10.1002/jor.22511.

137. DiFelice GS, Villegas C, Taylor SA. Anterior cruciate ligament preservation: early results of a novel arthroscopic technique for suture anchor primary anterior cruciate ligament repair. Arthroscopy. 2015;31:2162–71.
138. DiFelice GS, van der List JP. Arthroscopic primary repair of proximal anterior cruciate ligament tears. Arthrosc Tech. 2016;5:e1057. https://doi.org/10.1016/j.eats.2016.05.009.
139. DiFelice GS, van der List JP. Clinical outcomes of arthroscopic primary repair of proximal anterior cruciate ligament tears are maintained at mid-term follow-up. Arthrosc J Arthrosc Relat Surg. 2018;34(4):1085–93. https://doi.org/10.1016/j.arthro.2017.10.028.
140. Bigoni M, Gaddi D, Gorla M, Munegato D, Pungitore M, Piatti M, Turati M. Arthroscopic anterior cruciate ligament repair for proximal anterior cruciate ligament tears in skeletally immature patients: Surgical technique and preliminary results. Knee. 2017;24(1):40–8. https://doi.org/10.1016/j.knee.2016.09.017.
141. Gagliardi AG, Carry PM, Parikh HB, Traver JL, Howell DR, Albright JC. ACL repair with suture ligament augmentation is associated with a high failure rate among adolescent patients. Am J Sports Med. 2019;47(3):560–6. https://doi.org/10.1177/0363546518825255.
142. Dabis J, Yasen SK, Foster AJ, Pace JL, Wilson AJ. Paediatric proximal ACL tears managed with direct ACL repair is safe, effective and has excellent short-term outcomes. Knee Surg Sports Traumatol Arthrosc. 2020;28(8):2551–6. https://doi.org/10.1007/s00167-020-05872-2.
143. Murray MM, Kalish LA, Fleming BC, BEAR Trial Team, Flutie B, Freiberger C, Henderson RN, Perrone GS, Thurber LG, Proffen BL, Ecklund K, Kramer DE, Yen YM, Micheli LJ. Bridge-enhanced anterior cruciate ligament repair: two-year results of a first-in-human study. Orthop J Sports Med. 2019;7(3):2325967118824356. https://doi.org/10.1177/2325967118824356.
144. Eggli S, Kohlhof H, Zumstein M, Henle P, Hartel M, Evangelopoulos DS, et al. Dynamic intraligamentary stabilization: novel technique for preserving the ruptured ACL. Knee Surg Sports Traumatol Arthrosc. 2015;23:1215–21.
145. Eggli S, Roder C, Perler G, Henle P. Five year results of the first ten ACL patients treated with dynamic intraligamentary stabilisation. BMC Musculoskelet Disord. 2016;17:105.
146. Smith JO, Yasen SK, Palmer HC, Lord BR, Britton EM, Wilson AJ. Paediatric ACL repair reinforced with temporary internal bracing. Knee Surg Sports Traumatol Arthrosc. 2016;24:1845–51.
147. Henle P, Roder C, Perler G, Heitkemper S, Eggli S. Dynamic Intraligamentary Stabilization (DIS) for treatment of acute anterior cruciate ligament ruptures: case series experience of the first three years. BMC Musculoskelet Disord. 2015;16:27.
148. Knapik DM, Voos JE. Anterior cruciate ligament injuries in skeletally immature patients: a meta-analysis comparing repair versus reconstruction techniques. J Pediatr Orthop. 2020;40(9):492–502. https://doi.org/10.1097/BPO.0000000000001569.
149. Kercher J, Xerogeanes J, Tannenbaum A, Al-Hakim R, Black JC, Zhao J. Anterior cruciate ligament reconstruction in the skeletally immature: an anatomical study utilizing 3-dimensional magnetic resonance imaging reconstructions. J Pediatr Orthop. 2009;29(2):124–9. https://doi.org/10.1097/BPO.0b013e3181982228.

12 Natural History and Prevention of ACL Injuries in Children

Ariana Lott, Zachary Herman, and Bryson P. Lesniak

12.1 Incidence of Pediatric ACL Injuries and Current Management Trends

Increased participation in youth sports has led to an associated increase in pediatric anterior cruciate ligament (ACL) injuries over the past several decades with injury rates rising significantly higher compared to adult populations [1]. This increase in injury correlates with greater participation in year-round training among pediatric athletes and single-sport specialization at an earlier age [2]. A recent review of ACL tears in children and adolescents over the past 20 years cited an increase in ACL injuries by 2.3% per year with the highest rate of pediatric injuries occurring during high school [3]. Female adolescents have the greatest risk with about a 10% risk of injury over a high school career when participating in multiple sports [4]. These injuries impact pediatric and adolescent athletes physically with respect to knee function, quality of life, and the psychological burden associated with injury [5]. In addition to primary ACL injury, many young athletes suffer contralateral ACL injuries or ipsilateral ACL reinjury with some studies reporting rates of up to 30% within the first 2 years after returning to sport [6–8] (Fig. 12.1). With the unique considerations and complications that can result from ACL reconstruction in the pediatric population related to the physis, it is essential to understand the natural history of these injuries to counsel patients and their families regarding return to sport expectations and concerns regarding long-term preservation of the knee joint.

The treatment algorithm of ACL injuries in pediatric patients continues to be debated. This includes indications for surgery and the timing of surgery, particularly skeletally immature patients [9]. In the United States, the incidence of ACL reconstruction has increased significantly in the past several decades especially among patients aged 10–14 years [1]. However, a recent survey of pediatric orthopedic surgeons in the Netherlands demonstrated that 78% of surgeons "tend to treat" children with open physes nonoperatively [9]. This geographic discrepancy was confirmed in a recent survey distributed to European Paediatric Orthopaedic Society (EPOS) and Pediatric Orthopaedic Society of North America (POSNA) members on management of pediatric ACL tears [10]. In this survey, 17% of EPOS members and 70% of POSNA members recommended ACL reconstruction within 3 months of a prepubescent pediatric ACL injury. However, when asked about pubescent patients (aged 13 years), 83% of EPOS members recommended ACL reconstruction [10].

A. Lott · Z. Herman · B. P. Lesniak (✉)
UPMC Freddie Fu Sports Medicine Center, Department of Orthopaedic Surgery, University of Pittsburgh Medical Center, Pittsburgh, PA, USA
e-mail: lotta@upmc.edu; hermanz@upmc.edu; lesniakbp@upmc.edu

M. Karahan et al. (eds.), *Anterior Cruciate Ligament Injuries Throughout the Life Span*,
https://doi.org/10.1007/978-3-031-84501-7_12

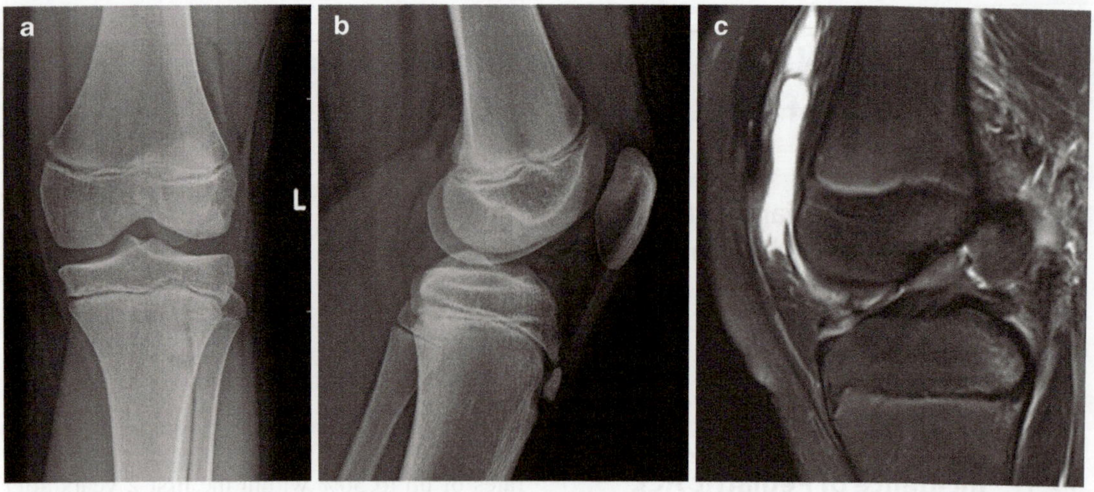

Fig. 12.1 Anteroposterior (AP) (**a**) and lateral (**b**) radiographs and MRI T2 sagittal image (**c**) of an 11-year-old male who sustained a left knee complete ACL tear while playing football. Abbreviations: *MRI* Magnetic resonance imaging, *ACL* Anterior cruciate ligament

12.2 Natural History of ACL Injuries

The natural history of ACL injuries in pediatric patients can be examined through evaluation of the outcome literature of nonoperative treatment of ACL injuries; however, physicians treating these patients must be mindful that nonoperative treatment and the associated injuries in these patients with ACL injuries vary [11]. As highlighted above, while ACL injuries themselves have been suggested to increase the likelihood of osteoarthritis development, it is injury to the associated meniscus and chondral surface that leads to further damage in the knee and loss of knee function [12].

While many children may sustain concomitant injuries at the time of initial injury, there are numerous studies that highlight the secondary injuries that can result in ACL deficient knees. A recent analysis by Pruneski et al. reported that 54% of pediatric and 71% of adolescent patients with ACL injuries had at least one concomitant soft tissue injury at the time of initial injury with increasing age predictive of sustaining a concomitant injury [13]. When examining delay of surgery, Lawrence et al. demonstrated that children 14 years of age or younger who underwent ACL reconstruction more than 12 weeks from injury had a 4.1 times increased risk of having a medial meniscus tear at the time of ACL reconstruction. Increased time to surgery (>12 weeks from injury) was also associated with a 5.6 times increased risk of medial compartment chondral injury and 11.3 times increased risk of lateral compartment chondral injury [14]. In patients with a sense of knee instability at the time of ACL reconstruction (defined as having one or more episodes of "giving way" or "shifting" with activity), the association was even greater between time to surgery and meniscal and chondral injuries [14]. The increase in meniscal pathology observed with delayed ACL reconstruction was also seen by Dumont et al. In an analysis of 370 pediatric patients who underwent ACL reconstruction, there was a 1.8 times increased rate of medial meniscal tears found in those patients who were treated with ACL reconstruction more than 150 days after initial ACL injury [15].

Literature also demonstrates that many patients treated nonoperatively will experience continued instability in their knee. When comparing operative and nonoperative treatment of ACL injuries, a meta-analysis by Ramski et al. concluded that 75% of patients reported instability/pathologic laxity with nonoperative treatment and that patients were 12 times more likely to have medial meniscal tears after nonoperative treatment [16]. Lastly, Vavken et al. in a systematic review of skeletally immature patients with ACL injuries, including 476 patients treated nonoperatively with an average follow-up of 52.7 months, demonstrated that, on average, 50.2% of patients had subjective instability with early severe meniscal degeneration and chondral defects that required surgical stabilization [17].

While the risk of secondary injury is present, there are studies that suggest patients can return to athletic activities with nonoperative treatment of their ACL injury particularly with the use of supplemental ACL braces and physical therapy. For example, Moksnes et al. reported that of 40 skeletally immature children who were treated nonoperatively for ACL injuries sustained at the age of 12 years or younger, 88% of them were able to participate in pivoting sports and/or physical education classes in school with use a custom-made and individually adjusted knee brace [18]. In those patients that required ACL reconstruction in the setting of continued instability symptoms or symptomatic meniscus injury, 46.2% required meniscal procedures with only 19.5% of patients having new meniscal injuries not seen on the first MRI [18].

With respect to return to sport, a subsequent study on a similar cohort of 46 children (12 years of age or younger) with nonoperatively treated ACL injuries reported that 91% of the patients returned to participation in pivoting sports and/or physical education in school. However, 38% of children changed their main activity to a level 2 activity from a level 1 activity [19]. Lastly, in a prospective case series of 44 patients with a mean 8-year follow-up who sustained ACL injury at the age of 13 years or younger and participated in an active rehabilitation program with the option of having delayed ACL reconstruction, only 55% of patients underwent ACL reconstruction. However, in those patients who were treated nonoperatively, 65% of them did not return to pivoting sports [20].

One example of a nonoperative rehabilitation program is described by Moksnes et al. [21] This structured and supervised rehabilitation program is divided into 4 phases that consist of range of motion, neuromuscular control, and muscle strengthening exercises [21]. Phase 1 focuses on active and passive knee range of motion using a stationary bike, dynamic extension, and prone extension hang exercises. Neuromuscular training during this phase focuses on quadriceps reactivation and single-leg stability with muscle strengthening forced on core and gluteus medius exercises. Phase 2 begins once unrestricted normal gait patterns have been re-established and transitions to single leg squats including on the BOSU ball with closed chain quadriceps, hamstring, and hip exercises. This phase is completed once patients return to activities of daily living including stairs with no instability nor swelling. Phase 3 incorporates open chain quadriceps strengthening and plyometric exercises including landing mechanics training to avoid dynamic valgus. Return to sport is allowed in a functional knee brace when patients can perform single-leg hop tests with quadriceps strength at least 90% of the contralateral side. Phase 4 progresses to secondary prevention and maintenance of neuromuscular training and functional stability [21].

In summary, while nonoperative treatment protocols can be considered for those patients who can comply with significant initial activity restriction and undergo a structured return to activity program, many may experience feelings of recurrent instability and inability to participate in pivoting activities. Furthermore, active patients have been shown to develop secondary meniscal and chondral injury with nonoperative treatment.

12.3 ACL Prevention Programs

12.3.1 ACL Risk Factors

Given the high rate of ACL injuries in young athletes and the consequences of these injuries on athletes both physically and mentally, there has been increased attention to ACL prevention programs. There are numerous risk factors for ACL injury including those that are modifiable and nonmodifiable. Nonmodifiable factors include age, sex, prior injury, ligamentous laxity, increased tibial slope, and decreased notch width [22, 23]. There are also multiple genetic markers that predispose pediatric patients to sustaining ACL injuries with a recent study demonstrating that children with an ACL injury had nearly three times higher odds of having a first degree relative with an ACL tear [24]. Modifiable factors for ACL injury include jump landing biomechanics and neuromuscular control/maturation. ACL prevention programs have focused on these modifiable factors to prevent primary and contralateral ACL injury [25].

An increased knee valgus moment during jump landing has been consistently demonstrated to be a risk factor for ACL injury particularly in adolescent female athletes [26, 27] (Fig. 12.2). In an analysis of adolescent female athletes performing side cutting maneuvers, Zebis et al. found that the hip flexion angle at initial contact, the internal knee rotation angle at initial contact, semitendinosus electromyography activity, and hip external rotator strength were significantly associated with future first time ACL injury [28]. An additional study of young female basketball and floorball players found that lower peak knee flexion angle and higher peak vertical ground-reaction forces were associated with increased risk of ACL injury over the subsequent 3 years [27].

The increased knee valgus motion during landing mechanics develops with the onset of puberty in female athletes [29, 30]. Interestingly, while preadolescent athletes have been shown to perform more poorly with respect to jump landing and balance skills when compared to collegiate-aged athletes, studies demonstrate that there are no sex differences in landing and balance at this earlier age [31]. This correlates with the significantly increased risk of ACL injury among female athletes after puberty [1].

While there has been significant study on landing mechanics, neuromuscular control and core strength have been increasingly studied with respect to ACL injury. Studies have shown that increased trunk displacement in response to sudden trunk force release is associated with an increase in knee injury risk [32]. With lateral trunk motion, the ground reaction forces move laterally, which can increase the abduction moment on the knee. Notably, when analyzing injury video from athletes who have sustained ACL injuries, female athletes landed with greater lateral trunk motion and knee abduction angles at the time of injury compared to male athletes [33].

Fig. 12.2 Proper knee landing mechanics (**a**) compared to excessive knee valgus (**b**) during jump landing

12.3.2 ACL Injury Prevention Programs

Numerous ACL injury prevention programs have been developed over the past several decades, and there are several common principles and ACL risk factors that the programs strive to modify. These include core stability, balance training, eccentric hamstring strengthening, and landing mechanics to limit the "dynamic knee valgus" moment during jump landing. In addition to these movement exercises, an emphasis on coach and athlete education on cutting techniques to avoid these high-risk positions is also essential. Many of these programs have been designed to use little to no additional equipment to assist in the effectiveness of these programs [25].

While there is variability in the treatment of ACL injuries in the pediatric population, ACL injury prevention programs are recommended in the most recent consensus statement on Pediatric ACL Injuries by the 2018 International Olympic Committee [34]. This recommendation is due to the breadth of evidence that highlights the success of these ACL prevention programs in pediatric populations specifically [25, 35–37]. In a recent meta-analysis of ACL injury prevention programs in young female athletes, the use of neuromuscular training programs reduced the risk of ACL injuries by nearly 50% [25]. In one of the largest studies, Walden et al. reported on the effectiveness of a neuromuscular training program in adolescent female soccer players (12–17 years old). Two-hundred and thirty Swedish soccer clubs were followed for one season and teams were cluster randomized into performing a 15 min neuromuscular warm up program twice a week throughout the season. In those teams that participated in the program, there was a 64% reduction in the rate of ACL injury [35]. Similarly, the use of the FIFA (Fédération Internationale de Football Association) "11+ Kids" program in pediatric soccer players has been demonstrated to reduce injury rates by 48% [36].

Improvements in motor performance and landing mechanics are also seen with participation in ACL injury prevention programs in pediatric athletes [38–40]. Kerman et al. reported that participation in the FIFA "11+ Kids" program in preadolescent female soccer players (ages 10–12 years) for 8 weeks improved knee flexion and hip internal rotation during single-leg drop landing [38]. Similar improvements have also been seen using the same program in children aged 7–12 years with beneficial effects observed in Y-balance and agility run [39] (Table 12.1). Lastly, isolated core strength training for 10 weeks has been shown to improve trunk endurance and to decrease in decrease knee valgus and hip adduction angles through improved activation of the vastus medialis and hamstring musculature during sports-related tasks [41].

In addition to being effective at preventing injuries and improving mechanics, ACL prevention programs are also cost-effective. Marshall et al. demonstrated that when implementing a neuromuscular training program to reduce injury in youth soccer, there was a reduction in health care costs by 43% [42]. Compliance with the neuromuscular training programs has been repeatedly demonstrated to correlate with injury risk reduction. Hägglund et al. performed a prospective cohort study based on a cluster randomized controlled trial on adolescent soccer players and reported that those players in the high compliance tertile had an ACL injury reduction of 88%. By contrast, the athletes in the low compliance tertile had similar ACL injury rates to that of the control group [43].

When analyzing specific program characteristics, a recent meta-analysis by Petushek et al. found that those programs that included more landing stabilization exercises (including drop landings and jump/hop and holds) and lower body strengthening exercises were most effective [25]. Lower body strength exercises included hamstring strengthening exercises, lunges, and heel-calf raises. However, those programs that included balance, core-strengthening, and agility work were no more effective than those programs that did not include these exercises. When looking at training timing and duration specifically, programs that were continued throughout the

Table 12.1 FIFA 11+ kids injury prevention program exercises and levels [35]

Exercises	Levels
Exercise 1: "Alertness" running game	Level 1: Listen for the stop command Level 2: Watch for the stop command Level 3: Ball in the hands and listen for the stop command Level 4: Ball in the hands and watch for the stop command Level 5: Dribble with the ball and listen for the stop command
Exercise 2: Skating jumps	Level 1: Learning how to land Level 2: Ball in both hands Level 3: Balancing with the ball in one hand Level 4: Dropping the ball onto the ground Level 5: Dynamic balancing with ball
Exercise 3: Single-leg stance	Level 1: Throwing the ball Level 2: Throwing the ball and moving it around the free leg Level 3: Passing game Level 4: Throwing the ball and passing back without touching the ground Level 5: Testing the partner's balance
Exercise 4: Press-ups	Level 1: Tunnel Level 2: Forearm support; shin resting on ball Level 3: Rolling the ball around the hands Level 4: Rolling the ball between hands and feet Level 5: Hands on the ball
Exercise 5: Single-leg jumps	Level 1: Forwards Level 2: Back and forth Level 3: Sideways Level 4: The coach indicates the direction Level 5: The coach indicates the direction; ball in both hands
Exercise 6: Spiderman	Level 1: Stroking the ball Level 2: A proper stretch Level 3: The crab Level 4: Dribbling Level 5: Ball bearing
Exercise 7: Sideways roll	Level 1: Squatting position Level 2: Slowing from a standing position Level 3: Dynamically from a standing position Level 4: From a slow walk Level 5: From a faster forward movement

Abbreviations: *FIFA* Fédération Internationale de Football Association

sport season as opposed to just in the pre-season were also found to be more effective at reducing the risk of ACL injury [25].

Lastly, the timing of when to begin an injury prevention program has been evaluated especially given that knee mechanics change with maturation particularly in female athletes. In an analysis of an ACL injury prevention program in female basketball players, Otsuki et al. found that starting a program in early puberty decreased the development of high knee abduction moments and high-risk knee mechanics. These findings support the idea that these prevention programs should be initiated in early puberty and continued through post-puberty [44].

While these programs have been present since the 1990s, there has not been widespread adoption of these programs in comparison to the number of patients who sustain or are exposed to ACL injuries [45]. In a recent study of parental awareness of ACL injury prevention programs, Sparagana et al. found that only 18% of parents had knowledge of ACL injury prevention programs and only 6% reported that their child had ever participated in a program. However, 12% of athletes in the study had experienced an ACL injury [46]. Continuing to encourage pediatric athletes to participate in these programs and to support these programs in our communities is essential as injury prevention efforts continue.

12.4 Conclusion

In conclusion, as children continue to remain active and participate in sports at younger ages, the prevalence of ACL injuries in the pediatric population can be expected to increase. With this high incidence of injury, the natural history of ACL injury should be examined to assess the risks of nonoperative treatment and provide expectations to patients and their families who wish to pursue nonoperative treatment as described in this chapter. Continued focus should remain on establishing and supporting injury prevention programs particularly in adolescent female athletes to not only prevent initial ACL injury, but also contralateral and recurrent ipsilateral injury.

References

1. Werner BC, Yang SD, Looney AM, et al. Trends in pediatric and adolescent anterior cruciate ligament injury and reconstruction. J Pediatr Orthop. 2016 Jul-Aug;36(5):447–52.
2. Johnsen MB, Guddal MH, Småstuen MC, et al. Sport participation and the risk of anterior cruciate ligament reconstruction in adolescents: a population-based prospective cohort study (The young-HUNT study). Am J Sports Med. 2016;44:2917–24.
3. Beck NA, Lawrence JTR, Nordin JD, et al. ACL tears in school-aged children and adolescents over 20 years. Pediatrics. 2017 Mar;139(3):e20161877.
4. Bram JT, Magee LC, Mehta NN, Patel NM, Ganley TJ. Anterior cruciate ligament injury incidence in adolescent athletes: a systematic review and meta-analysis. Am J Sports Med. 2021;49:1962–72.
5. Barenius B, Ponzer S, Shalabi A, et al. Increased risk of osteoarthritis after anterior cruciate ligament reconstruction: a 14-year follow-up study of a randomized controlled trial. Am J Sports Med. 2014;42(5):1049–57.
6. Allen MM, Pareek A, Krych AJ, et al. Are female soccer players at an increased risk of second anterior cruciate ligament injury compared with their athletic peers? Am J Sports Med. 2016;44(10):2492–8.
7. Paterno MV, Rauh MJ, Schmitt LC, et al. Incidence of second ACL injuries 2 years after primary ACL reconstruction and return to sport. Am J Sports Med. 2014 Jul;42(7):1567–73.
8. Dekker TJ, Godin JA, Dale KM, et al. Return to sport after pediatric anterior cruciate ligament reconstruction and its effect on subsequent anterior cruciate ligament injury. J Bone Joint Surg. 2017;99(11):897–904.
9. Dietvorst M, Reijman M, van Zutven R, van den Bekerom MP, Meuffels DE, Somford MP, Janssen RP. Current state of care for pediatric ACL ruptures in The Netherlands: a survey. J Knee Surg. 2021 Apr;34(05):520–5.
10. Accadbled F, Gracia G, Laumonerie P, Thevenin-Lemoine C, Heyworth BE, Kocher MS. Paediatric anterior cruciate ligament tears: management and growth disturbances. A survey of EPOS and POSNA membership. J Child Orthop. 2019 Oct 1;13(5):522–8.
11. Dingel A, Aoyama J, Ganley T, et al. Pediatric ACL tears: natural history. J Pediatr Orthop. 2019 Jul;39(Issue 6, Supplement 1 Suppl 1):S47–9.
12. Whittaker J, Woodhouse L, Nettel-Aguirre A, Emery C. Outcomes associated with early post-traumatic osteoarthritis and other negative health consequences 3–10 years following knee joint injury in youth sport. Osteoarthr Cartil. 2015;23(7):1122–9.
13. Pruneski JA, Heyworth BE, Kocher MS, Tavabi N, Milewski MD, Kramer DE, Christino MA, Yen YM, Micheli LJ, Murray MM, Gilreath L, Kim A, Murray J, Kiapour AM. Prevalence and predictors of concomitant meniscal and ligamentous injuries associated with ACL surgery: an analysis of 20 years of ACL reconstruction at a tertiary care children's hospital. Am J Sports Med. 2024 Jan;52(1):77–86.
14. Lawrence JT, Argawal N, Ganley TJ. Degeneration of the knee joint in skeletally immature patients with a diagnosis of an anterior cruciate ligament tear: is there harm in delay of treatment? Am J Sports Med. 2011 Dec;39(12):2582–7.
15. Dumont GD, Hogue GD, Padalecki JR, Okoro N, Wilson PL. Meniscal and chondral injuries associated with pediatric anterior cruciate ligament tears: relationship of treatment time and patient-specific factors. Am J Sports Med. 2012 Sep;40(9):2128–33.
16. Ramski DE, Kanj WW, Franklin CC, Baldwin KD, Ganley TJ. Anterior cruciate ligament tears in children and adolescents: a meta-analysis of nonoperative versus operative treatment. Am J Sports Med. 2014;42(11):2769–76.
17. Vavken P, Murray MM. Treating anterior cruciate ligament tears in skeletally immature patients. Arthroscopy. 2011;27:704–16.
18. Moksnes H, Engebretsen L, Risberg MA. Prevalence and incidence of new meniscus and cartilage injuries after a nonoperative treatment algorithm for ACL tears in skeletally immature children: a prospective MRI study. Am J Sports Med. 2013;41(8):1771–9.

19. Moksnes H, Engebretsen L, Eitzen I, Risberg MA. Functional outcomes following a non-operative treatment algorithm for anterior cruciate ligament injuries in skeletally immature children 12 years and younger. A prospective cohort with 2 years follow-up. Br J Sports Med. 2013 May 1;47(8):488–94.
20. Ekås GR, Moksnes H, Grindem H, Risberg MA, Engebretsen L. Coping with anterior cruciate ligament injury from childhood to maturation: a prospective case series of 44 patients with mean 8 Years' follow-up. Am J Sports Med. 2019 Jan;47(1):22–30.
21. Moksnes H, Engebretsen L, Risberg MA. Management of anterior cruciate ligament injuries in skeletally immature individuals. J Orthop Sports Phys Ther. 2012 Mar;42(3):172–83.
22. Freychet B, Lakhal W, Daggett M, Bonnard C. Intercondylar notch dysplasia in open-physis anterior cruciate ligament injuries: a case-control study. Orthop Traumatol Surg Res. 2016 Apr;102(2):203–6.
23. Edwards TC, Naqvi AZ, Dela Cruz N, Gupte CM. Predictors of pediatric anterior cruciate ligament injury: the influence of steep lateral posterior Tibial slope and its relationship to the lateral meniscus. Arthroscopy. 2021 May;37(5):1599–609.
24. Bram JT, Pascual-Leone N, Patel NM, DeFrancesco CJ, Talathi NS, Ganley TJ. Do pediatric patients with anterior cruciate ligament tears have a higher rate of familial anterior cruciate ligament injury? Orthop J Sports Med. 2020 Oct 30;8(10):2325967120959665.
25. Petushek EJ, Sugimoto D, Stoolmiller M, Smith G, Myer GD. Evidence-based best-practice guidelines for preventing anterior cruciate ligament injuries in young female athletes: a systematic review and meta-analysis. Am J Sports Med. 2019 Jun;47(7):1744–53.
26. Hewett TE, Myer GD, Ford KR, Heidt RS Jr, Colosimo AJ, McLean SG, van den Bogert AJ, Paterno MV, Succop P. Biomechanical measures of neuromuscular control and valgus loading of the knee predict anterior cruciate ligament injury risk in female athletes: a prospective study. Am J Sports Med. 2005 Apr;33(4):492–501.
27. Leppänen M, Pasanen K, Kujala UM, Vasankari T, Kannus P, Äyrämö S, Krosshaug T, Bahr R, Avela J, Perttunen J, Parkkari J. Stiff landings are associated with increased ACL injury risk in young female basketball and floorball players. Am J Sports Med. 2017 Feb;45(2):386–93.
28. Zebis MK, Aagaard P, Andersen LL, Hölmich P, Clausen MB, Brandt M, Husted RS, Lauridsen HB, Curtis DJ, Bencke J. First-time anterior cruciate ligament injury in adolescent female elite athletes: a prospective cohort study to identify modifiable risk factors. Knee Surg Sports Traumatol Arthrosc. 2022 Apr;30(4):1341–51.
29. Ford KR, Shapiro R, Myer GD, Van Den Bogert AJ, Hewett TE. Longitudinal sex differences during landing in knee abduction in young athletes. Med Sci Sports Exerc. 2010 Oct;42(10):1923–31.
30. Hewett TE, Myer GD, Kiefer AW, Ford KR. Longitudinal increases in knee abduction moments in females during adolescent growth. Med Sci Sports Exerc. 2015 Dec;47(12):2579–85.
31. Hutchinson AB, Yao P, Hutchinson MR. Single-leg balance and core motor control in children: when does the risk for ACL injury occurs? BMJ Open Sport Exerc Med. 2016;2(1):e000135.
32. Zazulak BT, Hewett TE, Reeves NP, Goldberg B, Cholewicki J. Deficits in neuromuscular control of the trunk predict knee injury risk: a prospective biomechanical-epidemiologic study. Am J Sports Med 2007 Jul;35(7):1123–1130. https://doi.org/10.1177/0363546507301585. Epub 2007 Apr 27.
33. Hewett TE, Torg JS, Boden BP. Video analysis of trunk and knee motion during non-contact anterior cruciate ligament injury in female athletes: lateral trunk and knee abduction motion are combined components of the injury mechanism. Br J Sports Med. 2009 Jun 1;43(6):417–22.
34. Ardern CL, Ekås G, Grindem H, Moksnes H, Anderson AF, Chotel F, et al. 2018 International Olympic Committee consensus statement on prevention, diagnosis, and management of pediatric anterior cruciate ligament injuries. Orthop J Sports Med. 2018;6(3):2325967118759953.
35. Waldén M, Atroshi I, Magnusson H, Wagner P, Hägglund M. Prevention of acute knee injuries in adolescent female football players: cluster randomised controlled trial. BMJ. 2012 May;3:344.
36. Rössler R, Junge A, Bizzini M, Verhagen E, Chomiak J, Aus der Fünten K, Meyer T, Dvorak J, Lichtenstein E, Beaudouin F, Faude O. A multinational cluster randomised controlled trial to assess the efficacy of '11+ kids': a warm-up programme to prevent injuries in children's football. Sports Med. 2018 Jun;48:1493–504.
37. Soligard T, Myklebust G, Steffen K, Holme I, Silvers H, Bizzini M, Junge A, Dvorak J, Bahr R, Andersen TE. Comprehensive warm-up programme to prevent injuries in young female footballers: cluster randomised controlled trial. BMJ. 2008 Dec 10;337
38. Taghizadeh Kerman M, Brunetti C, Yalfani A, Atri AE, Sforza C. The effects of FIFA 11+ kids prevention program on kinematic risk factors for ACL injury in preadolescent female soccer players: a randomized controlled trial. Children. 2023 Jul 12;10(7):1206.
39. Rössler R, Donath L, Bizzini M, Faude O. A new injury prevention programme for children's football–FIFA 11+ kids–can improve motor performance: a cluster-randomised controlled trial. J Sports Sci. 2016 Mar 18;34(6):549–56.
40. Gatterer H, Lorenzi D, Ruedl G, et al. The "FIFA 11+" injury prevention program improves body stability in child (10 year old) soccer players. Biol Sport. 2018;35:153–8.

41. Jeong J, Choi DH, Shin CS. Core strength training can alter neuromuscular and biomechanical risk factors for anterior cruciate ligament injury. Am J Sports Med. 2021 Jan;49(1):183–92.
42. Marshall DA, Lopatina E, Lacny S, Emery CA. Economic impact study: neuromuscular training reduces the burden of injuries and costs compared to standard warm-up in youth soccer. Br J Sports Med. 2016 Nov 1;50(22):1388–93.
43. Hägglund M, Atroshi I, Wagner P, Waldén M. Superior compliance with a neuromuscular training programme is associated with fewer ACL injuries and fewer acute knee injuries in female adolescent football players: secondary analysis of an RCT. Br J Sports Med. 2013 Oct 1;47(15):974–9.
44. Otsuki R, Benoit D, Hirose N, Fukubayashi T. Effects of an injury prevention program on anterior cruciate ligament injury risk factors in adolescent females at different stages of maturation. J Sports Sci Med. 2021 Apr 15;20(2):365–72.
45. Arundale AJH, Silvers-Granelli HJ, Myklebust G. ACL injury prevention: where have we come from and where are we going? J Orthop Res. 2022;40(1):43–54.
46. Sparagana PR, Selee B, Ellis HB Jr, Ellington M, Beck JJ, Carsen S, Crepeau A, Cruz AI Jr, Heyworth B, Mayer SW, Niu EL. Parental awareness and attitudes towards ACL injury prevention programs in youth athletes. J Pediatr Orthop Soc North Am. 2023 Nov 1;5(4):788.

Revision of Pediatric and Early Adolescent ACL Reconstruction After Re-injury

13

Hazem A. Farouk, Ashraf Abdelkafy, and Hatem Galal Said

Abbreviations

ACL	Anterior cruciate ligament
BTB	Bone tendon bone.
MARS	Multicenter ACL Revision Study.
ESSKA	European Society of Sports Traumatology, Knee Surgery and Arthroscopy.
PAMI	Pediatric Anterior Cruciate Ligament Monitoring Initiative.

13.1 Epidemiology

Anterior cruciate ligament (ACL) injury in the young skeletally immature population is gaining more popularity in the literature in the recent decades. This is accompanied by more inclination toward reconstructive intervention rather than non-operative treatment options to lower the risks of further meniscal and chondral damage [1]. This trend is driven by the fact that sports participation either on the recreational or the professional level is highly encouraged in children and adolescents. That is why the rates of ACL reconstruction in this population have been increasing along with the advances in sports injuries industry which encourage the surgeons to perform ACL reconstructive surgeries more efficiently and more safely [2, 3].

With the rising numbers of ACL reconstructions in the skeletally immature, the number of failed surgeries rises as well. This draws attention to further studying the failures as well as performing more and more revision surgeries. Incidence of ACL graft ruptures or failures have been under reported in the literature in the past, but recently studies have been reporting the true incidence of failures among ACL reconstruction procedures in the young and adolescent groups [1, 4]. Despite the fact that growth plate damage has been discussed as the most important complication after ACL reconstruction in the skeletally immature population, data from recent studies showed that incidence of ACL graft rupture or failure is the more common complication [3].

H. A. Farouk
Orthopedic & Trauma Surgery, Faculty of Medicine, Cairo University, Giza, Egypt
e-mail: hfarouk@kasralainy.edu.eg

A. Abdelkafy
Orthopedic & Trauma Surgery, Faculty of Medicine, Suez Canal University, Ismailia, Egypt
e-mail: ashraf.abdelkafy@med.suez.edu.eg

H. G. Said (✉)
Orthopedic & Trauma Surgery, Faculty of Medicine, Assiut University, Asyut, Egypt
e-mail: said.hgz@aun.edu.eg

M. Karahan et al. (eds.), *Anterior Cruciate Ligament Injuries Throughout the Life Span*,
https://doi.org/10.1007/978-3-031-84501-7_13

Previous studies reported ACL failure rates up to 30% [5–10]. Frosch et al. performed a meta-analysis of 935 skeletally immature patients from 55 articles. They reported an overall ACL graft rupture rate of 4.8% [11]. Wong et al. in a recent meta-analysis of 160 studies with a total of 1329 patients reported a failure rate of 8.7% in the form of ACL graft rupture [12]. These studies included different techniques for ACL reconstruction, either transphyseal or epiphyseal, with different graft types including either autografts (mainly hamstrings or patella tendon (BTB)) or allograft [6].

Re-injury mostly took place with sporting activities either in the adult or pediatric population. Wiggins et al. [13] stated that the risk of re-injury of ACL graft rises up to 30–40 times higher than the normal ACL with the return to sports. Timing of ACL graft ruptures ranged from as early as 6 weeks post-operatively up to 99 weeks. Hamrin et al. worryingly stated that one in every three to four skeletally immature patients with a reconstructed ACL will sustain a re-injury after return to sports and within the first 2 years post-operatively [14].

As the rate of ACL graft ruptures increases, the rates of the revision procedures increase as well. Ho et al. in 2016 stated that the rate of revision ACL surgeries in the skeletally immature population ranged from 6.5% up to 34% [2].

In a cohort of 534 skeletally immature patients who underwent transphyseal ACL reconstruction, Nelson et al. reported an overall revision rate of 8.2%, mirroring the current literature [20]. Wong et al. in their meta-analysis reported a similar rate of ACL graft ruptures (8.7%, 115/1329 cases) and among those cases with an ACL graft rupture, 94.6% went on to a revision surgery (70/74 cases) [12].

13.2 Risk Factors for ACL Graft Failure

Several risk factors have been associated with higher ACL graft failure rates in children and adolescents. Among those factors are: Age, time to return to sports, graft type, size, and diameter along with the surgical technique in ACL reconstruction [15]. Age is considered one of the primary unmodifiable risk factors for ACL graft failure [4]. Ho et al. reported that patients under the age of 20 show significantly higher ACL graft failure rates [2]. In 2013, Maletis et al. [12] reported the rates of aseptic revision among a cohort of 9817 ACL reconstructions. They noted that for each year increase in age, the risk of revision decreased by 7%. Aggregated results from reviews suggest that the younger patients who return to sports have a higher re-injury rate than those who do not [13, 14].

A report from the Norwegian National Knee Ligament Register confirms that age is a significant risk factor for revision with a hazard ratio of 4.0 for the 15 to 19-year age group, compared with patients over 30 years of age [16]. Additionally, the Swedish ACL Registry has reported the highest rates of early revision in adolescent patients 13–19 years of age [17, 18]. These increased risks of ACL graft failure could be attributed partially by the surgeon's choice of a technique that might not be anatomical or biomechanical but safer with less potential hazards to the physeal plate. The higher functional demands and higher impact activities exerted by the younger age groups also play a role in higher risks of ACL graft failure. Additionally, the time to return to sports influences the outcomes of ACL reconstruction. Early return to sports and engagement in high-pivoting activities are correlated with higher ACL graft failure rates, and this shall be discussed in further details later within this chapter.

The type of graft used is another risk factor associated with the failure of ACL reconstruction. Allografts have been popularly used in the past, but they have yielded more potential for ACL graft failure in contrast to using autografts [15]. Despite Cruz et al.'s findings in 2015 that no statistically significant correlation could be established between ACL graft failures and type of graft used; other studies clearly reported up to 15 times higher failure rates associated with the use of allografts [2, 19]. Several studies have highlighted the likelihood of ACL graft failures when an allograft is used especially in highly active candidates [2, 20, 21]. "Soft tissue grafts are twice as likely to fail compared to BTB," stated Ho et al. in 2016 in his retrospective analysis of 561 cases of ACL reconstruction in the children and adolescent population [2]. Rates of ACL graft failure with hamstring autograft was 12.5% compared to 5.5% when a BTB was used ($p = 0.007$). The most recent data from large American and Scandinavian registries consisting of thousands of patients have shown that compared with hamstring grafts, BTB autografts are less likely to require a revision surgery with a relative ratio ranging from 1.4 to 2.3. In addition, these studies have shown that these differences become magnified with younger, high-risk patients [2, 19–21]. Ultimately, this lower failure rate must be weighed against the risk of using a BTB autograft especially with open physes, as there is a potential for growth plate injury and subsequent deformities.

Graft size and diameter have also been implicated as risk factors for success or failure of ACL reconstruction especially in the skeletally immature population. An earlier study by Conte et al. [22] suggested a minimum graft diameter of 8 mm as a potential predictor for ACL graft failure. Following studies failed to affirm such claims and could not demonstrate a statistically significant correlation. Ho et al. [1] suggested the use of a hybrid graft (allograft augment to a hamstring autograft) to increase the graft diameter and reported excellent results, yet they stated that current literature did not support their conclusions. Current evidence falls short of proving the impact of graft size in rates of ACL graft failure [2, 19].

Surgical technique of ACL reconstruction in the pediatric population affects the outcomes and the failure rates, similar to the adult population. Several reviews have implicated non-anatomic graft placement as a significant cause for revision and ACL graft failure [20, 23]. Nelson et al. [20] reported higher failure rates with transtibial drilling as it results in more vertical tunnels with persistent rotatory knee instability that puts the ACL graft at higher stresses during daily activities. Historically, vertical tunnels were advocated in the skeletally immature population, as it lessens the risks of physeal damage and this might jeopardize the biomechanical efficiency of the graft. It is unclear how much influence the surgical approach has on ACL graft rupture rates compared with the graft choice. Buckle et al. [23] could not establish a significant difference in the frequency of ACL graft rupture ($p = 0.88$) observed between extraphyseal and transphyseal groups. Ho et al. [2] in their review of the risk factors reported many variables that influenced the ACL graft failure rates including graft choice without mentioning the drilling technique. On the contrary, the Danish Knee Ligament Reconstruction Registry [24] found that the revision rates for ACL reconstructions with 4 years follow-up with an anteromedial technique were higher than cases with a transtibial technique. This might have been attributed to the significant shortening of femoral tunnel lengths with anteromedial techniques, which can lead to problems with graft choice and type of fixation used. Also, the learning curve associated with transitioning from a non-anatomic technique to an anatomic technique may influence the outcomes and earlier suboptimal results may be justified. Finally, future studies should focus more on directly comparing long-term patient reported outcomes and complications using randomization of different ACL reconstruction techniques which may provide a greater understanding of their relative efficacy and safety. Also, studies failed to prove that a single method of fixation affects revision rates and failures. Again, the abundant confounding variables with limited number of cases could not establish a correlation between outcomes and fixation methods [2].

Finally, failure of ACL reconstruction in the pediatric population can also be attributed to concomitant pathologies that are either missed or not managed properly [25]. A meniscus root tear predisposes to either a primary injury to the ACL or a further injury to the reconstructed graft if not diagnosed or managed correctly. Missed or ignored medial or lateral side laxity may also contribute to ACL graft failure as these factors result in an increase in load and stress on the ACL graft during the postoperative period, leading to laxity and early failure. Generalized ligamentous laxity among certain patients is another intrinsic,

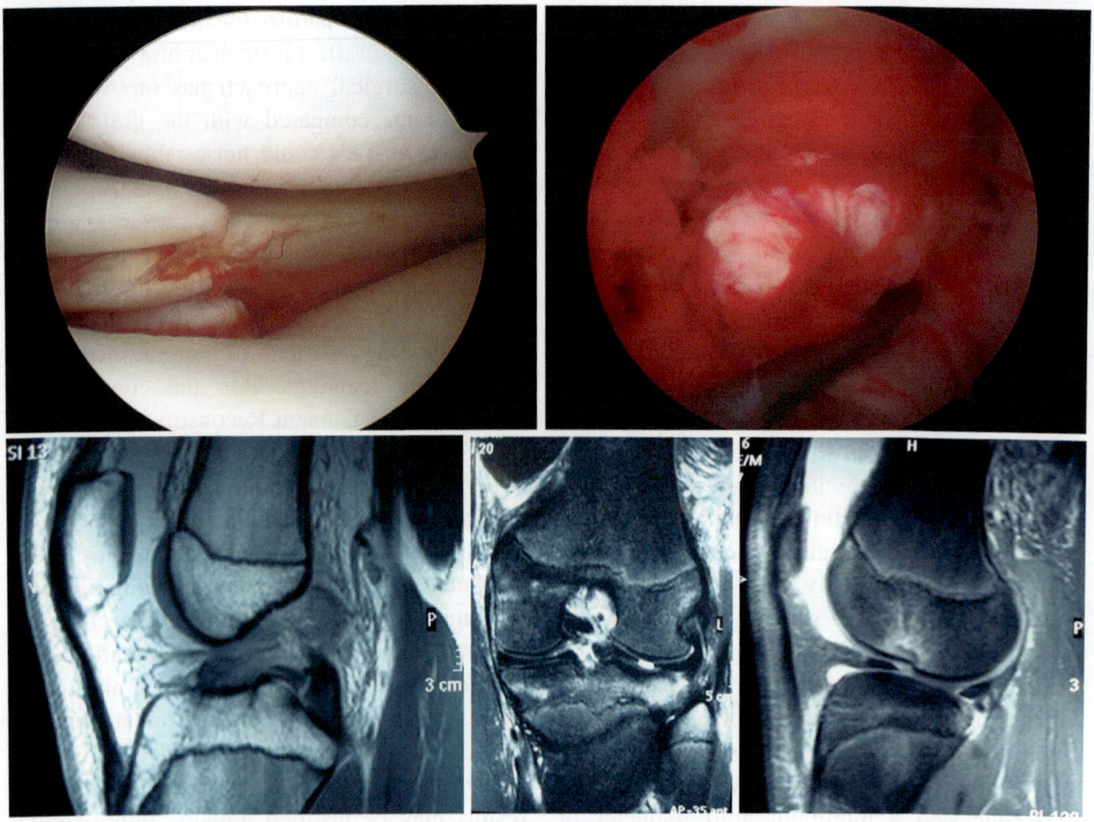

Fig. 13.1 Lateral meniscus root tear associated with anterior cruciate ligament injury (ACL) in a 15-year-old male (Top Left), completely torn ACL (Top Right) and a magnetic resonance imaging (MRI) showing a chondral injury of the lateral femoral condyle along with the complete ACL tear and root tear of the lateral meniscus (Bottom Row)

unmodifiable factor. Lower limb malalignment like genu valgum or varum along with other rotational deformities make an individual prone to primary ACL injury and subsequent failure of a previously reconstructed ACL graft. However, literature has not been able to identify with any certainty individuals at risk for a re-injury after a primary ACL reconstruction (Fig. 13.1).

Furthermore, preventative measures that decrease the risk of injury of the native ACL also play a role in decreasing the ACL graft failure rates. The lack of implementation and the difficulty to adhere to these protocols play in both primary ACL injury and re-injury rates [26]. The "FIFA 11 + for Kids" has shown to reduce football knee injuries by 48% and early implementation in the post-surgical rehabilitation protocol shall contribute to reducing the rates of re-injuries in the high-risk young athletic population [27].

13.3 Concerns in Revision Procedures

Revision procedures after primary ACL reconstruction are challenging especially in the pediatric population. Despite the higher failure rates that have been reported in younger patients after primary ACL reconstruction, outcomes and characteristics of revision ACL reconstructions in the pediatric and adolescent patient population have not been well studied. Yet, suboptimal outcomes have been reported along with increased risks of complications and a relatively high ACL graft failure rate. Also, the risk of contra-lateral ACL injury is high, and undoubtedly compromises return to sports rates. This requires meticulous planning and a detail-oriented approach during revision ACL reconstruction procedures to improve the outcomes and lower the risks of complications, either growth plate injury or ACL graft failures for the second time. Here we discuss the aspects that need to be addressed while attempting a revision ACL reconstruction in the pediatric skeletally immature population. These include (a) preoperative assessments like the morphological features of the knee and patient counselling, (b) surgical details such as graft choice, graft size, diameter, and tunnel placement, (c) post-surgical such as the post-operative rehabilitation program and time to return to sports.

First, the anatomical morphology of the patients should be further studied to identify any predisposing factors of injury. The revision procedure in itself requires a highly experienced surgeon with familiarity to the anatomy of the pediatric group. Changes in the width of the walls of the condyles, the location of the physeal plate in three dimensions, and time of its closure should be kept in mind prior to attempting a revision surgery. Three-dimensional modeling studies are useful in evaluating tunnel placement in primary ACL reconstruction techniques and should be mandatory in revision surgeries.

Patient counseling is mandatory prior to scheduling the patient for the revision surgery. The second procedure may potentially jeopardize these patients' ability and willingness to return to the same level of sport. In other words, a clear message should be communicated to the young patients, their parents, and their coaches with realistic expectations of returning to sport and the opportunity to reach the pre injury level. The literature tells us that with the return to pivoting sports at a high level, the risk of a second ACL injury is considerably high. This could hinder their participation at a high competitive level [10].

Graft options are limited in revision procedures, especially if an autograft was used in the primary procedure. Despite the high initial strength gained with the use of BTB autografts compared with other hamstrings, the risk of growth disturbances and physeal damage is higher [1]. Therefore, BTB autograft use in revision procedures should be considered wisely. Christiano et al., in their review of 90 knees from the pediatric population undergoing revision ACL reconstruction, found that allograft was used most commonly over the 16 years period of their study despite the literature that associates higher rates of ACL graft failure when allografts are utilized. Limited graft options and the refrain of using BTB due to its associated growth disturbance risks has led surgeons to use allograft as a primary option in revision ACL reconstruction surgeries [28]. The revision graft failure rates are reported to be 20%. An earlier Multicenter ACL Revision Study (MARS) with 460 revision patients from all age groups also utilized allografts most frequently and reported an ACL graft failure rate of 3.3%. This difference has been attributed to higher level of post-operative activity of younger patients, placing them potentially at a higher risk for re-injury compared with an older cohort. Despite higher outcome scores associated with autografts in the Christiano et al.'s study, no significant difference in ACL graft failure could be detected between autografts and allografts [28]. Further studies are needed on this topic, as autograft tissue may still be the preferred graft choice for revisions in young active patients to minimize failure rates [29]. Other studies advocated the use of a hybrid graft during the primary procedure to increase the thickness of the graft. This could be adopted as well in the revision procedure especially when larger graft size is needed to match the tunnel diameter drilled in the primary procedure, while others have suggested combining lateral extra-articular tenodesis with quadriceps tendon autograft in high risk adolescents undergoing primary ACL reconstruction [30].

Tunnel dilatation is another complication that has not been studied in the pediatric population but is a common sequela in the adult age groups. Thus, a CT scan is mandatory to assess the tunnel diameter prior to surgery and prepare for several options during surgery [28].

Other technical considerations include correct tunnel placement. Anatomical tunnel position adds more stability (especially rotatory knee stability), lowers the stresses on the revised ACL grafts, and decreases the rates of a new injury to the meniscus or the cartilage. This has not been studied in the pediatric population but rather a more sought after goal in revision procedures in adults. The risk of physeal damage stands as a barrier in aiming for an anatomical tunnel position primarily and more evidenced in revision surgeries. This means that in case where transphyseal tunnel placement is intended in the revision procedure, the cross-sectional areas of both tunnels (primary and revision) or the modified (potentially enlarged) tunnel will pose more risks for physeal damage. Moreover, the possibility of a focal bony bridge across the physis still exists if the non-anatomic primary tunnel is not filled with a soft tissue graft.

In addition, the diameter of the graft used in the revision procedure is another concern. This should be taken into consideration before attempting another surgery and while planning for the graft choice for revision surgeries. A hybrid graft has been utilized in primary ACL reconstruction with good outcomes combining hamstring autograft with allograft. Also, quadriceps tendon graft is receiving attention with not enough available data to recommend its utilization in revision surgeries in the pediatric population. Methods of fixation have not been reported to affect the outcomes of the revision procedure nor the ACL graft failure rates but it shall be taken in consideration based on the diameter of the old tunnel and if there is associated tunnel widening [15]. Prior acquaintance with the previous fixation methods is mandatory and whether it is planned to be removed during the revision procedure or not.

Associated injuries is another concern in revision surgeries as incidence of chondral or meniscal injuries is fairly high in the pediatric population as well as the adult population. Christiano et al. compared the incidence rate of associated injuries in their series with the MARS revision series and reported fairly close results (74.4% vs. 90% in the MARS group) [28, 31]. This high incidence rate necessitates proper assessment of the case prior to and during the surgery with preparation of different surgical modalities to deal with any associated injuries intra-operatively.

Another concern in the young population is the high possibility of sustaining a contralateral ACL injury after the primary injury. This can either be after the primary procedure or after the revision procedure. This is attributed to the level of sports participation that the young are desiring especially those who were willing to go through the primary and the revision surgery as well. In the Christiano et al. series of revision patients, a significant number of patients (33%) sustained contralateral ACL tears among those who went on to have revision ACL reconstructions. Thus, young athletes with more than one ACL tear continue to be at risk for further injury [28].

As mentioned earlier, the time to return to sports and the expectations of the young athlete/coach/parents trio should be thoroughly discussed and wisely decided upon. As much as this has been reported to influence the outcomes of the primary procedure, it is also known to influence the success of the revision procedure. Therefore, proper counseling and following a strict rehabilitation protocol is highly essential. Moreover, development of clear guidelines and criteria of when and how to return to sports is of the utmost importance.

ACL graft failure can also result from multiple technical and biological factors, secondary instability (malalignment, meniscal deficiency, or unappreciated additional ligamentous injury), and inadequate post-operative rehabilitation. Although detailed analysis of these variables was outside the scope of this review, these factors are important to acknowledge and assess for optimal surgical planning in a revision situation.

13.4 Return to Sports After Revision ACL Reconstruction in Children and Adolescents

Christino et al. studied 90 revision ACL reconstructions in skeletally immature patients and among the 69% who could return to competitive sports following revision ACL reconstruction, the average time to return to sport was 8.9 months (range 3–36 months). In this series, the authors concluded that ACL revisions in this particular age group were associated with compromised return to sports rates compared to 72% in adults [28, 32].

Return to sports criteria play an important role, if not the most important, to influence the rates of ACL graft failure. Data from the Norwegian National Knee Ligament Registry [16] and the Swedish ACL Registry [33] suggested that improving rehabilitation protocols, integrating neuromuscular training with a more cautious return to sports criteria improved outcomes and reduced the risk of re-injury [10]. These considerations after the surgery influence the outcomes of revision surgery particularly in children and adolescents with high activity demands [28]. Modifications to return to sport guidelines have the potential to reduce the re-injury risk and hopefully improve future sports performance in the young athlete after ACL injury. Although various return-to-sport guidelines exist, there is still no consensus about, or validation of, these guidelines. This means that return to sport decisions are often based on a combination of time since surgery and personal experience. Several factors related to training and competition load have been identified that influence the decision-making especially in the young. For instance, Malisoux et al. showed that an increase in weekly training intensity was associated with an increased risk of ACL injury in younger athletes, similar to what is found in their older counterparts [34]. This suggests that monitoring young athletes is beneficial in terms of reducing, and potentially preventing, injuries. The European Society of Sports Traumatology, Knee Surgery and Arthroscopy (ESSKA) that launched the ESSKA Pediatric Anterior Cruciate Ligament Monitoring Initiative (PAMI) aimed at better understanding of the specific risk factors in the young population and to allow an unbiased decision-making for a safe return to sport after ACL injury [35]. Consequently, the decision of the time to return to sports should be taken very cautiously in this young population. Delayed return to sports after a revision surgery is crucial to lower the risks of further injury and ACL graft failure. This in turn has to be communicated to the young athlete undergoing a revision surgery, as well as his parents and coaches. Their high expectations to early return to sports has to be lowered. Also, the possibility of not being able to perform at the same pre-injury level should be understood. In a cohort revision series by Christiano et al. [28], the return to sport rate was 69%; however, only 38.1% of these patients were able to return to sports at their pre-injury level. In other terms, 59.6% of patients were unable to return to pre-injury abilities. Return to sport rates following revision ACL reconstruction have demonstrated suboptimal return to sport rates in another study as well [28]. These rates match the return to sport rates in the adult population, with 46% rate of return to pre-injury levels as published by Anand et al. in 2015 [36]. A recent meta-analysis reported that rates of returning to the same level of sports participation ranges between 55% and 65% following revision ACL reconstruction [36]. That is why patient counseling is crucial before revision surgeries and expectations need to be tempered regarding time to return to sports and ability to participate at the pre-injury level, particularly in children and adolescents with high activity demands.

13.5 Authors' Preferred Approach to Revision Cases

There are many challenges surgeons face when attempting a revision ACL reconstruction in skeletally immature patients. Due to the complexity of the procedure, the keenness to avoid growth plate injury and the suboptimal outcomes that have been previously reported, the revision procedure should be well planned and the surgeon should be attentive to every detail before, during, and after the surgery.

13.5.1 Preoperative Planning

To increase success rates of the surgery and improve outcomes, planning the surgery preoperatively is of utmost importance. Thus each case should be analyzed individually to identify the causes of failure of the primary procedure. Routine thorough clinical assessment is mandatory in addition to a detailed report of the previous procedures with preoperative radiographs, in addition to new radiographs including a CT with 3D reconstruction to measure the tunnel diameters and to determine tunnel location. Morphological and anatomical risk factors that might have been overlooked before the primary procedure should be investigated using plain radiographs and CT. Full-length alignment films need to be obtained to assess for coronal and sagittal plane malalignment that may be necessary to address during the revision procedure. MRI is also important to confirm the re-tear and to identify any associated injuries. Any signs of generalized ligamentous laxity or syndromic features should be noted during thorough physical examination. Pivot shift testing to assess for the degree of rotatory knee instability may guide us to use an additional procedures such as extra-articular tenodesis.

Patient counseling as previously discussed shall be done with clear genuine communication about the patient's post-surgical expectations, planned activity level, and especially regarding return to competitive sports. The surgeon, patient, and parents (and maybe the coaches) should participate in shared decision-making about the course of action that starts with the timing of the revision surgery, technique, graft selection, and post-operative rehabilitation protocols. The concerns and possible complications shall be openly discussed as well.

Finally, a decision has to be made on whether the old hardware needs to be removed, and what instruments shall be used in the procedure. Many of these devices can require specialized tools for removal, and bone overgrowth, scar tissue, and implant migration can make removal of the old hardware even more difficult.

13.5.2 Operative Technical Consideration

In most cases, ACL reconstructions in the young and adolescent population can be revised in a single-stage surgery. Single stage surgery can be advocated with either anatomic, semi-anatomic (close to the anatomic sites), or non-anatomic primary tunnels. In the case of non-anatomic tunnels, new tunnels are drilled. If the tunnels were semi-anatomic, using same tunnels is preferred to avoid further damaging the physis with a new tunnel. If these tunnels are widened, either changing the direction of drilling to have a new tunnel direction exiting intra-articularly at the same aperture or using large-sized augmented grafts and a suspensory fixation with large sized buttons is used (Fig. 13.2). Thus, we advise a retrograde femoral tunnel drilling with only an epiphyseal socket which avoids further damage to the growth plate.

We advocate the use of contralateral quadrupled hamstring tendon autograft sized at least 8 mm in diameter. This should be communicated to the parents prior to surgery and decision is taken after counseling. Other alternative graft options include allografts, quadriceps tendon, or BTB autografts. Allograft as previously discussed is not advocated due to higher failure rates and sometimes unavailability in certain countries. Quadriceps is not recommended in young athletes as it carries the risk of weakness of the extensor mechanism but still remains a safe option. BTB use is not recommended due to the risk it carries to cause growth disturbances at the growth plate.

Fixation methods are numerous. Suspensory fixation is safer to use, both tibial and femoral, and care should be taken when using intra-tunnel fixation through the physis, and in such case, bioabsorbable material is preferred. An added extra-articular procedure (tenodesis) should be considered to increase the rotatory knee stability and relieve high stresses on the ACL graft especially in cases of high pivot shift (Fig. 13.3).

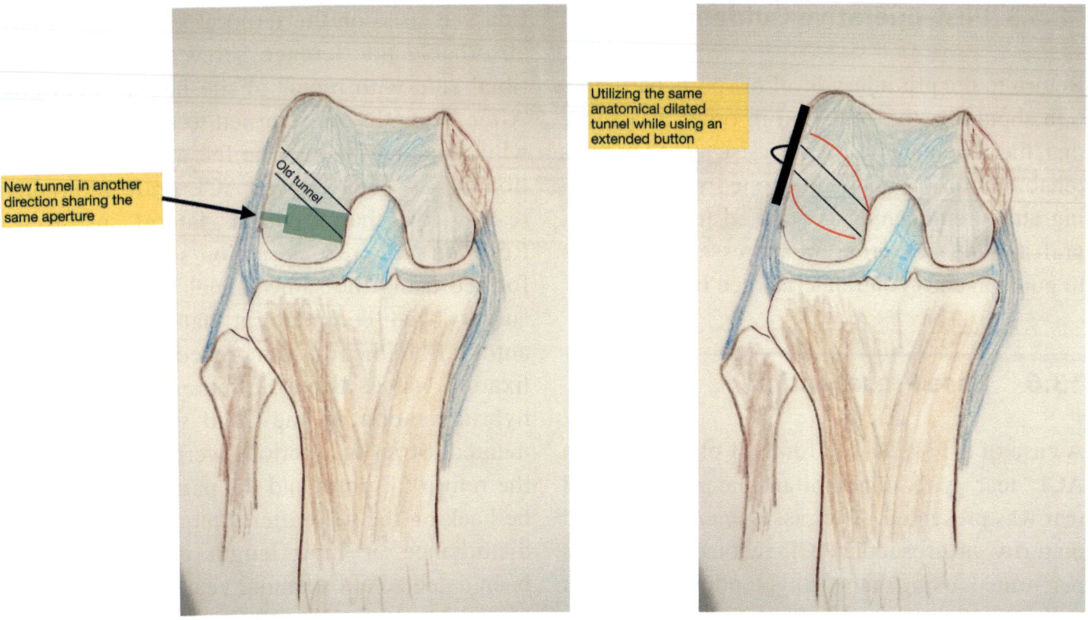

Fig. 13.2 In case of anatomic/semi-anatomic tunnels, the direction of the tunnel can be changed as in Fig. 13.2a, or using the same widened tunnel with an extended button and a larger-sized graft as in Fig. 13.2b

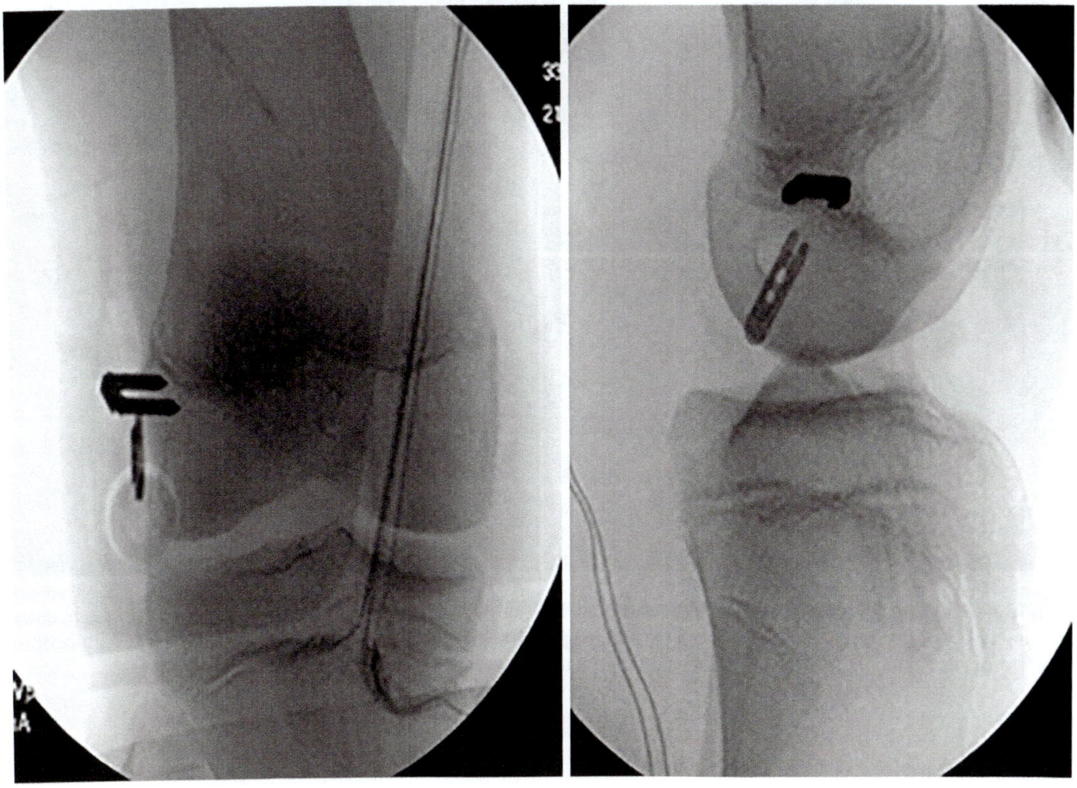

Fig. 13.3 A case of revision anterior cruciate ligament (ACL) reconstruction in a skeletally immature male with epiphyseal ACL reconstruction with hamstrings tendon autograft and iliotibial band extra-articular tenodesis (modified Lemaire procedure). (With permission from Prof. Shital Parikh, Cincinnati Children's Hospital Medical Center, University of Cincinnati College of Medicine, USA)

13.5.3 Post-operative Guidelines

Return to the pre-injury level of sports participation should be the goal, and in order to achieve that, the patient has to follow a strict 9–12 months rehabilitation program. Integrated muscle training and strengthening is adopted to avoid ACL graft failure. The program addresses both knees to guard against contralateral knee injury.

13.6 Case Presentation

A case of a 13-year-old football player with an ACL tear with concomitant lateral meniscal tear was presented. After assessment of skeletal maturity, arthroscopic ACL reconstruction was performed using hamstring tendons autograft with a transphyseal technique. Suspensory fixation was used on the femoral side while a bioabsorbable interference screw fixation on the tibial side with repair of the lateral meniscus. After 4 months, he returned to sports early and while kicking the ball, he felt a popping sensation. A new MRI showed ACL graft failure and re-tear of the repaired lateral meniscus. Revision ACL reconstruction surgery was performed using the same tunnels of the primary surgery and contralateral hamstring tendons autograft. In the revision procedure, suspensory fixation was used on the femoral side while a hybrid fixation on the tibial side was implemented. Strict instructions were given to delay the return to sports and the patient was able to be back on the field after 9 months. No growth disturbance or limb length discrepancy has been detected at one and 2 years follow-up (see Figs. 13.4, 13.5, 13.6, 13.7 and 13.8).

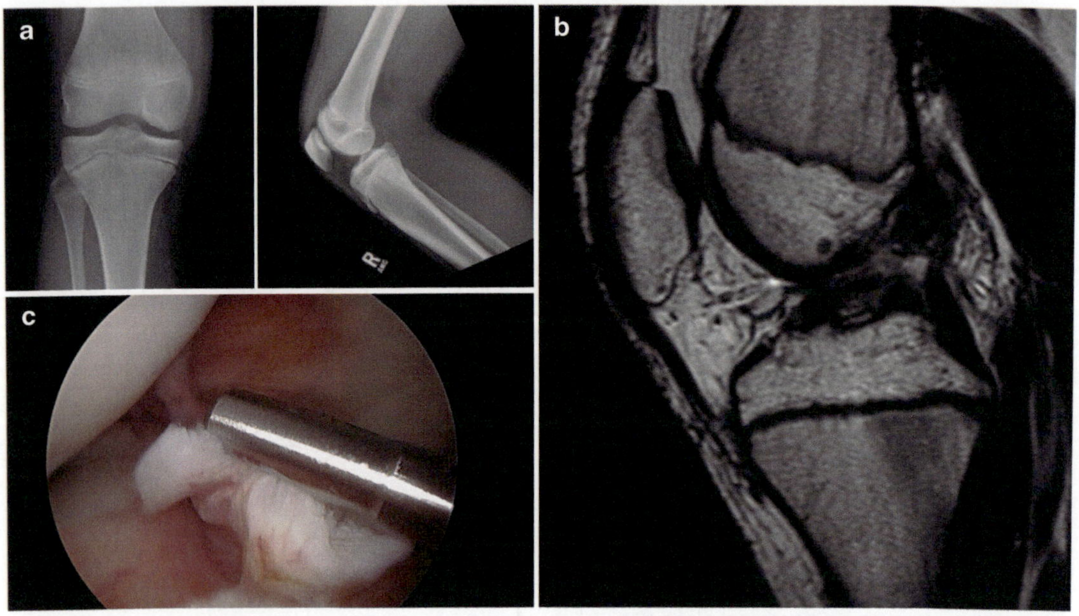

Fig. 13.4 13-year-old male footballer who suffered a twisting injury of his right knee. Radiograph (**a**) and magnetic resonance imaging (MRI) (**b**) showed complete anterior cruciate ligament (ACL) tear. Patient scheduled for an arthroscopic ACL reconstruction (**c**). (With permission from Prof. Shital Parikh)

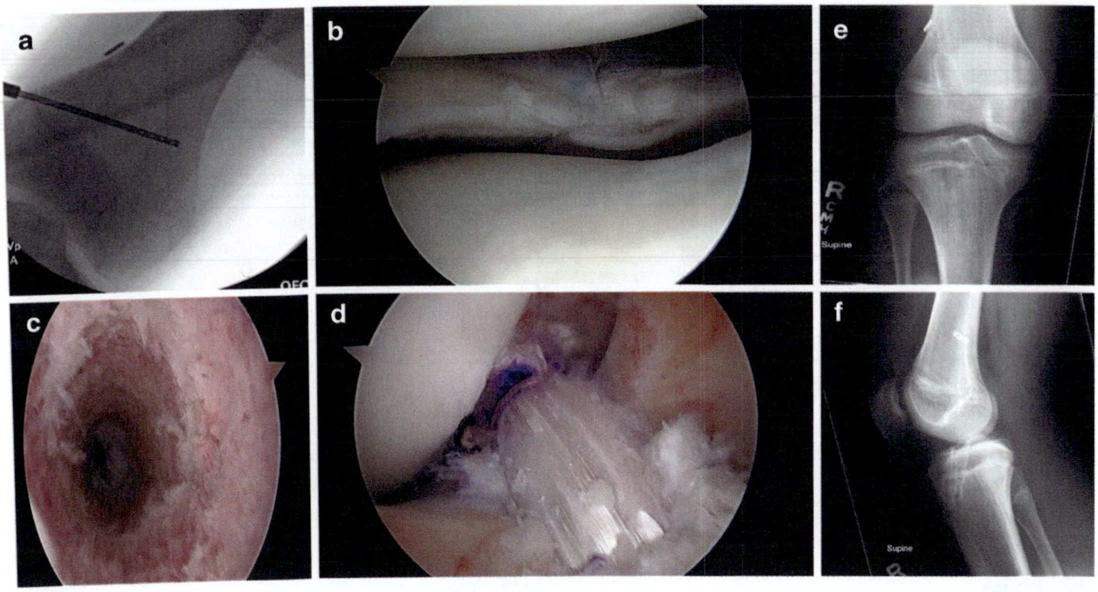

Fig. 13.5 Transphyseal anterior cruciate ligament (ACL) reconstruction with hamstring tendons autograft and fixation using suspensory fixation on the femoral side and interference screw on the tibial side as shown in the radiograph (**a**, **c**, **e**, and **f**) along with repair of a torn lateral meniscus (**b**). Arthroscopic picture showing the final ACL reconstruction (**d**). (With permission from Prof. Shital Parikh)

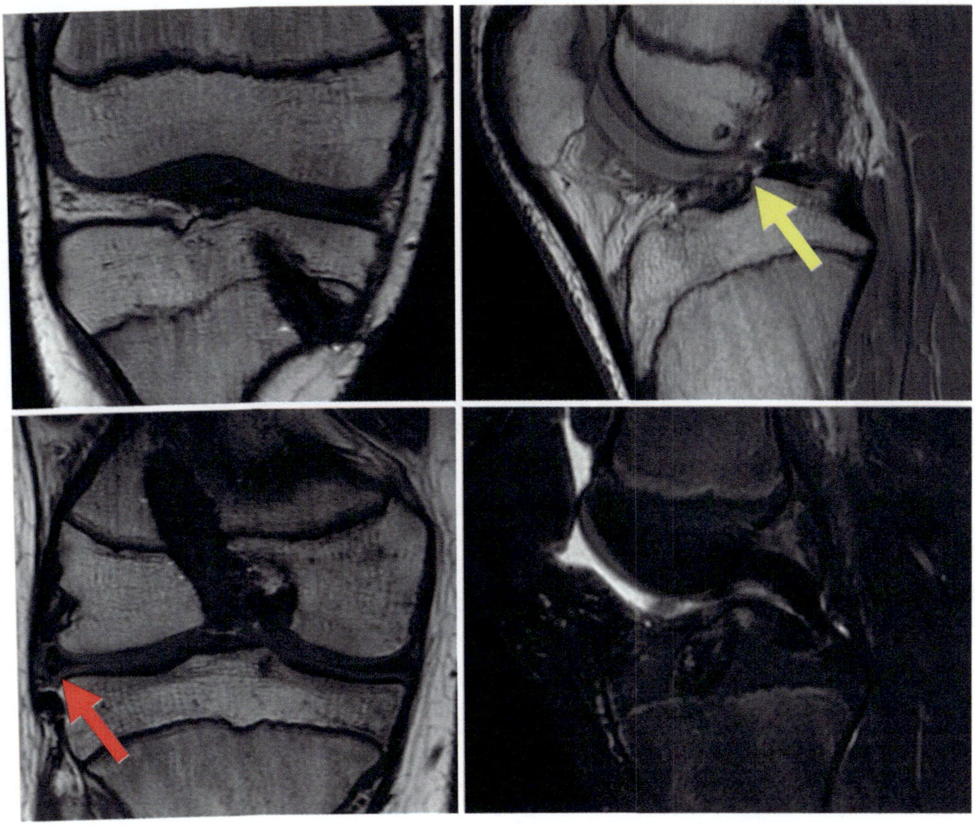

Fig. 13.6 Four months after the primary anterior cruciate ligament (ACL) reconstruction, the patient returned early to sports and felt a pop while trying to kick the ball with complete rupture of the reconstructed graft and re-tear of the lateral meniscus as shown in the MRI. (With permission from Prof. Shital Parikh)

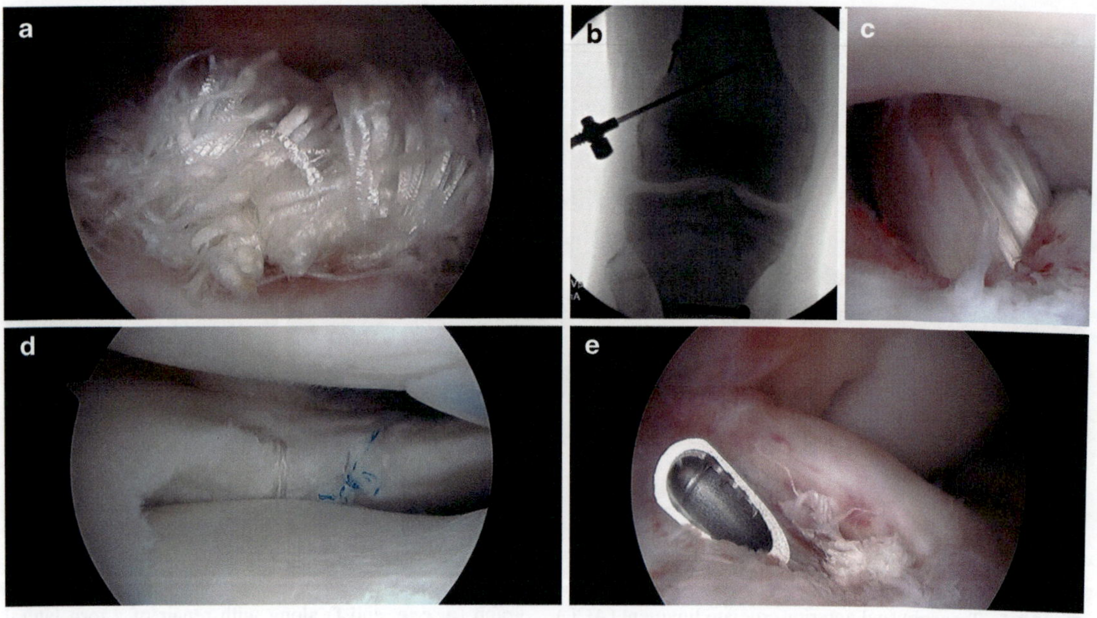

Fig. 13.7 Revision anterior cruciate ligament (ACL) reconstruction using the same tunnels with contralateral hamstring tendons autograft and repair of the re-torn lateral meniscus. (**a**) An arthroscopic picture of the torn graft, (**b**) Fluoroscopy picture during the ACL revision procedure, (**c**) the revised graft showing no impingement, (**d**) the revised repair of the lateral meniscus re-tear, (**e**) Tibial tunnel debrided with a shaver prior to graft passage. (With permission from Prof. Shital Parikh)

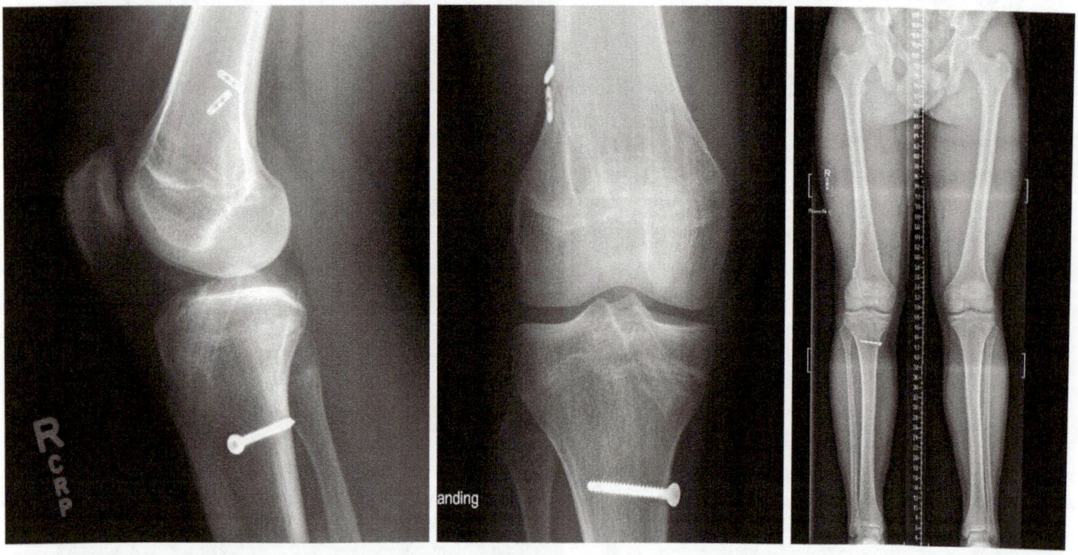

Fig. 13.8 One year follow-up radiographs. Patient back to competitive football 9 months postoperative. No limb length discrepancy. (With permission from Prof. Shital Parikh)

13.7 Conclusion

Revision ACL reconstructions in skeletally immature patients is challenging. Patient counseling is crucial before the surgery and expectations need to be tempered regarding time to return to sports and ability to participate at the pre-injury level, particularly in children and adolescents with high activity demands. We recommend single stage procedure, the use of contralateral quadrupled hamstring tendon autograft sized at least 8 mm in diameter, the use of retrograde drilling creating sockets rather than tunnels. Integrated muscle training and strengthening are adopted to avoid ACL graft failure. The program must address both knees to guard against contralateral knee injury.

References

1. Migliorini F, Cocconi F, Schäfer L, Memminger MK, Giorgino R, Maffulli N. Anterior cruciate ligament reconstruction in skeletally immature patients is effective: A systematic review. Knee Surg Sports Traumatol Arthrosc. 2024;32(2):418–31.
2. Ho B, Edmonds EW, Chambers HG, Bastrom TP, Pennock AT. Risk factors for early ACL reconstruction failure in pediatric and adolescent patients: a review of 561 cases. J Pediatr Orthop. 2016;38(7):388–92.
3. Jildeh TR, Bowen E, Bedi A. Revision anterior cruciate ligament reconstruction in skeletally immature patients. J Am Acad Orthop Surg. 2023;31(21):e920–9.
4. Rugg CM, Pitcher AA, Allen C, Pandya NK. Revision ACL reconstruction in adolescent patients. Orthop J Sports Med. 2020;8(9):2325967120953337.
5. Csintalan RP, Inacio MC, Desmond JL, Funahashi TT. Anterior cruciate ligament reconstruction in patients with open physes: early outcomes. J Knee Surg. 2013;26(4):225–32.
6. Engelman GH, Carry PM, Hitt KG, Polousky JD, Vidal AF. Comparison of allograft versus autograft anterior cruciate ligament reconstruction graft survival in an active adolescent cohort. Am J Sports Med. 2014;42(10):2311–8.
7. Fuchs R, Wheatley W, Uribe JW, Hechtman KS, Zvijac JE, Schurhoff MR. Intra-articular anterior cruciate ligament reconstruction using patellar tendon allograft in the skeletally immature patient. Arthroscopy. 2002;18(8):824–8.
8. Kim S-J, Shim D-W, Park K-W. Functional outcome of transphyseal reconstruction of the anterior cruciate ligament in skeletally immature patients. Knee Surg Relat Res. 2012;24(3):173–9.
9. Kohl S, Stutz C, Decker S, Ziebarth K, Slongo T, Ahmad SS, Kohlhof H, Eggli S, Zumstein M, Evangelopoulos DS. Mid-term results of transphyseal anterior cruciate ligament reconstruction in children and adolescents. Knee. 2014;21(1):80–5.
10. Senorski EH, Seil R, Svantesson E, Feller JA, Webster KE, Engebretsen L, Spindler K, Siebold R, Karlsson J, Samuelsson K. "I never made it to the pros…" Return to sport and becoming an elite athlete after pediatric and adolescent anterior cruciate ligament injury—current evidence and future directions. Knee Surg Sports Traumatol Arthrosc. 2018;26(4):1011–8.
11. Frosch K-H, Stengel D, Brodhun T, Stietencron I, Holsten D, Jung C, Reister D, Voigt C, Niemeyer P, Maier M, Hertel P, Jagodzinski M, Lill H. Outcomes and risks of operative treatment of rupture of the anterior cruciate ligament in children and adolescents. J Arthrosc Relat Surg. 2010;26(11):1539–50.
12. Wong SE, Feeley BT, Pandya NK. Complications after pediatric ACL reconstruction: a meta- analysis. J Pediatr Orthop. 2017;39(8):e566–71.
13. Wiggins AJ, Grandhi RK, Schneider DK, Stanfield D, Webster KE, Myer GD. Risk of secondary injury in younger athletes after anterior cruciate ligament reconstruction: a systematic review and meta-analysis. Am J Sports Med. 2016;44(7):1861–76.
14. Webster KE, Feller JA, Leigh WB, Richmond AK. Younger patients are at increased risk for graft rupture and contralateral injury after anterior cruciate ligament reconstruction. Am J Sports Med. 2014;42(3):641–7.
15. Zhao D, Pan J, Lin F, et al. Risk factors for revision or Rerupture after anterior cruciate ligament reconstruction: a systematic review and meta-analysis. Am J Sports Med. 2023;51(11):3053–75.
16. Persson A, Fjeldsgaard K, Gjertsen J-E, Kjellsen AB, Engebretsen L, Hole RM, Fevang JM. Increased risk of revision with hamstring tendon grafts compared with patellar tendon grafts after anterior cruciate ligament reconstruction: a study of 12,643 patients from the Norwegian Cruciate Ligament Registry, 2004–2012. Am J Sports Med. 2014;42(2):285–91.
17. Andernord D, Desai N, Björnsson H, Ylander M, Karlsson J, Samuelsson K. Patient predictors of early revision surgery after anterior cruciate ligament reconstruction: a cohort study of 16,930 patients with 2-year follow- up. Am J Sports Med. 2015;43(1):121–7.
18. Andernord D, Desai N, Björnsson H, Gillén S, Karlsson J, Samuelsson K. Predictors of contralateral anterior cruciate ligament reconstruction: a cohort study of 9061 patients with 5-year follow-up. Am J Sports Med. 2015;43(2):295–302.
19. Cruz Jr AI, Fabricant PD, McGraw M, Rozell JC, Ganley TJ, Wells L. All-epiphyseal ACL reconstruction in children: review of safety and early complications. J Pediatr Orthop. 2015;37(3):204–9.
20. Nelson IR, Chen J, Love R, Davis BR, Maletis GB, Funahashi TT. A comparison of revision and rerupture rates of ACL reconstruction between autografts and allografts in the skeletally immature. Knee Surg Sports Traumatol Arthrosc. 2016;24(3):773–9.

21. Pennock A, Murphy MM, Mark W. Anterior cruciate ligament reconstruction in skeletally immature patients. Curr Rev Musculoskelet Med. 2016;9(4):445–53.
22. Conte EJ, Hyatt AE, Gatt Jr CJ, Dhawan A. Hamstring autograft size can be predicted and is a potential risk factor for anterior cruciate ligament reconstruction failure. Arthroscopy. 2014;30(7):882–90.
23. Buckle C, Wainwright AM. A systematic review of long-term patient reported outcomes for the treatment of anterior cruciate ligament injuries in the skeletally immature. J Child Orthop. 2018;12(3):251–61.
24. Rahr-Wagner L, Thillemann TM, Pedersen AB, Lind MC. Increased risk of revision after anteromedial compared with transtibial drilling of the femoral tunnel during primary anterior cruciate ligament reconstruction: results from the Danish Knee ligament reconstruction register. Arthroscopy. 2013;29(1):98–105.
25. Phillips T, Ronna B, Sullivan B, Terner Z, Goldenberg N, Kushare I, Shybut T. Skeletal maturity is associated with increased meniscal and chondral pathology in patients under 21 years of age undergoing primary anterior cruciate ligament reconstruction within 6 months of injury. Arthroscopy. 2024;S0749-8063(24):00066–5.
26. Cancino B, Muñoz C, Tuca MJ, Birrer EA, Sepúlveda MF. Anterior cruciate ligament rupture in skeletally immature patients. JAAOS. 2022;6(5):e21.00166.
27. Thorborg K, Krommes KK, Esteve E, Clausen MB, Bartels EM, Rathleff MS. Effect of specific exercise-based football injury prevention programmes on the overall injury rate in football: a systematic review and meta-analysis of the FIFA 11 and 11+ programmes. Br J Sports Med. 2017;51:562–71.
28. Christino MA, Tepolt FA, Sugimoto D, Micheli LJ, Kocher MS. Revision ACL reconstruction in children and adolescents. J Pediatr Orthop. 2020;40(3):129–34.
29. Group M, Group M. Effect of graft choice on the outcome of revision anterior cruciate ligament reconstruction in the Multicenter ACL Revision Study (MARS) Cohort. Am J Sports Med. 2014;42:2301–10.
30. Green DW, Hidalgo Perea S, Brusalis CM, Chipman DE, Asaro LA, Cordasco FA. A modified lemaire lateral extra-articular tenodesis in high-risk adolescents undergoing anterior cruciate ligament reconstruction with quadriceps tendon autograft: 2-year clinical outcomes. Am J Sports Med. 2023;51(6):1441–6.
31. MARS Group; Wright RW, Huston LJ et al. Descriptive epidemiology of the Multicenter ACL Revision Study (MARS) cohort. Am J Sports Med. 2010;38(10):1979–86.
32. Manojlovic M, Ninkovic S, Matic R, Versic S, Modric T, Sekulic D, Drid P. Return to play and performance after anterior cruciate ligament reconstruction in soccer players: a systematic review of recent evidence. Sports Med. 2024; https://doi.org/10.1007/s40279-024-02035-y. Online ahead of print
33. Sandon A, Engström B, Forssblad M. High risk of further anterior cruciate ligament injury in a 10-year follow-up study of anterior cruciate ligament-reconstructed soccer players in the Swedish national knee ligament registry. Arthroscopy. 2020;36(1):189–95.
34. Malisoux L, et al. Monitoring of sport participation and injury risk in young athletes. J Sci Med Sport. 2013;16(6):504–508.
35. Moksnes H, et al. The ESSKA paediatric anterior cruciate ligament monitoring initiative. Knee Surg Sports Traumatol Arthrosc. 2016;24(3):680–7.
36. Anand BS, Feller JA, Richmond AK, Webster KE. Return-to-sport outcomes after revision anterior cruciate ligament reconstruction surgery. AJSM. 2015;44(3):580–4.

Part III

Adult ACL Injuries

Management of Adult ACL Injuries

14

Philipp W. Winkler, Rafael Buerba Siller, Jan-Dierk Clausen, Adrian Deichsel, Benjamin B. Rothrauff, and Volker Musahl

Abbreviations

AAOS	American Academy of Orthopaedic Surgeons
ACL	anterior cruciate ligament
ACLR	anterior cruciate ligament reconstruction
ALC	anterolateral complex
AM	anteromedial
AP	anteroposterior
BTB	bone-patellar-tendon-bone
CSA	cross-sectional area
CT	computed tomography
HT	hamstring tendons
IKDC	International Knee Documentation Committee
LET	lateral extra-articular tenodesis
OA	osteoarthritis
PL	posterolateral
PROs	patient reported outcomes
QT	quadriceps tendon
ROM	range-of-motion
RTS	return-to-sport

P. W. Winkler (✉)
Department of Orthopaedic Surgery, UPMC Freddie Fu Sports Medicine Center, University of Pittsburgh, Pittsburgh, PA, USA

Department for Orthopaedics and Traumatology, Kepler University Hospital GmbH, Johannes Kepler University Linz, Linz, Austria
e-mail: Philipp.Winkler@kepleruniklinikum.at

R. B. Siller · B. B. Rothrauff · V. Musahl
Department of Orthopaedic Surgery, UPMC Freddie Fu Sports Medicine Center, University of Pittsburgh, Pittsburgh, PA, USA

J.-D. Clausen
Orthopedic Trauma Department, Hannover Medical School, Hannover, Germany

A. Deichsel
Department of Trauma-, Hand- and Reconstructive Surgery, University Hospital Münster, Münster, Germany

14.1 Background

Anterior cruciate ligament (ACL) injuries inevitably affect the biomechanical properties of the knee joint and therefore impair knee kinematics given by nature. ACL deficiency leads to impaired knee-related quality of life compared to uninjured populations [1]. Patients undergoing ACL reconstruction (ACLR) also experience impaired knee-related quality of life [2]. In a randomized controlled trial, no statistical significant difference was observed between ACL-deficient and ACL-reconstructed groups [3]. Young athletes are most commonly affected by ACL injuries [4–6]. Return-to-sport (RTS) is essential for athletes, however, in one study only 65% of athletes return to their preinjury level [7], which is dependent on the desired sport [8]. Multiple modifiable and nonmodifiable risk factors [5, 6, 9, 10] affect treatment decision-making following ACL

M. Karahan et al. (eds.), *Anterior Cruciate Ligament Injuries Throughout the Life Span*,
https://doi.org/10.1007/978-3-031-84501-7_14

injuries and require comprehensive clinical and radiological diagnostics. An evidence-based approach to the management of adult ACL injuries is of the utmost importance to maintain social, professional, and sports activities and to prevent long-term adverse effects like knee joint osteoarthritis (OA).

14.2 Anatomy

Comprehensive anatomical knowledge is required for an individualized approach to restore the native ACL anatomy. With a mean length of 32 mm [11, 12], the ACL is an intra-articular, extra-synovial structure generally considered to be composed of the anteromedial (AM) and posterolateral (PL) bundles (Fig. 14.1) [12]. Connecting the femoral and tibial insertion site, the ACL follows an oblique orientation. The length of the ACL increases by almost 19% in knee extension compared to 90° of flexion, indicating anisometric characteristics of the native ACL [13]. The femoral insertion of the ACL is located on the medial facet of the lateral femoral condyle in between the lateral intercondylar ridge (resident's ridge) and the apex of the deep cartilage [14]. The isthmus that represents the section of the mid-substance with the smallest cross-sectional area (CSA) depends on the flexion

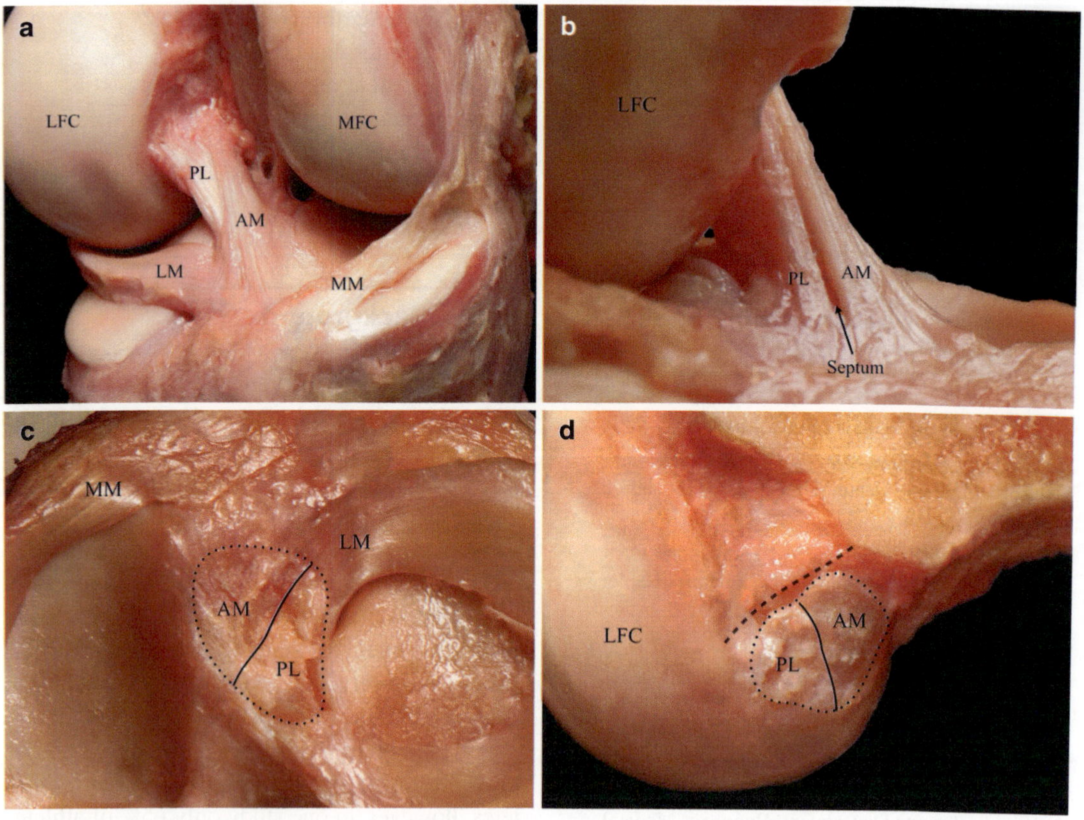

Fig. 14.1 Anatomy of the ACL and its femoral and tibial insertion sites in a human cadaveric knee specimen. (**a**) Anteromedial view with illustration of the two distinct bundles and the oblique orientation of the ACL in the intercondylar notch. (**b**) Separation of the AM and PL bundle by a membranous septum. (**c**) Tibial insertion site of the ACL adjacent to the anterolateral meniscal root. (**d**) Femoral insertion site of the ACL in between the lateral intercondylar ridge (resident's ridge) and the apex of the deep cartilage; *AM* anteromedial bundle, *LFC* lateral femoral condyle, *LM* lateral meniscus, *MFC* medial femoral condyle, *MM* medial meniscus, *PL* posterolateral bundle. Reprinted from [95], with permission from Elsevier

angle and loading condition of the knee [13]. Although the debate about the anatomical shape of the ACL is still ongoing, it is widely accepted that the ACL function consists of two distinct bundles (AM and PL bundles) [12]. Different strain patterns of thc AM and PL bundles lead to different restraining forces of the two bundles throughout the physiologic range-of-motion (ROM). The PL bundle has been proven to be the major restraint to anterior tibial translation in early flexion angles and to combined rotatory loads, whereas the AM bundle is the major restraint to anterior tibial translation in higher flexion angles with a negligible effect to combined rotatory loads [15]. It is imperative that the reconstructed ACL respects the biomechanical behavior of the native ACL as outlined above. Therefore, anatomic tunnel placement, not isometric (i.e., on the femur wrongly high and wrongly anterior) is necessary.

Evidence has demonstrated that the ACL acts as the primary restraint against anterior tibial translation and rotatory knee laxity [16]. However, diagnostic and treatment strategies also need to focus on concomitant injuries, since the menisci and the peripheral capsulo-ligamentous structures have been shown to be secondary restraints to translational and rotatory knee laxity [17, 18].

14.3 Epidemiology

The incidence of first-time non-contact ACL injury among college students is 0.15 per 1000 person-days and is more than twice as high as for high school students (0.061 per 1000 person-years). Additionally, female athletes have more than twice the risk of male athletes [6]. The peak incidence of males sustaining an ACL injury is between the ages of 19–25 years, whereas females aged between 14 and 18 years have a slightly left-sided shift of the peak [19]. Over the last years, a growing number of ACL reconstructions (ACLRs) have been recorded, with an increase of almost 50% between 1994 and 2006 [20]. The high number of ACLRs performed annually results in costs exceeding $7 billion per year in the United States of America [21].

14.4 Clinical & Radiological Presentation

A detailed diagnostic assessment consisting of the patient's history, a standardized clinical examination, and a radiological work-up is essential for an individualized approach in the treatment of ACL injuries.

Patients often describe a "popping" sensation, followed by subsequent swelling (hemarthrosis) and loss of function of the injured knee [22]. The physical examination specific to ACL injuries consists of the anterior drawer test, the Lachman test, and the pivot-shift test [21]. Given the importance of the peripheral knee structures and the menisci as secondary restraints, it is crucial to detect and if necessary to address combined injuries [18, 23]. Therefore it is recommended to follow a standardized examination protocol like the International Knee Documentation Committee (IKDC) form [24].

A standard radiological work-up for primary ACL injuries contains anteroposterior (AP) and lateral radiographs, native MRI scans, and—if necessary—hip-knee-ankle and stress radiographs for the assessment of limb alignment and concomitant injuries, respectively. Computed tomography (CT) scans are appropriate when bony deformities or fractures are suspected [25]. Radiographic signs indicating an ACL injury are the Segond fracture [26] or the lateral femoral notch sign [27]. However, magnetic resonance imaging (MRI) is considered the gold standard for imaging of ACL injuries (Fig. 14.2), with a pooled sensitivity of 87% and specificity of 90% [28]. The course of the ACL in sagittal MRI slices is approximately parallel (<15°) to the roof of the femoral notch, corresponding to Blumensaat's line on lateral plain radiographs [29]. The tibial inclination angle of the ACL is about 47° in the sagittal plane and 73° in the respective coronal plane [30]. Injury to the ACL and the associated loss of tension can result in a wavy appearance of the ACL, with the stumps not being parallel to the roof of the femoral notch [31]. Loss of continuity with elevated intraligamentary signal intensity is a clear sign of ACL injuries [31]. On coronal or axial images, femoral-sided tears present as the "empty notch sign," meaning that there are no

ACL fibers at the expected femoral insertion point of the ACL [32]. Bone bruise patterns on the femoral condyles and tibial plateau are common secondary signs of ACL injuries. Bone bruise patterns are observable in over 80% of patients with acute ACL injuries. The most frequently observed pattern is a bone bruise on the posterolateral tibial plateau with a corresponding bone bruise on the lateral femoral condyle [33]. Increased translational and rotatory knee laxity and the associated change in tibio-femoral position can lead to a buckled conformation of the posterior cruciate ligament [34] or the "uncovered lateral meniscus" sign (Fig. 14.2) [32].

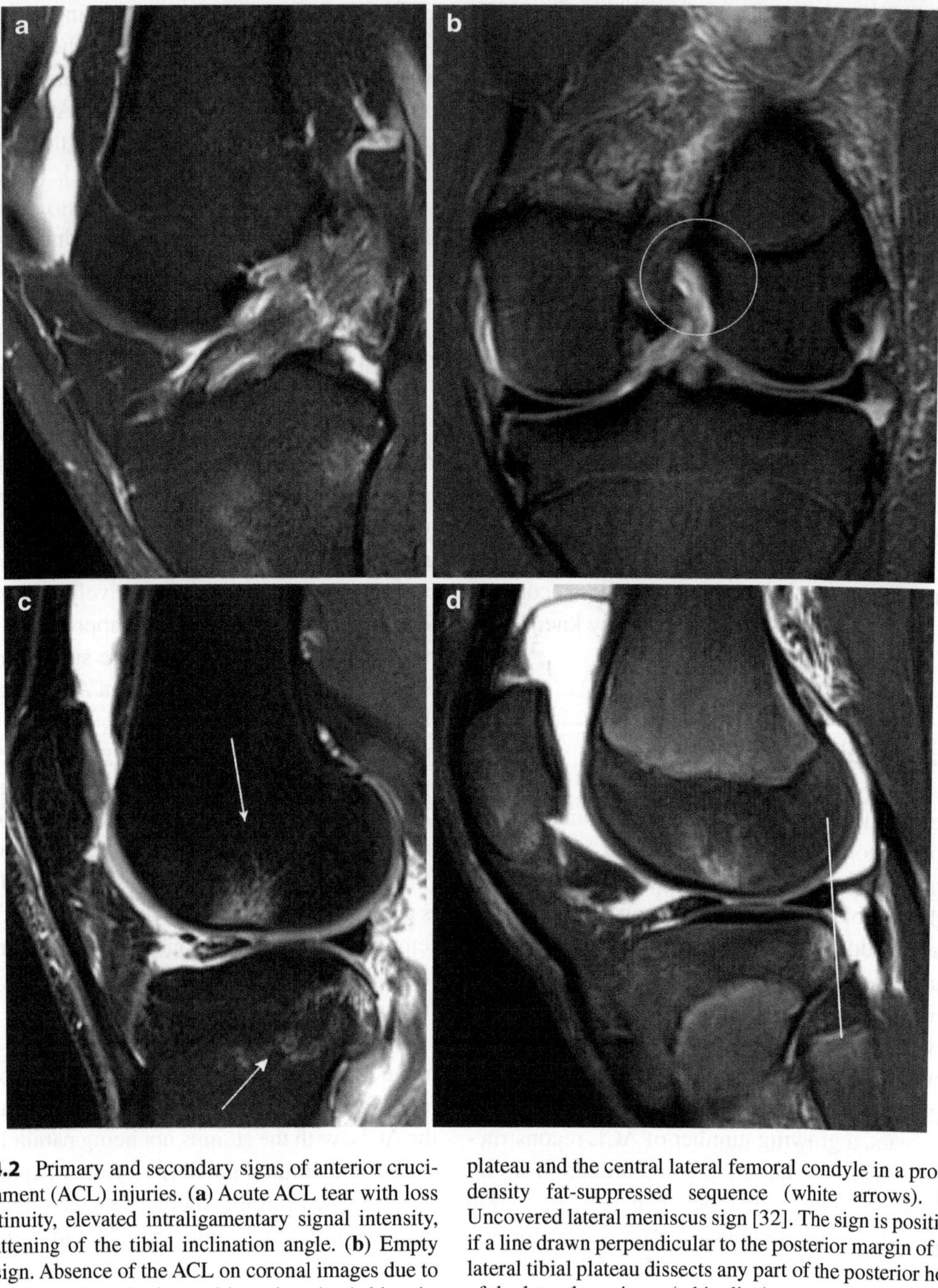

Fig. 14.2 Primary and secondary signs of anterior cruciate ligament (ACL) injuries. (**a**) Acute ACL tear with loss of continuity, elevated intraligamentary signal intensity, and flattening of the tibial inclination angle. (**b**) Empty notch sign. Absence of the ACL on coronal images due to a proximal tear from its femoral insertion site (white circle). (**c**) Bone bruise pattern on the posterolateral tibial plateau and the central lateral femoral condyle in a proton density fat-suppressed sequence (white arrows). (**d**) Uncovered lateral meniscus sign [32]. The sign is positive, if a line drawn perpendicular to the posterior margin of the lateral tibial plateau dissects any part of the posterior horn of the lateral meniscus (white line)

14.5 Nonoperative Treatment

Nonoperative treatment of ACL injuries usually consists of pain reduction by the use of nonsteroidal anti-inflammatory drugs as well as physical therapy structured in three phases [35]. In the acute phase, the primary goal is to reduce joint effusion and to restore ROM and strength of the injured leg. Combinations of cryotherapy, compression, stretching, and gentle strength training can be utilized to reach these goals. Strength training should encompass the whole lower extremity and the core, with emphasis on the quadriceps muscle [36]. Evidence exists for both the use of open and closed kinetic chain exercises, as well as neuromuscular electrical stimulation to restore quadriceps strength [37, 38]. Following the achievement of the aforementioned goals, rehabilitation turns toward neuromuscular training. In this phase, different dynamic weight-bearing exercises, including perturbation training, can be utilized [35]. Perturbation training is performed with the patient standing on unstable surfaces like rocker boards or roller boards. The therapist challenges the patient with unanticipated perturbations to the unstable surface. This type of training yields positive adaptations, in addition to a classic rehabilitation program [39].

The question of whether patients with ACL insufficiency should initially be treated operatively or nonoperatively has been debated. In one comparative study, statistically significant higher patient-reported outcomes (PROs; subjective IKDC) and less anterior tibial laxity (KT-1000) were reported for the ACLR group (n = 17; median age, 31 years) compared to the nonoperative control group (n = 15; median age, 33 years) at a median follow-up of 10.3 years. However, no difference was observed in terms of OA [40]. A second randomized controlled trial showed no statistically significant difference in PROs between patients with rehabilitation plus early ACLR (n = 62; mean age, 26.3 years) and patients with initial rehabilitation plus optional delayed ACLR (n = 59; mean age, 25.8 years) at two– and later 5–year follow-up [3, 41]. However, patients who were randomized to delayed surgery experienced more meniscal lesions, as well as cartilage damage [40]. Additionally, meniscus tears that occur secondary to chronic instability are often irreparable [42]. Furthermore, especially younger and more active patients are more prone to secondary adverse events, necessitating ACLR for this patient group [43].

14.6 Operative Treatment

14.6.1 Timing of Surgery

Although there is evidence supporting no differences in subjective and objective outcome measures regarding the timing of ACL surgery [44], delayed surgery has been associated with the worsening of concomitant injuries and with the development of new soft-tissue lesions [45, 46]. After ACLR, it has been hypothesized that the restoration of anterior-posterior and rotatory tibial laxity can prevent further damage to the meniscus, articular cartilage, or both due to the increased knee stability [21]. Currently, the American Academy of Orthopaedic Surgeons (AAOS) recommends initial nonoperative management of acute ACL injuries for 12 weeks followed by re-evaluation for the necessity of surgery. If surgery is to be performed, the AAOS recommends that it occurs within 5 months from injury in order to reduce recurrent instability and its associated sequelae [47]. Clinical studies have demonstrated that acute ACLR is not associated with negative outcomes (PROs, ROM, and stability testing), but might shorten the time period between injury and RTS [48, 49].

14.6.2 Graft Choices

There are many graft choices for ACLR available, including autografts and allografts. The most commonly used autografts include hamstring tendons (HT), bone-patellar-tendon-bone (BTB), and quadriceps tendon (QT) with or without a patellar bone block. For allografts, choices include BTB, tibialis anterior or posterior, HT, and Achilles tendon [50, 51]. The ideal graft should have similar properties to those of the native ligament, limit donor site morbidity, and allow for secure fixation and rapid incorporation [52]. The choice of graft should also be individualized to the patient's anatomy, sport, level of competition, age, and risk factors for failure [21, 50, 52]. Recent prospective cohort studies have identified risk factors for ACLR graft failure such as younger patient age, increased activity level, and the use of allograft; however, the only modifiable risk factor of these is graft choice [50]. It is thus critical that the operating surgeons have a lengthy discussion with the patient regarding which graft option(s) would be most suitable for the patient's individualized reconstruction, particularly given that the surgeon's recommendation has been shown to be one of the most important factors behind patient's graft choice [53].

14.6.2.1 Autograft

Bone-Patellar-Tendon-Bone (BTB) Autograft

The BTB autograft has been historically considered the "gold standard" for ACLR [50]. However, it is important to highlight that the paradigm is changing. Today, what is considered the current "gold standard" is the anatomic ACLR, regardless of graft, as it has been shown that anatomic ACLR reduces the risk of post-traumatic osteoarthritis at long-term follow-up [54]. Research has shown that BTB autografts have lower failure rates compared to HT autografts [55, 56]. Some studies also found less residual anterior knee laxity by the use of BTB autograft compared to HT autograft [57]. Proponents of BTB also cite the advantages of bone-to-bone healing compared to soft tissue-to-bone healing. It has been shown that bone grafts can be healed to the host bone within 6 weeks whereas soft tissue grafts take 8–12 weeks to fully incorporate [58]. Although outcomes with BTB have overall been good, there are complications associated with BTB that are almost exclusively related to the graft harvesting technique such as patellar fractures, patellar tendon rupture, and patellar tendinitis [59, 60]. One of the most common complications after ACLR that has been associated with BTB autograft use is anterior knee pain. In a meta-analysis of 21 studies, BTB autografts had an incidence of anterior knee pain of 17.4 versus 11.5% in HT autografts [57].

Hamstring Tendon (HT) Autograft

The advantages of HT autografts include greater CSA, no need to harm the extensor mechanism (i.e., quadriceps tendon, patellar tendon), and its option as a graft for the skeletally immature patients [61]. Regarding complications, HT autografts have been known to cause bone tunnel enlargement due to a "windshield wiper" effect from the suspensory fixation [62]. There is also a risk for residual hamstring weakness [61–63], making the graft a relative contraindication for athletes who heavily rely on their hamstrings for their athletic performance (i.e., sprinters). HT autografts have also been shown to have higher infection rates than BTB autografts [64]. As stated previously, recent studies and meta-analyses have found that HT autografts have higher failure rates compared to BTB autografts [55, 56]. Interestingly, a recent randomized controlled trial showed that combined ACLR and lateral extra-articular tenodesis (LET) by the use of single-bundle HT autograft resulted in a clinically relevant and statistically significant reduction in graft failure (4% (ACLR + LET) vs. 11% (ACLR)) and persistent rotatory knee laxity at 2 years follow-up [23].

Quadriceps Tendon (QT) Autograft

The QT autograft is a promising graft option for ACLR due to its versatility. The QT can be harvested as a full-thickness or a partial-thickness graft with or without a patellar bone block and is suitable for primary and revision ACLR [65]. It is associated with less donor-site morbidity, is longer, wider, has a higher tensile strength, and has about 50% more mass compared to BTB autograft with similar PROs at follow-up [21, 66, 67]. Additionally, the QT autograft is an excellent option for the skeletally immature patient [68]. Complications of QT harvesting include donor-site pain, bleeding (QT has a rich blood supply), retraction of the rectus femoris, and fractures of the patella after additional bone block excision [69, 70]. In a systematic review comparing QT, BTB, and HT autografts, there were no differences in knee laxity, PROs, or patient satisfaction between QT and either BTB or HT [69].

14.6.2.2 Allograft

The main advantages of allografts over autografts are shorter operative times, predictable graft size, no harvest-site morbidity, and easier recovery in the postoperative period [51]. Disadvantages of allografts include concerns for disease transmission and weakening of graft tissues that may occur during graft procurement and processing techniques and delayed incorporation [50]. Other disadvantages are that allografts are more costly than autografts [52] and that they are not as widely available in other countries outside of the United States [71]. Regarding outcomes after ACLR with allografts, studies have consistently shown that allografts have a higher re-rupture rate than autografts in young, athletic individuals [72, 73]. However, there are special circumstances in which allografts can be useful. These include multi-ligament knee injuries, revision ACLRs, patient preference, and cases in which autograft tissue is inadequate [74].

14.6.3 Individualized Anatomical ACL Reconstruction

In the past decade, the ACLR technique has evolved significantly with a shift away from transtibial drilling of the femoral tunnel to independent, transportal drilling, in an attempt to achieve lower failure and higher RTS rates [75]. The success of ACLR as measured by postoperative laxity testing, PROs, and failure rates is of tremendous importance regarding the development of future degenerative changes. Given the inability of nonanatomical ACLR to restore rotatory knee laxity [76], an individualized anatomical approach is recommended. A key component of anatomical ACLR is that there is variation between individuals in the size and shape of the native ACL and its tibial and femoral insertion site [77]. The basis of the individualized anatomical ACLR is a meticulous preoperative planning combined with intraoperative measurements of the femoral and tibial insertion site areas as well as the height and width of the intercondylar notch by using an arthroscopic ruler (Fig. 14.3). Preoperative determination of the CSA of the available grafts is performed on MRI scans since the goal is to achieve a mid-substance CSA of the graft that at minimum exceeds 50% of the tibial insertion area [13, 78]. Therefore, precise pre- and intraoperative measurements may support graft selection and thus minimize the risk of postoperative notch impingement. Understanding each patient's unique anatomy is crucial for individualized anatomical ACLR as the morphology of the native ACL and of the knee should direct the surgeon toward the most appropriate procedure for the patient (i.e., single- or double-bundle reconstruction, graft type, and tunnel size) [79].

14.6.4 Technical Aspects

Besides the graft choice, which best fits the patient's individual demands, many more technical aspects have to be considered to achieve optimal clinical results. Starting with precise anatomical femoral and tibial bone tunnel placement, many different types of graft fixation techniques are available. ACL graft fixation techniques can be divided into compression, expansion, suspension, cortical, extra-cortical, intra-tunnel, and extra-tunnel fixation depending on the mechanism and site of the technique [80]. There is no clear consensus regarding the superiority of any of the different graft fixation techniques. An in-depth discussion of all the methods of precise bone tunnel placement, the differences in the available fixation techniques, and the sometimes required concurrent interventions are beyond the scope of this chapter. However, the

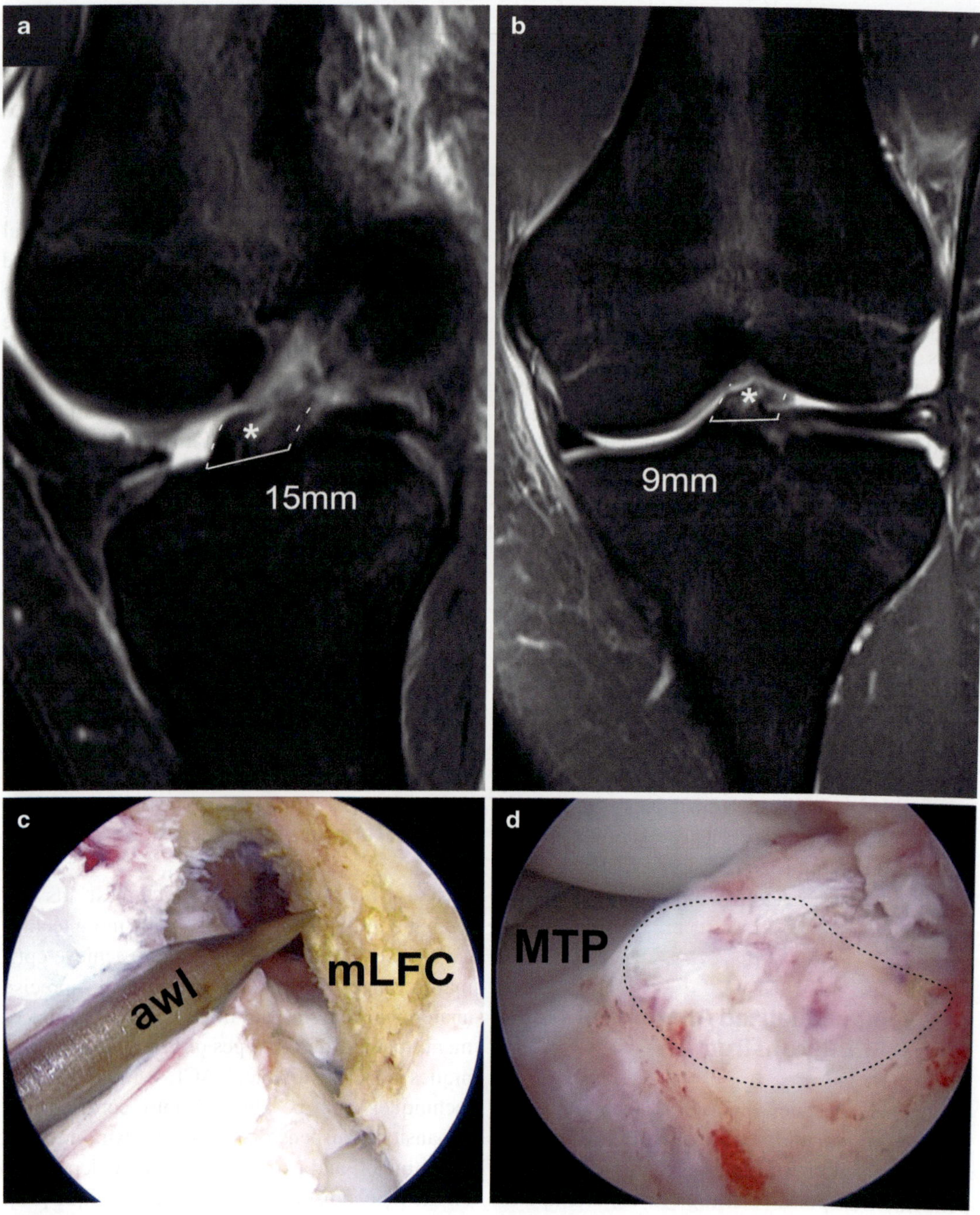

Fig. 14.3 Individualized anatomical anterior cruciate ligament (ACL) surgery (left knee). (**a**) Measurement of the length (15 millimeters (mm); mid-sagittal T2 weighted MRI layer) and (**b**) width (9 mm; coronal T2 weighted magnetic resonance imaging (MRI) layer) of the tibial insertion area of the ACL. (**c**) Marking the center of the femoral insertion area of the ACL using a microfracture awl (arthroscopic view). (**d**) Arthroscopic view of the tibial insertion area of the ACL (dotted line). (**e**) Arthroscopic length and (**f**) width measurement of the tibial insertion area of the ACL using an arthroscopic ruler (black lines); *, tibial ACL stump; **, arthroscopic ruler; mLFC, medial facet of the lateral femoral condyle; MTP, medial tibial plateau

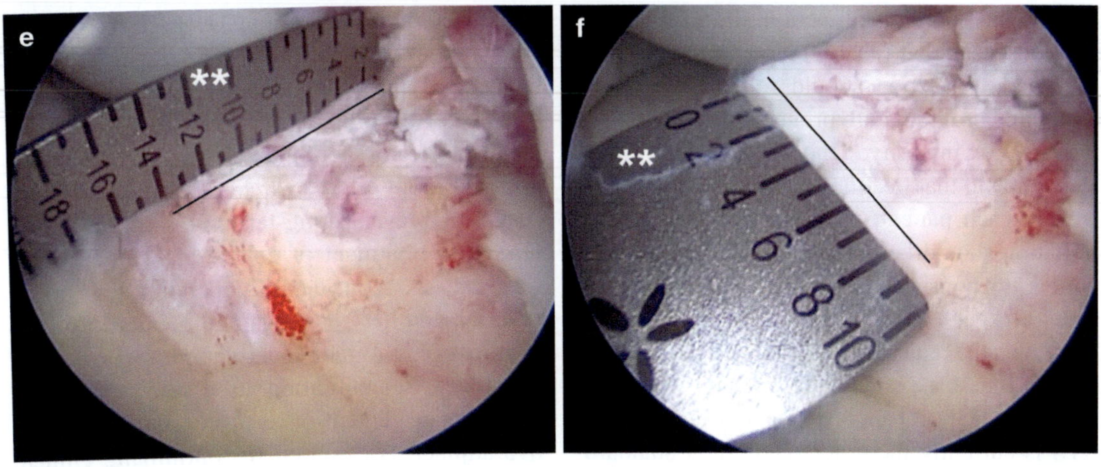

Fig. 14.3 (continued)

armamentarium of an ACLR surgeon should include the ability to perform physeal-sparing, physeal-respecting, and over-the-top ACLR techniques as well as concurrent procedures like LET, meniscal allograft transplantation, tibial slope-changing, and alignment-correcting osteotomies.

14.7 Combined Injuries

ACL tears are often accompanied by concomitant injuries, which remain unrecognized in many cases. These include meniscus, chondral, or bony injuries as well as lesions to the peripheral capsulo-ligamentous structures [81]. In addition to their unique functions in maintaining knee integrity, these structures are secondary stabilizers aiding the ACL in resisting anterior tibial translation and rotatory knee laxity. The lateral meniscus is a secondary stabilizer to the ACL in resisting anterolateral rotatory knee instability, with a concomitant lateral meniscal tear being a risk factor for a high-grade pivot shift [82, 83]. Similarly, the medial meniscus is a secondary stabilizer against anterior tibial translation, becoming a primary stabilizer with ACL insufficiency. Chronic ACL injury then predisposes the medial meniscus to degenerative tears when ACL function is not restored [84]. Therefore, repair of the menisci should be performed concomitantly with ACLR whenever possible, with the biological milieu of ACLR enhancing meniscus healing [85]. On the other hand, not all concomitant injuries require surgical treatment. Concomitant low-grade medial or lateral collateral ligament tears can often be managed nonoperatively when performing ACLR, with comparable outcomes to isolated ACL injury. Conversely, high-grade collateral ligament tears may require concomitant repair or reconstruction [86]. Operative treatment of the anterolateral complex (ALC) of the knee, including the capsule-osseous layer of the iliotibial band (i.e., anterolateral ligament) also remains controversial. Several cadaveric biomechanical studies have suggested that the anterolateral capsule of the knee lacks a distinct ligament that contributes substantively to the resistance of anterior tibial translation or internal tibial rotation [87, 88]. However, the multilayered ALC is known to serve an important role in rotatory knee stability [89], with LET likely warranted in certain cases, although the indications for its use with concurrent ACLR remain to be defined [23, 90].

14.8 Rehabilitation

Beginning in the 1990s, there was increasing emphasis on accelerated rehabilitation and increasingly early RTS [91]. While many questions remain on how to best regain or exceed the patient's preinjury function, there is a growing consensus that rehabilitation following ACL injury/reconstruction should follow a carefully structured program with progressively increasing movement intensity and complexity [92]. Recognizing that treatment should be individualized to each patient, progression through rehabilitation should be guided by the ability of the patient to consistently achieve objective measures of knee function before transitioning further down the rehabilitation pathway. To that end, purely time-based decision making regarding RTS should be abandoned and criterion-based RTS is currently the gold standard. A standardized, validated, and peer-reviewed RTS test, including an assessment of psychological readiness for RTS, should be passed prior to returning to full activity. For more specific guidelines on rehabilitation following ACL injury, please refer to recently published statements from the Panther Symposium ACL Consensus Meeting [92].

There is also emerging evidence to support delaying RTS until at least 9 months following ACL injury and/or ACLR, as side-to-side asymmetries in knee kinematics may remain at this time and imaging shows incomplete graft remodeling corresponding to compromised mechanical properties consistent with ongoing graft maturation [93, 94]. Taken together, the literature increasingly supports a progressive rehabilitation program following ACL injury/reconstruction with RTS testing performed no sooner than 9–12 months following injury/reconstruction.

14.9 Key Messages

ACL tears have an increasing incidence secondary to early sports participation and specialization and also continued sporting activity of an aging population. Nonoperative management of the injured ACL may be an initial evidence-based treatment strategy but is associated with an increased risk of damage to additional knee structures, namely the menisci and articular cartilage. Operative management, due to persistent knee instability or a desire to return to demanding, pivoting sports, should entail anatomic ACLR individualized to each patient. Autografts derived from the HT, BTB, and QT, possess unique advantages and disadvantages and are generally preferred over allografts. Repair of concomitant meniscus tears should be performed whenever possible in an effort to maintain meniscal function and prevent long-term OA of the knee. Operative treatment of collateral ligament injuries may be considered depending on injury severity, and LET has regained popularity in the context of high-grade anterolateral rotatory knee instability. Rehabilitation, used solely in nonoperative management or as a pre- and postoperative adjunct, is essential for optimizing clinical outcomes. RTS testing is useful and successful completion of a standardized, validated, and peer-reviewed RTS test is generally recommended prior to returning to full activity.

References

1. Filbay SR, Culvenor AG, Ackerman IN, Russell TG, Crossley KM. Quality of life in anterior cruciate ligament-deficient individuals: a systematic review and meta-analysis. Br J Sports Med. 2015;49(16):1033–41. https://doi.org/10.1136/bjsports-2015-094864.
2. Filbay SR, Ackerman IN, Russell TG, Macri EM, Crossley KM. Health-related quality of life after anterior cruciate ligament reconstruction: a systematic review. Am J Sports Med. 2014;42(5):1247–55. https://doi.org/10.1177/0363546513512774.
3. Frobell RB, Roos EM, Roos HP, Ranstam J, Lohmander LS. A randomized trial of treatment for acute anterior cruciate ligament tears. N Engl J Med. 2010;363(4):331–42. https://doi.org/10.1056/NEJMoa0907797.
4. Kaeding CC, Leger-St-Jean B, Magnussen RA. Epidemiology and diagnosis of anterior cruciate ligament injuries. Clin Sports Med. 2017;36(1):1–8. https://doi.org/10.1016/j.csm.2016.08.001.
5. Gornitzky AL, Lott A, Yellin JL, Fabricant PD, Lawrence JT, Ganley TJ. Sport-specific yearly risk and incidence of anterior cruciate ligament tears in high school athletes: a systematic review and meta-analysis. Am J Sports Med. 2016;44(10):2716–23. https://doi.org/10.1177/0363546515617742.
6. Beynnon BD, Vacek PM, Newell MK, Tourville TW, Smith HC, Shultz SJ, et al. The effects of level of competition, sport, and sex on the incidence of first-time noncontact anterior cruciate ligament injury. Am J Sports Med. 2014;42(8):1806–12. https://doi.org/10.1177/0363546514540862.

7. Ardern CL, Taylor NF, Feller JA, Webster KE. Fifty-five per cent return to competitive sport following anterior cruciate ligament reconstruction surgery: an updated systematic review and meta-analysis including aspects of physical functioning and contextual factors. Br J Sports Med. 2014;48(21):1543–52. https://doi.org/10.1136/bjsports-2013-093398.
8. Mai HT, Chun DS, Schneider AD, Erickson BJ, Freshman RD, Kester B, et al. Performance-based outcomes after anterior cruciate ligament reconstruction in professional athletes differ between sports. Am J Sports Med. 2017;45(10):2226–32. https://doi.org/10.1177/0363546517704834.
9. Alentorn-Geli E, Myer GD, Silvers HJ, Samitier G, Romero D, Lazaro-Haro C, et al. Prevention of non-contact anterior cruciate ligament injuries in soccer players. Part 1: mechanisms of injury and underlying risk factors. Knee Surg Sports Traumatol Arthrosc. 2009;17(7):705–29. https://doi.org/10.1007/s00167-009-0813-1.
10. Stijak L, Kadija M, Djulejic V, Aksic M, Petronijevic N, Markovic B, et al. The influence of sex hormones on anterior cruciate ligament rupture: female study. Knee Surg Sports Traumatol Arthrosc. 2015;23(9):2742–9. https://doi.org/10.1007/s00167-014-3077-3.
11. Amis AA, Dawkins GP. Functional anatomy of the anterior cruciate ligament. Fibre bundle actions related to ligament replacements and injuries. J Bone Joint Surg Br. 1991;73(2):260–7.
12. Zantop T, Petersen W, Sekiya JK, Musahl V, Fu FH. Anterior cruciate ligament anatomy and function relating to anatomical reconstruction. Knee Surg Sports Traumatol Arthrosc. 2006;14(10):982–92. https://doi.org/10.1007/s00167-006-0076-z.
13. Fujimaki Y, Thorhauer E, Sasaki Y, Smolinski P, Tashman S, Fu FH. Quantitative in situ analysis of the anterior cruciate ligament: length, Midsubstance cross-sectional area, and insertion site areas. Am J Sports Med. 2016;44(1):118–25. https://doi.org/10.1177/0363546515611641.
14. Hart A, Han Y, Martineau PA. The apex of the deep cartilage: a landmark and new technique to help identify femoral tunnel placement in anterior cruciate ligament reconstruction. Arthroscopy. 2015;31(9):1777–83. https://doi.org/10.1016/j.arthro.2015.03.026.
15. Zantop T, Herbort M, Raschke MJ, Fu FH, Petersen W. The role of the anteromedial and posterolateral bundles of the anterior cruciate ligament in anterior tibial translation and internal rotation. Am J Sports Med. 2007;35(2):223–7. https://doi.org/10.1177/0363546506294571.
16. Willinger L, Athwal KK, Williams A, Amis AA. An anterior cruciate ligament in vitro rupture model based on clinical imaging. Am J Sports Med. 2021;49(9):2387–95. https://doi.org/10.1177/03635465211017145.
17. Hughes JD, Rauer T, Gibbs CM, Musahl V. Diagnosis and treatment of rotatory knee instability. J Exp Orthop. 2019;6(1):48. https://doi.org/10.1186/s40634-019-0217-1.
18. Musahl V, Citak M, O'Loughlin PF, Choi D, Bedi A, Pearle AD. The effect of medial versus lateral meniscectomy on the stability of the anterior cruciate ligament-deficient knee. Am J Sports Med. 2010;38(8):1591–7. https://doi.org/10.1177/0363546510364402.
19. Sanders TL, Maradit Kremers H, Bryan AJ, Larson DR, Dahm DL, Levy BA, et al. Incidence of anterior cruciate ligament tears and reconstruction: a 21-year population-based study. Am J Sports Med. 2016;44(6):1502–7. https://doi.org/10.1177/0363546516629944.
20. Mall NA, Chalmers PN, Moric M, Tanaka MJ, Cole BJ, Bach BR Jr, et al. Incidence and trends of anterior cruciate ligament reconstruction in the United States. Am J Sports Med. 2014;42(10):2363–70. https://doi.org/10.1177/0363546514542796.
21. Musahl V, Karlsson J. Anterior cruciate ligament tear. N Engl J Med. 2019;380(24):2341–8. https://doi.org/10.1056/NEJMcp1805931.
22. Colombet P, Allard M, Bousquet V, DeLavigne C, Flurin P. L'histoire de la chirurgie du LCA. Maitrise Orthópedique. 1999;87:24–7.
23. Getgood AMJ, Bryant DM, Litchfield R, Heard M, McCormack RG, Rezansoff A, et al. Lateral extra-articular Tenodesis reduces failure of hamstring tendon autograft anterior cruciate ligament reconstruction: 2-year outcomes from the STABILITY study randomized clinical trial. Am J Sports Med. 2020;48(2):285–97. https://doi.org/10.1177/0363546519896333.
24. Hefti F, Müller W, Jakob RP, Stäubli HU. Evaluation of knee ligament injuries with the IKDC form. Knee Surg Sports Traumatol Arthrosc. 1993;1(3–4):226–34. https://doi.org/10.1007/bf01560215.
25. Naraghi AM, White LM. Imaging of athletic injuries of knee ligaments and menisci: sports imaging series. Radiology. 2016;281(1):23–40. https://doi.org/10.1148/radiol.2016152320.
26. Albers M, Shaikh H, Herbst E, Onishi K, Nagai K, Musahl V, et al. The iliotibial band and anterolateral capsule have a combined attachment to the Segond fracture. Knee Surg Sports Traumatol Arthrosc. 2018;26(5):1305–10. https://doi.org/10.1007/s00167-017-4549-z.
27. Cobby MJ, Schweitzer ME, Resnick D. The deep lateral femoral notch: an indirect sign of a torn anterior cruciate ligament. Radiology. 1992;184(3):855–8. https://doi.org/10.1148/radiology.184.3.1509079.
28. Li K, Du J, Huang LX, Ni L, Liu T, Yang HL. The diagnostic accuracy of magnetic resonance imaging for anterior cruciate ligament injury in comparison to arthroscopy: a meta-analysis. Sci Rep. 2017;7(1):7583. https://doi.org/10.1038/s41598-017-08133-4.
29. Gentili A, Seeger LL, Yao L, Do HM. Anterior cruciate ligament tear: indirect signs at MR imaging. Radiology. 1994;193(3):835–40. https://doi.org/10.1148/radiology.193.3.7972834.

30. Reid JC, Yonke B, Tompkins M. The angle of inclination of the native ACL in the coronal and sagittal planes. Knee Surg Sports Traumatol Arthrosc. 2017;25(4):1101–5. https://doi.org/10.1007/s00167-017-4419-8.
31. Lee JK, Yao L, Phelps CT, Wirth CR, Czajka J, Lozman J. Anterior cruciate ligament tears: MR imaging compared with arthroscopy and clinical tests. Radiology. 1988;166(3):861–4. https://doi.org/10.1148/radiology.166.3.3340785.
32. Tung GA, Davis LM, Wiggins ME, Fadale PD. Tears of the anterior cruciate ligament: primary and secondary signs at MR imaging. Radiology. 1993;188(3):661–7. https://doi.org/10.1148/radiology.188.3.8351329.
33. Patel SA, Hageman J, Quatman CE, Wordeman SC, Hewett TE. Prevalence and location of bone bruises associated with anterior cruciate ligament injury and implications for mechanism of injury: a systematic review. Sports Med. 2014;44(2):281–93. https://doi.org/10.1007/s40279-013-0116-z.
34. Robertson PL, Schweitzer ME, Bartolozzi AR, Ugoni A. Anterior cruciate ligament tears: evaluation of multiple signs with MR imaging. Radiology. 1994;193(3):829–34. https://doi.org/10.1148/radiology.193.3.7972833.
35. Paterno MV. Non-operative Care of the Patient with an ACL-deficient knee. Curr Rev Musculoskelet Med. 2017;10(3):322–7. https://doi.org/10.1007/s12178-017-9431-6.
36. Lewek M, Rudolph K, Axe M, Snyder-Mackler L. The effect of insufficient quadriceps strength on gait after anterior cruciate ligament reconstruction. Clin Biomech (Bristol, Avon). 2002;17(1):56–63. https://doi.org/10.1016/s0268-0033(01)00097-3.
37. Risberg MA, Lewek M, Snyder-Mackler L. A systematic review of evidence for anterior cruciate ligament rehabilitation: how much and what type? Phys Ther Sport. 2004;5(3):125–45. https://doi.org/10.1016/j.ptsp.2004.02.003.
38. Ediz L, Ceylan MF, Turktas U, Yanmis I, Hiz O. A randomized controlled trial of electrostimulation effects on effussion, swelling and pain recovery after anterior cruciate ligament reconstruction: a pilot study. Clin Rehabil. 2012;26(5):413–22. https://doi.org/10.1177/0269215511421029.
39. Fitzgerald GK, Axe MJ, Snyder-Mackler L. The efficacy of perturbation training in nonoperative anterior cruciate ligament rehabilitation programs for physical active individuals. Phys Ther. 2000;80(2):128–40.
40. Tsoukas D, Fotopoulos V, Basdekis G, Makridis KG. No difference in osteoarthritis after surgical and non-surgical treatment of ACL-injured knees after 10 years. Knee Surg Sports Traumatol Arthrosc. 2016;24(9):2953–9. https://doi.org/10.1007/s00167-015-3593-9.
41. Frobell RB, Roos HP, Roos EM, Roemer FW, Ranstam J, Lohmander LS. Treatment for acute anterior cruciate ligament tear: five year outcome of randomised trial. BMJ. 2013;346:f232. https://doi.org/10.1136/bmj.f232.
42. Mehl J, Otto A, Baldino JB, Achtnich A, Akoto R, Imhoff AB, et al. The ACL-deficient knee and the prevalence of meniscus and cartilage lesions: a systematic review and meta-analysis (CRD42017076897). Arch Orthop Trauma Surg. 2019;139(6):819–41. https://doi.org/10.1007/s00402-019-03128-4.
43. Dunn WR, Lyman S, Lincoln AE, Amoroso PJ, Wickiewicz T, Marx RG. The effect of anterior cruciate ligament reconstruction on the risk of knee reinjury. Am J Sports Med. 2004;32(8):1906–14. https://doi.org/10.1177/0363546504265006.
44. Andernord D, Karlsson J, Musahl V, Bhandari M, Fu FH, Samuelsson K. Timing of surgery of the anterior cruciate ligament. Arthroscopy. 2013;29(11):1863–71. https://doi.org/10.1016/j.arthro.2013.07.270.
45. Magnussen RA, Pedroza AD, Donaldson CT, Flanigan DC, Kaeding CC. Time from ACL injury to reconstruction and the prevalence of additional intra-articular pathology: is patient age an important factor? Knee Surg Sports Traumatol Arthrosc. 2013;21(9):2029–34. https://doi.org/10.1007/s00167-013-2380-8.
46. Sri-Ram K, Salmon LJ, Pinczewski LA, Roe JP. The incidence of secondary pathology after anterior cruciate ligament rupture in 5086 patients requiring ligament reconstruction. Bone Joint J. 2013;95-b(1):59–64. https://doi.org/10.1302/0301-620x.95b1.29636.
47. Shea KG, Carey JL. Management of anterior cruciate ligament injuries: evidence-based guideline. J Am Acad Orthop Surg. 2015;23(5):e1–5. https://doi.org/10.5435/jaaos-d-15-00094.
48. Herbst E, Hoser C, Gföller P, Hepperger C, Abermann E, Neumayer K, et al. Impact of surgical timing on the outcome of anterior cruciate ligament reconstruction. Knee Surg Sports Traumatol Arthrosc. 2017;25(2):569–77. https://doi.org/10.1007/s00167-016-4291-y.
49. von Essen C, Eriksson K, Barenius B. Acute ACL reconstruction shows superior clinical results and can be performed safely without an increased risk of developing arthrofibrosis. Knee Surg Sports Traumatol Arthrosc. 2019;28:2036. https://doi.org/10.1007/s00167-019-05722-w.
50. Duchman KR, Lynch TS, Spindler KP. Graft selection in anterior cruciate ligament surgery: who gets what and why? Clin Sports Med. 2017;36(1):25–33. https://doi.org/10.1016/j.csm.2016.08.013.
51. Kuhn MA, Ross G. Allografts in the treatment of anterior cruciate ligament injuries. Sports Med Arthrosc Rev. 2007;15(3):133–8. https://doi.org/10.1097/JSA.0b013e318134ecf6.
52. West RV, Harner CD. Graft selection in anterior cruciate ligament reconstruction. J Am Acad Orthop Surg. 2005;13(3):197–207. https://doi.org/10.5435/00124635-200505000-00006.
53. Cheung SC, Allen CR, Gallo RA, Ma CB, Feeley BT. Patients' attitudes and factors in their selection of grafts for anterior cruciate ligament reconstruction.

Knee. 2012;19(1):49–54. https://doi.org/10.1016/j.knee.2010.11.009.
54. Rothrauff BB, Jorge A, de Sa D, Kay J, Fu FH, Musahl V. Anatomic ACL reconstruction reduces risk of post-traumatic osteoarthritis: a systematic review with minimum 10-year follow-up. Knee Surg Sports Traumatol Arthrosc. 2020;28(4):1072–84. https://doi.org/10.1007/s00167-019-05665-2.
55. Persson A, Fjeldsgaard K, Gjertsen JE, Kjellsen AB, Engebretsen L, Hole RM, et al. Increased risk of revision with hamstring tendon grafts compared with patellar tendon grafts after anterior cruciate ligament reconstruction: a study of 12,643 patients from the Norwegian cruciate ligament registry, 2004–2012. Am J Sports Med. 2014;42(2):285–91. https://doi.org/10.1177/0363546513511419.
56. Laboute E, James-Belin E, Puig PL, Trouve P, Verhaeghe E. Graft failure is more frequent after hamstring than patellar tendon autograft. Knee Surg Sports Traumatol Arthrosc. 2018;26(12):3537–46. https://doi.org/10.1007/s00167-018-4982-7.
57. Freedman KB, D'Amato MJ, Nedeff DD, Kaz A, Bach BR Jr. Arthroscopic anterior cruciate ligament reconstruction: a metaanalysis comparing patellar tendon and hamstring tendon autografts. Am J Sports Med. 2003;31(1):2–11. https://doi.org/10.1177/03635465030310011501.
58. Rodeo SA, Arnoczky SP, Torzilli PA, Hidaka C, Warren RF. Tendon-healing in a bone tunnel. A biomechanical and histological study in the dog. J Bone Joint Surg Am. 1993;75(12):1795–803. https://doi.org/10.2106/00004623-199312000-00009.
59. Almekinders LC, Moore T, Freedman D, Taft TN. Post-operative problems following anterior cruciate ligament reconstruction. Knee Surg Sports Traumatol Arthrosc. 1995;3(2):78–82. https://doi.org/10.1007/bf01552379.
60. Marder RA, Raskind JR, Carroll M. Prospective evaluation of arthroscopically assisted anterior cruciate ligament reconstruction. Patellar tendon versus semitendinosus and gracilis tendons. Am J Sports Med. 1991;19(5):478–84. https://doi.org/10.1177/036354659101900510.
61. Cerulli G, Placella G, Sebastiani E, Tei MM, Speziali A, Manfreda F. ACL reconstruction: choosing the graft. Joints. 2013;1(1):18–24.
62. Samuelsson K, Andersson D, Karlsson J. Treatment of anterior cruciate ligament injuries with special reference to graft type and surgical technique: an assessment of randomized controlled trials. Arthroscopy. 2009;25(10):1139–74. https://doi.org/10.1016/j.arthro.2009.07.021.
63. Mohtadi NG, Chan DS, Dainty KN, Whelan DB. Patellar tendon versus hamstring tendon autograft for anterior cruciate ligament rupture in adults. Cochrane Database Syst Rev. 2011;2011(9):Cd005960:CD005960. https://doi.org/10.1002/14651858.CD005960.pub2.
64. Maletis GB, Inacio MC, Reynolds S, Desmond JL, Maletis MM, Funahashi TT. Incidence of postoperative anterior cruciate ligament reconstruction infections: graft choice makes a difference. Am J Sports Med. 2013;41(8):1780–5. https://doi.org/10.1177/0363546513490665.
65. Diermeier T, Tisherman R, Hughes J, Tulman M, Baum Coffey E, Fink C, et al. Quadriceps tendon anterior cruciate ligament reconstruction. Knee Surg Sports Traumatol Arthrosc. 2020;28:2644. https://doi.org/10.1007/s00167-020-05902-z.
66. Fulkerson JP, Langeland R. An alternative cruciate reconstruction graft: the central quadriceps tendon. Arthroscopy. 1995;11(2):252–4. https://doi.org/10.1016/0749-8063(95)90078-0.
67. Lund B, Nielsen T, Faunø P, Christiansen SE, Lind M. Is quadriceps tendon a better graft choice than patellar tendon? A prospective randomized study. Arthroscopy. 2014;30(5):593–8. https://doi.org/10.1016/j.arthro.2014.01.012.
68. Sheean AJ, Musahl V, Slone HS, Xerogeanes JW, Milinkovic D, Fink C, et al. Quadriceps tendon autograft for arthroscopic knee ligament reconstruction: use it now, use it often. Br J Sports Med. 2018;52(11):698–701. https://doi.org/10.1136/bjsports-2017-098769.
69. Slone HS, Romine SE, Premkumar A, Xerogeanes JW. Quadriceps tendon autograft for anterior cruciate ligament reconstruction: a comprehensive review of current literature and systematic review of clinical results. Arthroscopy. 2015;31(3):541–54. https://doi.org/10.1016/j.arthro.2014.11.010.
70. Fu FH, Rabuck SJ, West RV, Tashman S, Irrgang JJ. Patellar fractures after the harvest of a quadriceps tendon autograft with a bone block: a case series. Orthop J Sports Med. 2019;7(3):2325967119829051. https://doi.org/10.1177/2325967119829051.
71. Chechik O, Amar E, Khashan M, Lador R, Eyal G, Gold A. An international survey on anterior cruciate ligament reconstruction practices. Int Orthop. 2013;37(2):201–6. https://doi.org/10.1007/s00264-012-1611-9.
72. Kaeding CC, Pedroza AD, Reinke EK, Huston LJ, Spindler KP. Risk factors and predictors of subsequent ACL injury in either knee after ACL reconstruction: prospective analysis of 2488 primary ACL reconstructions from the MOON cohort. Am J Sports Med. 2015;43(7):1583–90. https://doi.org/10.1177/0363546515578836.
73. Wasserstein D, Sheth U, Cabrera A, Spindler KP. A systematic review of failed anterior cruciate ligament reconstruction with autograft compared with allograft in young patients. Sports Health. 2015;7(3):207–16. https://doi.org/10.1177/1941738115579030.
74. Tejwani SG, Chen J, Funahashi TT, Love R, Maletis GB. Revision risk after allograft anterior cruciate ligament reconstruction: association with graft processing techniques, patient characteristics, and graft type. Am J Sports Med. 2015;43(11):2696–705. https://doi.org/10.1177/0363546515589168.
75. Tibor L, Chan PH, Funahashi TT, Wyatt R, Maletis GB, Inacio MC. Surgical technique trends in primary

ACL reconstruction from 2007 to 2014. J Bone Joint Surg Am. 2016;98(13):1079–89. https://doi.org/10.2106/jbjs.15.00881.
76. Kondo E, Merican AM, Yasuda K, Amis AA. Biomechanical comparison of anatomic double-bundle, anatomic single-bundle, and nonanatomic single-bundle anterior cruciate ligament reconstructions. Am J Sports Med. 2011;39(2):279–88. https://doi.org/10.1177/0363546510392350.
77. Kopf S, Pombo MW, Szczodry M, Irrgang JJ, Fu FH. Size variability of the human anterior cruciate ligament insertion sites. Am J Sports Med. 2011;39(1):108–13. https://doi.org/10.1177/0363546510377399.
78. Offerhaus C, Albers M, Nagai K, Arner JW, Höher J, Musahl V, et al. Individualized anterior cruciate ligament graft matching: in vivo comparison of cross-sectional areas of hamstring, patellar, and quadriceps tendon grafts and ACL insertion area. Am J Sports Med. 2018;46(11):2646–52. https://doi.org/10.1177/0363546518786032.
79. Karlsson J, Irrgang JJ, van Eck CF, Samuelsson K, Mejia HA, Fu FH. Anatomic single- and double-bundle anterior cruciate ligament reconstruction, part 2: clinical application of surgical technique. Am J Sports Med. 2011;39(9):2016–26. https://doi.org/10.1177/0363546511402660.
80. Zeng C, Lei G, Gao S, Luo W. Methods and devices for graft fixation in anterior cruciate ligament reconstruction. Cochrane Database Syst Rev. 2018;2018(6):CD010730. https://doi.org/10.1002/14651858.CD010730.pub2.
81. Gillquist J, Messner K. Anterior cruciate ligament reconstruction and the long-term incidence of gonarthrosis. Sports Med. 1999;27(3):143–56. https://doi.org/10.2165/00007256-199927030-00001.
82. Tanaka M, Vyas D, Moloney G, Bedi A, Pearle AD, Musahl V. What does it take to have a high-grade pivot shift? Knee Surg Sports Traumatol Arthrosc. 2012;20(4):737–42. https://doi.org/10.1007/s00167-011-1866-5.
83. Musahl V, Rahnemai-Azar AA, Costello J, Arner JW, Fu FH, Hoshino Y, et al. The influence of meniscal and anterolateral capsular injury on knee laxity in patients with anterior cruciate ligament injuries. Am J Sports Med. 2016;44(12):3126–31. https://doi.org/10.1177/0363546516659649.
84. Nakamae A, Adachi N, Deie M, Ishikawa M, Nakasa T, Ikuta Y, et al. Risk factors for progression of articular cartilage damage after anatomical anterior cruciate ligament reconstruction: a second-look arthroscopic evaluation. Bone Joint J. 2018;100-b(3):285–93. https://doi.org/10.1302/0301-620x.100b3.Bjj-2017-0837.R1.
85. Westermann RW, Wright RW, Spindler KP, Huston LJ, Wolf BR. Meniscal repair with concurrent anterior cruciate ligament reconstruction: operative success and patient outcomes at 6-year follow-up. Am J Sports Med. 2014;42(9):2184–92. https://doi.org/10.1177/0363546514536022.
86. Spang RC 3rd, Getgood A, Strickland SM, Amendola AN, Gomoll AH. Optimizing anterior cruciate ligament (ACL) outcomes: what Else needs fixing besides the ACL? Instr Course Lect. 2020;69:653–60.
87. Dombrowski ME, Costello JM, Ohashi B, Murawski CD, Rothrauff BB, Arilla FV, et al. Macroscopic anatomical, histological and magnetic resonance imaging correlation of the lateral capsule of the knee. Knee Surg Sports Traumatol Arthrosc. 2016;24(9):2854–60. https://doi.org/10.1007/s00167-015-3517-8.
88. Musahl V, Herbst E, Burnham JM, Fu FH. The anterolateral complex and anterolateral ligament of the knee. J Am Acad Orthop Surg. 2018;26(8):261–7. https://doi.org/10.5435/jaaos-d-16-00758.
89. Getgood A, Brown C, Lording T, Amis A, Claes S, Geeslin A, et al. The anterolateral complex of the knee: results from the international ALC consensus group meeting. Knee Surg Sports Traumatol Arthrosc. 2019;27(1):166–76. https://doi.org/10.1007/s00167-018-5072-6.
90. Herbst E, Arilla FV, Guenther D, Yacuzzi C, Rahnemai-Azar AA, Fu FH, et al. Lateral extra-articular Tenodesis has no effect in knees with isolated anterior cruciate ligament injury. Arthroscopy. 2018;34(1):251–60. https://doi.org/10.1016/j.arthro.2017.08.258.
91. Shelbourne KD, Nitz P. Accelerated rehabilitation after anterior cruciate ligament reconstruction. Am J Sports Med. 1990;18(3):292–9. https://doi.org/10.1177/036354659001800313.
92. Meredith SJ, Rauer T, Chmielewski TL, Fink C, Diermeier T, Rothrauff BB, et al. Return to sport after anterior cruciate ligament injury: panther symposium ACL injury return to sport consensus group. Knee Surg Sports Traumatol Arthrosc. 2020;28:2403. https://doi.org/10.1007/s00167-020-06009-1.
93. Read PJ, Michael Auliffe S, Wilson MG, Graham-Smith P. Lower limb kinetic asymmetries in professional soccer players with and without anterior cruciate ligament reconstruction: nine months is not enough time to restore "functional" symmetry or return to performance. Am J Sports Med. 2020;0(0):1365. https://doi.org/10.1177/0363546520912218.
94. Hofbauer M, Soldati F, Szomolanyi P, Trattnig S, Bartolucci F, Fu F, et al. Hamstring tendon autografts do not show complete graft maturity 6 months postoperatively after anterior cruciate ligament reconstruction. Knee Surg Sports Traumatol Arthrosc. 2018;27:27. https://doi.org/10.1007/s00167-018-5033-0.
95. Irarrázaval S, Albers M, Chao T, Fu FH. Gross, arthroscopic, and radiographic anatomies of the anterior cruciate ligament: foundations for anterior cruciate ligament surgery. Clin Sports Med. 2017;36(1):9–23.

15 Age-Based Outcomes After Anterior Cruciate Ligament Reconstruction

Richard Danilkowicz, Nathan L. Grimm, Jeffrey Liles, and Annunziato Amendola

15.1 Introduction

Anterior cruciate ligament (ACL) reconstruction is a commonly performed procedure in young adults between the ages of 14 and 25, with the bulk of injuries sustained during athletic participation [1]. However, in recent years there has been a trend toward increased incidence of both younger and older patients undergoing ACL reconstruction (ACLR), as sports participation and specialization have occurred earlier in life and older individuals seek to continue their athletic endeavors [2–5]. With the changing landscape of individuals undergoing ACLR, it is important to understand how outcomes may differ among the various age cohorts undergoing ACL reconstruction for proper patient counseling and expectation setting. Currently, the bulk of the literature has focused on the young adult population; however, this chapter serves to highlight both the technical considerations and post-operative outcomes of ACL reconstruction as they pertain to adolescent, young adult, and adult populations.

R. Danilkowicz · J. Liles
Department of Orthopaedic Surgery, Division of Sports Medicine, Duke University Medical Center, Durham, NC, USA

A. Amendola (✉)
Department of Orthopaedic Surgery, Duke University Medical Center, Durham, NC, USA
e-mail: ned.amendola@duke.edu

N. L. Grimm
Idaho Sports Medicine Institute, Boise, ID, USA

15.2 Considerations for Skeletally Immature Patients

The surgical technique utilized for ACL reconstruction is based on the age and physeal status of the patient. Commonly, the bone age of the patient is used to determine the amount of estimated growth remaining, using age 14 for girls and 16 for boys as the age of growth cessation, which can help guide the choice of surgical technique. There are a number of published techniques to determine skeletal bone age, with hand radiographs being the most commonly employed [6]. In patients with open physes, a physeal sparing approach is utilized, unless the patient is nearing the end of their growth phase as determined on radiographs. For patients with an estimated 3 years or more of growth remaining, an all epiphyseal approach is commonly employed and has been shown to be both safe and efficacious [7]. Further stratification within the all epiphyseal cohort consists of the Micheli procedure, which utilizes an autologous IT band graft, and the all epiphyseal technique is more similar to the standard transphyseal approach by creating drill tunnels and utilizing a graft of the surgeon's

M. Karahan et al. (eds.), *Anterior Cruciate Ligament Injuries Throughout the Life Span*,
https://doi.org/10.1007/978-3-031-84501-7_15

choice [8, 9]. In patients with 2 to 3 years of growth remaining, a partial transphyseal approach may be utilized. And in patients with one or less years of growth remaining, a standard transphyseal approach is generally recommended as is used in adult patients.

Long-term outcome studies on the physeal sparing approaches have shown positive results. In the largest study of its kind, encompassing 240 Micheli-style ACL reconstructions, outcomes included 96.8% grade A Lachman and 98.8% grade A pivot-shift tests, with pedi-International Knee Documentation Committee (IKDC) and Lysholm scores both >93 and no limb length or angular deformities. There was a graft re-rupture rate of 6.6% among this cohort [10]. In the all epiphyseal technique utilizing drill tunnels, outcomes have also been positive. A 2020 systematic review by Gupta et al. reported IKDC scores >90, a return to sport rate of 93.2%, and a complication rate of 9.8% overall [11]. These findings were corroborated in a 2024 systematic review by Migliorini et al. who compared all epiphyseal with trans-physeal approaches in skeletally immature and found no difference in IKDC, Lysholm, and Tegner scores or complication rates, but an improved return to sport rate in the all epiphyseal group [12]. In patients with open physes who underwent trans-physeal reconstruction, the literature is mixed but does report potential risks of limb shortening on the femoral side up to 10 mm and angulation of 2 degrees in up to 24 and 82%, respectively, and should be taken into consideration [13, 14].

15.3 Age and ACL Reconstruction Failure

Due to the prevalence of ACLR performed annually, studies attempting to identify patient characteristics that could contribute to failure are common, particularly in light of re-tear incidence up to 7.7% [15]. Age is consistently cited as a key variable as younger patients tend to participate more in organized sports and have a goal to return to these activities. In one large study of 21,304 patients within the Kaiser database, age of <21 years was associated with a 7.76x risk of revision, with all graft types included, and approximately 31% of all revisions occurring in patients under 21 years old [16, 17]. A recent systematic review and meta-analysis found age < 25 to be associated with an odds of revision 4x that of patients >25 years old, with odds increasing 7% per year decrease in age [15]. While these figures appear alarmingly high, they are corroborated by other large national database studies out of Denmark, Norway, and Sweden where patients younger than 20 years of age had significantly higher rates of revision than older cohorts [18–20]. In a predictive model created by the MOON group, patients aged 10–19 years had the highest percentage of graft failures, with the odds of graft rupture increasing 2.3 times for each 10-year decrease in age among cohorts [21]. While seemingly high, these data are not altogether surprising as younger patients tend to be those who return to strenuous activities at higher rates than their older counterparts and will likely stress their reconstructed ACL for longer aggregate time periods, which increases the risk of re-injury.

15.4 Graft Choice

ACLR outcomes by graft choice is another area of significant study. Due to the numerous available options between autografts and allografts, with their associated benefits and drawbacks, there is a plethora of research available that compares the various graft choices. The recent trend has been away from the use of allografts, particularly in the younger and more active populations as allografts have been found to have 5x the odds of rupture compared to autografts, with a meta-analysis finding a 5% rupture rate in autografts compared with a 14% in allografts [21–24]. Despite the higher failure rates, allografts continue to be routinely utilized in older populations due primarily to the lack of harvest morbidity, surgical postoperative pain, and easier early rehabilitation, particularly in patients who do not plan to return to overly strenuous activity postoperatively. An important caveat to the use of allograft is consideration of their preparation technique, which should be non-irradiated and fresh-frozen without the use of chemical processing, each of which has been associated with worse outcomes [25]. While age is seldom the primary variable of interest in the graft survivorship studies, regression analysis often allows for the stratification of outcomes among varying age groups.

Hamstring Hamstring autograft has historically been a popular graft choice for younger patients undergoing ACLR, with a track record of successful outcomes and widespread use worldwide. While this graft is considered to have less donor site morbidity than other autograft options, the issue of patient hamstring size has brought into question the applicability among all comers [26]. Despite the concerns, large studies have shown hamstring autografts to have similar graft failure rates as hamstring autografts supplemented with allograft tissue, often referred to as hybrid grafts, with a failure rate of 7.5 and 8.0%, respectively, and no difference in Lysholm, Tender, or subjective IKDC scores [27]. Regarding the age of patients, Magnussen et al. found that the odds ratio (OR) for revision in patients <20 years of age vs > 20 was 18.97, which was far greater than the OR comparing patients with smaller grafts vs larger grafts (OR 2.20) [28]. Similarly, in a study of 98 patients, Kamien et al. found failure rates in hamstring autografts to be higher in patients under 25 years of age [29]. In a study of hamstring autografts vs allografts in a population with an average age of 27 and 31 respectively, no difference in failure rate was found [30]. Naot et al. investigated potential molecular level explanations for the age-related outcomes by comparing hamstring tendon samples from patients <20 and > 20 years of age and found younger tendons to have a higher elastic modulus and overrepresentation of certain genes that may in part play into the higher re-rupture rates [31].

Bone Patellar Tendon Bone Bone patellar tendon bone (BPTB) grafts have been a popular choice among highly active patients, particularly those who participate in collision sports, with closed physes due to their consistent size and strength in addition to the reliable bone to bone healing. There have been a number of studies that have compared autograft to allograft for outcome measures, many of which also cite age as a contributing demographic in the analysis. In a cohort of 90 patients aged 18 and younger with closed physes, Ellis et al. found a 15x higher revision rate in BPTB allografts (35%) in comparison to autografts (3%), with the majority of the failures attributed to premature return to sports [32]. Similarly, both Barrett et al. and Maletis et al. found an increased risk of failure in allografts in patients under 40 years of age, particularly in high-activity patients [16, 33]. The failure rate discrepancy does not appear as clear cut in older patients, however. Barber et al. compared patients under and over 40 years of age and found no difference in outcomes on physical exam and patient-reported outcome measures [34]. Barrett et al. found no significant difference in failure or clinical outcomes in patients over 40 who received an allograft vs autograft BPTB ACLR [35]. Within cohorts of young patients under 21 years old undergoing autograft reconstruction, Mittal et al. found a 2.9% graft rupture rate which is in line with New Zealand registry data of 2.7% failure rate for patients 20 and older, with each study showing a lower rate in comparison to hamstring autograft that persisted into older age cohorts [36, 37].

Quadriceps Tendon Use of the quadriceps tendon (QT) as an ACL graft choice, with or without a bone plug, has been gaining in popularity over the recent years in both the primary and revision settings [38, 39]. While long-term outcome data are less prevalent than the other graft options, the early results are promising. A study by Gagliardi et al. of 81 patients between the ages of 10–18 showed a failure rate of only 1.2% [40]. In a study of nearly 1000 patients with an average age of 20 years old, there was a reported failure rate of 4.2% [41]. A recent meta-analysis that included 1362 autograft QT ACLR with an average age of 29 years at surgery presented a failure rate of only 3.45%, which was statistically lower than a similar cohort of hamstring grafts [42]. The quad tendon graft also appears to be a viable option in older patients, with Meena et al. reporting prospectively on 57 patients with a mean age of 54.9 years old undergoing quad tendon autograft ACLR with each patient returning to desired sports and no failures [43]. Similarly, systematic reviews that include patients up to age 57 show a failure rate of 2.8% [44].

Tibial Slope In addition to graft choice considerations, the role of posterior tibial slope and ACLR failure should be considered, particularly in the younger population who are more prone to re-rupture. Multiple studies, clinical and biomechanical, have outlined the connection between reconstruction failure and increased posterior tibial slope, particularly as it pertains to increased forces across the ACL [45–48]. Waiwaiole et al. looked at the association between ACL injury, tibial slope, and various demographic parameters and found increased slope and decreased age to both be significant factors for injury in their study cohort [49]. Salmon et al. presented very strong 20-year follow-up data on a single surgeon series of 200 patients, finding ACL graft failure was significantly more common in adolescent males, with the strongest predictor of failure being a posterior tibial slope > 12 degrees [50]. Based on the clear connection between both age and tibial slope on ACLR failure, the addition of a slope-correcting osteotomy as part of a single or staged procedure has gained popularity in patients with increased tibial slope [51–55].

15.5 ACL Repair

A treatment option that has recently regained significant interest is ACL repair, primarily due to surgical and scientific advances. ACL repair is not a new procedure but in fact, was the primary mode of ACL tear treatment prior to the advances in arthroscopy technology. The field started to move away from repair due to the high failure rates reported in the literature among active patients, as evidenced in a paper by Feagin and Curl which documented repair failure in 17 of 32 West Point cadets who underwent the procedure at 5 years follow-up, results which were later confirmed in the cohort by Taylor et al. at 30 years follow-up [56, 57]. After the theorizing failure of direct repair is due to the dissolution of the fibrin clot that provides a healing scaffold, Murray et al. published their results of the BEAR trial in 2020, which reported on the outcomes of a primary repair augmented with a protein implant. The results concluded a non-inferiority in comparison to autograft reconstruction at 2 years follow-up, as evidenced by IKDC subjective scores, side-to-side knee AP laxity in addition to greater hamstring muscle strength in cohorts of young, active patients with mid-substance tears [58]. Additional randomized trials are currently being performed to compare repair and reconstruction as well [59]. At the minimum, these non-inferiority results and increased interest provide evidence to support additional research into the development of technical and biological advances that further validate the repair procedure.

15.6 Return to Sport/Activity

The ability to return to sport (RTS) or activity after ACLR is arguably one of the most important metrics tracked. More specifically, the ability to return to preinjury levels of activity is of utmost concern to many patients and families when being counseled on their options for ACLR. It is important to note that a patient may report significantly improved pain and function when comparing pre and post-operative time points, but it does not correlate directly to the ability to resume prior activities, a point that may not be represented in all standardized patient-reported outcome measures. A systematic review of level 1 studies by Harris et al. looked into published return to sport guidelines and found that only 5 of 49 included studies reported on whether a patient was able to successfully return to sport, highlighting the variability in the quality of available literature [60]. Overall return to sport rates is also variable in the literature, with one recent systematic review of 13 studies using a minimum of 1 year follow-up reporting RTS of 56–100%, and return to preinjury levels at 13–69%, which is in line with prior published studies [61, 62]. Despite the consensus thought that younger patients heal better after surgery, the literature is less convincing on the effect of age on the ability to return to sport after ACLR. There are a number of studies in the literature that do not identify age as a factor in the ability to either return to sport or return at prior levels of activity, while others cite age as a key demographic factor [63–65]. Despite the heterogenicity in the literature, it is still useful to attempt and stratify RTS by age.

In a study of 250 patients under 14 years of age, Chicorelli et al. found that 96% of patients were able to return to sport at their prior level of activity after ACLR based on survey response, with 85% of those returning within 12 months of surgery [66]. Similarly, Cordasco et al. reported on RTS rates in patients 20 years old or less at the time of surgery and found the ability to RTS ranging from 85 to 100% among three cohorts, with a lower rate of return for partial transphyseal and complete transphyseal with hamstring autograft and the highest return in all-epiphyseal technique with hamstring autograft patients [67]. Edwards et al. utilized a cohort of level I/II athletes and found that age 25 was a statistically significant inflection point of ability to return to sport, which is a similar age (25 vs 28) reported by Ardern et al. [64, 68] Finally, a large meta-analysis comprising 1156 ACLR with a mean age of 14.3 (range 6–19) found a RTS rate of 92% with return to preinjury levels of 81.0%, further indicating the success among this age group [69].

In patients older than 20 years of age, the ability to return to sport and the desired level of activity is less promising. Webster et al. reported only 53% of patients >20 years old were able to return to "strenuous activity." [70] Deakon and Zarnett reported on a cohort of patients over 40 years old who underwent ACLR with bone-patella-bone graft and found that only 15% of patients returned to competitive sports involving pivoting, 65% returned to recreational sports that included pivoting maneuvers, in comparison to 60% of patients returning to competitive pivoting sports under the age of 40 [71]. Similarly, Osti et al. studied 20 patients over 50 years of age after ACLR and found that 60% were able to return to sport at a preinjury level, in comparison to 90% of patients younger than 30 years old in the other arm of the study [72]. While age is clearly a factor in the ability to return to prior levels of activity after ACLR, it is important to note that other confounding factors undoubtedly play a role in addition to age that may not be reflected in the data as there are a multitude of reasons why individuals may not return.

15.7 Conclusion

In conclusion, the patterns of ACL injury and reconstruction are evolving, as well as the specific reconstructive techniques. Patients at both ends of the age spectrum are sustaining ACL injuries in higher numbers, although younger patients continue to sustain higher re-injury rates but also a greater ability to return to their prior activities. Both technique and graft choice should also be considered when planning for ACLR, with a preference toward physeal sparing techniques in skeletally immature patients and autografts in younger patients.

References

1. Sanders TL, Maradit Kremers H, Bryan AJ, et al. Incidence of anterior cruciate ligament tears and reconstruction: a 21-year population-based study. Am J Sports Med. 2016;44(6):1502–7. https://doi.org/10.1177/0363546516629944.
2. Beck NA, Lawrence JTR, Nordin JD, DeFor TA, Tompkins M. ACL tears in school-aged children and adolescents over 20 years. Pediatrics. 2017;139(3):e20161877. https://doi.org/10.1542/peds.2016-1877.
3. Werner BC, Yang S, Looney AM, Gwathmey FW Jr. Trends in pediatric and adolescent anterior cru-

ciate ligament injury and reconstruction. J Pediatr Orthop. 2016;36(5):447–52. https://doi.org/10.1097/bpo.0000000000000482.
4. Nicholls M, Aspelund T, Ingvarsson T, Briem K. Nationwide study highlights a second peak in ACL tears for women in their early forties. Knee Surg Sports Traumatol Arthrosc. 2018;26(2):648–54. https://doi.org/10.1007/s00167-017-4807-0.
5. Mall NA, Chalmers PN, Moric M, et al. Incidence and trends of anterior cruciate ligament reconstruction in the United States. Am J Sports Med. 2014;42(10):2363–70. https://doi.org/10.1177/0363546514542796.
6. Cavallo F, Mohn A, Chiarelli F, Giannini C. Evaluation of bone age in children: a mini-review. Front Pediatr. 2021;9:580314. https://doi.org/10.3389/fped.2021.580314.
7. Cruz AI Jr, Fabricant PD, McGraw M, Rozell JC, Ganley TJ, Wells L. All-epiphyseal ACL reconstruction in children: review of safety and early complications. J Pediatr Orthop. 2017;37(3):204–9. https://doi.org/10.1097/bpo.0000000000000606.
8. Anderson AF, Anderson CN. Transepiphyseal anterior cruciate ligament reconstruction in pediatric patients: surgical technique. Sports Health. 2009;1(1):76–80. https://doi.org/10.1177/1941738108326703.
9. Kocher MS, Garg S, Micheli LJ. Physeal sparing reconstruction of the anterior cruciate ligament in skeletally immature prepubescent children and adolescents. Surgical technique. J Bone Joint Surg Am. 2006;88(Suppl 1 Pt 2):283–93. https://doi.org/10.2106/jbjs.F.00441.
10. Kocher MS, Heyworth BE, Fabricant PD, Tepolt FA, Micheli LJ. Outcomes of Physeal-sparing ACL reconstruction with iliotibial band autograft in skeletally immature prepubescent children. J Bone Joint Surg Am. 2018;100(13):1087–94. https://doi.org/10.2106/jbjs.17.01327.
11. Gupta A, Tejpal T, Shanmugaraj A, Horner NS, Gohal C, Khan M. All-epiphyseal anterior cruciate ligament reconstruction produces good functional outcomes and low complication rates in pediatric patients: a systematic review. Knee Surg Sports Traumatol Arthrosc. 2020;28(8):2444–52. https://doi.org/10.1007/s00167-020-06085-3.
12. Migliorini F, Pilone M, Memminger MK, Eschweiler J, Giorgino R, Maffulli N. All-epiphyseal anterior cruciate ligament reconstruction yields superior sports performances than the trans-epiphyseal technique in skeletally immature patients: a systematic review. J Orthop Traumatol. 2024;25(1):7. https://doi.org/10.1186/s10195-024-00751-9.
13. Faunø P, Rømer L, Nielsen T, Lind M. The risk of transphyseal drilling in skeletally immature patients with anterior cruciate ligament injury. Orthop J Sports Med. 2016;4(9):2325967116664685. https://doi.org/10.1177/2325967116664685.
14. Calvo R, Figueroa D, Gili F, et al. Transphyseal anterior cruciate ligament reconstruction in patients with open physes: 10-year follow-up study. Am J Sports Med. 2015;43(2):289–94. https://doi.org/10.1177/0363546514557939.
15. Zhao D, Pan JK, Lin FZ, et al. Risk factors for revision or Rerupture after anterior cruciate ligament reconstruction: a systematic review and meta-analysis. Am J Sports Med. 2023;51(11):3053–75. https://doi.org/10.1177/03635465221119787.
16. Maletis GB, Chen J, Inacio MC, Funahashi TT. Age-related risk factors for revision anterior cruciate ligament reconstruction: a cohort study of 21,304 patients from the Kaiser Permanente anterior cruciate ligament registry. Am J Sports Med. 2016;44(2):331–6. https://doi.org/10.1177/0363546515614813.
17. Maletis GB, Inacio MC, Funahashi TT. Analysis of 16,192 anterior cruciate ligament reconstructions from a community-based registry. Am J Sports Med. 2013;41(9):2090–8. https://doi.org/10.1177/0363546513493589.
18. Lind M, Menhert F, Pedersen AB. Incidence and outcome after revision anterior cruciate ligament reconstruction: results from the Danish registry for knee ligament reconstructions. Am J Sports Med. 2012;40(7):1551–7. https://doi.org/10.1177/0363546512446000.
19. Persson A, Fjeldsgaard K, Gjertsen JE, et al. Increased risk of revision with hamstring tendon grafts compared with patellar tendon grafts after anterior cruciate ligament reconstruction: a study of 12,643 patients from the Norwegian cruciate ligament registry, 2004-2012. Am J Sports Med. 2014;42(2):285–91. https://doi.org/10.1177/0363546513511419.
20. Ahldén M, Sernert N, Karlsson J, Kartus J. Outcome of anterior cruciate ligament reconstruction with emphasis on sex-related differences. Scand J Med Sci Sports. 2012;22(5):618–26. https://doi.org/10.1111/j.1600-0838.2011.01306.x.
21. Kaeding CC, Aros B, Pedroza A, et al. Allograft versus autograft anterior cruciate ligament reconstruction: predictors of failure from a MOON prospective longitudinal cohort. Sports Health. 2011;3(1):73–81. https://doi.org/10.1177/1941738110386185.
22. Pallis M, Svoboda SJ, Cameron KL, Owens BD. Survival comparison of allograft and autograft anterior cruciate ligament reconstruction at the United States military academy. Am J Sports Med. 2012;40(6):1242–6. https://doi.org/10.1177/0363546512443945.
23. Krych AJ, Jackson JD, Hoskin TL, Dahm DL. A meta-analysis of patellar tendon autograft versus patellar tendon allograft in anterior cruciate ligament reconstruction. Arthroscopy. 2008;24(3):292–8. https://doi.org/10.1016/j.arthro.2007.08.029.
24. Prodromos C, Joyce B, Shi K. A meta-analysis of stability of autografts compared to allografts after anterior cruciate ligament reconstruction. Knee Surg Sports Traumatol Arthrosc. 2007;15(7):851–6. https://doi.org/10.1007/s00167-007-0328-6.
25. Hulet C, Sonnery-Cottet B, Stevenson C, et al. The use of allograft tendons in primary ACL recon-

struction. Knee Surg Sports Traumatol Arthrosc. 2019;27(6):1754–70. https://doi.org/10.1007/s00167-019-05440-3.
26. Conte EJ, Hyatt AE, Gatt CJ Jr, Dhawan A. Hamstring autograft size can be predicted and is a potential risk factor for anterior cruciate ligament reconstruction failure. Arthroscopy. 2014;30(7):882–90. https://doi.org/10.1016/j.arthro.2014.03.028.
27. Wang HD, Gao SJ, Zhang YZ. Hamstring autograft versus hybrid graft for anterior cruciate ligament reconstruction: a systematic review. Am J Sports Med. 2020;48(4):1014–22. https://doi.org/10.1177/0363546519849483.
28. Magnussen RA, Lawrence JT, West RL, Toth AP, Taylor DC, Garrett WE. Graft size and patient age are predictors of early revision after anterior cruciate ligament reconstruction with hamstring autograft. Arthroscopy. 2012;28(4):526–31. https://doi.org/10.1016/j.arthro.2011.11.024.
29. Kamien PM, Hydrick JM, Replogle WH, Go LT, Barrett GR. Age, graft size, and Tegner activity level as predictors of failure in anterior cruciate ligament reconstruction with hamstring autograft. Am J Sports Med. 2013;41(8):1808–12. https://doi.org/10.1177/0363546513493896.
30. Edgar CM, Zimmer S, Kakar S, Jones H, Schepsis AA. Prospective comparison of auto and allograft hamstring tendon constructs for ACL reconstruction. Clin Orthop Relat Res. 2008;466(9):2238–46. https://doi.org/10.1007/s11999-008-0305-5.
31. Naot D, Choi AJ, Street M, et al. Age-related differences in hamstring tendon used as autograft in reconstructive anterior cruciate ligament surgery. Int Orthop. 2022;46(4):845–53. https://doi.org/10.1007/s00264-021-05285-x.
32. Ellis HB, Matheny LM, Briggs KK, Pennock AT, Steadman JR. Outcomes and revision rate after bone-patellar tendon-bone allograft versus autograft anterior cruciate ligament reconstruction in patients aged 18 years or younger with closed physes. Arthroscopy. 2012;28(12):1819–25. https://doi.org/10.1016/j.arthro.2012.06.016.
33. Barrett GR, Luber K, Replogle WH, Manley JL. Allograft anterior cruciate ligament reconstruction in the young, active patient: Tegner activity level and failure rate. Arthroscopy. 2010;26(12):1593–601. https://doi.org/10.1016/j.arthro.2010.05.014.
34. Barber FA, Aziz-Jacobo J, Oro FB. Anterior cruciate ligament reconstruction using patellar tendon allograft: an age-dependent outcome evaluation. Arthroscopy. 2010;26(4):488–93. https://doi.org/10.1016/j.arthro.2009.08.022.
35. Barrett G, Stokes D, White M. Anterior cruciate ligament reconstruction in patients older than 40 years: allograft versus autograft patellar tendon. Am J Sports Med. 2005;33(10):1505–12. https://doi.org/10.1177/0363546504274202.
36. Marx RG, Hsu J, Fink C, Eriksson K, Vincent A, van der Merwe WM. Graft choices for paediatric anterior cruciate ligament reconstruction: state of the art. J isakos. 2023;8(3):145–52. https://doi.org/10.1016/j.jisako.2023.01.001.
37. Mittal A, Allahabadi S, Coughlan M, Pandya N. Graft failure in adolescent patients undergoing adult-type ACL reconstruction with bone patellar tendon bone or hamstring autograft. Orthop J Sports Med. 2021;9(7_suppl3):2325967121S00106. https://doi.org/10.1177/2325967121s00106.
38. Winkler PW, Vivacqua T, Thomassen S, et al. Quadriceps tendon autograft is becoming increasingly popular in revision ACL reconstruction. Knee Surg Sports Traumatol Arthrosc. 2022;30(1):149–60. https://doi.org/10.1007/s00167-021-06478-y.
39. Meena A, D'Ambrosi R, Runer A, et al. Quadriceps tendon autograft with or without bone block have comparable clinical outcomes, complications and revision rate for ACL reconstruction: a systematic review. Knee Surg Sports Traumatol Arthrosc. 2023;31(6):2274–88. https://doi.org/10.1007/s00167-022-07281-z.
40. Gagliardi AG, Carry PM, Parikh HB, Albright JC. Outcomes of quadriceps tendon with patellar bone block anterior cruciate ligament reconstruction in adolescent patients with a minimum 2-year follow-up. Am J Sports Med. 2020;48(1):93–8. https://doi.org/10.1177/0363546519885371.
41. Xerogeanes JW. Quadriceps tendon graft for anterior cruciate ligament reconstruction: THE GRAFT OF THE FUTURE! Arthroscopy. 2019;35(3):696–7. https://doi.org/10.1016/j.arthro.2019.01.011.
42. Nyland J, Collis P, Huffstutler A, et al. Quadriceps tendon autograft ACL reconstruction has less pivot shift laxity and lower failure rates than hamstring tendon autografts. Knee Surg Sports Traumatol Arthrosc. 2020;28(2):509–18. https://doi.org/10.1007/s00167-019-05720-y.
43. Meena A, Farinelli L, Hoser C, et al. Quadriceps autograft is a viable graft choice for arthroscopic ACL reconstruction in patients over 50 years of age. Knee Surg Sports Traumatol Arthrosc. 2023;31(8):3284–90. https://doi.org/10.1007/s00167-023-07367-2.
44. Belk JW, Kraeutler MJ, Marshall HA, Goodrich JA, McCarty EC. Quadriceps tendon autograft for primary anterior cruciate ligament reconstruction: a systematic review of comparative studies with minimum 2-year follow-up. Arthroscopy. 2018;34(5):1699–707. https://doi.org/10.1016/j.arthro.2018.01.047.
45. Feucht MJ, Mauro CS, Brucker PU, Imhoff AB, Hinterwimmer S. The role of the tibial slope in sustaining and treating anterior cruciate ligament injuries. Knee Surg Sports Traumatol Arthrosc. 2013;21(1):134–45. https://doi.org/10.1007/s00167-012-1941-6.
46. Shultz SJ, Schmitz RJ. Tibial plateau geometry influences lower extremity biomechanics during landing. Am J Sports Med. 2012;40(9):2029–36. https://doi.org/10.1177/0363546512453295.
47. Todd MS, Lalliss S, Garcia E, DeBerardino TM, Cameron KL. The relationship between posterior tibial slope and anterior cruciate ligament injuries.

Am J Sports Med. 2010;38(1):63–7. https://doi.org/10.1177/0363546509343198.
48. Bernhardson AS, Aman ZS, Dornan GJ, et al. Tibial slope and its effect on force in anterior cruciate ligament grafts: anterior cruciate ligament force increases linearly as posterior Tibial slope increases. Am J Sports Med. 2019;47(2):296–302. https://doi.org/10.1177/0363546518820302.
49. Waiwaiole A, Gurbani A, Motamedi K, et al. Relationship of ACL injury and posterior Tibial slope with patient age, sex, and race. Orthop J Sports Med. 2016;4(11):2325967116672852. https://doi.org/10.1177/2325967116672852.
50. Salmon LJ, Heath E, Akrawi H, Roe JP, Linklater J, Pinczewski LA. 20-year outcomes of anterior cruciate ligament reconstruction with hamstring tendon autograft: the catastrophic effect of age and posterior Tibial slope. Am J Sports Med. 2018;46(3):531–43. https://doi.org/10.1177/0363546517741497.
51. DePhillipo NN, Kennedy MI, Dekker TJ, Aman ZS, Grantham WJ, LaPrade RF. Anterior closing wedge proximal Tibial osteotomy for slope correction in failed ACL reconstructions. Arthrosc Tech. 2019;8(5):e451–7. https://doi.org/10.1016/j.eats.2019.01.001.
52. Akoto R, Alm L, Drenck TC, Frings J, Krause M, Frosch K-H. Slope-correction osteotomy with lateral extra-articular Tenodesis and revision anterior cruciate ligament reconstruction is highly effective in treating high-grade anterior knee laxity. Am J Sports Med. 2020;48(14):3478–85. https://doi.org/10.1177/0363546520966327.
53. Dejour D, Saffarini M, Demey G, Baverel L. Tibial slope correction combined with second revision ACL produces good knee stability and prevents graft rupture. Knee Surg Sports Traumatol Arthrosc. 2015;23(10):2846–52. https://doi.org/10.1007/s00167-015-3758-6.
54. Imhoff FB, Mehl J, Comer BJ, et al. Slope-reducing tibial osteotomy decreases ACL-graft forces and anterior tibial translation under axial load. Knee Surg Sports Traumatol Arthrosc. 2019;27(10):3381–9. https://doi.org/10.1007/s00167-019-05360-2.
55. Gupta A, Tejpal T, Shanmugaraj A, et al. Surgical techniques, outcomes, indications, and complications of simultaneous high Tibial osteotomy and anterior cruciate ligament revision surgery: a systematic review. HSS J. 2019;15(2):176–84. https://doi.org/10.1007/s11420-018-9630-8.
56. Feagin JA Jr, Curl WW. Isolated tear of the anterior cruciate ligament: 5-year follow-up study. Am J Sports Med. 1976;4(3):95–100. https://doi.org/10.1177/036354657600400301.
57. Taylor DC, Posner M, Curl WW, Feagin JA. Isolated tears of the anterior cruciate ligament: over 30-year follow-up of patients treated with arthrotomy and primary repair. Am J Sports Med. 2009;37(1):65–71. https://doi.org/10.1177/0363546508325660.
58. Murray MM, Fleming BC, Badger GJ, et al. Bridge-enhanced anterior cruciate ligament repair is not inferior to autograft anterior cruciate ligament reconstruction at 2 years: results of a prospective randomized clinical trial. Am J Sports Med. 2020;48(6):1305–15. https://doi.org/10.1177/0363546520913532.
59. van der List JP, Vermeijden HD, Sierevelt IN, et al. Repair versus reconstruction for proximal anterior cruciate ligament tears: a study protocol for a prospective multicenter randomized controlled trial. BMC Musculoskelet Disord. 2021;22(1):399. https://doi.org/10.1186/s12891-021-04280-y.
60. Harris JD, Abrams GD, Bach BR, et al. Return to sport after ACL reconstruction. Orthopedics. 2014;37(2):e103–8. https://doi.org/10.3928/01477447-20140124-10.
61. Glogovac G, Schumaier AP, Grawe BM. Return to sport following revision anterior cruciate ligament reconstruction in athletes: a systematic review. Arthroscopy. 2019;35(7):2222–30. https://doi.org/10.1016/j.arthro.2019.01.045.
62. Ardern CL, Webster KE, Taylor NF, Feller JA. Return to sport following anterior cruciate ligament reconstruction surgery: a systematic review and meta-analysis of the state of play. Br J Sports Med. 2011;45(7):596–606. https://doi.org/10.1136/bjsm.2010.076364.
63. Rodríguez-Roiz JM, Caballero M, Ares O, Sastre S, Lozano L, Popescu D. Return to recreational sports activity after anterior cruciate ligament reconstruction: a one- to six-year follow-up study. Arch Orthop Trauma Surg. 2015;135(8):1117–22. https://doi.org/10.1007/s00402-015-2240-8.
64. Edwards PK, Ebert JR, Joss B, et al. Patient characteristics and predictors of return to sport at 12 months after anterior cruciate ligament reconstruction: the importance of patient age and postoperative rehabilitation. Orthop J Sports Med. 2018;6(9):2325967118797575. https://doi.org/10.1177/2325967118797575.
65. Hamrin Senorski E, Svantesson E, Beischer S, et al. Low 1-year return-to-sport rate after anterior cruciate ligament reconstruction regardless of patient and surgical factors: a prospective cohort study of 272 patients. Am J Sports Med. 2018;46(7):1551–8. https://doi.org/10.1177/0363546518765120.
66. Chicorelli AM, Micheli LJ, Kelly M, Zurakowski D, MacDougall R. Return to sport after anterior cruciate ligament reconstruction in the skeletally immature athlete. Clin J Sport Med. 2016;26(4):266–71. https://doi.org/10.1097/jsm.0000000000000275.
67. Cordasco FA, Black SR, Price M, et al. Return to sport and reoperation rates in patients under the age of 20 after primary anterior cruciate ligament reconstruction: risk profile comparing 3

patient groups predicated upon skeletal age. Am J Sports Med. 2019;47(3):628–39. https://doi.org/10.1177/0363546518819217.
68. Ardern CL, Taylor NF, Feller JA, Whitehead TS, Webster KE. Psychological responses matter in returning to preinjury level of sport after anterior cruciate ligament reconstruction surgery. Am J Sports Med. 2013;41(7):1549–58. https://doi.org/10.1177/0363546513489284.
69. Kay J, Memon M, Marx RG, Peterson D, Simunovic N, Ayeni OR. Over 90% of children and adolescents return to sport after anterior cruciate ligament reconstruction: a systematic review and meta-analysis. Knee Surg Sports Traumatol Arthrosc. 2018;26(4):1019–36. https://doi.org/10.1007/s00167-018-4830-9.
70. Webster KE, Feller JA, Leigh WB, Richmond AK. Younger patients are at increased risk for graft rupture and contralateral injury after anterior cruciate ligament reconstruction. Am J Sports Med. 2014;42(3):641–7. https://doi.org/10.1177/0363546513517540.
71. Deakon RTZM. ACL reconstruction in patients over 40 years of age using autogenous bone-patellar tendon-bone. Arthroscopy. 1996;12(3):388.
72. Osti L, Papalia R, Del Buono A, Leonardi F, Denaro V, Maffulli N. Surgery for ACL deficiency in patients over 50. Knee Surg Sports Traumatol Arthrosc. 2011;19(3):412–7. https://doi.org/10.1007/s00167-010-1242-x.

Management of Accompanying Pathologies with ACL Rupture in the Adult

16

Volker Musahl, Abigail Boduch, Joseph D. Giusto, Ehab M. Nazzal, and Yoshiyuki Yahagi

16.1 ACL Reconstruction in the Setting of Lateral Meniscus Injury

Injuries to the lateral or medial meniscus are commonly associated with anterior cruciate ligament (ACL) tears, which can significantly impact translational and rotatory knee stability in addition to the risk for post-traumatic osteoarthritis. Specifically, a combined ACL and meniscus injury can increase the odds for a total knee arthroplasty fourfold compared to an isolated ACL injury [1]. The prevalence of any concomitant meniscus injury in those undergoing ACL reconstruction (ACL-R) ranges from 45 to 73% and 25–31% for the lateral meniscus [2–5].

Biomechanically, the lateral meniscus acts as a secondary stabilizer of the knee. A prior cadaveric study demonstrated this function after finding further increased anterior tibial translation during the pivot shift exam in ACL-deficient knees with a lateral meniscectomy [6]. Lateral meniscus injuries are more commonly observed in the acute setting because of the biomechanical contributions during the pivot shift mechanism of acute ACL injury. Furthermore, delaying ACL-R by 12 months has been found to have no additional risk for a lateral meniscus tear despite a nearly twofold increased risk of developing a medial meniscus tear [4].

16.1.1 Radial, Longitudinal, and Complex Lateral Meniscus Tears

A recently developed classification system found that the most common type of lateral meniscus tear is a posterior horn lateral meniscal oblique radial tear (LMORT), which occurs in 12% of ACL-R cases [2]. Radial tears, although less common, also occur significantly more often in the lateral meniscus (15.8%) than the medial meniscus (1.5%) [3]. Radial meniscus tears are characterized by a perpendicular split in the circumferentially oriented collagen fibers that create hoop stress forces (Fig. 16.1a). Conversely, longitudinal tears involve a split of the meniscus in the same direction as the fibers (Fig. 16.1b). However, the biomechanical consequences of a radial tear may be greater than that of a longitudinal tear given the disruption of the hoop stress forces. For example, a 100% radial tear in the lateral meniscus significantly decreased joint

V. Musahl (✉) · A. Boduch · J. D. Giusto · E. M. Nazzal
Department of Orthopaedic Surgery, University of Pittsburgh, Pittsburgh, PA, USA
e-mail: musahlv@Upmc.edu

Y. Yahagi
Department of Orthopaedic Surgery, University of Pittsburgh, Pittsburgh, PA, USA

Department of Orthopaedic Surgery, Nihon University School of Medicine, Tokyo, Japan

M. Karahan et al. (eds.), *Anterior Cruciate Ligament Injuries Throughout the Life Span*, https://doi.org/10.1007/978-3-031-84501-7_16

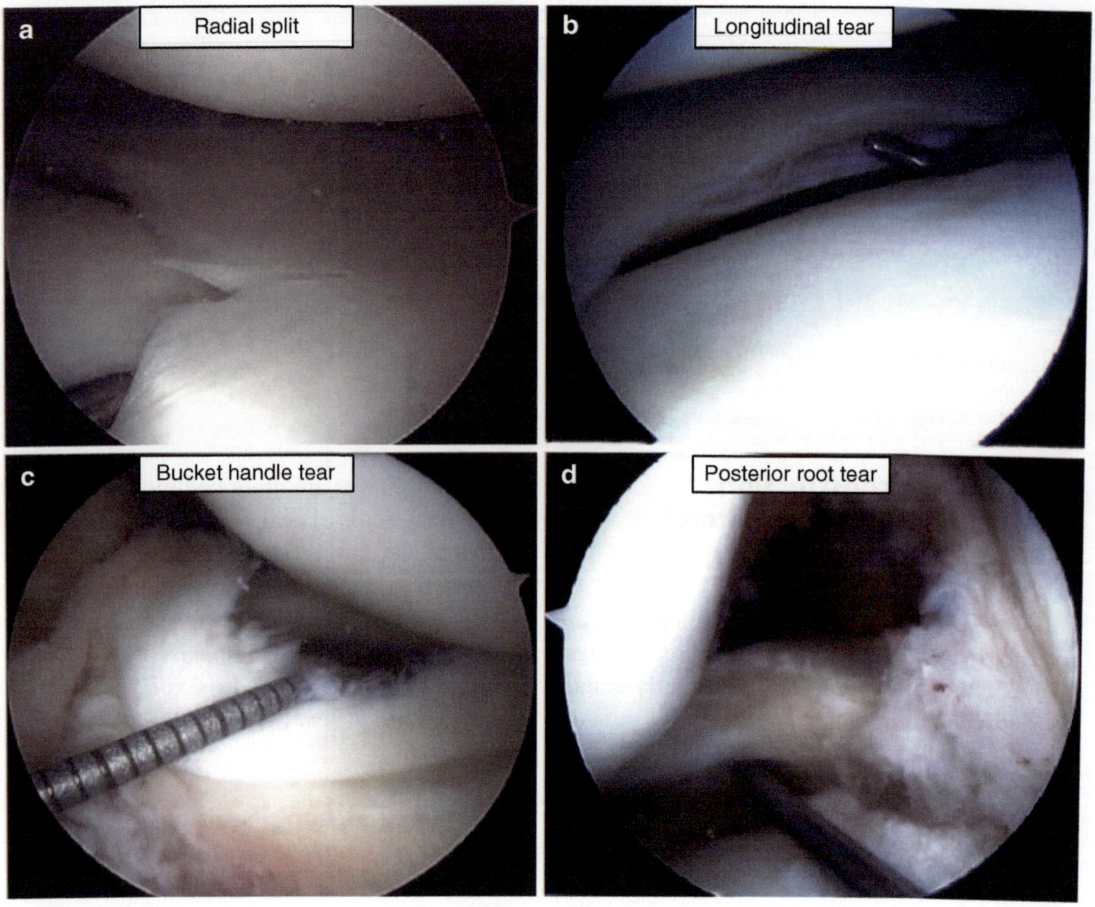

Fig. 16.1 Different meniscus tear patterns. Example arthroscopic images of commonly encountered meniscus tear patterns (all lateral meniscus tears). (**a**) Small mid-body radial tear. (**b**) A 1.5 cm horizontal tear of posterior horn in the red-white zone. (**c**) Displaced bucket-handle vertical split tear in the red-white zone. (**d**) Unstable posterior root tear.

contact area and increased peak contact forces in a cadaveric model [7].

Surgical decision-making and meniscus tear morphology can influence postoperative outcomes. For example, stable lateral meniscus tears have been shown to be successfully left in situ with a low reoperation rate of 2.4% at a mean follow-up of 6.6 years [8]. However, unstable lateral meniscus tears observed at the time of ACL-R are generally treated operatively with either meniscus resection or repair. Undergoing combined ACL-R and meniscectomy is associated with worse Knee Injury and Osteoarthritis Outcome Score Quality of Life (KOOS-QoL) subscale scores and an increased risk of osteoarthritis at long-term follow-up compared to combined ACL-R and meniscus repair [9, 10]. Meniscus repair is, therefore, recommended for the management of unstable meniscal tears and becoming increasingly favorable compared to meniscectomy. However, repair may not always be possible. Complex meniscus tears, which involve tears in multiple planes and are commonly seen as secondary injuries in ACL-deficient knees, are more likely to be treated with meniscectomy than repair [11].

Concomitant lateral meniscus tears can be repaired using either inside-out, outside-in, or all-inside suture techniques. All-inside repairs are becoming increasingly popular due the limited invasiveness and low failure rate, which has been reported to be 16% for lateral meniscus tears at a minimum follow-up of 10 years [12]. Lateral meniscus repairs with the all-inside technique also survive on average 3 years longer than medial meniscus repairs with the same technique [12]. Both radial and longitudinal tears can be successfully repaired with either all-inside or inside-out techniques with second-look arthroscopy demonstrating complete healing in 50–60% of cases and partial healing in 30–37.5% of cases [13, 14]. Additionally, meniscus tears are more likely to heal when repaired concomitantly with ACL-R compared to isolated repair, possibly due to the release of growth factors during bone tunnel drilling [15, 16]. Healing rates are also influenced by tear location as tears within the well-vascularized red-red zone are more likely to heal than those in the less vascularized zones [17].

16.1.2 Lateral Meniscus Bucket-Handle Tears

Bucket-handle meniscus tears involve a large longitudinal split of the meniscus and occur concomitantly with ACL injuries at a rate of approximately 13% (Fig. 16.1c) [18]. Surgical timing can also significantly affect the incidence of bucket-handle meniscus tears. Delayed ACL-R beyond 12 months from initial injury has been reported to increase the rate of concomitant bucket-handle meniscus tears [19, 20]. However, the incidence of concomitant lateral meniscus bucket-handle tears with ACL injury (1.3–2.8%) is lower than medial meniscus bucket-handle tears (5.6–9.4%) [20]. Bucket-handle meniscus tears carry a risk of displacement within the femoral intercondylar notch, which can result in patients presenting with a "locked" knee and restricted range of motion. More expedient surgical treatment may be warranted in such cases. A previous study has also found an increased risk of arthrolysis for loss of motion within 12 months in patients with concomitant ACL and bucket-handle meniscus tears compared to patients with isolated ACL tears, which further highlights the severity of this meniscus injury [21].

Treatment of lateral meniscus bucket-handle tears with repair is also recommended whenever possible, although repair may be more challenging. Patients who undergo resection have been shown to have worse postoperative pain at long-term follow-up compared to those undergoing repair [22]. A matched cohort analysis of radial and bucket-handle meniscus tears also found similar reoperation rates with improvements in visual analog scale (VAS) for pain and International Knee Documentation Committee (IKDC) scores, although some included patients did not have a concomitant ACL injury [23]. The failure rate for any bucket-handle repair has been reported to be 14.8% according to one systematic review; however, undergoing an isolated repair nearly doubled the risk of failure compared to repairs with a concomitant ACL-R [24].

16.1.3 Lateral Meniscus Root Tears

Lateral meniscus root tears are more commonly encountered with concomitant ACL injuries compared to medial meniscus root tears, likely due to the pivot shift mechanism of ACL injury [25]. The incidence of concomitant lateral root tears is less common compared to other tear patterns, occurring in only 14% of cases [25]. However, the biomechanical effects of a root tear can be determinantal due to the disruption of hoop stress forces, similar to the effects of a radial split tear (Fig. 16.1d). Prior cadaveric studies have shown that lateral meniscus root tears increase peak contact pressures and increase rotatory knee instability in ACL-deficient knees [26–28]. Patients with lateral meniscus root tears at the time of ACL-R have been found to have a 2.1 degrees higher lateral posterior tibial slope on average compared to controls, adding another important biomechanical consideration [29].

Repair of the lateral meniscus root can improve stability and potentially restore native knee kinematics in conjunction with anatomic ACL-R [28]. Two repair techniques are primarily used for root repair: the side-to-side repair and the trans-osseous pullout repair using a tibial tunnel. Clinical studies have found satisfactory improvements in IKDC scores with one study reporting that 96% of patients achieved the minimal clinically important difference at a minimum follow-up of 2 years [30, 31]. A magnetic resonance imaging (MRI) study investigating outcomes 1 year following combined ACL-R and lateral meniscus root repair also showed maintained reduction of lateral meniscus extrusion [32]. However, long-term clinical results for combined ACL-R and lateral meniscus root repair remain unclear.

16.2 ACL Reconstruction in the Setting of Medial Meniscus Tear

Medial meniscal tears are important to consider in the setting of ACL rupture, and have been estimated to occur in up to 40% of ACL injuries [33]. While lateral meniscus tears frequently occur at the time of ACL rupture, medial meniscal injuries are the result of abnormal kinematics in the ACL-deficient knee [34]. The following section will cover different types of medial meniscal tears, their respective rates, the biomechanical and clinical impacts they have on the knee, and management options with outcomes.

16.2.1 Management of Radial Medial Meniscal Tears

Radial tears are cleavage tears, and can occur at any part of the meniscus, from the anterior horn, the body, and all the way to the posterior body. These tears arise from the central region to the periphery, or from the white-white to the red-red zone [35]. As discussed before, radial tears play an important biomechanical role as they alter the meniscus's ability to convert loads into hoop stresses [36].

Radial tears can be separated into partial and full-thickness tears, and these are important to distinguish as they each have a different effect on subsequent knee biomechanics. A robotics study evaluating contact forces after medial meniscal radial tears found that in cases of larger radial tears (up to 90%), peak contact pressures shifted posterocentrally [37]. However, unlike previous studies evaluating the lateral meniscus, which found that >66% radial tears resulted in a loss of meniscal function, it was found that the residual medial meniscus continues to provide load transmission across the joint [38].

Given the wide circumferential involvement of a radial meniscal tear, nonoperative management may not be feasible, as the altered hoop stresses would result in abnormal kinematics [39]. Operative management includes meniscectomy versus meniscal repair. Meniscectomy is used for symptom management, with the assumption that this will ultimately lead to a progression of knee osteoarthritis. However, meniscal repair techniques for radial tears have been especially hampered by the low healing rates given their involvement of the avascular region of the meniscus [40].

Traditionally, radial meniscal tears were repaired with an inside-out approach with horizontal sutures, providing the ability to pass multiple stitches but coming with the risk of neurovascular injury with a posterior approach [41, 42]. However, recent studies have been more in favor of all-inside repair. In one study, porcine models were used to compare a vertical all-inside technique with horizontal inside-out repair and found that the all-inside repair resulted in lower displacement and higher load to failure [43]. A subsequent study in 2015 confirmed these findings not just for vertical all-inside repair, but for any all-inside repair in comparison with inside-out repair [44].

Overall, radial medial meniscal repair results in favorable patient-reported outcomes. In a 2021 systematic review, radial meniscal repairs demonstrated high patient-reported outcomes (Lysholm and Tegner Activity scores) compared with preoperative scores, as well as an overall low rate of failure [45]. Studies more specific to medial radial meniscal repair also demonstrated low revision rates and improved patient-reported outcomes, especially when trans-osseous fixation was utilized [46, 47].

16.2.2 Management of Horizontal Medial Meniscal Tears

Medial meniscal horizontal tears are typically associated with chronic ACL deficiency, aging, and osteoarthritis [48, 49]. With horizontal tears, the lesion divides the meniscus into upper and lower leaflets. Biomechanically, even larger horizontal meniscal tears can maintain relatively intact meniscal functions given maintenance of the circumferential fibers [50].

Unlike radial tears, horizontal tears of the medial meniscus are more amenable to nonoperative management given their association with chronic degeneration rather than an acute injury. Operative management of horizontal medial meniscal tears is indicated in the setting of unremitting pain refractory to nonoperative management, or with any definitive mechanical symptoms such as catching or locking.

Operative strategies for horizontal medial meniscal tears typically involve some level of meniscectomy, whether it is a total meniscectomy or a partial meniscectomy of one of the leaflets of the tear. Subtotal meniscectomy is a viable option to preserve the biomechanical function of the meniscus and preserve its axial load distribution, as demonstrated in a 2016 cadaveric study involving resection of the inferior leaflet for horizontal medial meniscal tears [51]. Total meniscectomy is an option for symptomatic relief and for mechanical symptoms, yet meniscectomy carries an increased risk of abnormal contact pressures on cartilage and progression of arthritis.

Traditionally, horizontal cleavage tears were not thought to be amenable to repair given its involvement of the avascular zone and a subsequent minimal capacity to heal. However, a 2014 systematic review looking at nine studies found a healing rate of nearly 80%, which is comparable to rates following repair of other tear patterns [52]. Repair strategies, like other meniscal tears, include open, inside-out, and all-inside techniques. Recently a "cerclage" suture meniscus repair technique has been described, in which circumferential compression stitches with an all-inside repair pattern stabilize a meniscus tear. A cerclage technique avoids the need for a posterior incision while also providing uniform compression to the superior and inferior leaflets [52].

16.2.3 Management of Meniscal Ramp Lesions

Meniscal ramp lesions are defined as tears that result in disruptions of the peripheral meniscocapsular attachments of the posterior horn of the medial meniscus [53]. These lesions are typically associated with ACL-deficient knees and have been reported in up to 25% of all ACL tears [54]. An early 2016 cadaveric study demonstrated that ACL-deficient knees with concomitant ramp lesions result in increased external rotation and anterior translation of the knee, resulting in altered knee biomechanics and a predisposition to meniscal and articular damage [55]. However, a more recent 2019 study looked at in situ forces after ramp lesion repair, and found that there were only significant decreases in the in situ forces with higher knee flexion angles [56].

Deciding whether to pursue operative management of meniscal ramp lesions depends on the size of the lesion. A tear smaller than 1 cm is considered to be "stable" and not requiring any specific intervention [57]. Instead, traditional nonoperative modalities such as physical therapy and symptomatic management are sufficient. However, studies have also described "biologic" treatment in these cases, especially if these lesions are encountered during arthroscopy, which includes such techniques as abrasion, trephination, and curettage of the lesion [58].

For larger tears or those that show translation of the posterior horn on arthroscopy (deemed to be "unstable"), operative intervention is indicated. Several repair techniques have been described, including all-inside repairs and suture hook techniques [57]. However, more unique to meniscal ramp lesions is the portal used for fixation. Other meniscal lesions are amenable to visualization and repair through two standard arthroscopic knee portals. While there have been some studies outlining repair strategies through the anteromedial portal [59], there is a theoretical risk of underestimating the true size of the ramp lesion, therefore necessitating establishment of a posteromedial portal for visualization and fixation.

The outcome data for management of meniscal ramp lesions is sparse, and overall does not overwhelmingly advocate for their repair. For stable lesions, little clinical benefit has been seen in repair, as one 2021 study comparing ACL-injured patients with and without ramp lesions found that there were no significant differences in various patient-reported outcomes even up to 3 years postoperatively [60]. However, substantial healing has been demonstrated after ramp repair, as one 2016 study evaluating 132 patients with 2-year follow-up demonstrated improvements in patient-reported outcomes, side-to-side knee laxity, and low rates of revision, demonstrating that repair is an effective method and can be used to restore knee biomechanics [61].

16.2.4 Management of Bucket-Handle Medial Meniscus Tears

Bucket-handle tears are a challenging subset of meniscal lesions, as they involve vertical or oblique longitudinal tears with a high propensity to displace into the intercondylar notch or around the femoral condyle and produce mechanical symptoms [62]. Though there are limited biomechanical studies on the effect of bucket-handle tears on the normal knee, the result of this tear type is not only mechanical symptoms, but a gradual degradation of the meniscus as it continues to displace, further leaving a patient susceptible to progression to osteoarthritis. This also means that unlike the other meniscal tear types, bucket-handle tears are mainly operative to prevent further symptoms and restore normal knee function.

Operative management of bucket-handle tears, like other tears, is comprised of meniscectomy versus meniscal repair with repair strategies like those outlined above. Repair is preferred given the risk of post-meniscectomy arthritis and loss of normal knee kinematics. While patient-reported outcomes improve after repair [63], bucket-handle tears have a high rate of failure postoperatively. Failure rates vary in the literature but are typically reported as ~15–30% [63, 64], especially in males, younger individuals, and those with previous meniscal injury or even early arthritic changes [65]. Despite these high failure rates, the recommendation of the senior author is to proceed with repair when encountered, especially with concomitant ACL-R.

16.2.5 Management of Medial Meniscal Root Tears

Just as has been discussed for other medial meniscal tears thus far, medial meniscal root tears are generally the result of degenerative changes and are seen in the older population. These tear types, defined as complete radial tears within 1 cm of the anterior or posterior meniscal attachments to the tibia [66], have been especially linked to degenerative changes in the knee, as biomechanical studies have shown that these medial meniscal root tears result in an increase in average contact pressure and decrease in contact area of the medial compartment relative to the intact knee state, especially at the posterior root [67].

Like radial tears, meniscal root tears are not typically amenable to nonoperative management given the disruption of the circumferential fibers and ability to dissipate hoop stresses, incurring increased articular degeneration [68]. Therefore, operative intervention should be attempted whenever possible in order to slow the progression of osteoarthritis. Similar techniques for repair discussed above are used for root tear repairs and include techniques such as all-inside repairs, utilization of suture anchors, and trans-osseous suture fixation in order to minimize root displacement and restore the anatomic footprint [69, 70].

Literature has demonstrated that restoring the meniscal root attachment site restores function of the meniscus and improves patient-reported outcomes. In a 2020 systematic review evaluating 19 studies, meniscal root repair led to the greatest improvement in function and lowest risk of conversion to arthroplasty, over nonoperative management and meniscectomy [71]. Another 2019 meta-analysis looking at medial meniscal root tears over a 10 year period reaffirmed these findings, and also found that meniscal root repair yielded the highest quality adjusted life years and lowest discounted 10-year cost [72]. The findings from these studies support the notion that for medial meniscal root tears, repair is the indicated intervention whenever possible.

16.3 Cartilage Lesions in the Setting of ACL Injury

Cartilage lesions represent an important pathology related to ACL rupture. Cartilage lesions occur from a combination of factors, including primary damage from the acute trauma and secondary lesions from chronic instability following ACL rupture [73]. The rate of cartilage lesions with ACL injury varies greatly in the literature with variations in grading systems. In one study, 100% of patients sustained an MRI-detectable cartilage lesion at the time of initial ACL injury [74]. Given that cartilage lesions are a frequent primary accompanying lesion in an acute ACL rupture, it is important to recognize those lesions and develop a plan for their management.

16.3.1 Types of Cartilage Lesions in the ACL Injured Knee

For ACL-R within the first 6 months after trauma, the mean incidence of cartilage lesion at the time of reconstruction was found to be 33% [75]. Cartilage lesions have also been reported to worsen over time in ACL injured knees [74–78]. Differences in sport related trauma also exist with a higher frequency of cartilage lesions occurring in soccer players than in alpine ski racers [79]. Cartilage lesions in the ACL injured knee have also been associated with older age and weaker quadriceps function [80]. Patients undergoing revision ACL-R or those with meniscal pathology in addition to primary ACL rupture have been reported to have a higher frequency of cartilage lesions. [81, 82] A recent systematic review reported the following frequencies for International Cartilage Research Society (ICRS) grades and locations; grades I–II: 75.7%, III–IV: 24.3%, location as medial femoral condyle 40.4%, lateral femoral condyle 17.5%, medial tibial plateau 6.8%, lateral tibial plateau 7.9%, patella 18.9%, and trochlea 8.5% [83].

A systematic review looking at the prevention of cartilage lesions after ACL-R included 40 cohort studies with a mean follow-up of 13.9 years found that patients who underwent ACL-R had fewer subsequent meniscal injuries and less need for further surgery as compared to those treated nonoperatively [84]. However, there was no significant development of radiographically evident osteoarthritis. Currently, there are few studies that are available on the protective effect of ACL-R on the development of secondary local cartilage lesions [73]. A recent study found cartilage lesions were encountered at the time of revision ACL-R: 43% in the medial femoral condyle, 29% in the lateral femoral condyle, 11% in the medial tibial plateau, 17% in the lateral tibial plateau, 30% in the patella, and 21% in the trochlea [85]. Figure 16.2 demonstrates cartilage damage in the setting of ACL injury. Articular cartilage injury likely has a greater impact as compared to concurrent meniscal pathology on long-term subjective outcomes after ACL-R [86].

16.3.2 Treatment of Cartilage Lesions with ACL Injury

There are multiple treatment pathways for the management of cartilage lesions in the setting of ACL injury. For small lesions (<2 cm), the most straightforward treatment is microfracture with chondroplasty. For lesions involving the underlying bone, osteochondral autograft transfers are typically used in the treatment of small (<2 cm) defects. Osteochondral allograft transfer and matrix-induced autologous chondrocyte implantation (ACI) are used for larger cartilage/bone and cartilage defects, respectively [87]. Outcomes after ACL-R with concurrent cartilage lesions in ACL injured knees are comparable to ACL-R

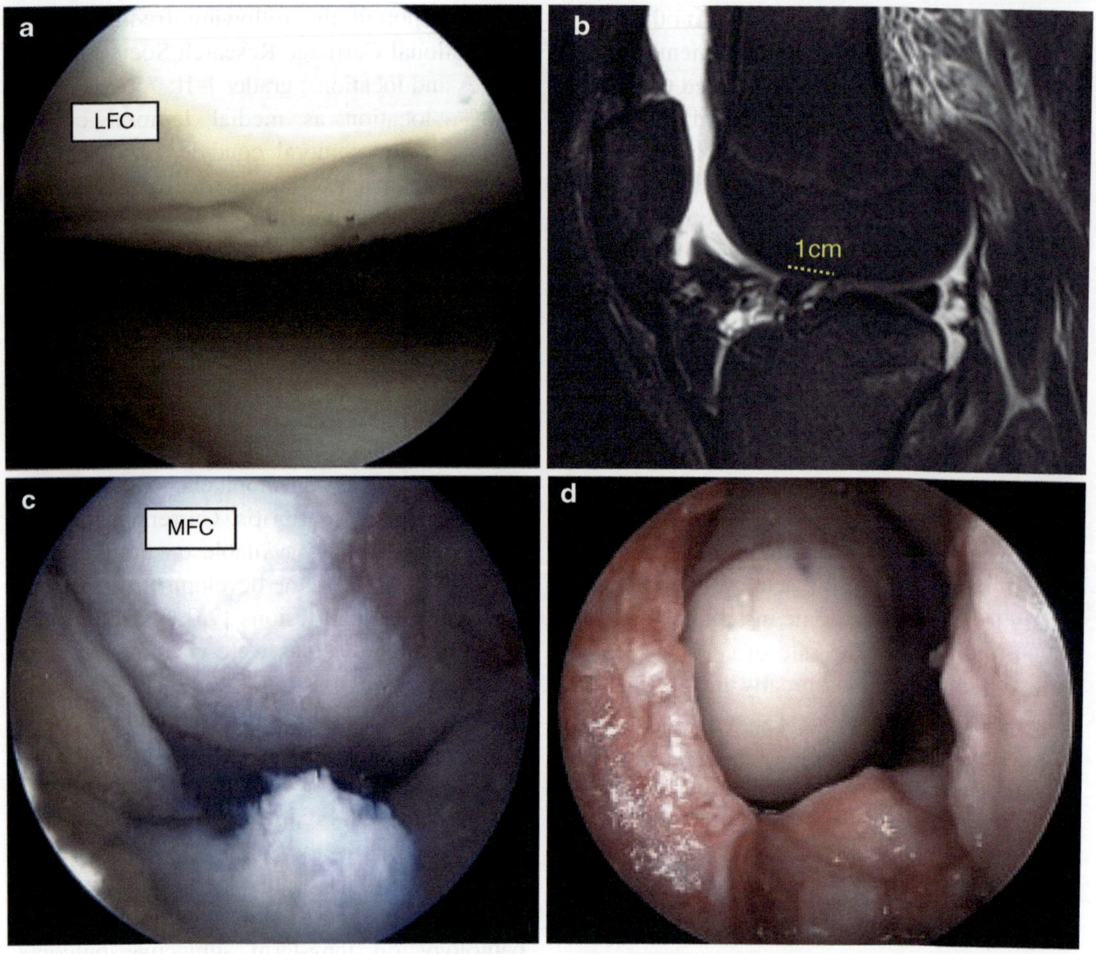

Fig. 16.2 Concomitant cartilage lesions treated during ACL-R. Lateral femoral condyle (LFC), medial femoral condyle (MFC). (**a**) Arthroscopic image of a 1x1cm Grade III LFC injury with a 0.5 cm flap (debrided) in a patient undergoing anterior cruciate ligament reconstruction (ACL-R) with concomitant chondroplasty. (**b**) Corresponding preoperative magnetic resonance imaging of patient in Fig. 16.2a demonstrating the condylar lesion on a sagittal T2 view. (**c**) Arthroscopic image of a 18x24mm Grade IV chondral defect on the weightbearing portion of the MFC. (**d**) Corresponding view from the patient in Fig. 16.2c of an osteochondral allograft implant (20 mm). Patient underwent a combined over-the-top revision ACL-R, medial meniscus transplantation, and MFC osteochondral allograft transplantation.

without concurrent cartilage injury. When considering long-term outcomes, ACL-reconstructed patients with a full-thickness cartilage lesion did not report worse outcomes at 10–15 years after surgery compared to patients with an isolated ACL rupture [88]. On the other hand, cartilage lesion size >2cm^2 and > 5 degrees of varus alignment were detected as major determinants leading to poorer outcomes in combined ACL-R and microfracture [89].

The clinical short-term outcomes after ACI at the knee joint in combination with ACL-R are good and are similar to the results after isolated ACI in ligament intact knees [90]. Recent reports have shown that in patients with an ACL injured knee, synovial fluid inflammatory biomarker concentrations at the time of injury can predict progression of early stage post-traumatic osteoarthritis [91]. Serum cartilage oligomeric matrix protein can be used to detect early cartilage change in patients with ACL injured knee [92]. In general, high-level studies comparing management techniques for the treatment of cartilage lesions in the setting of ACL injury are lacking [93].

16.4 ACL Reconstruction in the Setting of Multiligamentous Knee Injury

Multiligament knee injury (MLKI) can occur after a high or low velocity knee injury. In either scenario this is usually associated with a knee dislocation with at least two of the four major ligaments of the knee disrupted [94]. Although considered an overall rare orthopedic injury with MLKI accounting for just 0.02% of all orthopedic injuries [95], when MLKI occurs they most commonly include the ACL [96]. In a recent review, the most common patterns were an ACL tear combined with either a medial-sided injury (20.7%) or lateral-sided injury (23.2%) injury; one-third (34.7%) of injuries were bicruciate injuries [97].

When considering management of an ACL rupture in the setting of MLKI, one must use a systematic approach and identify all accompanying pathologies and designate a plan for each. Operative management including ACL-R is generally accepted as providing favorable outcomes in the setting of MLKI [98–100]. The operative plan must take into consideration timing, rehabilitation, and the patient's goals in terms of return to sport or preinjury level of function. Currently, there is ongoing research with the Surgical Timing and Rehabilitation for Multiligament Knee Injuries (STaR) trial looking at surgical timing and rehabilitation. It is also important to understand the specific risk factors for complications related to the management of ACL-R in the setting of MLKI. High-quality studies evaluating the relationship between surgical and rehabilitation timing and patient outcomes are lacking. However, ongoing randomized controlled trials may provide more information to aid clinical decision-making in the future. The following section will focus on the management of ACL injuries in the setting of associated pathologies including posterior cruciate ligament injury, posterolateral corner (PLC) injuries, medial collateral ligament injuries, posteromedial corner (PMC) injuries, and injuries involving the antero lateral complex of the knee.

16.4.1 ACL Reconstruction with Anterolateral Complex Injury

The anterolateral complex (ALC) has been shown to be a secondary stabilizer to anterolateral rotatory knee instability with the knee in flexion [101–103]. The ALC has also been shown to be a secondary stabilizer to the pivot shift in ACL-deficient knees [104]. Lateral extra-articular tenodesis (LET) is recommended after ACL-R in a patient with a high-grade pivot (lateral compartment translation >5 mm; tibial acceleration > 10 m/sec2) [104, 105]. A multicenter randomized control trial, STABILITY I, found at 2 years follow-up patients with LET showed a relative risk reduction in 0.38 in clinical failure and 0.67 in graft rupture when compared to ACL-R with hamstring autograft alone [106]. Additionally, anterolateral ligament reconstruction (ALL-R) can be performed and has shown reduced ACL graft failure rates compared to isolated ACL-R [107, 108].

16.4.2 ACL Reconstruction with Posterolateral Corner Injury

The PLC of the knee includes the lateral collateral ligament (LCL), popliteus muscle-tendon unit, and popliteofibular ligament (PFL) [109, 110]. Biomechanically the PLC protects the knee by acting as the primary restraint to varus stress and external rotation of the tibia [111]. Early cadaveric data demonstrated that sectioning of the PLC components significantly increased the force across the ACL graft after ACL-R [112]. The incidence of PLC injury based on MRI diagnosis is as high as 16% when considering all knee ligament injuries [113].

Operative management of PLC injuries in the setting of ACL injury is generally accepted. Failure to treat PLC injuries has been associated with increased failure of ACL-R with continued instability [112, 114, 115]. When considering management strategies, controversy exists in regard to repair versus reconstruction of the PLC. Currently, acute PLC repair for bony avulsions is recommended and PLC reconstruction is recommended for mid substance tears [116]. Acute repair (<3 weeks) of the PLC when concurrent ACL-R is performed results in similar outcomes at 6 years as compared to reconstruction of the PLC when considering isolated ACL-PLC injuries [117]. There is a higher rate of failure of PLC repair in the setting of ACL-R in the chronic setting (>3 weeks) and with staging of cruciate ligament reconstruction. In this setting, PLC repairs were revised in 37–40% of patients as compared to 6–9% revision rate in PLC reconstruction [100, 118].

Multiple techniques for PLC reconstruction exist. No functional or radiographic differences have been found when directly comparing PLC reconstruction techniques [119]. When combined with ACL injuries, PLC injuries have been shown to negatively impact ACL-R. In a case–control study comparing a matched cohort of isolated ACL-R patients to combined ACL-PLC reconstruction, patient-reported outcome measures including the KOOS and Lysholm scores were significantly lower in the combination ACL-PLC cohort [120]. In addition to lower patient-reported outcomes in the postoperative period when considering combination ACL-PLC injuries, it is important to recognize the increased complication of common peroneal nerve (CPN) palsy. CPN palsy presents in 23.4% of patients with MLKI and in 100% of patients with CPN palsy there was an associated ACL injury in addition to the PLC injury [121].

16.4.3 ACL Reconstruction with Medial Ligamentous Knee Injuries (Medial Collateral Ligament and Posteromedial Corner)

Medial-sided ligamentous injuries include both the medial collateral ligament (MCL) and posteromedial corner (PMC). When discussing medial-sided knee injuries, most of the focus remains on the superficial and deep MCL. However, when considering medial-sided knee injuries in conjunction with ACL injuries, it is important to recognize the implications of all the medial structures not just the MCL. Combined ACL, MCL, and PMC injuries are estimated to make up 6.7% of all knee ligamentous injuries. [122]

Controversy exists regarding the optimal treatment of medial-sided ligamentous injury in the setting of ACL injury. Biomechanically the medial side of the knee provides valgus, anteroposterior (AP), and rotatory stability to the knee. The superficial MCL is a primary valgus stabilizer of the knee in flexion and the deep MCL is considered a secondary valgus stabilizer [123]. AP translation is primarily controlled by the central compartment structures; however, in ACL-deficient knees the medial knee structures have been shown to have a significant contribution to stability [124]. Given the biomechanical contributions, recognizing and effectively managing medial-sided ligamentous knee injuries is an important factor in the setting of MLKI with ACL injury.

Traditionally nonsurgical management of MCL injuries in the setting of combined ACL injury has been the mainstay of treatment. Current treatment algorithms for combined ACL and MCL injury recommend nonsurgical treatment

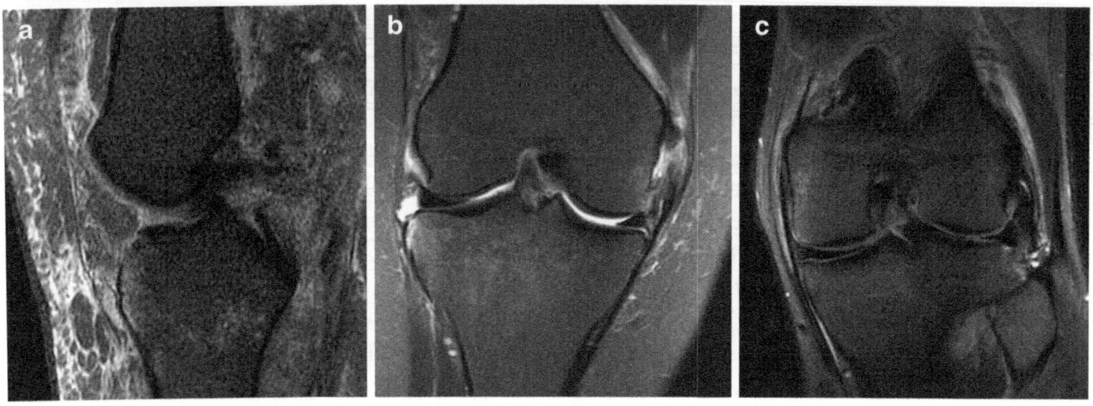

Fig. 16.3 MRI images of concomitant ligament injuries. Magnetic resonance imaging (MRI). (**a**) Posterior cruciate ligament tear on a sagittal T2 view. (**b**) Femoral-sided partial medial collateral ligament tear on a coronal T2 view. (**c**) Lateral collateral ligament tear on a coronal T2 view.

for grade I MCL injury and grade II femoral-sided MCL injury [123, 125]. However, unaddressed medial-sided laxity has been shown to increased primary ACL-R graft failure by 13 times [126]. Thus, controversy exists on the management of grade III MCL injury in combination with ACL injury. In the setting of combined ACL and MCL injury with a grade III MCL injury, no clear consensus on treatment exists. Some authors advocate for acute ACL-R and MCL repair (< 3 weeks from injury) when there is increased medial joint space opening with valgus stress in extension, a significant meniscotibial deep MCL injury (high-riding medial meniscus), or a displaced tibial-sided superficial MCL avulsion (Stener lesion of the knee) [127–129]. Figure 16.3 describes MRI findings for patients with multiligament knee injuries, including injury to the MCL.

Initial nonoperative management with delayed MCL reconstruction if continued valgus laxity exists has the advantage of avoiding surgical management of the MCL in some cases; however, there is a disadvantage with continued AP and anteromedial laxity in the setting of ACL and MCL injury [129]. There is some evidence to suggest equivocal outcomes [130] but also evidence to suggest knee stiffness with a delayed reconstruction approach in the ACL and MCL injured knee [130–132]. When considering repair versus reconstruction of the MCL, the tissue quality is most important regardless of timing.

The posteromedial corner (PMC) of the knee garners less attention especially when compared to the previously discussed PLC of the knee. The aptly termed "neglected corner" of the knee contains the posterior oblique ligament (POL), the semimembranosus tendon and its expansions, the oblique popliteal ligament (OPL), the posteromedial joint capsule, and the posterior horn of the medial meniscus [133]. In patients with MLKI including PMC injury, there is limited data on the management of the PMC. In an evaluation of consecutive knee dislocations, a cohort of 73 PMC injuries was available for evaluation. Of that cohort, 25 had PMC repairs and 48 had PMC reconstruction with either autograft or allograft. In this cohort, there was a 20% failure rate in the PMC repair group as compared to 4% failure rate in the PMC reconstruction group; however, given the small population, this was not found to be significant [134]. When considering PMC injury in the setting of ACL injury, it is important to recognize the implications on stability of the PMC; however, there is limited clinical data to support surgical management of the PMC in the isolated ACL and PMC injured patient.

16.4.4 ACL Reconstruction with PCL Injury

Isolated PCL injury is rare with an incidence of 1–6% [135]. Frequently, PCL injuries are found with concurrent injury to other structures of the knee. ACL/PCL combined injuries have a prevalence of 4–7% [5, 136, 137]. Multiple case studies have shown promising outcomes after single stage ACL and PCL reconstruction [138, 139]. The patient-reported outcomes after ACL and PCL reconstruction have a large variability in the reported return to sport rate. Optimal management of bicruciate injuries remains controversial and needs to be individualized to the patient's demographics and goals.

Bicruciate injury alone is rare and usually occurs with damage to other knee structures. Thus, there is limited data on bicruciate ligament damage alone. In a recent systematic review of bicruciate injuries, patient reports outcomes were higher at 2 years follow-up in the PCL reconstruction group as compared to combined ACL and PCL reconstruction. Decreased outcomes after ACL and PCL reconstruction have been theorized to be related to the varying degree of trauma and heterogenous patient population.

16.4.5 Complications of ACL Reconstruction in Multiligament Knee Injury

When discussing surgical management of ACL injuries in the setting of multiple ligamentous pathology, it is important to recognize the associated risk factors and specific complications. In a recent case–control study looking at risk factors for complications after surgically managed MLKI, smoking and planned staged surgery significantly increased the risk for postoperative complications. Increased time from injury to surgery decreased the risk of postoperative complications, specifically delayed surgery was associated with a decreased rate of knee stiffness [140, 141].

16.5 Conclusions

Management of ACL tears can become routine; however, there are many accompanying pathologies that need to be recognized. These pathologies need to be treated appropriately to provide the best patient outcomes. Each combination of pathologies and each patient is unique. A thorough plan should be developed to address each injury with consideration to the patient population, patient goals, and patient's ability to complete the appropriate rehabilitation.

References

1. Khan T, Alvand A, Prieto-Alhambra D, et al. ACL and meniscal injuries increase the risk of primary total knee replacement for osteoarthritis: a matched case-control study using the clinical practice research datalink (CPRD). Br J Sports Med. 2019;53(15):965. https://doi.org/10.1136/bjsports-2017-097762.
2. Krych AJ, LaPrade MD, Cook CS, et al. Lateral meniscal oblique radial tears are common with ACL injury: A classification system based on arthroscopic tear patterns in 600 consecutive patients. Orthop J Sports Med. 2020;8(5):2325967120921737. https://doi.org/10.1177/2325967120921737.
3. Tandogan RN, Taşer Ö, Kayaalp A, et al. Analysis of meniscal and chondral lesions accompanying anterior cruciate ligament tears: relationship with age, time from injury, and level of sport. Knee Surg Sports Traumatol Arthrosc. 2004;12(4):262. https://doi.org/10.1007/s00167-003-0398-z.
4. Brambilla L, Pulici L, Carimati G, et al. Prevalence of associated lesions in anterior cruciate ligament reconstruction. Am J Sports Med. 2015;43(12):2966. https://doi.org/10.1177/0363546515608483.
5. Zsidai B, Horvath A, Winkler PW, et al. Different injury patterns exist among patients undergoing operative treatment of isolated PCL, combined PCL/ACL, and isolated ACL injuries: a study from the Swedish National Knee Ligament Registry. Knee Surg Sports Traumatol Arthrosc. 2022;30(10):3451–60. https://doi.org/10.1007/s00167-022-06948-x.
6. Musahl V, Citak M, O'Loughlin PF, Choi D, Bedi A, Pearle AD. The effect of medial versus lateral meniscectomy on the stability of the anterior cruciate ligament-deficient knee. Am J Sports Med. 2010;38(8):1591. https://doi.org/10.1177/0363546510364402.

7. Ode GE, Van Thiel GS, McArthur SA, et al. Effects of serial sectioning and repair of radial tears in the lateral meniscus. Am J Sports Med. 2012;40(8):1863. https://doi.org/10.1177/0363546512453291.
8. Shelbourne KD, Heinrich J. The long-term evaluation of lateral meniscus tears left in situ at the time of anterior cruciate ligament reconstruction. Arthroscopy. 2004;20(4):346. https://doi.org/10.1016/j.arthro.2004.01.029.
9. Barenius B, Ponzer S, Shalabi A, Bujak R, Norlén L, Eriksson K. Increased risk of osteoarthritis after anterior cruciate ligament reconstruction: A 14-year follow-up study of a randomized controlled trial. Am J Sports Med. 2014;42(5):1049. https://doi.org/10.1177/0363546514526139.
10. Chalatsis G, Mitrousias V, Siouras A, et al. Long-term quality of life in patients after ACL reconstruction with concomitant meniscal injury treatment: patient-reported outcomes at minimum 10-year follow-up. Orthop J Sports Med. 2023;11(6):23259671231177279. https://doi.org/10.1177/23259671231177279.
11. Hagmeijer MH, Hevesi M, Desai VS, et al. Secondary meniscal tears in patients with anterior cruciate ligament injury: relationship among operative management, osteoarthritis, and arthroplasty at 18-year mean follow-up. Am J Sports Med. 2019;47(7):1583. https://doi.org/10.1177/0363546519844481.
12. Wright RW, Huston LJ, Haas AK. Ten-year outcomes of second-generation, all-inside meniscal repair in the setting of ACL reconstruction. J Bone Joint Surg. 2023;105(12):908. https://doi.org/10.2106/JBJS.22.01196.
13. Tsujii A, Yonetani Y, Kinugasa K, et al. Outcomes more than 2 years after meniscal repair for radial/flap tears of the posterior lateral meniscus combined with anterior cruciate ligament reconstruction. Am J Sports Med. 2019;47(12):2888. https://doi.org/10.1177/0363546519869955.
14. Tsujii A, Yonetani Y, Kinugasa K, et al. Outcomes more than 2 years after meniscal repair for longitudinal tears of the lateral meniscus combined with anterior cruciate ligament reconstruction. Am J Sports Med. 2021;49(3):684. https://doi.org/10.1177/0363546520981976.
15. de Girolamo L, Galliera E, Volpi P, et al. Why menisci show higher healing rate when repaired during ACL reconstruction? Growth factors release can be the explanation. Knee Surg Sports Traumatol Arthrosc. 2015;23(1):90. https://doi.org/10.1007/s00167-013-2712-8.
16. Wasserstein D, Dwyer T, Gandhi R, Austin PC, Mahomed N, Ogilvie-Harris D. A matched-cohort population study of reoperation after meniscal repair with and without concomitant anterior cruciate ligament reconstruction. Am J Sports Med. 2013;41(2):349. https://doi.org/10.1177/0363546512471134.
17. Gerritsen LM, van der Lelij TJN, van Schie P, et al. Higher healing rate after meniscal repair with concomitant ACL reconstruction for tears located in vascular zone 1 compared to zone 2: a systematic review and meta-analysis. Knee Surg Sports Traumatol Arthrosc. 2022;30(6):1976. https://doi.org/10.1007/s00167-022-06862-2.
18. Kawakami Y, Takayama K, Matsumoto T, et al. Anterior cruciate ligament-derived stem cells transduced with BMP2 accelerate graft-bone integration after ACL reconstruction. Am J Sports Med. 2017;45(3):584. https://doi.org/10.1177/0363546516671707.
19. Kawashima I, Tsukahara T, Sakai T, et al. Delayed anterior cruciate ligament reconstruction increases the incidence of medial meniscal bucket handle tears and medial compartment chondral injuries in patients aged 40 years and older. Arch Orthop Trauma Surg. 2021;141(6):971. https://doi.org/10.1007/s00402-020-03745-4.
20. Erard J, Cance N, Shatrov J, et al. Delaying ACL reconstruction is associated with increased rates of medial meniscal tear. Knee Surg Sports Traumatol Arthrosc. 2023;31(10):4458–66. https://doi.org/10.1007/s00167-023-07516-7.
21. Deroche E, Batailler C, Swan J, Lustig S, Servien E. Significant risk of arthrolysis after simultaneous anterior cruciate ligament reconstruction and treatment of dislocated bucket-handle meniscal tear. Orthop Traumatol Surg Res. 2022;108(3):103252. https://doi.org/10.1016/j.otsr.2022.103252.
22. Shelbourne KD, Dersam MD. Comparison of partial meniscectomy versus meniscus repair for bucket-handle lateral meniscus tears in anterior cruciate ligament reconstructed knees. Arthroscopy. 2004;20(6):581–5. https://doi.org/10.1016/j.arthro.2004.03.009.
23. Wu IT, Hevesi M, Desai VS, et al. Comparative outcomes of radial and bucket-handle meniscal tear repair: A propensity-matched analysis. Am J Sports Med. 2018;46(11):2653. https://doi.org/10.1177/0363546518786035.
24. Costa GG, Grassi A, Zocco G, et al. What is the failure rate after arthroscopic repair of bucket-handle meniscal tears? A systematic review and meta-analysis. Am J Sports Med. 2022;50(6):1742. https://doi.org/10.1177/03635465211015425.
25. Forkel P, Reuter S, Sprenker F, et al. Different patterns of lateral meniscus root tears in ACL injuries: application of a differentiated classification system. Knee Surg Sports Traumatol Arthrosc. 2015;23(1):112. https://doi.org/10.1007/s00167-014-3467-6.
26. LaPrade CM, Jansson KS, Dornan G, Smith SD, Wijdicks CA, LaPrade RF. Altered tibiofemoral contact mechanics due to lateral meniscus posterior horn root avulsions and radial tears can be restored with in situ pull-out suture repairs. J Bone Joint Surg. 2014;96(6):471. https://doi.org/10.2106/JBJS.L.01252.
27. Shybut TB, Vega CE, Haddad J, et al. Effect of lateral meniscal root tear on the stability of

the anterior cruciate ligament-deficient knee. Am J Sports Med. 2015;43(4):905. https://doi.org/10.1177/0363546514563910.
28. Tang X, Marshall B, Wang JH, et al. Lateral meniscal posterior root repair with anterior cruciate ligament reconstruction better restores knee stability. Am J Sports Med. 2019;47(1):59. https://doi.org/10.1177/0363546518808004.
29. Bernholt D, DePhillipo NN, Aman ZS, Samuelsen BT, Kennedy MI, LaPrade RF. Increased posterior tibial slope results in increased incidence of posterior lateral meniscal root tears in ACL reconstruction patients. Knee Surg Sports Traumatol Arthrosc. 2021;29(11):3883–91. https://doi.org/10.1007/s00167-021-06456-4.
30. Shekhar A, Tapasvi S, Williams A. Outcomes of combined lateral meniscus posterior root repair and anterior cruciate ligament reconstruction. Orthop J Sports Med. 2022;10(3):23259671221083318. https://doi.org/10.1177/23259671221083318.
31. Zheng T, Song G, Li Y, et al. Clinical, radiographic, and arthroscopic outcomes of surgical repair for radial and avulsed lesions on the lateral meniscus posterior root during ACL reconstruction: A systematic review. Orthop J Sports Med. 2021;9(3):2325967121989678. https://doi.org/10.1177/2325967121989678.
32. Cuvillier M, Marot V, Bukvić F, et al. Evaluation of the all-inside technique for the repair of lateral meniscus root tears at 1 year after ACL reconstruction. Orthop J Sports Med. 2023;11(4):23259671221149716. https://doi.org/10.1177/23259671221149716.
33. Levy AS, Meier SW. Approach to cartilage injury in the anterior cruciate ligament-deficient knee. Orthop Clin N Am. 2003;34(1):149. https://doi.org/10.1016/S0030-5898(02)00065-2.
34. Hoshino Y, Salzler M, Jiang K, Musahl V. ACL injuries with concomitant meniscal pathologies. In: Doral M, Karlsson J, editors. Sports injuries. Berlin/Heidelberg: Springer; 2014. https://doi.org/10.1007/978-3-642-36801-1_88-1.
35. Mameri ES, Jackson GR, Gonzalez F, et al. Meniscus radial tears: Current concepts on management and repair techniques. Curr Rev Musculoskelet Med. 2023;16(5):182. https://doi.org/10.1007/s12178-023-09831-5.
36. Mameri ES, Dasari SP, Fortier LM, et al. Review of meniscus anatomy and biomechanics. Curr Rev Musculoskelet Med. 2022;15:323–35. https://doi.org/10.1007/s12178-022-09768-1.
37. Tachibana Y, Mae T, Fujie H, et al. Effect of radial meniscal tear on in situ forces of meniscus and tibiofemoral relationship. Knee Surg Sports Traumatol Arthrosc. 2017;25(2):355. https://doi.org/10.1007/s00167-016-4395-4.
38. Bedi A, Kelly NH, Baad M, et al. Dynamic contact mechanics of the medial meniscus as a function of radial tear, repair, and partial meniscectomy. J Bone Joint Surg. 2010;92(6):1398. https://doi.org/10.2106/JBJS.I.00539.
39. Kim JG, Lee DW. Editorial commentary: efforts to heal meniscal radial tears are ongoing. Arthroscopy. 2021;37(3):941. https://doi.org/10.1016/j.arthro.2020.12.002.
40. Barber-Westin SD, Noyes FR. Clinical healing rates of meniscus repairs of tears in the central-third (red-white) zone. Arthroscopy. 2014;30(1):134. https://doi.org/10.1016/j.arthro.2013.10.003.
41. Vint H, Quartley M, Robinson JR. All-inside versus inside-out meniscal repair: A systematic review and meta-analysis. Knee. 2021;28:28. https://doi.org/10.1016/j.knee.2020.12.005.
42. Nelson CG, Bonner KF. Inside-out meniscus repair. Arthrosc Tech. 2013;2:4. https://doi.org/10.1016/j.eats.2013.07.006.
43. Beamer BS, Masoudi A, Walley KC, et al. Analysis of a new all-inside versus inside-out technique for repairing radial meniscal tears. Arthroscopy. 2015;31(2):293. https://doi.org/10.1016/j.arthro.2014.08.011.
44. Branch EA, Milchteim C, Aspey BS, Liu W, Saliman JD, Anz AW. Biomechanical comparison of arthroscopic repair constructs for radial tears of the meniscus. Am J Sports Med. 2015;43(9):2270. https://doi.org/10.1177/0363546515591994.
45. Milliron EM, Magnussen RA, Cavendish P, Quinn J, AC DB, Flanigan DC. Repair of radial meniscus tears results in improved patient-reported outcome scores: A systematic review. Arthrosc Sports Med Rehabil. 2021;3(3):e967. https://doi.org/10.1016/j.asmr.2021.03.002.
46. LaPrade RF, Matheny LM, Moulton SG, James EW, Dean CS. Posterior meniscal root repairs: outcomes of an anatomic Transtibial pull-out technique. Am J Sports Med. 2017;45(4):884. https://doi.org/10.1177/0363546516673996.
47. Cinque ME, Geeslin AG, Chahla J, Dornan GJ, Laprade RF. Two-tunnel Transtibial repair of radial meniscus tears produces comparable results to inside-out repair of vertical meniscus tears. Am J Sports Med. 2017;45(10):2253. https://doi.org/10.1177/0363546517704425.
48. Tiftikçi U, Serbest S. Repair of isolated horizontal meniscal tears with all-inside suture materials using the overlock method: outcome study with a minimum 2-year follow-up. J Orthop Surg Res. 2016;11(1):131. https://doi.org/10.1186/s13018-016-0466-y.
49. Li L, Yang X, Yang L, et al. Biomechanical analysis of the effect of medial meniscus degenerative and traumatic lesions on the knee joint. Am J Transl Res. 2019;11(2):744.
50. Lee JK, Lee MC, Il KJ, Lim S. Prognostic factors for the treatment of meniscus horizontal tear. Sci Rep. 2022;12(1):17293. https://doi.org/10.1038/s41598-022-21599-1.
51. Brown MJ, Farrell JP, Kluczynski MA, Marzo JM. Biomechanical effects of a horizontal medial meniscal tear and subsequent leaflet resection.

Am J Sports Med. 2016;44(4):850. https://doi.org/10.1177/0363546515623782.
52. Woodmass JM, Johnson JD, Wu IT, Saris DBF, Stuart MJ, Krych AJ. Horizontal cleavage meniscus tear treated with all-inside circumferential compression stitches. Arthrosc Tech. 2017;6(4):e1329. https://doi.org/10.1016/j.eats.2017.05.016.
53. Ozeki N, Seil R, Krych AJ, Koga H. Surgical treatment of complex meniscus tear and disease: state of the art. J ISAKOS. 2021;6(1):35. https://doi.org/10.1136/jisakos-2019-000380.
54. DePhillipo NN, Moatshe G, Brady A, et al. Effect of Meniscocapsular and Meniscotibial lesions in ACL-deficient and ACL-reconstructed knees: A biomechanical study. Am J Sports Med. 2018;46(10):2422. https://doi.org/10.1177/0363546518774315.
55. Stephen JM, Halewood C, Kittl C, Bollen SR, Williams A, Amis AA. Posteromedial Meniscocapsular lesions increase Tibiofemoral joint laxity with anterior cruciate ligament deficiency, and their repair reduces laxity. Am J Sports Med. 2016;44(2):400. https://doi.org/10.1177/0363546515617454.
56. Naendrup JH, Pfeiffer TR, Chan C, et al. Effect of meniscal ramp lesion repair on knee kinematics, bony contact forces, and in situ forces in the anterior cruciate ligament. Am J Sports Med. 2019;47(13):3195. https://doi.org/10.1177/0363546519872964.
57. Marin F, Soto J, Barahona M, Negrin R. Searching for the best treatment for ramp lesions: A systematic review and network meta-analysis. Cureus. Published online. 2023;15:e41651. https://doi.org/10.7759/cureus.41651.
58. Yang J, Guan K, Wang J. Clinical study on the arthroscopic refreshing treatment of anterior cruciate ligament injury combined with stable medial meniscus ramp injury. J Musculoskelet Neuronal Interact. 2017;17(2):108–13.
59. Balazs GC, Greditzer HG, Wang D, et al. Non-treatment of stable ramp lesions does not degrade clinical outcomes in the setting of primary ACL reconstruction. Knee Surg Sports Traumatol Arthrosc. 2020;28(11):3576. https://doi.org/10.1007/s00167-020-06017-1.
60. Albayrak K, Buyukkuscu MO, Kurk MB, Kaya O, Kulduk A, Misir A. Leaving the stable ramp lesion unrepaired does not negatively affect clinical and functional outcomes as well as return to sports rates after ACL reconstruction. Knee Surg Sports Traumatol Arthrosc. 2021;29(11):3773. https://doi.org/10.1007/s00167-020-06402-w.
61. Thaunat M, Jan N, Fayard JM, et al. Repair of meniscal ramp lesions through a posteromedial portal during anterior cruciate ligament reconstruction: outcome study with a minimum 2-year follow-up. Arthroscopy. 2016;32(11):2269. https://doi.org/10.1016/j.arthro.2016.02.026.
62. Muench LN, Achtnich A, Krivec L, et al. Clinical outcome and healing rate after meniscal bucket handle tear repair. BMC Musculoskelet Disord. 2022;23(1):1063. https://doi.org/10.1186/s12891-022-06037-7.
63. Saltzman BM, Cotter EJ, Wang KC, et al. Arthroscopically repaired bucket-handle meniscus tears: patient demographics, postoperative outcomes, and a comparison of success and failure cases. Cartilage. 2020;11(1):77. https://doi.org/10.1177/1947603518783473.
64. El Helou A, Gousopoulos L, Shatrov J, et al. Failure rates of repaired bucket-handle tears of the medial meniscus concomitant with anterior cruciate ligament reconstruction: A cohort study of 253 patients from the SANTI study group with a mean follow-up of 94 months. Am J Sports Med. 2023;51(3):585. https://doi.org/10.1177/03635465221148497.
65. Hagmeijer MH, Kennedy NI, Tagliero AJ, et al. Long-term results after repair of isolated meniscal tears among patients aged 18 years and younger: an 18-year follow-up study. Am J Sports Med. 2019;47(4):799. https://doi.org/10.1177/0363546519826088.
66. Floyd ER, Rodriguez AN, Falaas KL, et al. The natural history of medial meniscal root tears: A biomechanical and clinical case perspective. Front Bioeng Biotechnol. 2021;9:9. https://doi.org/10.3389/fbioe.2021.744065.
67. Padalecki JR, Jansson KS, Smith SD, et al. Biomechanical consequences of a complete radial tear adjacent to the medial meniscus posterior root attachment site: in situ pull-out repair restores derangement of joint mechanics. Am J Sports Med. 2014;42(3):699. https://doi.org/10.1177/0363546513499314.
68. Hein CN, Deperio JG, Ehrensberger MT, Marzo JM. Effects of medial meniscal posterior horn avulsion and repair on meniscal displacement. Knee. 2011;18(3):189. https://doi.org/10.1016/j.knee.2010.04.006.
69. Laprade CM, Laprade MD, Turnbull TL, Wijdicks CA, Laprade RF. Biomechanical evaluation of the transtibial pull-out technique for posterior medial meniscal root repairs using 1 and 2 transtibial bone tunnels. Am J Sports Med. 2015;43(4):899. https://doi.org/10.1177/0363546514563278.
70. Pache S, Aman ZS, Kennedy M, et al. Current concepts review meniscal root tears: current concepts review. Arch Bone Jt Surg. 2018;6:250–9.
71. Stein JM, Yayac M, Conte EJ, Hornstein J. Treatment outcomes of meniscal root tears: A systematic review. Arthrosc Sports Med Rehabil. 2020;2(3):e251. https://doi.org/10.1016/j.asmr.2020.02.005.
72. Faucett SC, Geisler BP, Chahla J, et al. Meniscus root repair vs Meniscectomy or nonoperative management to prevent knee osteoarthritis after medial meniscus root tears: clinical and economic effectiveness. Am J Sports Med. 2019;47(3):762. https://doi.org/10.1177/0363546518755754.

73. Petersen W, Guenther D, Imhoff AB, et al. Management after acute rupture of the anterior cruciate ligament (ACL). Part 1: ACL reconstruction has a protective effect on secondary meniscus and cartilage lesions. Knee Surg Sports Traumatol Arthrosc. 2023;31(5):1665. https://doi.org/10.1007/s00167-022-06960-1.
74. Potter HG, Jain SK, Ma Y, Black BR, Fung S, Lyman S. Cartilage injury after acute, isolated anterior cruciate ligament tear: immediate and longitudinal effect with clinical/MRI follow-up. Am J Sports Med. 2012;40(2):276–85. https://doi.org/10.1177/0363546511423380.
75. Mehl J, Otto A, Baldino JB, et al. The ACL-deficient knee and the prevalence of meniscus and cartilage lesions: a systematic review and meta-analysis (CRD42017076897). Arch Orthop Trauma Surg. 2019;139(6):819–41. https://doi.org/10.1007/s00402-019-03128-4.
76. Prodromidis AD, Drosatou C, Mourikis A, Sutton PM, Charalambous CP. Relationship between timing of anterior cruciate ligament reconstruction and chondral injuries: A systematic review and meta-analysis. Am J Sports Med. 2022;50(13):3719–31. https://doi.org/10.1177/03635465211036141.
77. Nakamae A, Miyamoto A, Kamei G, et al. An older age, a longer duration between injury and surgery, and positive pivot shift test results increase the prevalence of articular cartilage injury during ACL reconstruction in all three compartments of the knee in patients with ACL injuries. Knee Surg Sports Traumatol Arthrosc. 2022;30(1):219–30. https://doi.org/10.1007/s00167-021-06461-7.
78. Sommerfeldt M, Raheem A, Whittaker J, Hui C, Otto D. Recurrent instability episodes and meniscal or cartilage damage after anterior cruciate ligament injury: A systematic review. Orthop J Sports Med. 2018;6(7):2325967118786507. https://doi.org/10.1177/2325967118786507.
79. Farinelli L, Csapo R, Meena A, Abermann E, Hoser C, Fink C. Concomitant injuries associated with ACL rupture in elite professional alpine ski racers and soccer players: A comparative study with propensity score matching analysis. Orthop J Sports Med. 2023;11(8):23259671231192127. https://doi.org/10.1177/23259671231192127.
80. Everhart JS, DiBartola AC, Swank K, et al. Cartilage damage at the time of anterior cruciate ligament reconstruction is associated with weaker quadriceps function and lower risk of future ACL injury. Knee Surg Sports Traumatol Arthrosc. 2020;28(2):576–83. https://doi.org/10.1007/s00167-019-05739-1.
81. Beynnon BD, Fiorentino N, Gardner-Morse M, et al. Combined injury to the ACL and lateral meniscus alters the geometry of articular cartilage and meniscus soon after initial trauma. J Orthop Res. 2020;38(4):759–67. https://doi.org/10.1002/jor.24519.
82. Pioger C, Claes S, Haidar I, et al. Prevalence and incidence of chondral and meniscal lesions in patients undergoing primary and subsequent revision anterior cruciate ligament reconstruction: an analysis of 213 patients from the SANTI group. Am J Sports Med. 2022;50(7):1798–804. https://doi.org/10.1177/03635465221094624.
83. Filardo G, de Caro F, Andriolo L, Kon E, Zaffagnini S, Marcacci M. Do cartilage lesions affect the clinical outcome of anterior cruciate ligament reconstruction? A systematic review. Knee Surg Sports Traumatol Arthrosc. 2017;25(10):3061–75. https://doi.org/10.1007/s00167-016-4097-y.
84. Chalmers PN, Mall NA, Moric M, et al. Does ACL reconstruction alter natural history?: A systematic literature review of long-term outcomes. J Bone Joint Surg. 2014;96(4):292–300. https://doi.org/10.2106/JBJS.L.01713.
85. Wright RW, Huston LJ, Haas AK, et al. Meniscal and articular cartilage predictors of outcome after revision ACL reconstruction: A 6-year follow-up cohort study. Am J Sports Med. 2023;51(3):605–14. https://doi.org/10.1177/03635465231151389.
86. Shelbourne KD, Gray T. Results of anterior cruciate ligament reconstruction based on meniscus and articular cartilage status at the time of surgery: five- to fifteen-year evaluations. Am J Sports Med. 2000;28(4):446. https://doi.org/10.1177/03635465000280040201.
87. Krych AJ, Saris DBF, Stuart MJ, Hacken B. Cartilage injury in the knee: assessment and treatment options. J Am Acad Orthop Surg. 2020;28(22):914–22. https://doi.org/10.5435/JAAOS-D-20-00266.
88. Wang K, Eftang CN, Ulstein S, Årøen A, Jakobsen RB. Concomitant full-thickness cartilage lesions do not affect patient-reported outcomes at minimum 10-year follow-up after ACL reconstruction. Knee Surg Sports Traumatol Arthrosc. 2022;30(5):1836–45. https://doi.org/10.1007/s00167-021-06757-8.
89. Sofu H, Gumussuyu G, Guler O, Ucpunar H, Duman S, Camurcu Y. Lesion size and varus malalignment are the major determinants leading to poorer clinical outcomes after combined microfracture treatment for focal cartilage lesions during anterior cruciate ligament reconstruction. Arch Orthop Trauma Surg. 2022;142(8):1941–9. https://doi.org/10.1007/s00402-021-04138-x.
90. Mehl J, Feucht M, Achtnich A, et al. Autologous chondrocyte implantation combined with anterior cruciate ligament reconstruction: similar short-term results in comparison with isolated cartilage repair in ligament intact joints. Knee Surg Sports Traumatol Arthrosc. 2022;30(10):3249–57. https://doi.org/10.1007/s00167-021-06764-9.
91. Markus DH, Hurley ET, Mojica ES, et al. Concentration of synovial fluid biomarkers on the day of anterior cruciate ligament (ACL)-reconstruction predict size and depth of cartilage lesions on 5-year follow-up. Knee Surg Sports Traumatol Arthrosc. 2023;31(5):1753–60. https://doi.org/10.1007/s00167-022-07045-9.

92. Nishida Y, Hashimoto Y, Orita K, Nishino K, Kinoshita T, Nakamura H. Serum cartilage Oligomeric matrix protein detects early osteoarthritis in patients with anterior cruciate ligament deficiency. Arthroscopy. 2022;38(3):873–8. https://doi.org/10.1016/j.arthro.2021.06.019.
93. Pike AN, Patzkowski JC, Bottoni CR. Meniscal and chondral pathology associated with anterior cruciate ligament injuries. J Am Acad Orthop Surg. 2019;27(3):75–84. https://doi.org/10.5435/JAAOS-D-17-00670.
94. Levy BA, Dajani KA, Whelan DB, et al. Decision making in the multiligament-injured knee: an evidence-based systematic review. Arthroscopy. 2009;25(4):430. https://doi.org/10.1016/j.arthro.2009.01.008.
95. Mook WR, Ligh CA, Moorman CT, Leversedge FJ. Nerve injury complicating multiligament knee injury: Current concepts and treatment algorithm. J Am Acad Orthop Surg. 2013;21(6):343. https://doi.org/10.5435/JAAOS-21-06-343.
96. Kaeding CC, Pedroza AD, Parker RD, Spindler KP, McCarty EC, Andrish JT. Intra-articular findings in the reconstructed multiligament-injured knee. Arthroscopy. 2005;21(4):424. https://doi.org/10.1016/j.arthro.2004.12.012.
97. Erratum. Presentation and surgical management of multiple ligament knee injuries. A multicenter study from the surgical timing and rehabilitation (STaR) Trial for MLKIs Network. J Bone Joint Surg. 2023;105(8):607–13. https://doi.org/10.2106/JBJS.20.02051.
98. Hendel MD, Liu JN, Levy BA, Jakobsen BW, Marx RG. ACL reconstruction in the multiligament knee. In: Controversies in the technical aspects of ACL reconstruction: an evidence-based medicine approach; 2017. https://doi.org/10.1007/978-3-662-52742-9_35.
99. Lynch AD, Chmielewski T, Bailey L, et al. Current concepts and controversies in rehabilitation after surgery for multiple ligament knee injury. Curr Rev Musculoskelet Med. 2017;10(3):328. https://doi.org/10.1007/s12178-017-9425-4.
100. Stannard JP, Bauer KL. Current concepts in knee dislocations: PCL, ACL, and medial sided injuries. J Knee Surg. 2012;25(4):287. https://doi.org/10.1055/s-0032-1326998.
101. Musahl V, Rahnemai-Azar AA, Costello J, et al. The influence of meniscal and anterolateral capsular injury on knee laxity in patients with anterior cruciate ligament injuries. Am J Sports Med. 2016;44(12):3126. https://doi.org/10.1177/0363546516659649.
102. Willinger L, Athwal KK, Holthof S, Imhoff AB, Williams A, Amis AA. Role of the anterior cruciate ligament, anterolateral complex, and lateral meniscus posterior root in anterolateral rotatory knee instability: A biomechanical study. Am J Sports Med. 2023;51(5):1136. https://doi.org/10.1177/03635465231161071.
103. Spencer L, Burkhart TA, Tran MN, et al. Biomechanical analysis of simulated clinical testing and reconstruction of the anterolateral ligament of the knee. Am J Sports Med. 2015;43(9):2189. https://doi.org/10.1177/0363546515589166.
104. Tanaka M, Vyas D, Moloney G, Bedi A, Pearle AD, Musahl V. What does it take to have a high-grade pivot shift? Knee Surg Sports Traumatol Arthrosc. 2012;20(4):737. https://doi.org/10.1007/s00167-011-1866-5.
105. Musahl V, Getgood A, Neyret P, et al. Contributions of the anterolateral complex and the anterolateral ligament to rotatory knee stability in the setting of ACL injury: a roundtable discussion. Knee Surg Sports Traumatol Arthrosc. 2017;25(4):997. https://doi.org/10.1007/s00167-017-4436-7.
106. Getgood AMJ, Bryant DM, Litchfield R, et al. Lateral extra-articular Tenodesis reduces failure of hamstring tendon autograft anterior cruciate ligament reconstruction: 2-year outcomes from the STABILITY study randomized clinical trial. Am J Sports Med. 2020;48(2):285. https://doi.org/10.1177/0363546519896333.
107. Sonnery-Cottet B, Thaunat M, Freychet B, Pupim BHB, Murphy CG, Claes S. Outcome of a combined anterior cruciate ligament and anterolateral ligament reconstruction technique with a minimum 2-year follow-up. Am J Sports Med. 2015;43(7):1598. https://doi.org/10.1177/0363546515571571.
108. Sonnery-Cottet B, Haidar I, Rayes J, et al. Long-term graft rupture rates after combined ACL and anterolateral ligament reconstruction versus isolated ACL reconstruction: A matched-pair analysis from the SANTI study group. Am J Sports Med. 2021;49(11):2889. https://doi.org/10.1177/03635465211028990.
109. Porrino J, Sharp JW, Ashimolowo T, Dunham G. An update and comprehensive review of the posterolateral corner of the knee. Radiol Clin North Am. 2018;56(6):935. https://doi.org/10.1016/j.rcl.2018.06.006.
110. Camarda L, Caradonna G, Grassedonio E, D'Arienzo A, D'Arienzo M. Update in the diagnosis of posterolateral knee injuries. Minerva Ortopedica e Traumatologica. 2011;62(6):471–7.
111. Maniar AR, White AE, Musahl V, Ranawat A. Posterolateral corner of the knee: an update on current evaluation and management strategies. J Am Acad Orthop Surg. Published online July 19. 2023;32:e13–23. https://doi.org/10.5435/jaaos-d-23-00278.
112. LaPrade RF, Resig S, Wentorf F, Lewis JL. The effects of grade III posterolateral knee complex injuries on anterior cruciate ligament graft force. A biomechanical analysis. Am J Sports Med. 1999;27(4):469–75. https://doi.org/10.1177/03635465990270041101.
113. LaPrade RF, Wentorf FA, Fritts H, Gundry C, Hightower CD. A prospective magnetic resonance imaging study of the incidence of posterolateral

and multiple ligament injuries in acute knee injuries presenting with a Hemarthrosis. Arthroscopy. 2007;23(12):1341. https://doi.org/10.1016/j.arthro.2007.07.024.

114. Noyes FR, Barber-Westin SD. Surgical reconstruction of severe chronic posterolateral complex injuries of the knee using allograft tissues. Am J Sports Med. 1995;23(1):2. https://doi.org/10.1177/036354659502300102.

115. Kannus P. Nonoperative treatment of grade II and III sprains of the lateral ligament compartment of the knee. Am J Sports Med. 1989;17(1):83. https://doi.org/10.1177/036354658901700114.

116. Chahla J, Murray IR, Robinson J, et al. Posterolateral corner of the knee: an expert consensus statement on diagnosis, classification, treatment, and rehabilitation. Knee Surg Sports Traumatol Arthrosc. 2019;27(8):2520–9. https://doi.org/10.1007/s00167-018-5260-4.

117. Westermann RW, Marx RG, Spindler KP, et al. No difference between posterolateral corner repair and reconstruction with concurrent ACL surgery: results from a prospective multicenter cohort. Orthop J Sports Med. 2019;7(7):2325967119861062. https://doi.org/10.1177/2325967119861062.

118. Levy BA, Dajani KA, Morgan JA, Shah JP, DiL D, Stuart MJ. Repair versus reconstruction of the fibular collateral ligament and posterolateral corner in the multiligament-injured knee. Am J Sports Med. 2010;38:804. https://doi.org/10.1177/0363546509352459.

119. van Gennip S, van der Wal WA, Heesterbeek PJC, Wymenga AB, Busch VJJF. Posterolateral corner reconstruction in combined injuries of the knee: improved stability with Larson's fibular sling reconstruction and comparison with LaPrade anatomical reconstruction. Knee. 2020;27(1):124–31. https://doi.org/10.1016/j.knee.2019.09.008.

120. Cartwright-Terry M, Yates J, Tan CK, Pengas IP, Banks JV, McNicholas MJ. Medium-term (5-year) comparison of the functional outcomes of combined anterior cruciate ligament and posterolateral corner reconstruction compared with isolated anterior cruciate ligament reconstruction. Arthroscopy. 2014;30(7):811. https://doi.org/10.1016/j.arthro.2014.02.039.

121. Essilfie AA, Alaia EF, Bloom DA, et al. Distal posterolateral corner injury in the setting of multiligament knee injury increases risk of common peroneal palsy. Knee Surg Sports Traumatol Arthrosc. 2022;30(1):239. https://doi.org/10.1007/s00167-021-06469-z.

122. Jari S, Shelbourne KD. Nonoperative or delayed surgical treatment of combined cruciate ligaments and medial side knee injuries. Sports Med Arthrosc Rev. 2001;9(3):185. https://doi.org/10.1097/00132585-200107000-00004.

123. Jiang KN, West RV. Management of chronic combined ACL medial posteromedial instability of the knee. Sports Med Arthrosc Rev. 2015;23(2):85. https://doi.org/10.1097/JSA.0000000000000056.

124. Haimes JL, Wroble RR, Grood ES, Noyes FR. Role of the medial structures in the intact and anterior cruciate ligament-deficient knee: limits of motion in the human knee. Am J Sports Med. 1994;22(3):402. https://doi.org/10.1177/036354659402200317.

125. Elkin JL, Zamora E, Gallo RA. Combined anterior cruciate ligament and medial collateral ligament knee injuries: anatomy, diagnosis, management recommendations, and return to sport. Curr Rev Musculoskelet Med. Published online. 2019;12:239. https://doi.org/10.1007/s12178-019-09549-3.

126. Ahn JH, Lee SH. Risk factors for knee instability after anterior cruciate ligament reconstruction. Knee Surg Sports Traumatol Arthrosc. 2016;24(9):2936. https://doi.org/10.1007/s00167-015-3568-x.

127. Bollier M, Smith P. Anterior cruciate ligament and medial collateral ligament injuries. Journal of Knee Surgery. 2014;27(05):359–68. https://doi.org/10.1055/s-0034-1381961.

128. Shelbourne KD, Porter DA. Anterior cruciate ligament-medial collateral ligament injury: nonoperative management of medial collateral ligament tears with anterior cruciate ligament reconstruction: A preliminary report. Am J Sports Med. 1992;20(3):283. https://doi.org/10.1177/036354659202000308.

129. van der List JP, Muscott RK, Parikh N, Waterman BR, Trasolini NA. Early anterior cruciate ligament treatment might be crucial for acute combined anterior cruciate ligament and medial collateral ligament injuries: A systematic review of the various treatment strategies. Arthroscopy. Published online January 17. 2024;40:2322. https://doi.org/10.1016/J.ARTHRO.2024.01.009.

130. Vermeijden HD, Yang XA, Rademakers MV, Kerkhoffs GMMJ, van der List JP, DiFelice GS. Early and delayed surgery for isolated ACL and multiligamentous knee injuries have equivalent results: A systematic review and meta-analysis. Am J Sports Med. 2023;51(4):1106. https://doi.org/10.1177/03635465211069356.

131. Deabate L, Previtali D, Grassi A, Filardo G, Candrian C, Delcogliano M. Anterior cruciate ligament reconstruction within 3 weeks does not increase stiffness and complications compared with delayed reconstruction: A meta-analysis of randomized controlled trials. Am J Sports Med. 2020;48(5):1263. https://doi.org/10.1177/0363546519862294.

132. Holuba K, Vermeijden HD, Yang XA, O'Brien R, van der List JP, DiFelice GS. Treating combined anterior cruciate ligament and medial collateral ligament injuries operatively in the acute setting is potentially advantageous. Arthroscopy. 2023;39(4):1099–107. https://doi.org/10.1016/j.arthro.2022.06.023.

133. Dold AP, Swensen S, Strauss E, Alaia M. The posteromedial corner of the knee: anatomy, pathology, and management strategies. J Am Acad Orthop Surg. 2017;25(11):752–61. https://doi.org/10.5435/JAAOS-D-16-00020.

134. Stannard JP, Black BS, Azbell C, Volgas DA. Posteromedial corner injury in knee dislocations. J Knee Surg. 2012;25(5):429. https://doi.org/10.1055/s-0032-1322605.
135. Bram JT, Magee LC, Mehta NN, Patel NM, Ganley TJ. Anterior cruciate ligament injury incidence in adolescent athletes: A systematic review and meta-analysis. Am J Sports Med. 2021;49(7):1962. https://doi.org/10.1177/0363546520959619.
136. Winkler PW, Zsidai B, Narup E, et al. Sports activity and quality of life improve after isolated ACL, isolated PCL, and combined ACL/PCL reconstruction. Knee Surg Sports Traumatol Arthrosc. 2023;31(5):1781. https://doi.org/10.1007/s00167-022-07060-w.
137. Mygind-Klavsen B, Nielsen TG, Lind MC. Outcomes after posterior cruciate ligament (PCL) reconstruction in patients with isolated and combined PCL tears. Orthop J Sports Med. 2017;5(4):2325967117700077. https://doi.org/10.1177/2325967117700077.
138. Fayed AM, Rothrauff BB, de Sa D, Fu FH, Musahl V. Clinical studies of single-stage combined ACL and PCL reconstruction variably report graft tensioning, fixation sequence, and knee flexion angle at time of fixation. Knee Surg Sports Traumatol Arthrosc. 2021;29(4):1238. https://doi.org/10.1007/s00167-020-06171-6.
139. Fayed AM, Kanto R, Price TM, et al. No difference in knee kinematics between anterior cruciate ligament–first and posterior cruciate ligament–first fixation during single-stage multiligament knee reconstruction: a biomechanical study. Orthop J Sports Med. 2022;10(9):23259671221118587. https://doi.org/10.1177/23259671221118587.
140. Patel NK, Lian J, Nickoli M, et al. Risk factors associated with complications after operative treatment of multiligament knee injury. Orthop J Sports Med. 2021;9(3):2325967121994203. https://doi.org/10.1177/2325967121994203.
141. Özbek EA, Dadoo S, Grandberg C, et al. Early surgery and number of injured ligaments are associated with postoperative stiffness following multi-ligament knee injury surgery: a systematic review and meta-analysis. Knee Surg Sports Traumatol Arthrosc. 2023;31(10):4448. https://doi.org/10.1007/s00167-023-07514-9.

ACL Reconstruction and the Development of Early Arthrosis

17

Richard Danilkowicz, Nathan L. Grimm, Jeffrey Liles, and Annunziato Amendola

17.1 Introduction

The anterior cruciate ligament (ACL) conveys substantial stability to the knee joint, predominantly in the prevention of excessive anterior tibial translation and internal rotation [1]. Without a functional ACL, the literature suggests a significant risk of progression to degenerative changes within the knee joint and subsequent development of osteoarthritis (OA). However, quantifying an incidence for this progression to osteoarthritis is difficult for a multitude of reasons including the numerous potential confounding variables and the length of time required to conduct an appropriate natural history study. One known contributing factor for early arthrosis is the high incidence of concomitant soft tissue and chondral injuries that often accompany an ACL tear, including meniscal and other ligamentous structures. Injuries to these structures further destabilize the joint and can expedite the progression of degenerative changes in an additive manner, particularly when those injuries cannot be repaired themselves. The activity level of patients post injury also likely plays a significant role in progression as highly active individuals will put a different level of stress on the knee joint over subsequent years than sedentary individuals, leading to advanced degenerative changes. Conversely, there is some literature to suggest that return to activities after ACL reconstruction (ACLR) may lead to lower rates of OA [2]. Perhaps most difficult for incidence tracking is the inherent issues in performing long-term studies to track OA progression based on the ability to physically track patients over a number of years as they change locations and providers. There is also debate as to whether ACLR with or without additional interventions can reduce the risk of OA progression after ACL rupture. This chapter focuses on the progression of OA after ACL injury, covering the basic science and biomechanical foundations as well as the surgical considerations that may impact the trajectory of OA progression in the injured knee.

R. Danilkowicz · J. Liles
Department of Orthopaedic Surgery, Division of Sports Medicine, Duke University Medical Center, Durham, NC, USA

A. Amendola (✉)
Department of Orthopaedic Surgery, Duke University Medical Center, Durham, NC, USA
e-mail: ned.amendola@duke.edu

N. L. Grimm
Idaho Sports Medicine Institute, Boise, ID, USA

17.2 Biomechanics and Natural History of the ACL-Deficient Knee

The primary role of the intact ACL is to prevent excessive anterior tibial translation and internal rotation. When the ACL is injured, the biomechanics of the knee are altered in both their resting

M. Karahan et al. (eds.), *Anterior Cruciate Ligament Injuries Throughout the Life Span*,
https://doi.org/10.1007/978-3-031-84501-7_17

positions and during motion and activity. Even after stability is restored with ACL reconstruction (ACLR), studies have shown up to 85% of knees will continue to display abnormal gait mechanics [3]. The ACL-deficient knee is more varus and internally rotated, which, combined with additional freedom of movement, increases mean contract stresses in both the medial and lateral compartments of the knee [4, 5]. The ACL-deficient knee also moves with more anterior–posterior freedom as well as increased internal rotation of the tibial relative to the femur. In addition to the static stabilizers being compromised, ACL deficiency results in a loss of proprioception within the knee that is critical for the coordination of the dynamic stabilizers surrounding the knee including the quadriceps and hamstring musculature that provide additional secondary stability to the knee [6]. The increased forces and lack of stability are then transferred to the cartilage and menisci, which can result in damage to those structures as well. Regarding the cartilage specifically, increased physiological forces can alter the homeostasis of the cartilage environment which leads to more wear and tear over time. Sutter et al. evidenced these changes in an magnetic resonance imaging (MRI)-based study on patients performing single leg hops on both an ACL competent and an ACL-deficient knee and found that the medial femoral condyle adjacent to the intercondylar notch in the ACL-deficient knee exhibited elevated cartilage strain and loss of cartilage thickness that were associated with time since injury [7]. These findings were later corroborated in a similar study focusing on patient's under normal physiologic stresses of walking, which showed significantly increased cartilage strain of the medial femoral condyle adjacent to the intercondylar notch and across the medial and lateral tibial plateaus relative to the patient's ACL competent contralateral knee [8]. Similarly, the increased stresses can have deleterious effects on the menisci which are secondary stabilizers of the knee. Biomechanical studies have established increased stresses across the menisci in ACL-deficient knees, particularly in the posterior horn of the medial meniscus [9, 10]. Perhaps more interesting, Arner et al. used a cadaveric model to show that ACL-deficient knees have increased meniscal mobility as well, particularly in the lateral meniscus, which may explain the increased stresses and degeneration in these tissues [11]. The change in biomechanics of the knee and the associated injuries to the chondral surfaces and meniscal tissues has been hypothesized as contributing factors for the development of post-traumatic arthritis in the joint.

The natural history of knees after ACL injury has been well documented in the literature, as a primary concern is the long-term potential for arthritis in the joint. With follow-up out to 32–37 years after ACL injury Kvist et al. found that patients who underwent early surgery at a mean of five days post injury had lower prevalence of radiographic OA as defined by Kellgren and Lawrence grade 2 or higher compared to those with delayed or no surgery [12]. Regardless of reconstruction, a large percentage of patients will go on to develop OA changes, which is likely in part to the prolonged inflammatory environment in the joint after injury [13, 14]. This is of particular concern as the majority of ACL injuries are sustained in younger patients. There is difficulty in determining the true prevalence of this population as the literature does not use one accepted definition for determining the presence of arthritis; therefore, the results should be interpreted with some caution. Despite these limitations, long-term follow-up studies after ACL injury estimate that the affected knee is aged by 30 years based on radiographic findings and functional outcomes and up to 87% of patients will eventually develop some degree of post-traumatic osteoarthritis (PTOA) [15–17]. However, these is a significant range of reported osteoarthritis rates in the literature, as evidenced by a 2019 systematic review that cites a prevalence range of 0–100% >10 years out from injury, which does not provide much guidance when counseling patients before surgery [18]. An updated literature review and meta-analysis on OA development after operative and non-operative management by Ferrero et al. in 2023 found a relative risk of OA of 1.72 in the non-operative cohort compared to operative.

In patients who specifically did not have a reconstruction performed the evidence is again mixed but appears to trend towards poor outcomes long term. Kannus and Jarvinen reported on 49 patients treated conservatively after ACL injury and found 70% of patients developed definitive arthritis only eight years after injury [19]. Similarly, Nebelung and Wuschech studied 19 Olympic level athletes treated conservatively after ACL rupture who all returned to high level activity and found 95% of the knees showed severe symptoms of OA and instability at 20-year follow-up with >50% progressing to total joint replacement [20]. Conversely, a 2020 systematic review and meta-analysis by Lien-Iverson et al. found OA prevalence in the non-surgical group to range from 11 to 68%, but significantly lower prevalence of radiographic knee OA in two studies when compared to a surgical cohort [21]. This is a different finding to a prior systematic review and meta-analysis that supports a surgical role in reducing the risk of progression to OA [22]. As mentioned previously, there is significant heterogenicity within the patient population, associated injuries, defining characteristics of OA, and overall quality of studies on this topic; therefore, these results should be interpreted with caution. Additionally, there is potential that surgery allows patients to return to prior levels of activity, which is a confounding factor in future development of OA over time with increased use and should also be considered in light of the literature.

17.3 Factors in the Development of Post-Traumatic Osteoarthritis After ACL Injury

A common misnomer is that surgery for ACL rupture will prevent the development of OA. As recent literature has further demonstrated, this may not be the case as Cinque et al. modeled the proportion of PTOA after ACL surgery at 5, 10, and 20 years after surgery to be 11.3% (6.4–19.1%), 20.6% (14.9–27.7%), and 51.6% (29.1–73.5%) [23]. The rationale behind these numbers is not as clear cut however. As briefly mentioned above, there are a number of potential risk factors in the development of PTOA after ACL injury that confound any analysis on prevalence alone, for example, patient specific factors, biologic factors, mechanical factors, and structural factors. Some patient-specific factors can be altered, such as body mass index, where a reduction can help offload the joint, or when and what level to return to back to activity; however, others such as age cannot [2, 24, 25]. The biologic considerations include immediate and long-term effects of the joint microenvironment after ACL injury, which include the presence of pro-inflammatory and cartilage degrading cytokines and matrix mellaoproteinases [26–28]. The mechanical factors including the surrounding joint musculature that helps in stabilizing the knee can be optimized with physical therapy to re-create normal mechanics; however, proprioception deficits are less amenable to intervention. The structural factors are an avenue that surgical intervention may provide long-term benefit from the patient and include ACLR, cartilage, and meniscal repair.

17.3.1 ACL Reconstruction Effects on Osteoarthritis Development

ACL reconstruction is more than a simple decision to proceed with surgery or not and the accompanying choices may play a role in the long-term development of osteoarthritis. These decisions include time from injury to surgery, choice of graft, and surgical technique. An understanding of how each may play a role in the long-term health of the joint is important for surgical counseling of patients.

Surgical timing may be the most clearly defined risk factor in the surgical algorithm in regard to OA development, primarily due to continued damage that instability incurs. The primary hypothesis for these time-related findings comes from the continued strain on the chondral and mensical surfaces of an unstable ACL deficient knee [29]. Cinque et al. evaluated 4108 ACL patients and amongst their significant findings was the increased time from injury to surgery that was associated with increased risk of OA [23]. Church and Keating evaluated 183 patients split into early (<12 months to surgery) and late (>12 months) and found a significantly higher incidence of meniscal tears in patients undergoing reconstruction after 12 months reconstruction, compared with early reconstruction (71.2 vs. 41.7%), which could lead to advanced degenerative changes in the knee [30]. As part of their review, Ajuied et al. cited three studies containing 151 total knees, 73 of which had an ACLR performed within 6 months of injury. In these patients, there was only Kellgren and Lawrence grade 0-I OA progression at minimum 10-year follow-up, versus a cohort of 187 procedures done between 24 and 34 months that contained grades of II, III, or IV [22]. Seon et al. evaluated 58 knees and found an odds ratio of 4.77 for the development of OA in knees where surgery was delayed more than 6 months compared to earlier to procedures [31]. Keays et al. utilized a cohort study of 56 patients and found a time difference before surgery between the patients who developed OA and those who did not, at 40 months and 24 months on average, respectively [32]. Lastly, as mentioned previously, Kvist et al. found that patients who underwent early surgery at a mean of five days post injury had lower prevalence of radiographic OA at 32–37 years follow-up compared to later or no surgery [12]. In consideration of these findings, there may be a benefit towards early surgical intervention within 6–12 months of injury before additional damage is done to the remaining knee structures, as the risk of additional lesions has been estimated to be approximately 0.6% per month [33]. However, despite a convincing association between time to surgery and reduction in the risk of arthritis progression, patients should still be optimized before surgery with appropriate pre-habilitation to maximize range of motion, decrease swelling, and improve muscular strengthening to further optimize postoperative outcomes. Postoperatively, considerations on return to activity and activity level also apply. In an MRI-based study of 111 patients one year after ACLR, those who returned to competitive sport within 10 months of surgery were found to have signs of early onset osteoarthritis, as defined by bone marrow lesions on MRI, when compared to those who waited longer than 10 months to return, suggesting too early of a return may be detrimental to long-term knee health [34].

A second consideration with ACLR and OA progression is the potential effect of graft choice, in which there is currently no consensus as reflected in the minimal, yet conflicting literature. In one large multicenter cohort study by Lecoq et al. encompassing 541 patients, 311 patellar tendon and 230 hamstring grafts were directly compared with no difference in the rate of OA at 12-year follow-up [35]. Conversely, a large meta-analysis of 1443 autograft patients with five years of follow-up found significantly increased radiographic evidence of OA in bone patellar bone graft patients compared with hamstring graft groups based on the IKDC system [36]. Similarly, the trend towards increased OA risk in bone patellar tendon bone autograft has been found in additional studies, yet no definitive conclusions can be made [37, 38]. There is also a paucity of studies that compare either allografts versus autografts or other comparative autograft studies that involve quad tendon use, likely due to the extended follow-up needed to detect the outcome of interest.

17.4 Chondral and Meniscal Injury and Their Effects on Osteoarthritis

Arguably more concerning for the progression of OA after ACL injury is the accompanying damage to the cartilage and meniscal tissues, which many studies have shown to be associated with earlier osteoarthritis than a knee with a simple ACL tear [39, 40]. In fact, Shelbourne et al. found that only 3% of patients with intact cartilage and menisci at time of ACLR had abnormal IKDC radiographic findings compared to 32% of patients with injuries to both at 7.5 years of follow-up [16]. In terms of cartilage specifically, studies have reported that to 42% of ACL injuries are associated with severe chondral damage, while others have cited that 100% of injuries have some degree of associated damage [41, 42]. As cartilage does not have the inherent healing capacity as bone or other tissues, damage is likely to progress and be a significant contributor to further degeneration of the knee [43]. Patient who sustain large, operable cartilage injuries at the time of ACL tear can undergo simultaneous procedures, which have shown acceptable outcomes, or the procedures can be staged depending on surgeon comfortability [44, 45]. Regarding the meniscus, isolated meniscal injuries are known to be independent factors associated with development of osteoarthritis, and therefore when combined with an ACL tear present an additive issue [46, 47]. In patient who sustained meniscal injury at the time of ACL tear, Suter et al. found that the lifetime risk of developing symptomatic OA was 34%, which was significantly higher than the 16% in isolated ACL injuries and 14% in uninjured controls [48]. Similarly, Poulson et al. cite odds of developing OA after concomitant ACL and meniscal tear at six times that of non-injured knees, compared to four times in isolated ACL [49]. Simultaneous repair of both ACL and meniscal tears had proven to be an effective method and is commonly performed, with good result [50–52]. In fact, studies have shown improved outcomes when the issues are treated in one procedure versus a staged approach [53]. It is the recommendation of the authors that when able, the two pathologies should be treated concurrently to both allow for a decreased overall recovery time and limit the risk for progression of the injuries.

17.5 Conclusion

The development of post-traumatic arthritis after ACL injury and repair is a highly debatable topic. While the ACL-deficient knee undoubtably undergoes abnormal biomechanics during the gait cycle, there is also a complex interplay of patient-specific factors and difficulty in conducting well designed, long-term follow-up studies that results in mixed literature. However, there is strong evidence to suggest that ACL injuries that are accompanied by chondral and meniscal pathology will progress to OA at higher rates than ACL injuries alone. Additionally, there is strong evidence to suggest that time to surgery of <12 months after injury is associated with a decreased prevalence of OA, likely due to the decreased risk of developing chondral and meniscal damage. Despite unconvincing data on the progression of OA in surgically repaired ACL patients, the outcomes are favorable for the procedure, and the preference is for repair of any concomitant meniscal injuries at the time of ACLR.

References

1. Domnick C, Raschke MJ, Herbort M. Biomechanics of the anterior cruciate ligament: Physiology, rupture and reconstruction techniques. World J Orthop. 2016;7(2):82–93. https://doi.org/10.5312/wjo.v7.i2.82.
2. Øiestad BE, Holm I, Risberg MA. Return to pivoting sport after ACL reconstruction: association with osteoarthritis and knee function at the 15-year follow-up. Br J Sports Med. 2018;52(18):1199. https://doi.org/10.1136/bjsports-2017-097718.
3. Scanlan SF, Chaudhari AM, Dyrby CO, Andriacchi TP. Differences in tibial rotation during walking in ACL reconstructed and healthy contralateral knees. J Biomech. 2010;43(9):1817–22. https://doi.org/10.1016/j.jbiomech.2010.02.010.
4. Imhauser C, Mauro C, Choi D, et al. Abnormal tibiofemoral contact stress and its association with altered kinematics after center-center anterior cruciate ligament reconstruction: an in vitro study. Am J Sports Med. 2013;41(4):815–25. https://doi.org/10.1177/0363546512475205.
5. Keays SL, Sayers M, Mellifont DB, Richardson C. Tibial displacement and rotation during seated knee extension and wall squatting: a comparative study of tibiofemoral kinematics between chronic unilateral anterior cruciate ligament deficient and healthy knees. Knee. 2013;20(5):346–53. https://doi.org/10.1016/j.knee.2012.07.005.
6. Reider B, Arcand MA, Diehl LH, et al. Proprioception of the knee before and after anterior cruciate ligament reconstruction. Arthroscopy. 2003;19(1):2–12. https://doi.org/10.1053/jars.2003.50006.
7. Sutter EG, Liu B, Utturkar GM, et al. Effects of anterior cruciate ligament deficiency on Tibiofemoral cartilage thickness and strains in response to hopping. Am J Sports Med. 2019;47(1):96–103. https://doi.org/10.1177/0363546518802225.
8. Crook BS, Collins AT, Lad NK, Spritzer CE, Wittstein JR, DeFrate LE. Effect of walking on in vivo tibiofemoral cartilage strain in ACL-deficient versus intact

knees. J Biomech. 2021;116:110210. https://doi.org/10.1016/j.jbiomech.2020.110210.
9. Yao J, Snibbe J, Maloney M, Lerner AL. Stresses and strains in the medial meniscus of an ACL deficient knee under anterior loading: a finite element analysis with image-based experimental validation. J Biomech Eng. 2006;128(1):135–41. https://doi.org/10.1115/1.2132373.
10. Guess TM, Razu S. Loading of the medial meniscus in the ACL deficient knee: a multibody computational study. Med Eng Phys. 2017;41:26–34. https://doi.org/10.1016/j.medengphy.2016.12.006.
11. Arner JW, Irvine JN, Zheng L, et al. The effects of anterior cruciate ligament deficiency on the Meniscus and Articular Cartilage: a novel dynamic in vitro pilot study. Orthop J Sports Med. 2016;4(4):2325967116639895. https://doi.org/10.1177/2325967116639895.
12. Kvist J, Filbay S, Andersson C, Ardern CL, Gauffin H. Radiographic and symptomatic knee osteoarthritis 32 to 37 years after acute anterior cruciate ligament rupture. Am J Sports Med. 2020;48(10):2387–94. https://doi.org/10.1177/0363546520939897.
13. Haslauer CM, Elsaid KA, Fleming BC, Proffen BL, Johnson VM, Murray MM. Loss of extracellular matrix from articular cartilage is mediated by the synovium and ligament after anterior cruciate ligament injury. Osteoarthr Cartil. 2013;21(12):1950–7. https://doi.org/10.1016/j.joca.2013.09.003.
14. Irie K, Uchiyama E, Iwaso H. Intraarticular inflammatory cytokines in acute anterior cruciate ligament injured knee. Knee. 2003;10(1):93–6. https://doi.org/10.1016/s0968-0160(02)00083-2.
15. Lohmander LS, Ostenberg A, Englund M, Roos H. High prevalence of knee osteoarthritis, pain, and functional limitations in female soccer players twelve years after anterior cruciate ligament injury. Arthritis Rheum. 2004;50(10):3145–52. https://doi.org/10.1002/art.20589.
16. Shelbourne KD, Stube KC. Anterior cruciate ligament (ACL)-deficient knee with degenerative arthrosis: treatment with an isolated autogenous patellar tendon ACL reconstruction. Knee Surg Sports Traumatol Arthrosc. 1997;5(3):150–6. https://doi.org/10.1007/s001670050043.
17. Nakata K, Shino K, Horibe S, et al. Arthroscopic anterior cruciate ligament reconstruction using fresh-frozen bone plug-free allogeneic tendons: 10-year follow-up. Arthroscopy. 2008;24(3):285–91. https://doi.org/10.1016/j.arthro.2007.09.007.
18. Lie MM, Risberg MA, Storheim K, Engebretsen L, Øiestad BE. What's the rate of knee osteoarthritis 10 years after anterior cruciate ligament injury? An updated systematic review. Br J Sports Med. 2019;53(18):1162. https://doi.org/10.1136/bjsports-2018-099751.
19. Kannus P, Järvinen M. Conservatively treated tears of the anterior cruciate ligament. Long-term results. J Bone Joint Surg Am. 1987;69(7):1007–12.
20. Nebelung W, Wuschech H. Thirty-five years of follow-up of anterior cruciate ligament-deficient knees in high-level athletes. Arthroscopy. 2005;21(6):696–702. https://doi.org/10.1016/j.arthro.2005.03.010.
21. Lien-Iversen T, Morgan DB, Jensen C, Risberg MA, Engebretsen L, Viberg B. Does surgery reduce knee osteoarthritis, meniscal injury and subsequent complications compared with non-surgery after ACL rupture with at least 10 years follow-up? A systematic review and meta-analysis. Br J Sports Med. 2020;54(10):592. https://doi.org/10.1136/bjsports-2019-100765.
22. Ajuied A, Wong F, Smith C, et al. Anterior cruciate ligament injury and radiologic progression of knee osteoarthritis: a systematic review and meta-analysis. Am J Sports Med. 2014;42(9):2242–52. https://doi.org/10.1177/0363546513508376.
23. Cinque ME, Dornan GJ, Chahla J, Moatshe G, LaPrade RF. High rates of osteoarthritis develop after anterior cruciate ligament surgery: an analysis of 4108 patients. Am J Sports Med. 2018;46(8):2011–9. https://doi.org/10.1177/0363546517730072.
24. Lebel B, Hulet C, Galaud B, Burdin G, Locker B, Vielpeau C. Arthroscopic reconstruction of the anterior cruciate ligament using bone-patellar tendon-bone autograft: a minimum 10-year follow-up. Am J Sports Med. 2008;36(7):1275–82. https://doi.org/10.1177/0363546508314721.
25. Kessler MA, Behrend H, Henz S, Stutz G, Rukavina A, Kuster MS. Function, osteoarthritis and activity after ACL-rupture: 11 years follow-up results of conservative versus reconstructive treatment. Knee Surg Sports Traumatol Arthrosc. 2008;16(5):442–8. https://doi.org/10.1007/s00167-008-0498-x.
26. Friel NA, Chu CR. The role of ACL injury in the development of posttraumatic knee osteoarthritis. Clin Sports Med. 2013;32(1):1–12. https://doi.org/10.1016/j.csm.2012.08.017.
27. Elsaid KA, Fleming BC, Oksendahl HL, et al. Decreased lubricin concentrations and markers of joint inflammation in the synovial fluid of patients with anterior cruciate ligament injury. Arthritis Rheum. 2008;58(6):1707–15. https://doi.org/10.1002/art.23495.
28. Riordan EA, Little C, Hunter D. Pathogenesis of post-traumatic OA with a view to intervention. Best Pract Res Clin Rheumatol. 2014;28(1):17–30. https://doi.org/10.1016/j.berh.2014.02.001.
29. Marks PH, Donaldson ML. Inflammatory cytokine profiles associated with chondral damage in the anterior cruciate ligament-deficient knee. Arthroscopy. 2005;21(11):1342–7. https://doi.org/10.1016/j.arthro.2005.08.034.

30. Church S, Keating JF. Reconstruction of the anterior cruciate ligament: timing of surgery and the incidence of meniscal tears and degenerative change. J Bone Joint Surg Br. 2005;87(12):1639–42. https://doi.org/10.1302/0301-620x.87b12.16916.
31. Seon JK, Song EK, Park SJ. Osteoarthritis after anterior cruciate ligament reconstruction using a patellar tendon autograft. Int Orthop. 2006;30(2):94–8. https://doi.org/10.1007/s00264-005-0036-0.
32. Keays SL, Newcombe PA, Bullock-Saxton JE, Bullock MI, Keays AC. Factors involved in the development of osteoarthritis after anterior cruciate ligament surgery. Am J Sports Med. 2010;38(3):455–63. https://doi.org/10.1177/0363546509350914.
33. Brambilla L, Pulici L, Carimati G, et al. Prevalence of associated lesions in anterior cruciate ligament reconstruction: correlation with surgical timing and with patient age, sex, and body mass index. Am J Sports Med. 2015;43(12):2966–73. https://doi.org/10.1177/0363546515608483.
34. Culvenor AG, Patterson BE, Guermazi A, Morris HG, Whitehead TS, Crossley KM. Accelerated return to sport after anterior cruciate ligament reconstruction and early knee osteoarthritis features at 1 year: an exploratory study. PM R. 2018;10(4):349–56. https://doi.org/10.1016/j.pmrj.2017.09.005.
35. Lecoq FA, Parienti JJ, Murison J, et al. Graft choice and the incidence of osteoarthritis after anterior cruciate ligament reconstruction: a causal analysis from a cohort of 541 patients. Am J Sports Med. 2018;46(12):2842–50. https://doi.org/10.1177/0363546518795137.
36. Xie X, Xiao Z, Li Q, et al. Increased incidence of osteoarthritis of knee joint after ACL reconstruction with bone-patellar tendon-bone autografts than hamstring autografts: a meta-analysis of 1,443 patients at a minimum of 5 years. Eur J Orthop Surg Traumatol. 2015;25(1):149–59. https://doi.org/10.1007/s00590-014-1459-3.
37. Magnussen RA, Carey JL, Spindler KP. Does autograft choice determine intermediate-term outcome of ACL reconstruction? Knee Surg Sports Traumatol Arthrosc. 2011;19(3):462–72. https://doi.org/10.1007/s00167-010-1277-z.
38. Murray JR, Lindh AM, Hogan NA, et al. Does anterior cruciate ligament reconstruction lead to degenerative disease?: Thirteen-year results after bone-patellar tendon-bone autograft. Am J Sports Med. 2012;40(2):404–13. https://doi.org/10.1177/0363546511428580.
39. Øiestad BE, Engebretsen L, Storheim K, Risberg MA. Knee osteoarthritis after anterior cruciate ligament injury: a systematic review. Am J Sports Med. 2009;37(7):1434–43. https://doi.org/10.1177/0363546509338827.
40. Li RT, Lorenz S, Xu Y, Harner CD, Fu FH, Irrgang JJ. Predictors of radiographic knee osteoarthritis after anterior cruciate ligament reconstruction. Am J Sports Med. 2011;39(12):2595–603. https://doi.org/10.1177/0363546511424720.
41. Brophy RH, Zeltser D, Wright RW, Flanigan D. Anterior cruciate ligament reconstruction and concomitant articular cartilage injury: incidence and treatment. Arthroscopy. 2010;26(1):112–20. https://doi.org/10.1016/j.arthro.2009.09.002.
42. Potter HG, Jain SK, Ma Y, Black BR, Fung S, Lyman S. Cartilage injury after acute, isolated anterior cruciate ligament tear: immediate and longitudinal effect with clinical/MRI follow-up. Am J Sports Med. 2012;40(2):276–85. https://doi.org/10.1177/0363546511423380.
43. Houck DA, Kraeutler MJ, Belk JW, Frank RM, McCarty EC, Bravman JT. Do Focal Chondral defects of the knee increase the risk for progression to osteoarthritis? a review of the literature. Orthop J Sports Med. 2018;6(10):2325967118801931. https://doi.org/10.1177/2325967118801931.
44. Pike AN, Bryant T, Ogura T, Minas T. Intermediate-to long-term results of combined anterior cruciate ligament reconstruction and autologous chondrocyte implantation. Orthop J Sports Med. 2017;5(2):2325967117693591. https://doi.org/10.1177/2325967117693591.
45. Gudas R, Gudaitė A, Mickevičius T, et al. Comparison of osteochondral autologous transplantation, microfracture, or debridement techniques in articular cartilage lesions associated with anterior cruciate ligament injury: a prospective study with a 3-year follow-up. Arthroscopy. 2013;29(1):89–97. https://doi.org/10.1016/j.arthro.2012.06.009.
46. Englund M, Guermazi A, Lohmander SL. The role of the meniscus in knee osteoarthritis: a cause or consequence? Radiol Clin North Am. 2009;47(4):703–12. https://doi.org/10.1016/j.rcl.2009.03.003.
47. Englund M, Guermazi A, Roemer FW, et al. Meniscal tear in knees without surgery and the development of radiographic osteoarthritis among middle-aged and elderly persons: the Multicenter Osteoarthritis Study. Arthritis Rheum. 2009;60(3):831–9. https://doi.org/10.1002/art.24383.
48. Suter LG, Smith SR, Katz JN, et al. Projecting lifetime risk of symptomatic knee osteoarthritis and total knee replacement in individuals sustaining a complete anterior cruciate ligament tear in early adulthood. Arthritis Care Res (Hoboken). 2017;69(2):201–8. https://doi.org/10.1002/acr.22940.
49. Poulsen E, Goncalves GH, Bricca A, Roos EM, Thorlund JB, Juhl CB. Knee osteoarthritis risk is increased 4-6 fold after knee injury - a systematic review and meta-analysis. Br J Sports Med. 2019;53(23):1454–63. https://doi.org/10.1136/bjsports-2018-100022.

50. Westermann RW, Wright RW, Spindler KP, Huston LJ, Group MK, Wolf BR. Meniscal repair with concurrent anterior cruciate ligament reconstruction: operative success and patient outcomes at 6-year follow-up. Am J Sports Med. 2014;42(9):2184–92. https://doi.org/10.1177/0363546514536022.
51. Melton JT, Murray JR, Karim A, Pandit H, Wandless F, Thomas NP. Meniscal repair in anterior cruciate ligament reconstruction: a long-term outcome study. Knee Surg Sports Traumatol Arthrosc. 2011;19(10):1729–34. https://doi.org/10.1007/s00167-011-1501-5.
52. Toman CV, Dunn WR, Spindler KP, et al. Success of meniscal repair at anterior cruciate ligament reconstruction. Am J Sports Med. 2009;37(6):1111–5. https://doi.org/10.1177/0363546509337010.
53. Gallacher PD, Gilbert RE, Kanes G, Roberts SN, Rees D. Outcome of meniscal repair prior compared with concurrent ACL reconstruction. Knee. 2012;19(4):461–3. https://doi.org/10.1016/j.knee.2011.04.004.

Risks and Complications of ACL Surgery

18

Sachin R. Tapasvi, Amit Meena, and Anshu Shekhar

18.1 Introduction

The occurrence of a complication in any surgical procedure impacts the overall patient outcome. There is no globally accepted definition of a surgical complication, and a number of terms like risk, sequela, medical error, adverse event, and negligence need to be differentiated for their cause and effect [1]. One of these many definitions is the one proposed by Sokol and Wilson and it reads as "any undesirable, unintended, and direct result of an operation affecting the patient, which would not have occurred had the operation gone as well as reasonably hoped." They further believed that a complication is not a fixed reality but varies according to the medical facilities available and a surgeon's skill set [2]. As a result of the ambiguity as to what may be labeled as a complication (Fig. 18.1), there may be unexpected errors in the accurate reporting of the data and its analysis. Ricketts et al., in their audit of an orthopedic computer database, found that any surgical procedure would have a complication rate of 10% or more and values lower than those may indicate poor data collection [3].

Anterior cruciate ligament (ACL) reconstruction (ACLR) is the most performed knee ligament surgery. As many as 120,000 ACLRs are performed annually in the United States, and this number is only likely to increase with greater participation in high school sports, especially among female athletes [3]. With advanced techniques and technology, arthroscopic ACLR is now a safe surgery. The rates of complications vary widely depending on how or what one considers a complication in the absence of a standard definition [4–6]. For example, re-tear of the ACL graft should not be labeled as a complication since it can happen for any number of reasons, several of which are not directly related to index surgery.

Complications in ACLR can be divided into three categories based on the timeline during which they occur: (1) pre-operative, (2) intra-operative, and (3) post-operative. This chapter will discuss each complication's cause, consequence, circumvention, and cure. This discussion does not include any complication(s) attributable to anesthesia or pharmacotherapy, such as adverse drug events.

S. R. Tapasvi (✉)
The Orthopaedic Speciality Clinic, Pune, India

A. Meena
Department of Orthopedics, Shalby Hospital, Jaipur, India

A. Shekhar
Shri Rawatpura Sarkar Institute of Medical Sciences and Research, Naya Raipur, India

M. Karahan et al. (eds.), *Anterior Cruciate Ligament Injuries Throughout the Life Span*,
https://doi.org/10.1007/978-3-031-84501-7_18

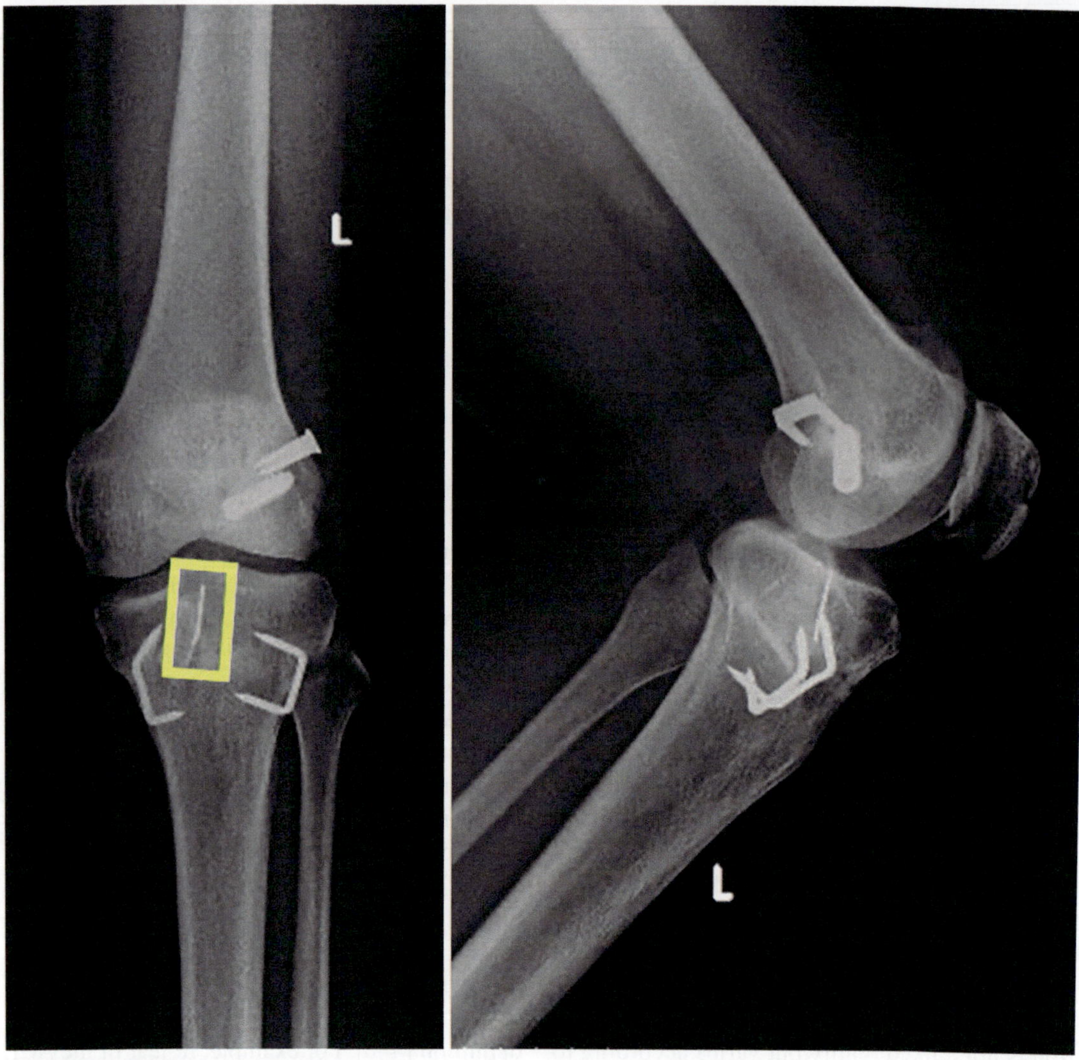

Fig. 18.1 Anteroposterior radiograph of a patient who had undergone a revision ACLR and tibia slope-reducing osteotomy. The broken screw guidewire in the tibia (yellow box) was noticed after surgery. ACLR anterior cruciate ligament reconstruction

18.2 Complications: Pre-Operative Origin

Complications during the pre-operative phase are usually due to improper patient selection and planning. Accurate pathology diagnosis, thorough clinical examination, and necessary investigations are the mainstays in decision-making before scheduling an ACLR. Any deficiencies in the above can be the forerunner of further undesirable outcomes that are later labeled as a *Complication*.

18.2.1 Incorrect Diagnosis: Intact ACL

18.2.1.1 Cause

A complete or partial ACL tear is often not a difficult diagnosis if careful history and examination are performed. A negative history of pivoting injury, popping sensation, normal Lachman test, and normal pivot shift test would make the diagnosis of an ACL tear highly unlikely [7]. However, there are instances when even the most astute clinician might have difficulties. A patient with traumatic patellar instability may need to give a clear history, which may be misleading to the physician. Similarly, partial ACL tears may not have frank instability symptoms. Magnetic resonance imaging (MRI) scans are used to confirm the diagnosis of ACL tears. They have a sensitivity of 87% and a specificity of 93% for diagnosing an ACL tear [8]. It needs to be clarified if the magnetic field strength alone considerably increases diagnostic accuracy, despite reports of 100% accuracy with a three-Tesla MRI scan compared to an arthroscopic examination [8, 9].

18.2.1.2 Consequence

The patient is subjected to an avoidable admission, anesthesia, and an unindicated surgery. This is an unnecessary burden for the patient and the healthcare system and may also have potential medicolegal consequences.

18.2.1.3 Circumvention

Correlation of the patient's history, clinical examination, and MRI scan can avoid this potential complication altogether. The Lachman test is the most sensitive (87.1%) with the highest negative likelihood ratio, while the pivot shift test is the most specific (97.5%) and has the highest positive likelihood ratio [10]. Both of these tests are best performed after relaxing the patient adequately. Additionally, MRI sequences in knee flexion and/or images in sagittal oblique planes improve diagnostic yield, especially if a partial tear is suspected [11] (Fig. 18.2). A thorough examination under anesthesia is essential, and this should be made as a standard operating procedure. If the doubt persists, diagnostic arthroscopy should be performed to confirm the diagnosis before any autograft is harvested.

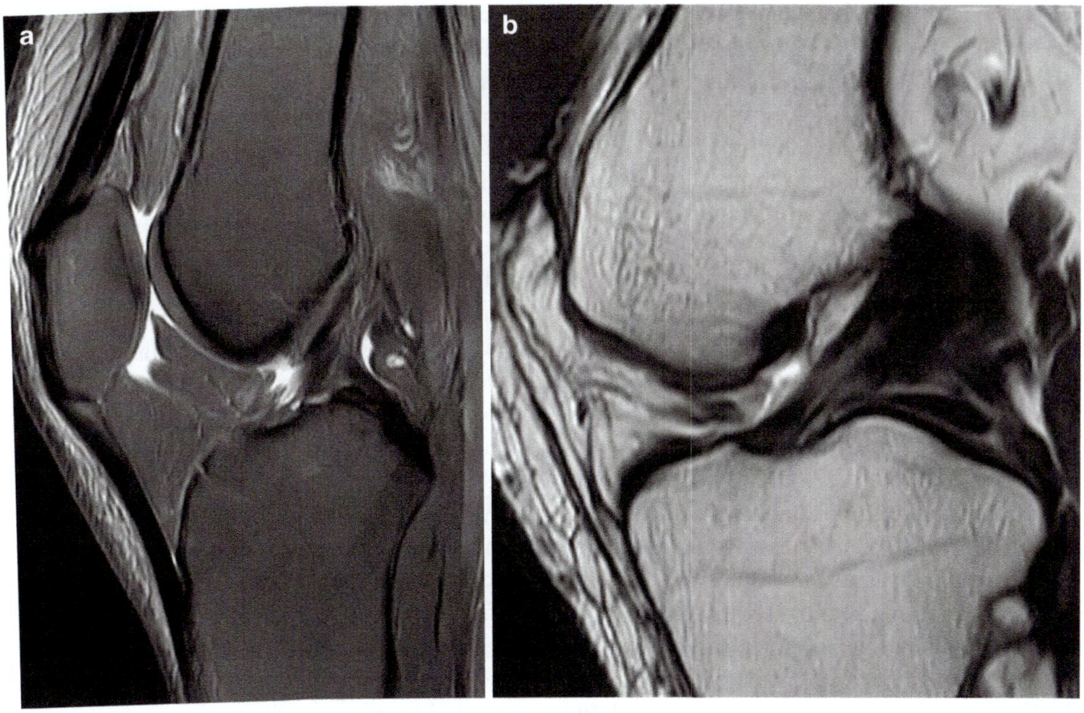

Fig. 18.2 (**a**) Proton-density fat-saturated sagittal MRI section showing the anterior cruciate ligament with poorly visualized tibial and femoral insertions. (**b**) The T2 sagittal oblique section of the same patient shows the entire ACL. ACL anterior cruciate ligament, MRI magnetic resonance imaging

18.2.1.4 Cure

One must abort the reconstruction/proposed procedure and treat any other pathology if detected.

18.2.2 Missed Diagnosis

18.2.2.1 Cause

It is common knowledge that ACL tears may be associated with concomitant injuries. Such injuries involving the medial collateral ligament and posterolateral corner (PLC), tear of the posterior root of the lateral meniscus, and posteromedial meniscus capsular separation (ramp lesion) are likely to be missed. The leading causes of missed diagnosis of these concomitant injuries are improper clinical history, inadequate preoperative clinical evaluation, failure to examine the injured joint under anesthesia, and poor correlation and reliability of the available MRI [12]. MRI for the above concomitant injuries in the setting of ACL deficiency has a reported sensitivity of 82–90% and a specificity of 89–92%, with high false-negative results [9, 13–16]. Maintaining high suspicion for these injuries in patients with a high velocity or contact injury, significant and abnormal bone bruise patterns, and a high-grade pivot shift test is necessary.

18.2.2.2 Consequence

Approximately 15% of ACLRs may fail due to missed diagnosis of an associated ligament, meniscus, or cartilage lesion [17]. Missed posteromedial corner (PMC) injuries have been implicated in anteromedial rotary instability, persistent pain, and valgus instability [14]. Although the medial collateral ligament (MCL) is the primary restraint to valgus force, the PMC and ACL increase their contribution to valgus restraint, especially with knee extension [18]. On the other hand, an unrecognized posterolateral corner (PLC) injury causes persistent varus instability and the development of a varus thrust, which may progress to the failure of the ACL graft [19, 20]. The posteromedial menisco-capsular junction plays a vital role in the posteromedial stability of the knee, as it limits the anterior tibial translation [15]. Failure to recognize and repair ramp lesions is associated with persistent anterior and rotatory knee laxity [21–23]. The lateral meniscus posterior root is an essential stabilizer against internal rotation at all knee flexion angles in an ACL-deficient knee [24]. Failure to identify and treat this lesion may lead to persistent rotatory knee instability after the ACLR and subsequent failure of the index procedure.

18.2.2.3 Circumvention

A thorough history regarding the mechanism of injury may provide clues to look for associated injuries. A detailed systematic clinical examination, which includes a varus–valgus stress test, dial test, pivot shift test, Slocum test, etc., is essential [12, 14, 19, 22]. Both lateral meniscus posterior root tear and ramp lesion can cause high-grade pivot shift phenomenon and must be suspected in those scenarios [25, 26]. MRI scans, especially the T2-weighted gradient-echo coronal and sagittal images, must be studied in detail. Bone bruises and posteromedial menisco-capsular separation suggest a ramp lesion, whereas posterolateral contusions may suggest a posterolateral corner injury [27]. A notch sign of the lateral femoral condyle of the femur should draw attention to possible anterolateral rotatory knee instability, while extrusion of the meniscus in the coronal sequence and the absence of an identifiable meniscus in the sagittal plane (ghost sign) are pathognomonic of medial meniscus posterior root tears [16]. Well-performed stress radiographs can reliably diagnose and assess collateral injuries. Careful examination under anesthesia and a thorough diagnostic arthroscopy, including inspection and probing of the menisci, posterior meniscus roots, and ramp lesion, are essential before proceeding to any graft harvest or other procedure (Fig. 18.3). A note should also be made of any arthroscopic drive-through sign indicating either a PLC injury or medial laxity.

18.2.2.4 Cure

Early recognition of the missed diagnosis of any concomitant injury and its confirmation with additional imaging and specific clinical tests at

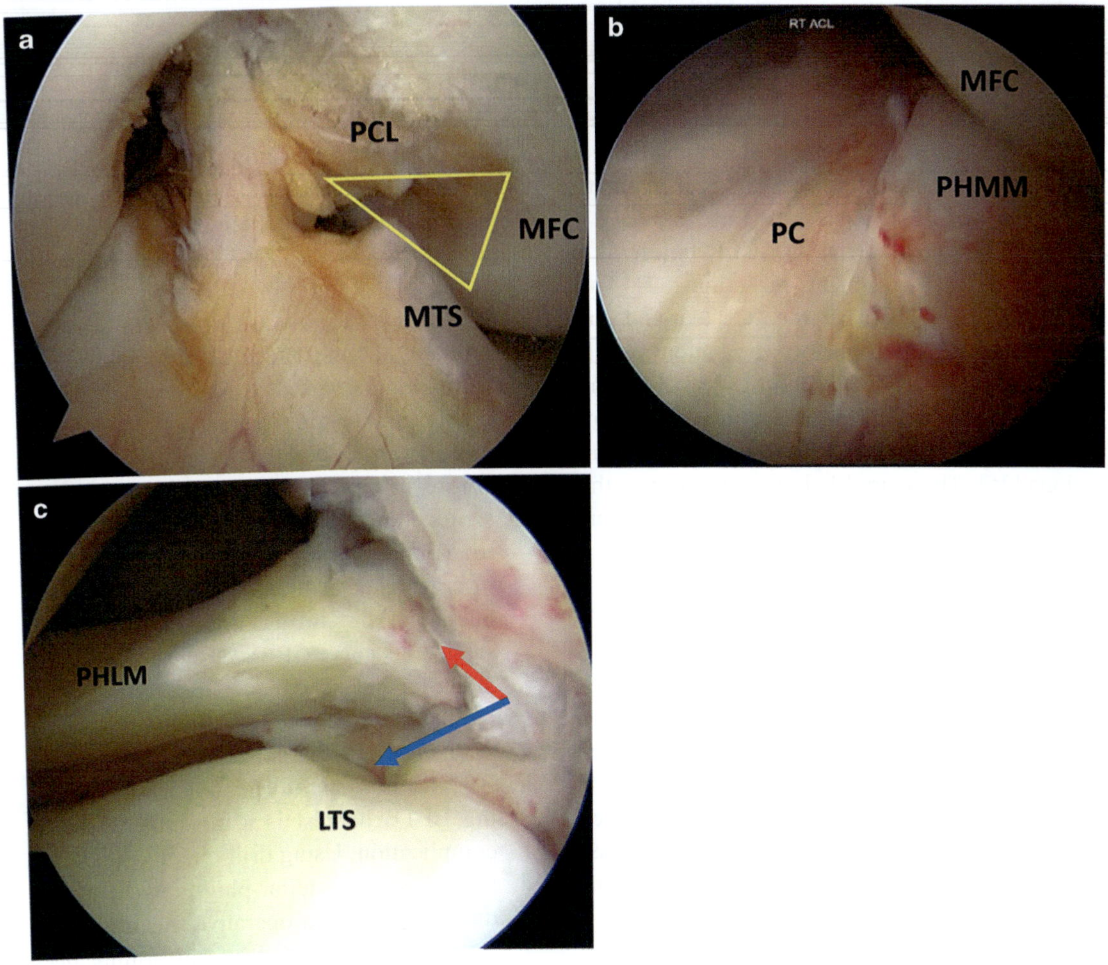

Fig. 18.3 Right knee arthroscopy. (**a**) The area in the interval between the posterior cruciate ligament, medial tibial spine, and medial femoral condyle (yellow triangle) is where the arthroscope is inserted from the anterolateral portal to visualize the ramp area. (**b**) A medial meniscus ramp lesion was seen with the arthroscope in the interval above, and the light source rotated. (**c**) An avulsed lateral meniscus posterior root (red arrow) from its insertion site (blue arrow) is visualized from the anterolateral portal and knee in the figure-of-4 position. LTS lateral tibial spine, MFC medial femoral condyle, MTS medial tibial spine, PC posterior capsule, PCL posterior cruciate ligament, PHLM posterior horn of lateral meniscus, PHMM posterior horn of medial meniscus

follow-up will be helpful in decision-making regarding further management. Timely intervention will contribute to a better outcome.

18.2.3 Timing of Surgery: Too Early or Too Late

18.2.3.1 Cause

It is conventional wisdom to perform an ACLR during the inflammation-free interval once the joint effusion has resolved and a good range of motion (ROM) and quadriceps strength has been achieved [28]. If ACLR is performed within 3 weeks of the injury, it is termed “acute.” If it is performed after 3 weeks, it is termed “delayed.” However, this distinction may not be scientifically accurate, and there is still no definite consensus regarding the timing of ACLR. Recent evidence suggests no significant differences in outcomes when surgery is performed during either time frame [29].

18.2.3.2 Consequence

Performing any arthroscopic surgery too early after an extensive injury may cause extravasation of the fluid through the capsular rent, leading to compartment-like syndrome and knee stiffness [30, 31]. Delay in performing ACLR in patients with medial or lateral meniscus tears may have a significantly increased risk of ACLR failure as a result of potential meniscus resection [32]. The longer the interval between the ACL tear and surgery, the higher the chance of the patient developing additional secondary injuries [33, 34]. The more the delay following the injury, the more likely the possibility of developing chondral lesions [35]. These are likely to compromise the outcomes and predispose to secondary osteoarthrosis [36, 37].

18.2.3.3 Circumvention

One has to be prudent about deciding the timing of surgery. Unnecessary delay is certainly deleterious in an unstable knee. Proponents of acute ACLR have reported non-inferior outcomes when the surgery is performed within 8 days or even 48 hours, especially in competitive athletes [38, 39]. However, the authors advise against rushing into surgery in the presence of gross effusion and poor quadriceps tone.

18.2.3.4 Cure

Knee stiffness as a consequence of early surgery that needs to be dealt with appropriately. This will be discussed in a later section. Any meniscal tears and chondral lesions that develop due to delayed surgery must be evaluated and managed judiciously.

18.3 Complications: Intra-Operative Origin

Complications during surgery are overwhelmingly due to technical errors or faulty instrumentation. They are best avoided by perfecting the skills required for an ACLR. These events require immediate recognition and remedial measures to prevent impacting the surgery's outcome.

18.3.1 Graft Harvest

18.3.1.1 Cause

The three most commonly used autografts for ACLR are hamstring tendons (HT), bone-patella tendon-bone (BPTB), and quadriceps tendon (QT), and harvesting each is fraught with complications.

(i) Hamstrings: Premature graft amputation is probably the most feared complication with hamstring harvest. It is likely to occur if the gracilis and semitendinosus tendons are not freed from fascial bands and attachments before stripping. The accessory insertion of semitendinosus, which blends with the gastrocnemius fascia, is present in about 77% of cases and must always be released [40]. Secondly, the infrapatellar branch of the saphenous nerve (IPBSN) is at risk from the anteromedial skin incision because its highly variable anatomy may be minimized with an oblique skin incision.

(ii) BPTB: Intra-operative patellar fracture during BPTB harvest is a rare but disastrous complication. Using drill bits or osteotomes to harvest the patellar plug is likely to predispose to an intra-operative fracture. Stress risers may be created by performing deep bone cuts in the patella. These may predispose to late patellar fractures. Shredding of tendon fibers and mismatch between the tendon and bone plugs are likely if the patellar tendon is not cut in its central third in a parallel fashion. Additionally, a failure to precisely measure the length of the bone plugs in a BPTB graft can lead to too long or too short of a graft. Lastly, the infrapatellar branch of the saphenous nerve (IPBSN) is also at risk during the BPTB harvest skin incision (Fig. 18.4).

(iii) QT: QT harvest can also be compromised by a short tendon harvest, either due to premature amputation or if the tendinous portion of the quadriceps is short in itself. Furthermore, if the incision in the tendon extends proximally into the muscle fibers, it is likely to

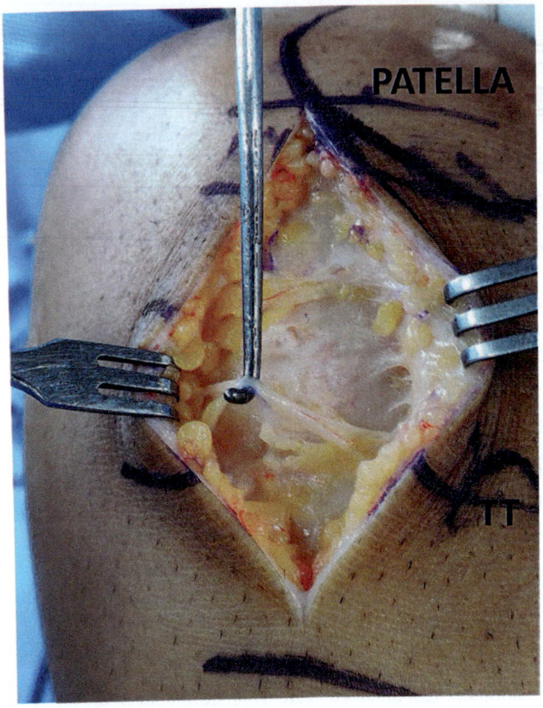

Fig. 18.4 The infrapatellar branch of the saphenous nerve isolated during exposure for the bone-patella tendon-bone graft harvest

cause significant bleeding. As with the BPTB, the patellar plug may fracture if a strict technique is not followed.

18.3.1.2 Consequence

(i) If the hamstring tendons are too short, preparing a 5/6 strand construct may not be possible, resulting in a thin-diameter graft. Preparing a 5- or 6-strand HT graft allows a graft diameter of >8 mm in almost all patients [41]. An HT graft diameter of >8 mm is associated with better functional outcomes and lower revision rates [42]. Further, every 0.5 mm increment in HT graft diameter from 7 mm to 10 mm reduces the likelihood of revision surgery by 0.86 times [43].

(ii) Patellar fractures may require internal fixation, prolonged splinting, and delayed rehabilitation after surgery. Graft–fiber mismatch can lead to weak BPTB grafts and may predispose to early graft re-rupture. Injury to IPBSN and its sequela are discussed in the next section.

18.3.1.3 Circumvention

(i) Hamstrings: The tendons are identified and separated proximally before graft harvest. This bifurcation point is located at a mean distance of 4.5 cm medial and 2.2 cm distal to the tibial tuberosity [44]. One should be able to release and retrieve the tendons almost 3 cm outside the skin incision. This indicates that the fascial bands have been released adequately [45]. Prodromos has described a posteromedial mini-incision in the popliteal fossa for HT harvest, allowing direct access to the fascial bands and significantly reducing graft amputation risk [46, 47].

(ii) BPTB: A small saw blade with incremental depth markings must always be used to harvest the patellar bone plug. This will allow controlled bone cuts not extending more than 8–9 mm in depth. The apex in the patella should be made triangular to reduce stress risers and to enable it to be fashioned like a bullet tip [48]. Parallel cuts in the patellar tendon can be ensured using the parallel tendon knife (Fig. 18.5). The combined length of the patellar bone plug and the patellar tendon is measured before the harvest of the tibial bone plug. If the tibial cortical length is to be kept short, a tongue of bone can be harvested to fill the tibial tunnel (Fig. 18.6). Horizontal incisions for BPTB harvesting may reduce, but not eliminate, the chances of injuring the IPBSN [49].

(iii) QT: The vastus medialis crosses over more toward the midline than the vastus lateralis. The central quadriceps tendon, hence, is more toward the lateral side rather than the central. This should be kept in mind before the QT harvest. An arthroscope introduced through the skin incision can determine the adequacy of the tendinous length and prevent proximal extension of the incision within the muscle belly. While harvesting the patellar bone plug with QT graft, its depth should not extend >30% of the patellar thickness and should not be harvested from too lateral. This helps to reduce stress risers and the risk of subsequent fracture [50].

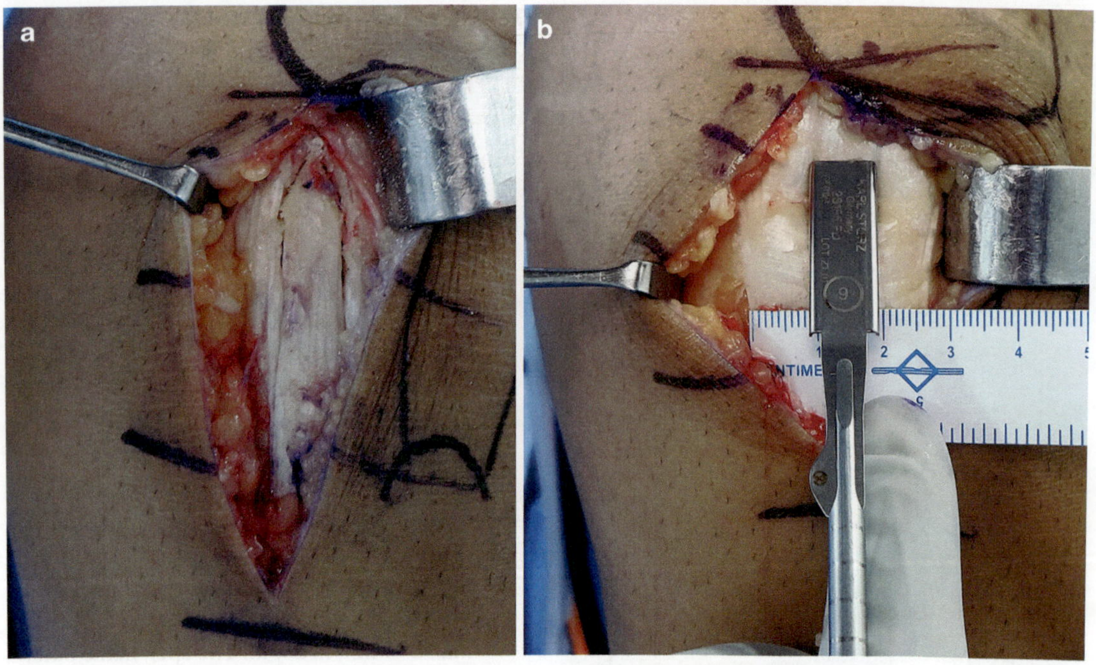

Fig. 18.5 Bone-patella tendon-bone harvest. (**a**) A triangular cut in the patella allows fashioning on the graft tip like a bullet. (**b**) A special knife used for quadriceps tendon harvest allows a perfectly parallel cut in the central third of the tendon

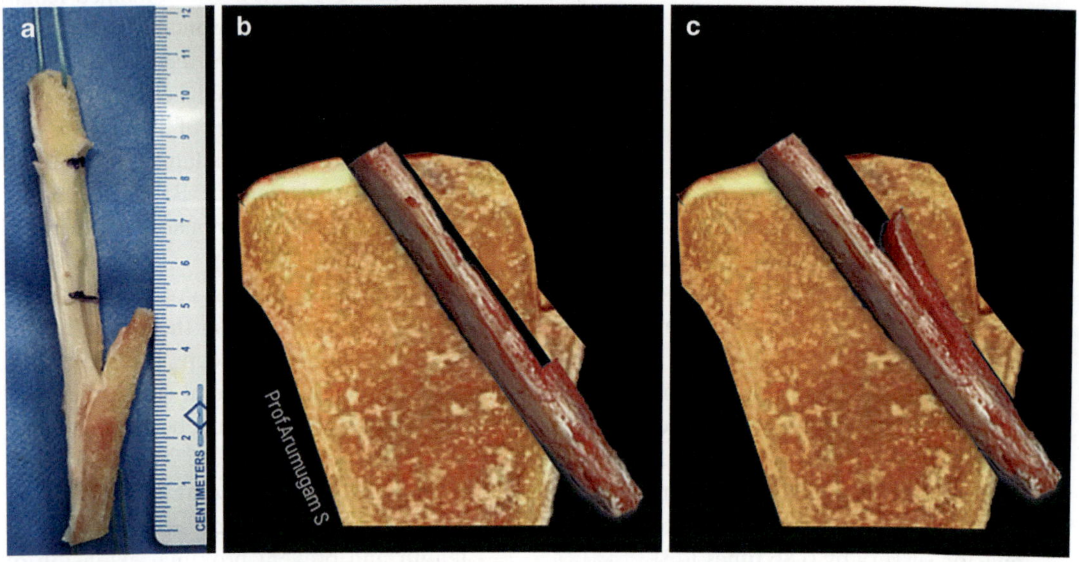

Fig. 18.6 (**a**) A bone-patella tendon-bone (BPTB) graft harvested with a tongue of bone on the tibial plug side. (**b**) Possible graft tunnel mismatch with traditional BPTB graft. (**c**) Graft tunnel mismatch negated by the bone tongue, allowing maximum bone to fill in the tibial tunnel by the modified technique. (Image courtesy Prof. S. Arumugam, SRMC, Chennai)

18.3.1.4 Cure

(i) Hamstrings: If the diameter of the HT graft is <8 mm, then one should consider a hybrid graft by adding an allograft to the existing graft and making the graft diameter >8 mm. However, hybrid grafts have higher failure rates and worse functional outcomes than autografts [51, 52]. A thin graft can also be protected by suture tape augmentation, which increases the construct's dynamic stiffness and ultimate strength. This tape must be fixed independently of the ACL graft in the tibia [53, 54]. If both hamstring tendons get amputated, harvesting the BPTB, QT, opposite knee hamstrings, or allograft (if available) is the better option.

(ii) BPTB: A long BPTB can be tackled by drilling a longer femoral tunnel (by the outside in technique) and a long tibial tunnel (by increasing the angle on the tibial jig). It is possible to save a thin or cracked bone plug by "piggybacking" the bone graft and inserting strong locking sutures through both plugs. When the tibial bone plug harvested is inadequate, the "tongue" of the bone can be folded onto the graft and used within the tunnel.

(iii) QT: A short QT is easily salvageable by carefully harvesting extra length on the patellar bone plug. The patellar defect should be grafted to avoid any stress risers. The bone plug is fixed in the femoral tunnel, and the free QT portion is fixed within the tibial tunnel.

18.3.2 Graft Preparation and Sizing

18.3.2.1 Cause

The purpose of accurate graft preparation and sizing is to facilitate easy passage and to optimize fixation. Preparation of a 5- or 6-strand HT graft construct with a suspensory loop button may be technically demanding. If the tendon suturing to the suspensory loop is not secure, the tendon may get dislodged from the loop. In the case of HT grafts fixed with interference screws in the tibia, suturing the graft with high-strength suture increases its ultimate failure load and ultimate stiffness [55]. The use of a weaker suture (suture type and diameter) will increase the possibility of graft laceration during interference screw fixation. If aperture fixation on the tibia is desired, the length of the prepared HT graft must be at least 75–80 mm. If the all-inside short HT technique is contemplated, the prepared HT graft should be at most 60–65 mm. This allows for firm graft security. Fracture through the bone plugs while drilling for suture holes in BPTB or QT grafts is possible. Inaccurate sizing or poor shaping of the patellar bone plug can lead to added difficulty during graft passage.

18.3.2.2 Consequence

The torque strain on the graft and the suture is very high when an interference screw is used for aperture fixation. Thin, flimsy, and weak suture preparation will predispose to graft laceration during graft fixation. Conversely, using a thick or large-diameter suture (e.g., No. 5 polyester) will lead to a graft tunnel mismatch, compromising fixation. Fracture of the patellar plug (either BPTB or QT) will render it inadequate for fixation in the femoral tunnel. Fracture of the tibial plug will lead to insecure fixation in the tibia. If the grafts are too short, it would compromise fixation at time zero itself. Additionally, long BPTB grafts are also not desirable.

18.3.2.3 Circumvention

Whip stitches should be taken with a high-strength suture like No. 2 polyester or ultra-high molecular weight polyethylene for HT graft preparation. This secures the attachment of the graft to the suspensory device and prevents graft laceration during screw insertion. The patellar bone plug for the BPTB or QT graft must be fashioned like a bullet tip and should be no longer than 16–18 mm to facilitate easy passage in the femoral tunnel. Using graft compaction pliers is recommended for shaping and sizing the bone plugs. The bone plugs should be placed in a "bone plug holder" and drilled. This technique is far safer than drilling the "freehand" bone plugs.

Alternatively, the bone plugs may be drilled "in situ" on the patella or the tibia before they are cut out with a saw.

18.3.2.4 Cure

Preparing the HT graft by using No. 2 high-strength suture in a locking or non-locking suture configuration will ensure graft security. Compression of HT graft using special tubes or graft sizers makes it less bulky, facilitates easy graft passage, and allows for compaction of the graft collagen [56]. Excess graft length in the case of BPTB may be managed in any of the following ways: (1) by outside-in drilling of the femoral tunnel, starting more proximally to increase the length of the femoral tunnel; (2) by increasing the length of the tibial tunnel by starting more distally and medially; or, (3) by folding the tibial plug onto the tendon by 180° and suturing them together [57]. When the BPTB graft length is short, the lengths of bone tunnels must also be kept short. If the graft length is too short for secure aperture fixation in the tibia, suspensory fixation with a washer or additional fixation with a post-fixation screw must be planned for. If the patellar plug of the QT or BPTB graft fractures, the graft may be reversed, and this fractured piece may be fixed in the tibia with an interference screw. Alternatively, an intact bone plug is "piggybacked" on the compromised bone plug and is prepared as a composite.

18.3.3 Tunnel Placement and Drilling

18.3.3.1 Cause

Anatomic ACLR necessitates graft placement at the native femoral and tibial insertions with sufficient footprint coverage to restore normal knee kinematics [58]. Inadequate visualization and failure to identify bony landmarks for the femoral footprint are usual causes of a non-anatomic tunnel placement. Trans-portal drilling helps achieve the correct femoral tunnel position and requires the knee to be flexed >90° to achieve adequate tunnel length. However, hyperflexion may not be possible in obese patients. Again, failure to sufficiently clear the fat pad may obscure the vision and compromise femoral tunnel placement [59]. A breach of the lateral femoral cortex or "lateral blow-out" is usually the result of aggressive uncontrolled reaming. Posterior wall blow-out is a complication that results from incorrect drilling trajectory (low knee flexion or high medial portal) or extreme posterior placement of the femoral tunnel [60]. ACL remnant preservation is desirable, but this must not be at the expense of inaccurate tunnel placement. Malposition of the tibial tunnel is less common and occurs if the tibial footprint is not identified accurately. An inaccurate tibial tunnel may violate the medial tibial articular surface or the anterior root of the lateral meniscus.

18.3.3.2 Consequence

Non-anatomic tunnel placement, more commonly on the femur, is the most common technical error causing ACLR graft rupture [61] (Fig. 18.7). Tunnel malposition may result in ACL graft impingement or inadequate ACL graft tension. Such a graft may impinge with either the roof of the notch, the posterior cruciate ligament (PCL), or the wall of the lateral femoral condyle. A "high noon" femoral or anterior tibial

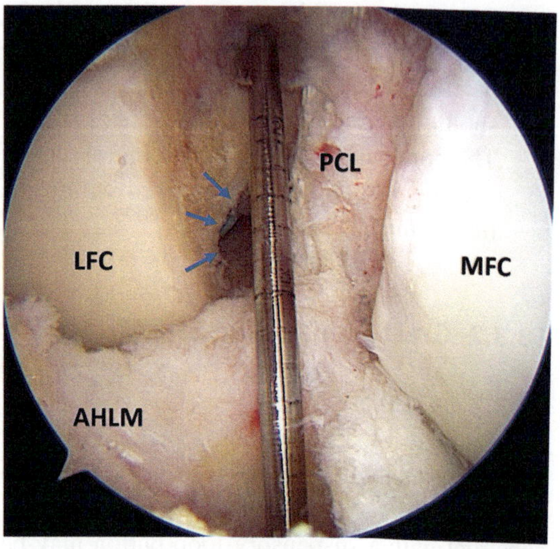

Fig. 18.7 The completely non-anatomic femoral tunnel in a failed ACL seen during revision surgery is far from the anatomic insertion site (blue arrows). ACL anterior cruciate ligament, AHLM anterior horn of lateral meniscus, LFC lateral femoral condyle, MFC medial femoral condyle, PCL posterior cruciate ligament

tunnel may lead to roof impingement. A high femoral tunnel may lead to impingement with the PCL; an excessively lateralized tibial tunnel may lead to impingement with the wall of the lateral femoral condyle [62]. A femoral tunnel that is placed in a "shallow" position may result in impingement with rotatory knee instability [60]. Howell et al. demonstrated higher failure rates and loss of extension after ACLR in grafts with roof impingement (and probable PCL impingement) [63]. For the lateral blow-out of >6 mm, additional salvage aperture fixation or a larger button construct (XTENDOBUTTON®, Smith Nephew, Andover, MA) is required [64]. An unrecognized posterior wall blow-out with aperture screw fixation in the femoral tunnel is likely to fail early due to inadequate graft containment [65].

18.3.3.3 Circumvention

Adequate visualization of the femoral footprint area can be improved by creating an additional central medial portal [66]. The tissues on the medial wall of the lateral cortex are cleared right up to the back of the knee to completely identify all the landmarks like the intercondylar ridge and the bifurcate ridge. This may be best achieved with a radiofrequency ablation device. A pilot hole is created with an awl, and the arthroscope is shifted to the anteromedial portal to reduce image distortion and obtain an "en-face" view of the femoral footprint. There are various landmarks described for optimal placement of the femoral tunnel. Pearle et al. detailed the "I.D.E.A.L. spot" for the most favorable femoral tunnel positioning [67]. The use of an arthroscopic ruler to identify this location has been validated by Bird et al. [68]. The sequential reaming technique permits the surgeon to make minor changes in the femoral tunnel position. To avoid posterior blow-out, the knee should be flexed to 110° when using flexible reamers and 125° when using rigid reamers [69]. Trans-portal drilling of the femoral tunnel via an accessory anteromedial portal will significantly reduce the possibility of posterior wall blow-outs [70]. Arthroscopic inspection of the femoral tunnel helps to rule out a posterior wall blow-out [71] (Fig. 18.8). Measurement of the aperture-cortex distance and controlled drilling to the appropriate depth will help to avoid a lateral wall blow-out. Flexible reamers or outside-in drilling may be used when hyperflexion is not possible, such as in obese patients. Tibial tunnel malposition

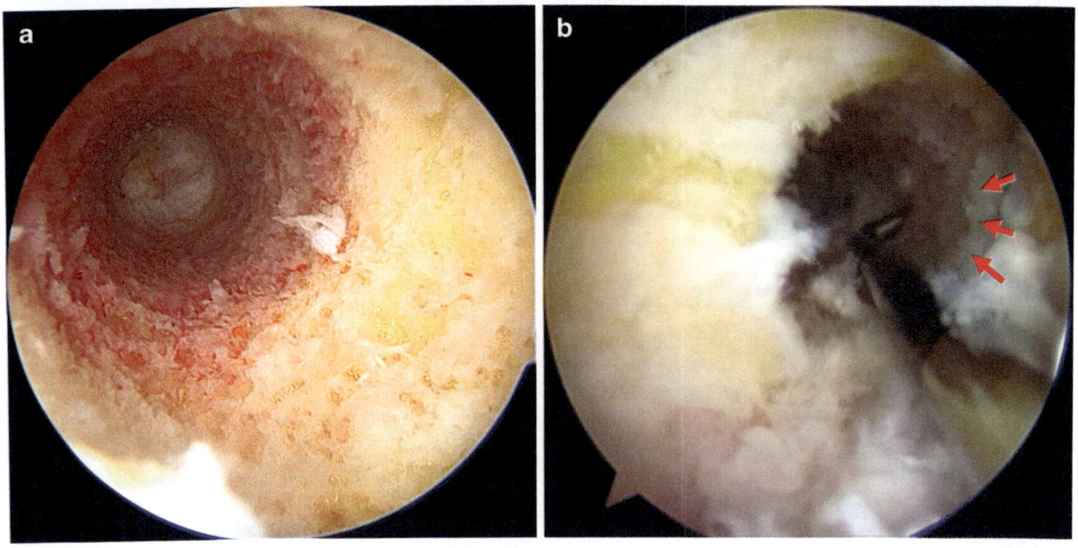

Fig. 18.8 (**a**) The standard femoral tunnel view from the anterolateral portal by positioning the arthroscope in the notch and rotating the light source to face a medial wall of the lateral femoral condyle. (**b**) Posterior blow-out (red arrows) with no bone in the posterior wall of the tunnel (hook probe)

may be avoided by carefully identifying anatomic landmarks like the anterior horn of the lateral meniscus, PCL, tibial spines, and ACL remnant [72]. Sequential drilling will allow for minor adjustments in tunnel position.

There has been a recent emphasis on retaining the ACL remnant whenever possible. Retention of the remnant allows for faster healing, ligamentization, and better proprioception. Obsession about remnant preservation may sacrifice anatomy for biology, as the presence of the remnant may misguide the anatomical tibial tunnel drilling.

18.3.3.4 Cure

Non-anatomic femoral or tibial tunnels make the situation non-salvageable. In these situations, one may attempt slight over-drilling and push the graft to the desired location by strategic screw placement. The resultant impingement (roof or lateral wall) may then be treated with notchplasty to avoid subsequent graft attrition and extension loss. Various fixation devices must be available to manage a lateral or posterior blow-out. A lateral blow-out requires an aperture fixation or a more oversized button like the XTENDOBUTTON® (Smith and Nephew, Andover, MA) with the conventional cortical button (Fig. 18.9). A posterior blow-out requires a suspensory, a hybrid (suspensory plus aperture), or an over-the-top fixation as a salvage procedure [60].

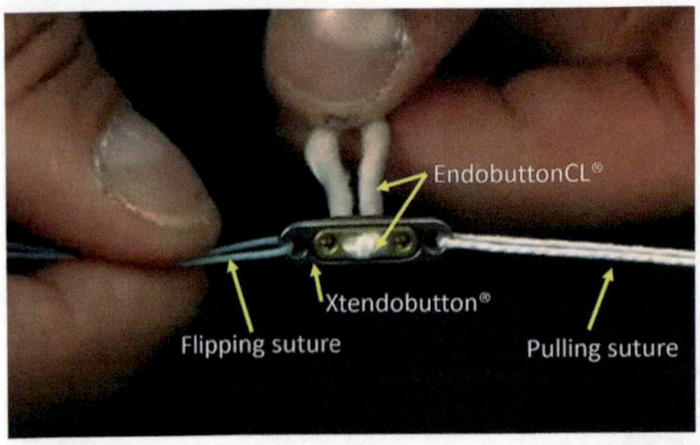

Fig. 18.9 The XTENDOBUTTON® (Smith and Nephew, Andover, MA) used with the regular suspensory cortical button in case of a lateral wall blow-out

18.3.4 Graft Passage and Fixation

18.3.4.1 Cause

Difficulty in passage of bone plugs of BPTB or QT grafts through the tibial tunnel may be attributed to inadequate soft tissue clearance at the mouth of the tibial tunnel or improper graft sizing. Problems with the femoral cortical suspensory buttons include the inability to flip within the tunnel or flip over the iliotibial band (ITB) (Fig. 18.10). Screw breakage, especially in the femoral tunnel, has been reported either due to plastic deformation of the screwdriver or failure of the screw biomaterial or improper surgical technique [73]. In patients with hard bone, the screw may get jammed in the femoral tunnel after partial insertion and break. At times, fatigue failure of the metal may occur and lead

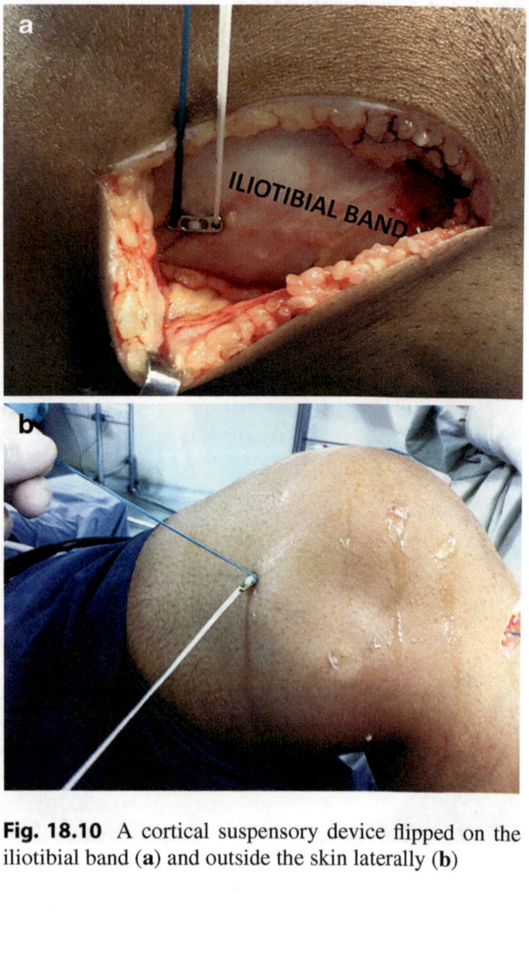

Fig. 18.10 A cortical suspensory device flipped on the iliotibial band (**a**) and outside the skin laterally (**b**)

to the breaking-off of the screwdriver tip. Screw divergence in the femoral tunnel is likely if the screw is not inserted from the same portal and with the same degree of knee flexion as while drilling the tunnel. Careless placement of the screw guidewire may result in divergence in the tibial tunnel. An anteriorly positioned tibial screw may push the graft posterolaterally away from the native insertion. This will place the graft more vertically, which certainly is not desirable [74]. Graft laceration during screw fixation is possible, especially in hard tibial bone. Finally, fixation of the ACL graft in high flexion angles leads to loss of knee extension due to abnormal graft tensioning [75].

18.3.4.2 Consequence

Application of excess force for graft passage can lead to breakage of sutures at the suspensory loop button or the bone plug. It may prove more disastrous if the suture lacerates the soft tissue graft or fractures through the bone plug of the BPTB or QT graft. An inappropriate flipping of the suspensory loop button (intra-tunnel or over the ITB) can lead to post-operative laxity. While using aperture fixation interference screws, a graft-screw divergence of >20° reduces the strength of fixation [76]. Lastly, if the graft lacerates and sufficient length is not salvaged, another graft may need to be harvested.

18.3.4.3 Circumvention

The soft tissue at the mouth of the tibial tunnel cortex must be cleared to permit the bone plug of the BPTB or QT graft to pass through easily. When using a fixed loop suspensory device, the femoral socket should be reamed to a sufficient depth to allow the button to flip on the lateral cortex of the femoral condyle. For an adjustable loop suspensory device, the aperture-cortex distance should be marked on the loop to identify that the button has passed outside the cortex. If one is unsure of the button flipping, inserting the arthroscope in the lateral gutter and confirming and/or performing the flip under visualization is advocated. A flipped button over the ITB should be suspected if there is a dimpling of the lateral thigh's skin during graft cycling. This should also be confirmed under an image intensifier. When femoral fixation with an interference screw is planned, the socket should be tapped, and a "notcher" should be used to create a slot superior to the graft (Fig. 18.11). This position is crucial for

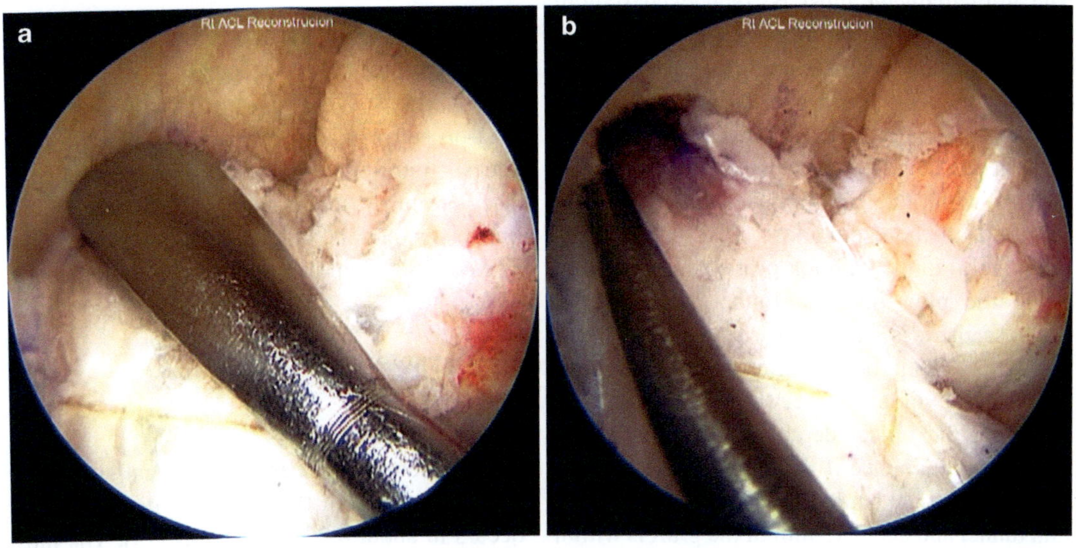

Fig. 18.11 (**a**) A tunnel notcher positioned superior to the bone-patella tendon-bone graft in the femur. (**b**) A guidewire placed in the socket thus created facilitates interference screw fixation of the graft

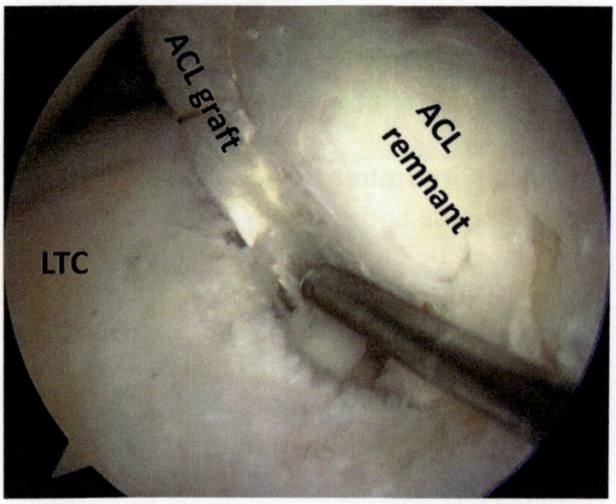

Fig. 18.12 The preferred interference screw position on the tibia is posterolateral (hook probe), which pushes the graft anteromedially

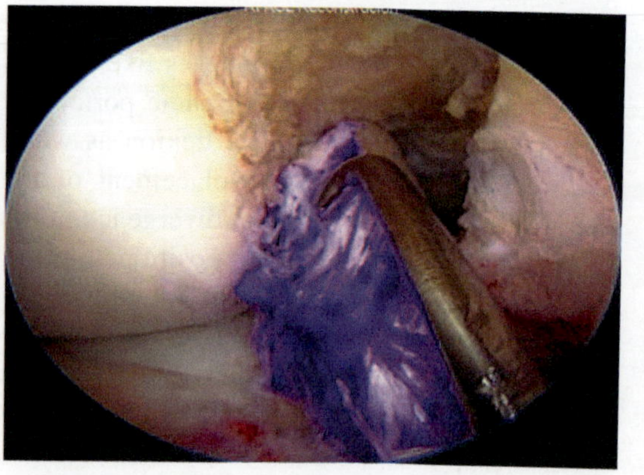

Fig. 18.13 A hook probe inserted from the accessory anteromedial portal is used to maneuver the bone plug of a quadriceps graft into the femoral socket

BPTB, where orienting the graft collagen inferiorly has been shown to provide better footprint coverage [77]. A technique for ensuring posterolateral screw position in the tibia is described wherein the guidewire is passed and clamped before graft passage [78] (Fig. 18.12). Suturing of the HT graft with a high-strength suture protects against graft laceration from a tibial screw.

18.3.4.4 Cure

The graft diameter must be rechecked if graft passage is difficult. If the bone plug is large, it can be carefully trimmed or compressed. For soft tissue grafts, compressing the graft in compression tubes usually helps. Alternatively, the graft is retrieved completely, and the tunnels are re-drilled with the appropriate larger-diameter reamer. If the bone plug is stuck at the mouth of the femoral tunnel, a hook probe can be used to guide the plug into the tunnel (Fig. 18.13). Suppose there is an inability to deploy a fixed suspensory loop button after several attempts successfully. In that case, the graft should be retrieved, the tunnel depth checked, and extra tunnel length reamed if found short. If a suspensory loop cortical button gets jammed in the femoral tunnel and about 15 mm of the graft is within the tunnel, the ACL graft tension is checked. If the graft is adequately tensioned, nothing may need to be done. A flipped button on the ITB will require a lateral incision, soft tissue release, and repositioning on the cortex. If this is recognized after tibial fixation of the graft, an additional interference screw in the femur is necessary with tensioning of the graft proximally in the femur. A broken screw in the femur or tibia has to be replaced. If the screw breaks partially, an additional backup fixation, like a fixation post, is advisable.

18.4 Complications: Post-Operative Origin

Post-operative complications usually have their genesis intra-operatively but are present later. They are most important because of their direct impact on patient outcomes and rehabilitation. The common complications are discussed below.

18.4.1 Infection: Early/Late

18.4.1.1 Cause

The incidence of superficial surgical site infection (SSI) is about 0.94%, while deep infection occurs in about 0.2% of patients [79]. The most common organisms are *Staphylococcus aureus* and coagulase-negative *Staphylococci* (CONS), although Methicillin-resistant *S. aureus* (MRSA), gram-negative bacilli, and anaerobic microorganisms have also been isolated [80]. The patient factors associated with a higher infection rate are males, tobacco use, morbid obesity, diabetes, younger age, pre-operative steroid use, increased operative time, atopic dermatitis, and immune-compromised state [79, 81, 82]. Fluid reservoir surgical drapes are used commonly during ACLR, and 22% of these drapes may get contaminated within 60 minutes of surgery. Thus, there is higher contamination with increasing surgical duration [83]. ACL graft harvest is also associated with bacterial contamination [84]. The type of graft has a definite bearing on infection rate, with HT autografts having a significantly higher incidence than BPTB autografts. Allografts do not demonstrate a higher rate of infection than autografts [85]. Probably, graft contamination and/or surgical field happens during the surgery.

18.4.1.2 Consequence

Infection following any orthopedic intervention is a disastrous complication. The spectrum of infection after ACLR ranges from minor SSI to septic arthritis. It leads to patient distress and increased expense due to the need for prolonged antimicrobial therapy, re-admission, and repeated surgical intervention. Infection also compromises outcomes, delays rehabilitation, and delays return to work or sports [86].

18.4.1.3 Circumvention

Strict adherence to aseptic precautions is essential within the operating room. Impermeable drapes are paramount when using the "foot off-table" or "hanging down" position [45]. Since prolonged surgical times adversely affect infection rates, the workflow must be optimized to reduce the overall duration of the procedure. A single dose of prophylactic antibiotic at induction of anesthesia, preferably a first-generation (Cefazolin) or second-generation (Cefuroxime) cephalosporin, is helpful as prophylaxis. These cephalosporins are safe and cost-effective and achieve good bone, muscle, and synovium concentration. They provide coverage against gram-positive bacteria, clinically significant aerobic gram-negative bacilli, and anaerobic gram-positive bacteria [87]. If the patient is allergic to penicillin or is an MRSA carrier, intravenous Vancomycin or Teicoplanin is recommended [88]. Another successful intervention has been using a Vancomycin solution (5 mg/mL) soaked wrap for the graft [89] (Fig. 18.14). Although initially described for hamstrings graft, any graft can be soaked in Vancomycin. This technique has been shown to reduce the infection rate ten times [90]. It decreases graft contamination without increasing the risk of graft rupture or knee stiffness [91, 92]. The use of Gentamicin (80 mg/L) in the irrigation solution has also been found to reduce infection (0.57%) when compared to normal saline alone (2.1%) [93]. When an allograft is preferred, the best sterilization method without compromising the mechanical strength is irradiation of <1.8 Mrad [88]. It is essential to discuss the choice of graft with the patient, especially when risk factors for infection are identified.

18.4.1.4 Detection

A high index of suspicion is necessary to diagnose infection. Although there is no consensus on terminology, early infection occurs within 2 weeks, and infection after 4 weeks is labeled late infection [80, 94]. A combination of clinical features, serology, and knee aspiration analysis has been recommended as the diagnostic algorithm for post-ACLR infection [88]. Clinical symptoms and signs suggestive of an infection are fever, sudden increase in pain or swelling, knee effusion, local erythema, loss of range of motion (ROM), and serous or purulent discharge from the surgical site [95]. A raised erythrocyte sedimentation rate (ESR) and C-reactive protein

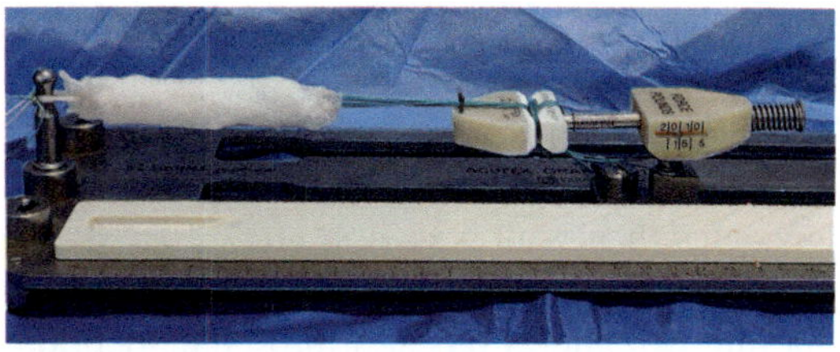

Fig. 18.14 Use of a Vancomycin-soaked gauge to wrap the graft while it is being pre-tensioned

(CRP) are essential to decision-making [88]. A CRP of >150 mg/dL on day 3 post-ACLR or >20 mg/dL on day 15 post-ACLR is highly suspicious of infection. Joint aspiration must be performed in every case of suspected infection. A positive result on culture or gram-stain, purulent nature of aspirate, total cell count >100,000, and polymorphonuclear cells >90% in the synovial fluid are diagnostic of infection [80]. MRI scan is a valuable adjunct that may reveal synovitis, marrow edema, sinus tracts, or soft-tissue abscesses [96].

18.4.1.5 Cure

S. aureus, or CONS, is known to form biofilms early in the infection. This makes their eradication difficult. Hence, it must be stressed here that early treatment initiation is paramount [97]. In the presence of superficial SSI without intra-articular involvement, antibiotic suppression alone may be considered. Empirical antimicrobial therapy in consultation with an infectious diseases expert may be initiated if the organism is not cultured. A more aggressive protocol has been proposed by some authors, with the preservation of articular cartilage and salvage of the ACL graft as the main aim [98]. This involves arthroscopic lavage, sampling from two to three areas for culture, synovectomy, and removing devitalized tissues and the fibrin layer formed over the graft. It must be removed if the graft is necrotic or its mechanical integrity is compromised. A repeat arthroscopy is advisable if the clinical features and CRP titer do not improve after 72 hours. Antibiotic therapy is initiated empirically until culture is available. After obtaining culture reports and sensitivity patterns, the appropriate antibiotic therapy is initiated as intravenous antibiotics for 2–3 weeks (until a favorable clinical or serological response is obtained). It is followed by 4 weeks of oral therapy [80, 99]. Graft retention must be attempted whenever possible. It allows for better functional outcomes than graft resection [100, 101]. If the signs, symptoms, and laboratory parameters do not improve after a repeat arthroscopic debridement, removal of the graft and all hardware is recommended [88].

18.4.2 Loss of Knee Motion

18.4.2.1 Cause

Loss of extension is the most frequently encountered complication after ACLR. The causative factors include graft impingement, tensioning of the ACL graft in high flexion angles, cyclops formation, and hamstring contracture [75]. The pathogenesis of graft impingement has been discussed in earlier sections. Cyclops syndrome is an entity causing loss of extension due to forming a fibrous nodule in the anterior portion of the tibial footprint and is symptomatic in almost 2–11% of cases [102]. A systematic review identified the risk factors for the development of cyclops as: (a) pre-operative—knee inflammation and poor ROM; (b) intra-operative—narrow intercondylar notch and excessively anterior tibial tunnel; and (c) post-operative—persistent hamstring spasm [103]. Loss of flexion can be the result of a high-tension ACL fixation. Arthrofibrosis is a more generic term to describe the loss of ROM due to the involvement of one or more compartments or extracapsular tissue [104]. The reported rate of arthrofibrosis following ACLR is about 5%. The independent risk factors for arthrofibrosis include early surgery, age <18 years, females, post-operative hematoma, infection, BPTB or QT autograft, and concomitant meniscus repair [105, 106]. Intermediate to low signal intensity on T1- and T2-weighted MRI images is characteristic of diffuse synovial hyperplasia seen in arthrofibrosis [96].

18.4.2.2 Consequence

A cyclops lesion after ACLR is associated with an abnormal gait pattern characterized by a higher first peak knee flexion moment and higher knee flexion moment impulse. These changes cause excessive loading of the medial compartment and subsequent cartilage degeneration. This may develop as early as 1 year post surgery [107]. A loss of extension of >10% also causes excessive patellofemoral joint loading and can contribute to anterior knee pain [108]. Though flexion loss is better tolerated, flexion <125° causes difficulty in squatting [104]. Loss of motion requires re-admission and procedures like manipulation under anesthesia (MUA) or arthroscopic adhesiolysis to gain ROM. Independent risk factors for these procedures in one series were using a QT autograft, infection, and meniscus repair [105].

18.4.2.3 Circumvention

The one factor that can prevent the onset of knee stiffness would be to delay surgery until the inflammatory response in the knee has settled. Prevention of technical intra-operative errors, which can lead to loss of ROM (flexion or extension), is critical. Previously, the identification of graft impingement and notch impingement has been discussed. Tensioning and securing the ACL graft in the tibia at or near full extension leads to significantly less loss of extension in the first 2 weeks [109]. Early initiation of ROM exercises and bracing the knee in full extension may also be implemented to prevent loss of extension [75]. One must achieve full passive and active knee extension within the first 2 weeks post surgery by instituting appropriate and aggressive physiotherapy [86].

18.4.2.4 Cure

The remedies for post-operative loss of ROM include physiotherapy, drop-casting, oral steroids, MUA, arthroscopic adhesiolysis, and open release. These therapies are most beneficial when performed within eight months of surgery. While drop-casting is most helpful for extension loss, MUA is most effective for achieving greater flexion [110]. Oral short-course corticosteroid therapy of 84 mg methylprednisolone tapered over six days has been reported beneficial for flexion loss without extension deficit [111]. Arthroscopic adhesiolysis effectively clears scar tissue and adhesions. The suprapatellar pouch may need distension before portal placement. After clearing the suprapatellar pouch and medial/lateral gutters, the infrapatellar fat pad may need to be released from the anterior tibia to establish the pre-tibial recess. In the intercondylar notch, it is crucial to identify and rectify factors like a cyclops lesion graft impingement or a narrow notch [104]. In exceptional cases of flexion contracture, a posterior capsular release may need to be performed. It can either be a posterolateral or a posteromedial release and is a safe and effective procedure [112].

18.4.3 Graft Donor Site Morbidity

18.4.3.1 Cause

Anterior knee and kneeling pain were reported after the HT and BPTB harvests. Although these symptoms are more common with BPTB harvest (about 46%), there appears to be no difference between graft types at long-term follow-up [5, 113–116]. BPTB harvest results in direct injury to the inferior pole of the patella, leading to inflammation, scarring, and prolonged pain. HT harvest-related pain is probably related to increased blood flow in the infrapatellar fat pad [117]. Sensory disturbances are due to injury to the infrapatellar branch of the saphenous nerve during the BPTB harvest and the sartorial branch of the saphenous nerve during the HT harvest. This sensory loss was experienced in about 77% of BPTB and 58% of HT grafts in one study [118]. Patellar fracture or patellar tendon rupture is a rare complication of BPTB harvest, with a reported incidence of 0.2% [119].

18.4.3.2 Consequence

Anterior knee pain resolves over time in almost all patients. When it is severe, it may limit kneeling activities. Sensory deficits are problematic for the patients but have no sinister consequences. They usually recover completely within 2 years in 25% of patients. Patients with HT grafts are more likely to encounter partial sensory recovery [118]. Patellar fracture or patellar tendon rupture is a serious consequence, requiring additional surgery, immobilization, rehabilitation, and loss of workdays.

18.4.3.3 Circumvention

A minimally invasive double-incision BPTB harvest technique has been found to significantly reduce the incidence of both anterior knee pain and hypoesthesia secondary to nerve injury [120]. An oblique incision for HT harvest significantly reduces the incidence of nerve injury and its sequelae [118, 121]. QT harvest is associated with the least donor site morbidity and better cosmesis [122].

18.4.3.4 Cure

Anterior knee pain has no specific treatment. Counseling and activity modification to avoid kneeling is advisable. A massage therapy protocol that includes lymphatic drainage, myofascial release, trigger point release, and cross-fiber friction has demonstrated successful outcomes [123]. Sensory deficits arising from injury to the saphenous nerve may result in patient dissatisfaction. This can be avoided by good pre-operative counseling. Rarely, a neuroma may develop in the injured nerve, requiring surgical excision [124]. Graft selection should be individualized. BPTB could be avoided in patients who kneel constantly for sports or jobs, and HT could be avoided in athletes like sprinters who require hamstring explosiveness.

18.4.4 Hardware Issues

18.4.4.1 Cause

Insecurely fixed tibial and femoral interference screws used for aperture fixation of the graft are known to migrate [125]. Screw divergence, graft-tunnel mismatch, and tunnel widening are possible causes of screw migration. Screw breakage, femoral tunnel widening, and prolonged knee effusion are more familiar with bioabsorbable screws, especially first- and second-generation biomaterials [126]. Formation of intra-osseous tibial cysts is a problem associated with bioabsorbable screws, more so with poly-*L*-lactic acid (PLLA) screws compared to PLLA-hydroxyapatite-coated screws [127, 128]. These cysts usually take about 2 to 3 years to develop. Suspensory cortical fixation buttons have also been reported to migrate [6]. The older designs of suspensory cortical buttons necessitated the creation of a closed loop intraoperatively using a Mersilene® (Ethicon, Somerville, NJ) tape. This would fail over a while and allow the button to migrate. The extracortical fixation-post screw used on the tibia may cause irritation symptoms in 10% of patients [5].

18.4.4.2 Consequence

Intra-articular migration of a screw or cortical button may cause pain, effusion, locking, loss of ROM, and even chondral damage [6]. A tibial cyst usually presents with pre-tibial pain or a palpable tender nodule [127]. These may mimic local infection or cellulitis. Clinical and functional outcomes or graft-rupture rates are not influenced by the material of the interference screw used for ACLR [129]. Computed tomography (CT) and MRI scans are helpful tools to diagnose implant migration and intra-osseous cyst formation [96, 127].

18.4.4.3 Circumvention

Prevention of screw divergence, use of titanium screws or newer-generation bio-absorbable screws, use of newer-generation suspensory loops, and avoiding prominent fixation posts are some ways in which these hardware-related complications can be minimized.

18.4.4.4 Cure

Issues resulting from extra-articular implant prominence may initially be treated symptomatically. If symptoms persist, hardware removal may be necessary. Intra-articular loose hardware always requires removal. Arthroscopic removal is often successful and involves the creation of additional portals depending on the location of the loose hardware [6]. Symptomatic tibial cysts frequently require open excision and curettage of the cyst and screw debris. If the defect is significant, grafting with bone substitutes or allograft may be necessary [127].

18.4.5 Persistent Laxity (Without Re-Tear)

18.4.5.1 Cause

Failure to adequately tension the graft during tibial fixation will likely lead to a loose graft. In an all-inside technique, an excessively long graft is likely to bottom out and result in a loose graft.

Another important cause of persistent laxity is the failure to address preexisting anterolateral rotatory instability (ALRI) adequately. The ACL graft cannot control the ALRI, which may necessitate an extra-articular procedure [130, 131]. Lengthening of the graft construct with an adjustable loop button has also been described in porcine models, although its clinical impact is unknown [132, 133]. Failure to recognize and treat concomitant ligament injuries (PLC/MCL/PCL) may result in late instability.

18.4.5.2 Consequence

The patient would continue to experience instability despite surgery. This may lead to kinesiophobia and may impact return to sport or pre-injury activity [134]. The worst-case possibility would be re-injury with an ACL graft tear or a contralateral ACL rupture.

18.4.5.3 Circumvention

As discussed in the previous section, it is mandatory to avoid missing concomitant injuries. All soft-tissue grafts are best pre-tensioned at 20 N for 5 minutes, followed by cyclic flexion-extension loading after femoral fixation, and then final tensioning of the graft with simultaneous tibial fixation at 20° knee flexion while applying a posterior drawer force. However, evaluating intra-articular graft tension based on extra-articular tension at time zero is challenging. It is commonly observed that the intra-articular tension is lower than the applied external tension [135]. An accurate assessment of graft length and tibial socket is necessary when an all-inside technique is performed. If the graft that has been prepared is too long, the tibial tunnel is drilled completely, and the fixation modality is changed to that of an interference screw. If the pivot shift persists after ACL graft fixation, it may be worthwhile to perform an extra-articular procedure. Although its indications are still evolving, in the STABILITY 1 trial, Getgood et al. found that the following patients may benefit from the lateral extra-articular procedure: (1) young age; (2) those who desire to return to contact pivoting sport; (3) a posterior tibial slope >12°; and (4) significant meniscus deficiency [130, 131, 136].

18.4.5.4 Cure

Missed ligament injuries may require revision surgery. The Porto KT MRI sequence is helpful in assessing persistent rotatory knee laxity when a negative Lachman test is found. This MRI would also confirm the integrity of the ACL graft. If such is the case, an extra-articular procedure should be performed, either an anterolateral ligament reconstruction or an iliotibial band tenodesis. If the graft is lax and functionally redundant, a revision ACLR may need to be considered.

18.4.6 Venous Thromboembolism (VTE)

18.4.6.1 Cause

The reported incidence of symptomatic VTE within 90 days of ACLR is 0.4–0.5%, and this includes deep vein thrombosis (DVT) as well as pulmonary thromboembolism (PE) [4, 137, 138]. However, since about 73% of patients may be asymptomatic, the actual rate of DVT may be as high as 8.4% [138]. The risk factors associated with VTE are age >35 years, prolonged duration of surgery (relative risk of 1.12 for every 15-minute increment), tobacco use, and concomitant procedures like an osteotomy [4, 82, 137, 138].

18.4.6.2 Consequence

VTE following ACLR has not been reported to be fatal. It may lead to re-admission within the first 90 days after surgery [4]. Patients with symptomatic VTE have poorer outcomes on patient-reported outcome measures [137].

18.4.6.3 Circumvention

The routine use of thromboprophylaxis is not recommended and has not been shown to reduce VTE in large databases [4, 137]. Mitigation of risk factors like reducing the duration of surgery and cessation of smoking is worth considering. Thromboprophylaxis (pharmacological and mechanical) in selected high-risk patients may be instituted, although current evidence does not support this.

18.4.6.4 Cure

The constantly evolving guidelines for diagnosis and treatment of VTE demand that they are best managed by specialty physicians. A duplex compression ultrasound scan and fibrin D-dimer test are used to diagnose DVT, while the best investigation for a PE is CT pulmonary angiography. Treatment strategies range from anticoagulant pharmacotherapy to thrombolysis to placement of inferior vena cava filters [139]. The rehabilitation program in these patients requires modification until the DVT or PE is resolved.

18.5 Conclusion

Arthroscopic ACLR is a technically demanding surgery with a definite learning curve. It is, however, a safe procedure if performed after adequate planning. Complications are inevitable, as with any surgery, and every attempt has to be made to avoid these complications. Strict adherence to a standard operating protocol can help reduce technical errors. A thorough knowledge of various grafts, drilling techniques, and implants is vital to salvage difficult situations. Anticipating complications and having a plan for managing these complications improve the quality of patient care and satisfaction.

Acknowledgments Professor S. Arumugam for allowing us to use his illustrations (Fig. 18.6) and Dr. Shantanu Patil for helping with the editing.

References

1. Ricketts D, Rogers RA, Roper T, Ge X. Recognising and dealing with complications in orthopaedic surgery. Ann R Coll Surg Engl. 2017;99(3):185–8.
2. Sokol DK, Wilson J. What is a surgical complication? World J Surg. 2008;32(6):942–4.
3. Kaeding CC, Léger-St-Jean B, Magnussen RA. Epidemiology and diagnosis of anterior cruciate ligament injuries. Clin Sports Med. 2017;36(1):1–8.
4. Jameson SS, Dowen D, James P, Serrano-Pedraza I, Reed MR, Deehan D. Complications following anterior cruciate ligament reconstruction in the English NHS. Knee. 2012;19(1):14–9.
5. Rousseau R, Labruyere C, Kajetanek C, Deschamps O, Makridis KG, Djian P. Complications after anterior cruciate ligament reconstruction and their relation to the type of graft: a prospective study of 958 cases. Am J Sports Med. 2019;47(11):2543–9.
6. Palazzolo A, Rosso F, Bonasia DE, Saccia F, Rossi R. Knee Committee SUncommon complications after anterior cruciate ligament reconstruction. Joints. 2018;6(3):188–203.
7. Décary S, Fallaha M, Belzile S, Martel-Pelletier J, Pelletier J-P, Feldman D, et al. Clinical diagnosis of partial or complete anterior cruciate ligament tears using patients' history elements and physical examination tests. PLoS One. 2018;13(6):e0198797-e.
8. Phelan N, Rowland P, Galvin R, O'Byrne JM. A systematic review and meta-analysis of the diagnostic accuracy of MRI for suspected ACL and meniscal tears of the knee. Knee Surg Sports Traumatol Arthrosc. 2015;24(5):1525–39.
9. Sampson MJ, Jackson MP, Moran CJ, Moran R, Eustace SJ, Shine S. Corrigendum to: three Tesla MRI for the diagnosis of meniscal and anterior cruciate ligament pathology: a comparison to arthroscopic findings [63 (10) 1106–1111]. Clin Radiol. 2009;64(7):747.
10. Zhang Y. Clinical examination of anterior cruciate ligament rupture: a systematic review and meta-analysis. Acta Orthop Traumatol Turc. 2016;50:22.
11. Simpfendorfer C, Polster J. MRI of the knee: what do we miss? Curr Radiol Rep. 2014;2(4):43.
12. Kurzweil PR, Kelley ST. Physical examination and imaging of the medial collateral ligament and posteromedial corner of the knee. Sports Med Arthrosc Rev. 2006;14(2):67–73.
13. Praz C, Vieira TD, Saithna A, Rosentiel N, Kandhari V, Nogueira H, et al. Risk factors for lateral meniscus posterior root tears in the anterior cruciate ligament–injured knee: an epidemiological analysis of 3956 patients from the SANTI Study Group. Am J Sports Med. 2019;47(3):598–605.
14. Craft JA, Kurzweil PR. Physical examination and imaging of medial collateral ligament and posteromedial corner of the knee. Sports Med Arthrosc Rev. 2015;23(2):e1–6.
15. Edgar C, Kumar N, Ware JK, Ziegler C, Reed DN, DiVenere J, et al. Incidence of posteromedial meniscocapsular separation and the biomechanical implications on the anterior cruciate ligament. J Am Acad Orthop Surg. 2019;27(4):e184–e92.
16. Wong KPL, Han AX, Wong JLY, Lee DYH. Reliability of magnetic resonance imaging in evaluating meniscal and cartilage injuries in anterior cruciate ligament-deficient knees. Knee Surg Sports Traumatol Arthrosc. 2016;25(2):411–7.
17. Samitier G, Marcano AI, Alentorn-Geli E, Cugat R, Farmer KW, Moser MW. Failure of anterior cruciate ligament reconstruction. Arch Bone Joint Surg. 2015;3(4):220–40.
18. Zaffagnini S, Bignozzi S, Martelli S, Lopomo N, Marcacci M. Does ACL reconstruction restore knee stability in combined lesions? Clin Orthop Relat Res. 2007;454:95–9.

19. Ricchetti ET, Sennett BJ, Huffman GR. Acute and chronic management of posterolateral corner injuries of the knee. Orthopedics. 2008;31(5):479–88.
20. Pacheco RJ, Ayre CA, Bollen SR. Posterolateral corner injuries of the knee. J Bone Joint Surg. 2011;93-B(2):194–7.
21. Peltier A, Lording TD, Lustig S, Servien E, Maubisson L, Neyret P. Posteromedial meniscal tears may be missed during anterior cruciate ligament reconstruction. Arthroscop J Arthroscop Relat Surg. 2015;31(4):691–8.
22. Sonnery-Cottet B, Serra Cruz R, Vieira TD, Goes RA, Saithna A. Ramp lesions. Clin Sports Med. 2020;39(1):69–81.
23. Bumberger A, Koller U, Hofbauer M, Tiefenboeck TM, Hajdu S, Windhager R, et al. Ramp lesions are frequently missed in ACL-deficient knees and should be repaired in case of instability. Knee Surg Sports Traumatol Arthroscop. 2020;28(3):840–54.
24. Forkel P, von Deimling C, Lacheta L, Imhoff FB, Foehr P, Willinger L, et al. Repair of the lateral posterior meniscal root improves stability in an ACL-deficient knee. Knee Surg Sports Traumatol Arthrosc. 2018;26(8):2302–9.
25. DePhillipo NN, Moatshe G, Brady A, Chahla J, Aman ZS, Dornan GJ, et al. Effect of meniscocapsular and meniscotibial lesions in ACL-deficient and ACL-reconstructed knees: a biomechanical study. Am J Sports Med. 2018;46(10):2422–31.
26. Song G-y, Zhang H, Liu X, Zhang J, Xue Z, Qian Y, et al. Complete posterolateral meniscal root tear is associated with high-grade pivot-shift phenomenon in noncontact anterior cruciate ligament injuries. Knee Surg Sports Traumatol Arthrosc. 2017;25(4):1030–7.
27. Temponi EF, de Carvalho Júnior LH, Saithna A, Thaunat M, Sonnery-Cottet B. Incidence and MRI characterization of the spectrum of posterolateral corner injuries occurring in association with ACL rupture. Skeletal Radiol. 2017;46(8):1063–70.
28. Irarrázaval S, Kurosaka M, Cohen M, Fu FH. Anterior cruciate ligament reconstruction. J ISAKOS Joint Disorders OrthopSports Med. 2016;1(1):38–52.
29. Ferguson D, Palmer A, Khan S, Oduoza U, Atkinson H. Early or delayed anterior cruciate ligament reconstruction: is one superior? A systematic review and meta-analysis. Eur J Orthop Surg Traumatol Orthop Traumatol. 2019;29(6):1277–89.
30. Brambilla L, Pulici L, Carimati G, Quaglia A, Prospero E, Bait C, et al. Prevalence of associated lesions in anterior cruciate ligament reconstruction. Am J Sports Med. 2015;43(12):2966–73.
31. Mohtadi NGH, Webster-Bogaert S, Fowler PJ. Limitation of motion following anterior cruciate ligament reconstruction. Am J Sports Med. 1991;19(6):620–5.
32. Church S, Keating JF. Reconstruction of the anterior cruciate ligament. J Bone Joint Surg. 2005;87-B(12):1639–42.
33. Anstey DE, Heyworth BE, Price MD, Gill TJ. Effect of timing of ACL reconstruction in surgery and development of meniscal and chondral lesions. Phys Sportsmed. 2012;40(1):36–40.
34. Tayton E, Verma R, Higgins B, Gosal H. A correlation of time with meniscal tears in anterior cruciate ligament deficiency: stratifying the risk of surgical delay. Knee Surg Sports Traumatol Arthrosc. 2008;17(1):30–4.
35. Filardo G, de Caro F, Andriolo L, Kon E, Zaffagnini S, Marcacci M. Do cartilage lesions affect the clinical outcome of anterior cruciate ligament reconstruction? A systematic review. Knee Surg Sports Traumatol Arthrosc. 2017;25(10):3061–75.
36. Fok AWM, Yau WP. Delay in ACL reconstruction is associated with more severe and painful meniscal and chondral injuries. Knee Surg Sports Traumatol Arthrosc. 2012;21(4):928–33.
37. Sommerfeldt M, Goodine T, Raheem A, Whittaker J, Otto D. Relationship between time to ACL reconstruction and presence of adverse changes in the knee at the time of reconstruction. Orthop J Sports Med. 2018;6(12):2325967118813917.
38. von Essen C, Eriksson K, Barenius B. Acute ACL reconstruction shows superior clinical results and can be performed safely without an increased risk of developing arthrofibrosis. Knee Surg Sports Traumatol Arthrosc. 2019;28:2036.
39. Herbst E, Hoser C, Gföller P, Hepperger C, Abermann E, Neumayer K, et al. Impact of surgical timing on the outcome of anterior cruciate ligament reconstruction. Knee Surg Sports Traumatol Arthrosc. 2016;25(2):569–77.
40. Pagnani MJ, Warner JJP, O'Brien SJ, Warren RF. Anatomic considerations in harvesting the semitendinosus and gracilis tendons and a technique of harvest. Am J Sports Med. 1993;21(4):565–71.
41. Tutkus V, Kluonaitis K, Silove S, Tutkuviene J. ACL reconstruction using 5- or 6-strand hamstring autograft provides graft's diameter bigger than 8 mm. Knee Surg Sports Traumatol Arthrosc. 2017;26(5):1349–56.
42. Mariscalco MW, Flanigan DC, Mitchell J, Pedroza AD, Jones MH, Andrish JT, et al. The influence of Hamstring Autograft size on patient-reported outcomes and risk of revision after anterior cruciate ligament reconstruction: a Multicenter Orthopaedic Outcomes Network (MOON) Cohort study. Arthroscop J Arthroscop Relat Surg. 2013;29(12):1948–53.
43. Snaebjörnsson T, Hamrin Senorski E, Ayeni OR, Alentorn-Geli E, Krupic F, Norberg F, et al. Graft diameter as a predictor for revision anterior cruciate ligament reconstruction and KOOS and EQ-5D values: a Cohort Study from the Swedish National Knee Ligament Register based on 2240 patients. Am J Sports Med. 2017;45(9):2092–7.
44. Tillett E, Madsen R, Rogers R, Nyland J. Localization of the semitendinosus-gracilis tendon bifurcation

point relative to the tibial tuberosity: an aid to hamstring tendon harvest. Arthroscop J Arthroscop Relat Surg. 2004;20(1):51–4.
45. Tjoumakaris FP, Herz-Brown AL, Bowers AL, Sennett BJ, Bernstein J. Complications in brief: anterior cruciate ligament reconstruction. Clin Orthop Relat Res. 2012;470(2):630–6.
46. Prodromos CC, Han YS, Keller BL, Bolyard RJ. Posterior mini-incision technique for hamstring anterior cruciate ligament reconstruction graft harvest. Arthroscop J Arthroscop Relat Surg. 2005;21(2):130–7.
47. Prodromos CC. Posterior Mini-incision Hamstring harvest. Sports Med Arthrosc Rev. 2010;18(1):12–4.
48. Steen H, Tseng KF, Goldstein SA, Carpenter JE. Harvest of Patellar Tendon (Bone–Tendon–Bone) autograft for ACL reconstruction significantly alters surface strain in the human Patella. J Biomech Eng. 1999;121(2):229–33.
49. Kerver ALA, Leliveld MS, den Hartog D, Verhofstad MHJ, Kleinrensink GJ. The surgical anatomy of the Infrapatellar Branch of the Saphenous nerve in relation to incisions for anteromedial knee surgery. J Bone Joint Surg. 2013;95(23):2119–25.
50. Ferrer GA, Miller RM, Murawski CD, Tashman S, Irrgang JJ, Musahl V, et al. Quantitative analysis of the patella following the harvest of a quadriceps tendon autograft with a bone block. Knee Surg Sports Traumatol Arthrosc. 2015;24(9):2899–905.
51. Burrus MT, Werner BC, Crow AJ, Brockmeier SF, Carson EW, Miller MD, et al. Increased failure rates after anterior cruciate ligament reconstruction with soft-tissue Autograft-Allograft Hybrid Grafts. Arthroscop J Arthroscop Relat Surg. 2015;31(12):2342–51.
52. Xu H, Lin W, Jin G, Xin D, Zhang J, Kang K, et al. Graft choice for anatomic anterior cruciate ligament reconstruction: the comparison between thin autograft and thick hybrid graft. An observational study. Medicine. 2018;97(30):e11597–e.
53. Noonan BC, Bachmaier S, Wijdicks CA, Bedi A. Independent Suture Tape reinforcement of tripled smaller-diameter and quadrupled grafts for anterior cruciate ligament reconstruction with tibial screw fixation: a biomechanical full construct model. Arthroscop J Arthroscop Relat Surg. 2020;36(2):481–9.
54. Frangie R, Warth RJ, Harner CD. Editorial commentary: will Suture Tape augmentation prove to be the answer to anterior cruciate ligament graft remodeling, ultimate strength, and safe return to play? Arthroscop J Arthroscop Relat Surg. 2020;36(2):490–1.
55. Höher J, Offerhaus C, Steenlage E, Weiler A, Scheffler S. Impact of tendon suturing on the interference fixation strength of quadrupled hamstring tendon grafts. Arch Orthop Trauma Surg. 2013;133(9):1309–14.
56. Cruz AI, Fabricant PD, Seeley MA, Ganley TJ, Lawrence JTR. Change in size of Hamstring Grafts during preparation for ACL reconstruction. J Bone Joint Surg. 2016;98(6):484–9.
57. Charalambous CP, Kwaees TA. Anatomical considerations in hamstring tendon harvesting for anterior cruciate ligament reconstruction. Muscles Ligaments Tendons J. 2012;2(4):253–7.
58. Araujo PH, Asai S, Pinto M, Protta T, Middleton K, Linde-Rosen M, et al. ACL Graft position affects in situ graft force following ACL reconstruction. J Bone Joint Surg. 2015;97(21):1767–73.
59. Basdekis G, Abisafi C, Christel P. Influence of knee flexion angle on femoral tunnel characteristics when drilled through the anteromedial portal during anterior cruciate ligament reconstruction. Arthroscop J Arthroscop Relat Surg. 2008;24(4):459–64.
60. Mitchell JJ, Dean CS, Chahla J, Menge TJ, Cram TR, LaPrade RF. Posterior wall blowout in anterior cruciate ligament reconstruction: a review of anatomic and surgical considerations. Orthop J Sports Med. 2016;4(6):2325967116652122.
61. Group M, Wright RW, Huston LJ, Spindler KP, Dunn WR, Haas AK, et al. Descriptive epidemiology of the Multicenter ACL Revision Study (MARS) cohort. Am J Sports Med. 2010;38(10):1979–86.
62. Iriuchishima T, Shirakura K, Fu FH. Graft impingement in anterior cruciate ligament reconstruction. Knee Surg Sports Traumatol Arthrosc. 2012;21(3):664–70.
63. Howell SM, Taylor MA. Failure of reconstruction of the anterior cruciate ligament due to impingement by the intercondylar roof. J Bone Joint Surg. 1993;75(7):1044–55.
64. Herbort M, Heletta S, Raschke MJ, Schliemann B, Osada N, Petersen W, et al. Accidental perforation of the lateral femoral cortex in ACL reconstruction: an investigation of mechanical properties of different fixation techniques. Arthroscop J Arthroscop Relat Surg. 2012;28(3):382–9.
65. Jiang X-D, Zheng H-L, Yang Y-P. Outcome of posterior wall blowout in anterior cruciate ligament (ACL) reconstruction via anteromedial portal approach: a retrospective research in 20 patients with 6 years follow-up. Chinese J Traumatol = Zhonghua chuang shang za zhi. 2019;22(1):24–8.
66. Cohen SB, Fu FH. Three-portal technique for anterior cruciate ligament reconstruction: use of a central medial portal. Arthroscop J Arthroscop Relat Surg. 2007;23(3):325.e1–.e5.
67. Pearle AD, McAllister D, Howell SM. Rationale for strategic graft placement in anterior cruciate ligament reconstruction: I.D.E.A.L. Femoral Tunnel Position. Am J Orthop (Belle Mead NJ). 2015;44(6):253–8.
68. Bird JH, Carmont MR, Dhillon M, Smith N, Brown C, Thompson P, et al. Validation of a new technique to determine midbundle femoral tunnel position in anterior cruciate ligament reconstruction using 3-Dimensional computed tomography analysis. Arthroscop J Arthroscop Relat Surg. 2011;27(9):1259–67.

69. Forsythe B, Collins MJ, Arns TA, Zuke WA, Khair M, Verma NN, et al. Optimization of anteromedial portal femoral tunnel drilling with flexible and straight reamers in anterior cruciate ligament reconstruction: a Cadaveric 3-Dimensional computed tomography analysis. Arthroscop J Arthroscop Relat Surg. 2017;33(5):1036–43.
70. Moon D-K, Jo H-S, Lee D-Y, Kang D-G, Byun J-H, Hwang S-C. Influence of the different anteromedial portal on femoral tunnel orientation during anatomic ACL reconstruction. Acta Orthop Traumatol Turc. 2017;51(3):227–32.
71. Temponi EF, Oliveira JNP, Soares LFM, Carvalho Júnior LH. The femoral tunnel view test during ACL reconstruction can ensure tunnel integrity. Rev Bras Ortop. 2018;53(4):427–31.
72. Shimodaira H, Tensho K, Akaoka Y, Takanashi S, Kato H, Saito N. Remnant-preserving Tibial Tunnel positioning using anatomic landmarks in double-bundle anterior cruciate ligament reconstruction. Arthroscop J Arthroscop Relat Surg. 2016;32(9):1822–30.
73. Watson JN, McQueen P, Kim W, Hutchinson MR. Bioabsorbable interference screw failure in anterior cruciate ligament reconstruction: a case series and review of the literature. Knee. 2015;22(3):256–61.
74. Mall NA, Matava MJ, Wright RW, Brophy RH. Relation between anterior cruciate ligament graft obliquity and knee laxity in elite athletes at the National Football League Combine. Arthroscop J Arthroscop Relat Surg. 2012;28(8):1104–13.
75. Petsche TS, Hutchinson MR. Loss of extension after reconstruction of the anterior cruciate ligament. J Am Acad Orthop Surg. 1999;7(2):119–27.
76. Steiner ME, Wing D, Walley KC, Manoukian O, Perez-Viloria M, Okajima S, et al. Hyperflexion and femoral interference screw insertion in ACL reconstruction. Orthop J Sports Med. 2018;6(8):2325967118788810.
77. Bedi A, Weber AE, Trasolini NA, Mayer EN, Bolia IK, Higgins J, et al. Does bone plug and graft orientation (inferior versus posterior) alter native femoral footprint coverage in bone Patellar Tendon bone anterior cruciate ligament reconstruction? Arthroscop J Arthroscop Relat Surg. 2020;36:1875.
78. Parate P, Chernchujit B. A surgical technique for posterolateral placement of interference screw accurately in Tibial Tunnel in single-bundle anterior cruciate ligament reconstruction. Arthrosc Tech. 2016;5(6):e1481–e6.
79. Kawata M, Sasabuchi Y, Taketomi S, Inui H, Matsui H, Fushimi K, et al. Atopic dermatitis is a novel demographic risk factor for surgical site infection after anterior cruciate ligament reconstruction. Knee Surg Sports Traumatol Arthrosc. 2018;26(12):3699–705.
80. Torres-Claramunt R, Gelber P, Pelfort X, Hinarejos P, Leal-Blanquet J, Pérez-Prieto D, et al. Managing septic arthritis after knee ligament reconstruction. Int Orthop. 2015;40(3):607–14.
81. Clement RC, Haddix KP, Creighton RA, Spang JT, Tennant JN, Kamath GV. Risk factors for infection after knee arthroscopy: analysis of 595,083 cases from 3 United States databases. Arthroscopy. 2016;32(12):2556–61.
82. Agarwalla A, Gowd AK, Liu JN, Garcia GH, Bohl DD, Verma NN, et al. Effect of operative time on short-term adverse events after isolated anterior cruciate ligament reconstruction. Orthop J Sports Med. 2019;7(2):2325967118825453.
83. Fuchs M, von Roth P, Pfitzner T, Kopf S, Sass FA, Hommel H. Contamination of irrigation fluid during primary total knee arthroplasty. J Am Acad Orthop Surg Glob Res Rev. 2018;2(5):e027–e.
84. Pérez-Prieto D, Torres-Claramunt R, Gelber PE, Shehata TMA, Pelfort X, Monllau JC. Autograft soaking in vancomycin reduces the risk of infection after anterior cruciate ligament reconstruction. Knee Surg Sports Traumatol Arthrosc. 2014;24(9):2724–8.
85. Bansal A, Lamplot JD, VandenBerg J, Brophy RH. Meta-analysis of the risk of infections after anterior cruciate ligament reconstruction by graft type. Am J Sports Med. 2018;46(6):1500–8.
86. Eckenrode BJ, Carey JL, Sennett BJ, Zgonis MH. Prevention and management of post-operative complications following ACL reconstruction. Curr Rev Musculoskelet Med. 2017;10(3):315–21.
87. Hansen E. Perioperative antibiotics. J Orthop Res. 2014;32(S1):S31–59.
88. Parvizi JGT. Proceedings of the second international consensus meeting on musculoskeletal infection. In: Parvizi JGT, editor. . Brooklandville: Data Trace; 2018.
89. Phegan M, Grayson JE, Vertullo CJ. No infections in 1300 anterior cruciate ligament reconstructions with vancomycin pre-soaking of hamstring grafts. Knee Surg Sports Traumatol Arthrosc. 2015;24(9):2729–35.
90. Baron JE, Shamrock AG, Cates WT, Cates RA, An Q, Wolf BR, et al. Graft preparation with intraoperative vancomycin decreases infection after ACL reconstruction. J Bone Joint Surg. 2019;101(24):2187–93.
91. Perez-Prieto D, Torres-Claramunt R, Gelber PE, Shehata TM, Pelfort X, Monllau JC. Autograft soaking in vancomycin reduces the risk of infection after anterior cruciate ligament reconstruction. Knee Surg Sports Traumatol Arthrosc. 2016;24(9):2724–8.
92. Offerhaus C, Balke M, Hente J, Gehling M, Blendl S, Höher J. Vancomycin pre-soaking of the graft reduces postoperative infection rate without increasing risk of graft failure and arthrofibrosis in ACL reconstruction. Knee Surg Sports Traumatol Arthrosc. 2019;27(9):3014–21.
93. Yazdi H, Moradi A, Herbort M. The effect of gentamicin in irrigating solutions on articular infection prophylaxis during arthroscopic ACL reconstruction. Arch Orthop Trauma Surg. 2014;134(2):257–61.

94. Fong SY, Tan JL. Septic arthritis after arthroscopic anterior cruciate ligament reconstruction. Ann Acad Med Singap. 2004;33(2):228–34.
95. Gobbi A, Karnatzikos G, Chaurasia S, Abhishek M, Bulgherhoni E, Lane J. Postoperative infection after anterior cruciate ligament reconstruction. Sports Health. 2016;8(2):187–9.
96. Papakonstantinou O, Chung CB, Chanchairujira K, Resnick DL. Complications of anterior cruciate ligament reconstruction: MR imaging. Eur Radiol. 2003;13(5):1106–17.
97. Lazzarini L, Lipsky BA, Mader JT. Antibiotic treatment of osteomyelitis: what have we learned from 30 years of clinical trials? Int J Infect Dis. 2005;9(3):127–38.
98. Komnos GA, Chalatsis G, Mitrousias V, Hantes ME. Postoperative infection after anterior cruciate ligament reconstruction: prevention and management. Microorganisms. 2022;10(12):2349.
99. Abdel-Aziz A, Radwan YA, Rizk A. Multiple arthroscopic debridement and graft retention in septic knee arthritis after ACL reconstruction: a prospective case-control study. Int Orthop. 2014;38(1):73–82.
100. Pogorzelski J, Themessl A, Achtnich A, Fritz EM, Wörtler K, Imhoff AB, et al. Septic arthritis after anterior cruciate ligament reconstruction: how important is graft salvage? Am J Sports Med. 2018;46(10):2376–83.
101. Waterman BR, Arroyo W, Cotter EJ, Zacchilli MA, Garcia ESJ, Owens BD. Septic arthritis after anterior cruciate ligament reconstruction: clinical and functional outcomes based on graft retention or removal. Orthop J Sports Med. 2018;6(3):2325967118758626.
102. Kambhampati SBS, Gollamudi S, Shanmugasundaram S, Josyula VVS. Cyclops lesions of the knee: a narrative review of the literature. Orthop J Sports Med. 2020;8(8):2325967120945671.
103. Noailles T, Chalopin A, Boissard M, Lopes R, Bouguennec N, Hardy A. Incidence and risk factors for cyclops syndrome after anterior cruciate ligament reconstruction: a systematic literature review. Orthop Traumatol Surg Res. 2019;105(7):1401–5.
104. Magit D, Wolff A, Sutton K, Medvecky MJ. Arthrofibrosis of the knee. J Am Acad Orthop Surg. 2007;15(11):682–94.
105. Huleatt J, Gottschalk M, Fraser K, Boden A, Dalwadi P, Xerogeanes J, et al. Risk factors for manipulation under anesthesia and/or lysis of adhesions after anterior cruciate ligament reconstruction. Orthop J Sports Med. 2018;6(9):2325967118794490.
106. Nwachukwu BU, McFeely ED, Nasreddine A, Udall JH, Finlayson C, Shearer DW, et al. Arthrofibrosis after anterior cruciate ligament reconstruction in children and adolescents. J Pediatr Orthop. 2011;31(8):811–7.
107. Samaan MA, Facchetti L, Pedoia V, Tanaka MS, Link TM, Souza RB, et al. Cyclops lesions are associated with altered gait patterns and medial knee joint cartilage degeneration at 1 year after ACL-reconstruction. J Orthop Res. 2017;35(10):2275–81.
108. Biggs A, Jenkins WL, Urch SE, Shelbourne KD. Rehabilitation for patients following ACL reconstruction: a knee symmetry model. N Am J Sports Phys Ther. 2009;4(1):2–12.
109. Yazdi H, Moradi A, Sanaie A, Ghadi A. Does the hyperextension maneuver prevent knee extension loss after arthroscopic anterior cruciate ligament reconstruction? J Orthop Traumatol. 2016;17(4):327–31.
110. Ekhtiari S, Horner NS, de Sa D, Simunovic N, Hirschmann MT, Ogilvie R, et al. Arthrofibrosis after ACL reconstruction is best treated in a step-wise approach with early recognition and intervention: a systematic review. Knee Surg Sports Traumatol Arthrosc. 2017;25(12):3929–37.
111. Rue J-PH, Ferry AT, Lewis PB, Bach BR. Oral Corticosteroid use for loss of flexion after primary anterior cruciate ligament reconstruction. Arthroscop J Arthroscop Relat Surg. 2008;24(5):554–9.e1.
112. Tardy N, Thaunat M, Sonnery-Cottet B, Murphy C, Chambat P, Fayard J-M. Extension deficit after ACL reconstruction: is open posterior release a safe and efficient procedure? Knee. 2016;23(3):465–71.
113. Kartus J, Movin T, Karlsson J. Donor-Site Morbidity after anterior cruciate ligament reconstruction using autografts. In: Anterior knee pain and patellar instability. Springer-Verlag. p. 305–19.
114. Xie X, Liu X, Chen Z, Yu Y, Peng S, Li Q. A meta-analysis of bone–patellar tendon–bone autograft versus four-strand hamstring tendon autograft for anterior cruciate ligament reconstruction. Knee. 2015;22(2):100–10.
115. Webster KE, Feller JA, Hartnett N, Leigh WB, Richmond AK. Comparison of Patellar Tendon and Hamstring Tendon anterior cruciate ligament reconstruction. Am J Sports Med. 2015;44(1):83–90.
116. Hardy A, Casabianca L, Andrieu K, Baverel L, Noailles T. Complications following harvesting of patellar tendon or hamstring tendon grafts for anterior cruciate ligament reconstruction: systematic review of literature. Orthop Traumatol Surg Res. 2017;103(8):S245–S8.
117. Kanamoto T, Tanaka Y, Yonetani Y, Kita K, Amano H, Kusano M, et al. Anterior knee symptoms after double-bundle ACL reconstruction with hamstring tendon autografts: an ultrasonographic and power Doppler investigation. Knee Surg Sports Traumatol Arthrosc. 2014;23(11):3324–9.
118. Haviv B, Bronak S, Rath E, Yassin M. Nerve injury during anterior cruciate ligament reconstruction: a comparison between patellar and hamstring tendon grafts harvest. Knee. 2017;24(3):564–9.
119. Lee GH, McCulloch P, Cole BJ, Bush-Joseph CA, Bach BR. The incidence of acute Patellar Tendon Harvest complications for anterior cruciate ligament reconstruction. Arthrosc J Arthroscop Relat Surg. 2008;24(2):162–6.

120. Beaufils P, Gaudot F, Drain O, Boisrenoult P, Pujol N. Mini-invasive technique for bone patellar tendon bone harvesting: its superiority in reducing anterior knee pain following ACL reconstruction. Curr Rev Musculoskelet Med. 2011;4(2):45–51.
121. Papastergiou SG, Voulgaropoulos H, Mikalef P, Ziogas E, Pappis G, Giannakopoulos I. Injuries to the infrapatellar branch(es) of the saphenous nerve in anterior cruciate ligament reconstruction with four-strand hamstring tendon autograft: vertical versus horizontal incision for harvest. Knee Surg Sports Traumatol Arthrosc. 2005;14(8):789–93.
122. Mouarbes D, Dagneaux L, Olivier M, Lavoue V, Peque E, Berard E, et al. Lower donor-site morbidity using QT autografts for ACL reconstruction. Knee Surg Sports Traumatol Arthrosc. 2020;28:2558.
123. Zalta J. Massage therapy protocol for post-anterior cruciate ligament reconstruction patellofemoral pain syndrome: a case report. Int J Ther Massage Bodywork. 2008;1(2):11–21.
124. Nelissen E, van Arkel E, Hazelbag H. Traumatic neuroma of the infrapatellar branch of the saphenous nerve after hamstring harvesting. J Knee Surg. 2010;23(04):233–6.
125. Fang CH, Li M, Zhang YF, Liu H. Extra-articular migration of PEEK interference screw after anterior cruciate ligament reconstruction: a report of two cases. BMC Musculoskelet Disord. 2021;22(1):498.
126. Mascarenhas R, Saltzman BM, Sayegh ET, Verma NN, Cole BJ, Bush-Joseph C, et al. Bioabsorbable versus metallic interference screws in anterior cruciate ligament reconstruction: a systematic review of overlapping meta-analyses. Arthroscop J Arthroscop Relat Surgery. 2015;31(3):561–8.
127. Chevallier R, Klouche S, Gerometta A, Bohu Y, Herman S, Lefevre N. Bioabsorbable screws, whatever the composition, can result in symptomatic intra-osseous tibial tunnel cysts after ACL reconstruction. Knee Surg Sports Traumatol Arthrosc. 2018;27(1):76–85.
128. Lee DW, Lee JW, Kim SB, Park JH, Chung KS, Ha JK, et al. Comparison of Poly-L-Lactic Acid and Poly-L-Lactic Acid/Hydroxyapatite bioabsorbable screws for tibial fixation in ACL reconstruction: clinical and magnetic resonance imaging results. Clin Orthop Surg. 2017;9(3):270–9.
129. Debieux P, Franciozi CES, Tamaoki MJS, Lenza M, Faloppa F, Belloti JC. Bioabsorbable versus metallic interference screws for graft fixation in anterior cruciate ligament reconstruction. Cochrane Database Syst Rev. 2012;7:CD009772. Wiley.
130. Mathew M, Dhollander A, Getgood A. Anterolateral ligament reconstruction or extra-articular tenodesis. Clin Sports Med. 2018;37(1):75–86.
131. Musahl V, Herbst E, Burnham JM, Fu FH. The anterolateral complex and anterolateral ligament of the knee. J Am Acad Orthop Surg. 2018;26(8):261–7.
132. Glasbrenner J, Domnick C, Raschke MJ, Willinghöfer T, Kittl C, Michel P, et al. Adjustable buttons for ACL graft cortical fixation partially fail with cyclic loading and unloading. Knee Surg Sports Traumatol Arthrosc. 2018;27(8):2530–6.
133. Smith PA, Piepenbrink M, Smith SK, Bachmaier S, Bedi A, Wijdicks CA. Adjustable- versus fixed-loop devices for femoral fixation in ACL reconstruction: an in vitro full-construct biomechanical study of surgical technique-based tibial fixation and graft preparation. Orthop J Sports Med. 2018;6(4):2325967118768743.
134. Cozzi AL, Dunn KL, Harding JL, McLeod TCV, Bacon CEW. Kinesiophobia after anterior cruciate ligament reconstruction in physically active individuals. J Sport Rehabil. 2015;24(4):434–9.
135. Nishizawa Y, Hoshino Y, Nagamune K, Araki D, Nagai K, Kurosaka M, et al. Comparison between intra- and extra-articular tension of the graft during fixation in anterior cruciate ligament reconstruction. Arthroscopy J Arthroscop Relat Surg. 2017;33(6):1204–10.
136. Firth AD, Bryant DM, Litchfield R, McCormack RG, Heard M, MacDonald PB, et al. Predictors of graft failure in young active patients undergoing hamstring autograft anterior cruciate ligament reconstruction with or without a lateral extra-articular tenodesis: the stability experience. Am J Sports Med. 2022;50(2):384–95.
137. Kraus Schmitz J, Lindgren V, Janarv PM, Forssblad M, Stålman A. Deep venous thrombosis and pulmonary embolism after anterior cruciate ligament reconstruction. Bone Joint J. 2019;101-B(1):34–40.
138. Erickson BJ, Saltzman BM, Campbell KA, Fillingham YA, Harris JD, Gupta AK, et al. Rates of deep venous thrombosis and pulmonary embolus after anterior cruciate ligament reconstruction: a systematic review. Sports health. 2015;7(3):261–6.
139. Pollak AW, McBane RD. Succinct review of the new VTE prevention and management guidelines. Mayo Clin Proc. 2014;89(3):394–408.

19 Management of Revision ACL Reconstruction

M. Süleyman Abul, Selim Ergün,
and Mustafa Karahan

19.1 Introduction

Two to ten percent rate of failure has been reported after anterior cruciate ligament (ACL) reconstruction (ACLR) [1]. Revision ACLR (R-ACLR) is a highly complex procedure with poorer outcomes and lower rates of return to sport compared to primary ACLR [2–4]. The most common cause of R-ACLR is malpositioned tunnels, which are often caused by surgeons who are inexperienced in ACLR surgery. Other common causes include omission of concomitant rotational ligament injuries and meniscus tears, increased posterior tibial slope or varus deformity in the coronal plane, new trauma, inappropriate rehabilitation, and infection. In R-ACLR management, first determining the specific etiology and mode of failure, then performing the necessary interventions as a single- or two-stage surgery will increase success.

The indication for revision surgery after failure will be necessary for the majority of patients, and the demand of the patient whose instability continues will be in the same direction. However, it should not be forgotten that this requirement is not met for every patient. Especially in patients with arthritis affecting two or three compartments, regional pain syndrome, multiple previous revisions, and a sedentary lifestyle, revision surgery can be avoided [5].

19.2 Diagnosis

Failed ACLR is defined as functional instability in daily life or sports activities and increased laxity in physical examination and tests performed with physical examination and instruments [6]. In another definition, failure is defined as a laxity of more than 6 mm on the graft in the KT-2000 arthrometer test or pivot-shift test [7]. A detailed anamnesis, physical examination, and magnetic resonance imaging (MRI) are essential to diagnose rerupture of the graft and to detect the presence of accompanying meniscal and ligamentous injuries. Computerized tomography (CT) will be helpful to evaluate tunnel locations and diameters. In order not to miss sagittal and coronal plane deformities, an ortho-roentgenogram and full lateral radiographs of the cruris should be taken.

M. S. Abul · S. Ergün
Kartal Dr. Lütfi Kırdar City hospital, Istanbul, Turkey

M. Karahan (✉)
Department of Orthopaedics, Acıbadem Mehmet Ali Aydinlar University, Istanbul, Turkey

M. Karahan et al. (eds.), *Anterior Cruciate Ligament Injuries Throughout the Life Span*,
https://doi.org/10.1007/978-3-031-84501-7_19

19.3 History and Physical Examination

A recent history of trauma or time of reappearance of instability complaint should be noted. While complaints that persist in the early postoperative period are often associated with loss of fixation and infection, midterm failures (3–12 months) might be caused by errors in surgical technique, aggressive physical therapy, and overlooked accompanying meniscoligamentous injuries. Recurrent trauma is the main cause of late failures (>12 months). An improper/failed graft implantation or accompanying ligamentous injuries, complaints increasing gradually with time may suggest a malpositioned tunnel. In general, indicators of ACLR failure include recurrent instability, or a stable but painful and stiff knee. Limited extension and flexion, range of motion between 10 and 120° of flexion, is a finding that should be taken seriously into account. The patient's physical examination may not match the patient's perception [8, 9]. Loss of extension may cause many symptoms in patients, whereas loss of flexion may be better tolerated [10, 11].

In the physical examination:

- On inspection, old incision scars give information about the autograft used in previous surgery. Redness and warmness over the incision may be evaluated in favor of infection.
- The presence of instability can be examined with the Lachman, anterior drawer, and pivot shift tests. When the pivot shift test gives high-grade findings, additional anterolateral stabilization would be necessary.
- Posterior drawer, varus–valgus stress tests, and rotatory instability tests, which include dial test for posterolateral and anterior drawer test with the knee in 15° of external rotation for anteromedial rotatory instability, should be performed to question any accompanying ligamentous injuries.
- Tests for meniscus pathologies along with radiological findings are necessary, especially for the diagnosis of posterolateral root tears or posteromedial ramp lesions that contribute to rotational deficiencies.
- The presence of antalgic gait, varus thrust pattern, or stiff knee gait in the gait analysis provides diagnostic clues [12].

19.4 Radiology and Laboratory Findings

Weightbearing anteroposterior (AP) and lateral radiographs of the knee provide information regarding the topography and diameters of the tunnels. Radiographs also provide information regarding the presence of arthrosis in any knee compartment [13]. The best assessment of tunnel location and diameter is provided by three-dimensional CT imaging [14]. MRI is useful to demonstrate graft failure, intraarticular effusion, bone marrow edema in the proximal tibia or distal femur, and associated chondral and meniscoligamentous injuries [15]. Laboratory tests should be ordered to rule out infection. Complete blood count, sedimentation, and C-reactive protein levels are useful to rule out infection. If laboratory tests are in favor of infection, contrast-enhanced MRI is used as a further investigation. Joint fluid aspiration for leucocyte count and culture should be ordered in highly suspected patients [16, 17].

19.5 Etiology-Oriented Management

The cause of ACLR failure can be grouped into three main categories (Fig. 19.1).

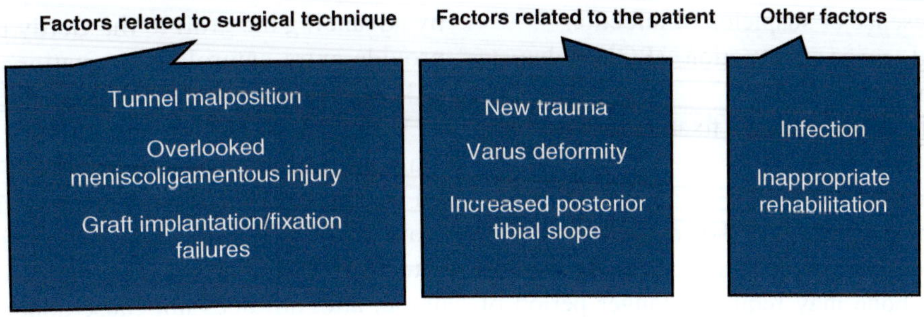

Fig. 19.1 Main causes for anterior cruciate ligament reconstruction failure

19.6 Tunnel Malposition and Enlargement, Bone Defects

Points to consider in preoperative evaluation:

- Is it possible to use the same tunnels, or are new tunnel/tunnels necessary?
- Diameters of preexisting tunnels need bone grafting/staged surgery or graft with a bone block?
- If new tunnels are necessary, can new tunnels be drilled without creating a bony defect (convergent tunnels)?
- Lastly, will stable fixation be achieved and is it necessary to remove previous implants? If necessary, will implant removal create a relevant bony defect?

Tunnel malposition may present as malposition of the tibial tunnel, femoral tunnel, or both. Femoral tunnel malposition is the most common, which is often placed vertically or anteriorly [18]. Tibial tunnel malposition is more often encountered with a posterior location [19]. Anteriorly located femoral tunnel will cause flexion limitation in the early stages, and over time, cyclic stretching will cause the graft to fail. Posteriorly drilled tibial tunnel causes the graft to be placed more centrally in the joint and fail may occur due to the rotational forces that the graft cannot resist. Iatrogenic cortical damage during drilling new tunnel or removal of previous implant may occur.

Tunnel enlargement, rather than being a radiological finding, indicates the presence of osteolysis and problem in bone–tendon healing. But even if tunnel enlargement is radiographically obvious after primary ACLR, it can also be asymptomatic [20]. CT images to evaluate tunnel geometry, bone sclerosis, graft integration, and implant position are critical in symptomatic patients. Enlarged tunnels with a diameter greater than 14 mm is a general set point for determining whether to stage the revision procedure or not [21]. But there is no absolute threshold for "critical tunnel diameter" with values ranging from 10 to 15 mm [22–25], and this threshold value may vary depending on graft selection, drilling and fixation technique.

The need for bone grafting and two-stage surgery may arise due to preoperatively those predictable reasons such as the previously mentioned tunnel convergence or the presence of an enlarged tunnel, or due to unpredictable bone loss occurring per-operatively while drilling or previous implant removal. However, specific techniques such as drilling with a different angle from a new cortical location by the outside-in method in convergent tunnels, the over-the-top technique, preference of graft with large bone blocks, and filling the defect with large interference screws can eliminate the need for bone grafting and two-stage surgery. Definitive indications for two-stage surgery can be listed as follows:

- The presence or suspicion of infection
- Limited range of motion (ROM) due to arthrofibrosis
- Inability to provide safe fixation due to insufficient bone stock [26]

A follow-up period of 3–6 months is recommended before the second stage [27]. The use of allogeneic bone may require a longer period of time. Adequate graft union can be determined by CT imaging [28]. Still, with specific techniques such as impacted bone grafting, it can also be performed as a one-stage procedure.

Since anatomic landmarks or native ACL remnants are no longer exist in revision surgeries, fluoroscopy can be used to determine the true tunnel location. While malpositioned tibial tunnels usually do not overlap with the new one, convergence with the previous tunnel is common for the femoral tunnel [19].

Loss of tunnel wall integrity is most commonly caused by iatrogenic injury of the posterior wall of the femoral tunnel [29]. Evaluation of the whole femoral tunnel 360° through the anteromedial portal can be useful [30]. While defects in posterior wall are less than 5 mm can be ignored, fixation with isolated cortical suspension implants should be avoided in defects larger than 5 mm. Suture anchor or post-fixation screw to the lateral cortex with an additional incision will be necessary [31, 32]. If the graft is long enough, the graft can be passed behind the posterolateral femoral cortex by the "over the top" method and can be fixed to the lateral cortex with a U-Nail. Depending on the length of the graft, additional rotational stability can be provided by advancing the graft to the Gerdy tubercle to provide extra anterolateral rotational stability [33, 34].

When two-stage surgery is decided, autograft, allograft, or synthetic grafts can be used to fill the tunnel opened in the previous surgery (Table 19.1). While autograft provides more osteoinduction and reduces the risk of nonunion, allograft does not cause donor site morbidity and provides a shorter surgical time. Synthetic grafts such as calcium phosphate can be preferred with osteoconductive effect. While some recent studies showed that nonirradiated allografts showed similar union rates with the autografts [35, 36], others showed that the failure rate with autograft was found to be much lower than the allograft (2.8 vs. 13%, respectively), [37, 38]. Regardless of the type of bone graft used, it is important to carefully prepare the tunnel before implanting the bone graft by removing all graft material, sutures, and disintegrating the sclerotic bone.

Bone–patellar tendon–bone graft preference is not recommended in patients where single-stage bone block graft would be appropriate instead of two-stage, due to donor site comorbidities such as iatrogenic patella fracture [39]. Quadriceps tendon has recently became popular. Quadriceps tendon graft is advantageous because its length and width can be adjusted according to the need and a portion can be harvested together with the patella (Fig. 19.2). When the outcomes of the quadriceps tendon graft, which has higher biomechanical strength compared to the hamstring autograft, were compared, no significant difference was found in terms of clinical results and failure [40].

Table 19.1 Properties of bone grafts used to fill tunnels

Allograft	Demineralized, mineralized frozen bone graft	Osteoconductive	Manufactured, ease of use, osteointegration slow, high risk of immunological reaction
Autograft	Fibula, tibial tubercle, iliac crest, Gerdy tubercle	Osteoconductive, osteoinductive	Better osteointegration, donor site issues
Synthetic grafts	Calcium phosphate, calcium sulfate, hydroxyapatite	Osteoconductive, scaphoid	Different characteristics of each graft

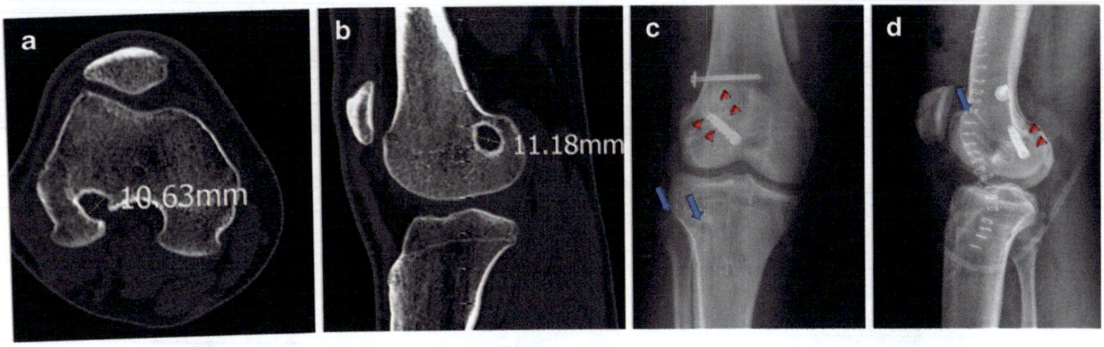

Fig. 19.2 Failed ACLR surgery in an 18-year-old male patient who underwent ACLR one year ago. Posterolateral corner (PLC) instability was detected on his physical examination. (**a**, **b**) Enlarged femoral tunnel with a diameter of 11 mm was measured on CT images. Same femoral and tibial tunnels were planned to be used in R-ACLR. (**c**, **d**) Postoperative knee AP (**c**) and lateral (**d**) radiographs show one-stage revision operation with quadriceps tendon autograft with a 12 mm bone block (red arrow heads) which was fixed with a metallic interference screw and sutures tied over a post-fixation screw on lateral cortex. Posterolateral corner reconstruction was performed simultaneously through tunnels on fibula, tibia, and femur (blue arrows) with contralateral hamstring tendons. ACLR anterior cruciate ligament reconstruction, CT computed tomography, R-ACLR revision anterior cruciate ligament reconstruction, AP anteroposterior

19.7 Sagittal and Coronal Malalignment

Posterior tibial slope (PTS) is a risk factor for primary ACL tear and retear [41–46]. Dejour et al. showed that every increase of 10° in PTS leads to a 6 mm increment of passive anterior tibial translation of the tibia, which increases the stress on the ACL [47]. It has been shown that the slope of the lateral plateau is more significant than the medial for the tension occurring in the ACL [48, 49]. The rate of ACL reinjury was found to be five times higher in patients with a PTS of 12° or more [50, 51] Dejour et al. reported no failure or instability in the two-year follow-up period when deflection osteotomy was performed simultaneously with R-ACLR surgery in patients with PTS over 12° [52].

To correct high PTS during R-ACLR, deflection osteotomy in the form of a closed wedge osteotomy can be performed superior, at the same level, or distal to the tibial tubercle. Osteotomy superior to the tibial tubercle (supra-tubercular) is more advantageous than the distal ones in terms of union because the osteotomy line is in the metaphyseal region [53, 54], (Fig. 19.3). However, in this osteotomy, there is a more limited area for the tibial tunnel. Osteotomy at the level of the tibial tubercle is safer in terms of the tibial tunnel, but it is recommended for tertiary revisions because it has some risks in terms of neurovascular injury, pseudoarthrosis, and stiffness [55]. In osteotomy distal to the tibial tubercle, patellofemoral biomechanics are preserved [56]. However, there is a disadvantage in terms of nonunion [57].

In the coronal plane, varus deformity increases the tension on the ACL fibers, creating a risk for graft failure [58–60]. Definition of varus deformity is the angle between the tibial and femoral mechanical axis more than 3°. Patients with varus or valgus deformity ≥5° (measurement of the angle between the femur and tibia mechanical axis) accompanied by osteoarthritis findings, significant cartilage damage and/or symptomatic meniscus insufficiency without any signs of osteoarthritis, and dynamic ligament insufficiency (varus-thrust) are indicated to correct coronal alignment. In recent years, it has become

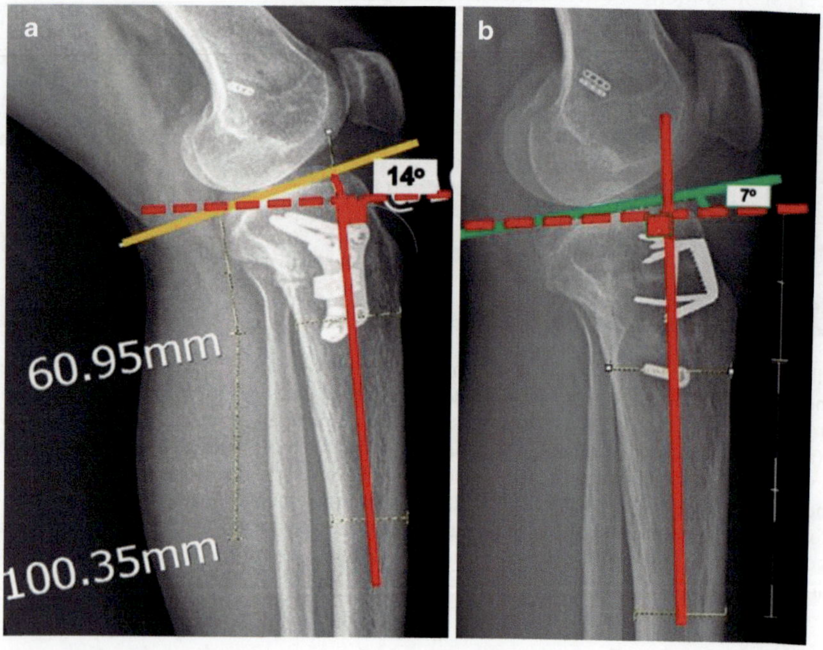

Fig. 19.3 (**a**) In a 30-year-old male patient who underwent R-ACLR with concomitant medial open wedge high tibial osteotomy due to varus deformity, the PTS angle measurement was 14° in the lateral radiograph. (**b**) After removing the medial plate and screws, performing supra tubercular deflection osteotomy, and performing ACL revision in the same session, the measured PTS angle decreased to 7°. (Orange and green lines are parallel to tibia plateau surface, red line is the axis of tibia, red dashed line is the line perpendicular to the tibia axis. R-ACLR Revision anterior cruciate ligament reconstruction, PTS posterior tibial slope, ACL anterior cruciate ligament)

indicated for even lower degree deformities. Therefore, correcting even a <5° deformity may be indicated in selected cases, such as patients undergoing concurrent meniscus transplantation, cartilage repair procedures, or collateral ligament reconstruction. However, an isolated varus malalignment without the mentioned conditions of arthrosis, chondral–meniscal damage is not an indication for osteotomy alone [4].

In a clinical study we conducted, we found that there was an increase in the PTS angle and varus deformities in the coronal plane of the affected knees of patients who were not reconstructed after ACL rupture and who were exposed to ACL deficiency for long time (more than 5 years), compared to the contralateral healthy knees [61]. For this reason, we believe that PTS and coronal plane deformity examination should be performed preoperatively, even when planning primary reconstruction, in patients with ongoing instability due to long-term exposure to ACL deficiency.

19.8 Rotational Instabilities

It has been reported that rotational instability persists in one out of every four patients who underwent ACLR surgery [62]. The anterolateral complex (ALC), which is an extraarticular structure, plays an important role in rotational stability together with the ACL. This structure is shown to be injured in 78% of patients with ACL injury [63]. However, it may be difficult to diagnose an ALC injury clinically. Studies have reported that instability observed in Lachman and pivot shift tests may not be always correlated with ACL injury and is more closely related to ALC injury.

MRI and ultrasonography can be utilized in the diagnosis [64]. Interventions for anterolateral instability can be planned in patients present with gross laxity (pivot shift+++, and/ or in pivoting sports or in hyperlaxity). Extra-articular tenodesis or anterolateral ligament (ALL) reconstruction can be performed according to the surgeon's preference [62, 65, 66].

A retrospective analysis of patients who underwent R-ACLR after ACLR revealed a higher number of associated meniscal and cartilage lesions compared to the normal population [67]. In patients with ACL insufficiency, the medial meniscus acts as one of the structures preventing anterior tibial translation [68]. Biomechanical studies have shown that the lateral meniscus has a role in preventing knee pivot shift under axial load [69]. Ramp lesions, defined as tears at the peripheral attachment site of the posterior horn of the medial meniscus, are effective in anteromedial rotatory instability [70]. Ramp lesions were detected in 16.6% of patients undergoing ACLR [71]. Therefore, these lesions should be carefully diagnosed and repaired in patients whom R-ACLR is planned. The risk of ACLR failure was found to be 4.5 times higher in patients with medial meniscal deficiency and 3.5 times higher in patients with lateral meniscal deficiency [72]. In the absence of meniscus (no posterior root or less than 40% is intact) with subjective joint line pain or symptomatic instability after a failed ACLR, meniscal transplantation with R-ACLR is considered [73].

It has been demonstrated that increased load on the graft occurs when there is a missed medial collateral ligament (MCL) or posterolateral corner (PLC) injury in patients undergoing ACLR [74, 75]. MCL injury is associated with 20–38% of patients with traumatic ACL injury [76]. Preoperative medial laxity is a risk factor for worse ACL revision outcomes. Accompanying medial collateral ligament (MCL) reconstruction may be considered for grade 2 and 3 MCL laxity, but there is no high level of evidence study to support this recommendation.

PLC injury alone is rare [77]. When PLC injury concomitant with ACL injury is neglected surgically, it will result in permanent instability and early degenerative changes [78]. Dial test or reverse–pivot shift test are helpful in diagnose. In lateral collateral ligament (LCL) and/or PLC injuries, unlike rotational instabilities such as anterolateral or anteromedial instability, more than normal opening in the joint space may be notable. Before starting arthroscopy, bilateral comparative AP radiographs taken under fluoroscopy, especially under varus stress at 10–15° of flexion, can provide strong evidence of diagnosis. Excessive opening in the joint space with noticeable "drive through sign" in the lateral compartment during arthroscopy should also make the surgeon consider PLC injury. After clarifying the diagnosis of PLC, tibiofibular-based PLC reconstruction technique (LaPrade technique) or fibula-based PLC reconstruction technique (Larson or Arciero) can be preferred to accompany R-ACLR.

19.9 Postoperative Rehabilitation

Postoperative rehabilitation is one of the most important steps in returning to sports activity and normal life cycle following the R-ACLR. Rehabilitation should progress more slowly compared to the primary reconstruction [79]. Since the graft, articular cartilage, and meniscal status, accompanying ligamentous injuries and patient expectations, are different in revision ACLR, an individualized rehabilitation process should be managed. With allograft use in revision surgeries, bone-graft integration will be slower than autograft. It has been shown that early aggressive rehabilitation after revision ACLR may cause early graft failure and late laxity [80]. In general, the first 4 weeks following ACLR revision surgery consists of pain modulation, effusion reduction, neutral extension, and quadriceps activation. Between weeks 2 and 8, it is recommended to continue with neuromuscular control exercises with the inclusion of weightbearing, mobility, and low-level conditioning exercises. Weeks 6–12 constitute the period when neuromuscular exercises are intensified and conditioning exercises are tightened. If patients do not have multiple ligament injuries or serious osteochondral defects, braces are not necessary or should be used for a short time [81].

References

1. Wright RW, Magnussen RA, Dunn WR, Spindler KP. Ipsilateral graft and contralateral ACL rupture at five years or more following ACL reconstruction: a systematic review. J Bone Joint Surg Am. 2011;93(12):1159–65.
2. Grassi A, Kim C, Marcheggiani Muccioli GM, et al. What is the mid-term failure rate of revision ACL reconstruction? A systematic review. Clin Orthop Relat Res. 2017;475(10):2484–99.
3. Wright RW, Johnson L, Brophy RH, et al. Revision anterior cruciate ligament reconstruction outcomes at a minimum of 5-year follow-up: a systematic review. J Knee Surg. 2019;32(3):218–21.
4. Tischer T, Beaufilis P, Becker R, Ahmad SS, Bonomo M, Dejour D, Eriksson K, Filardo G, Feucht MJ, Grassi A, Wilson A, Menetrey J, Pujol N, Rathcke M, Seil R, Strauss MJ, Condello V. Management of anterior cruciate ligament revision in adults: the 2022 ESSKA consensus part I-diagnostics and preoperative planning. Knee Surg Sports Traumatol Arthrosc. 2023;31(11):4642–51. https://doi.org/10.1007/s00167-022-07214-w. Epub 2022 Nov 2. PMID: 36322180.
5. Osti L, Buda M, Osti R, Massari L, Maffulli N. Preoperative planning for ACL revision surgery. Sports Med Arthrosc Rev. 2017;25(1):19–29. https://doi.org/10.1097/JSA.0000000000000140. PMID: 28045870.
6. Safran MR, Harner CD. Technical considerations of revision anterior cruciate ligament surgery. Clin Orthop Relat Res. 1996;325:50–64.
7. Noyes FR, Barber-Westin SD. Revision anterior cruciate surgery with use of bone-patellar tendon-bone autogenous grafts. J Bone Joint Surg Am. 2001;83(8):1131–43. https://doi.org/10.2106/00004623-200108000-00001. PMID: 11507120.
8. Harter RA, Osternig LR, Singer KM, James SL, Larson RL, Jones DC. Long-term evaluation of knee stability and function following surgical reconstruction for anterior cruciate ligament insufficiency. Am J Sports Med. 1988;16:434–43.
9. Lephart SM, Perrin DH, Fu FH, Gieck JH, McCue FC, Irrgang JJ. Relationship between selected physical characteristics and functional capacity in the anterior cruciate ligament-insufficient athlete. J Orthop Sports Phys Ther. 1992;16:174–81.
10. Sachs RA, Daniel DM, Stone ML. Patellofemoral problems after anterior cruciate ligament reconstruction. Am J Sports Med. 1989;17:760.
11. Irrgang JJ, Harner CD. Loss of motion following knee ligament reconstruction. Sports Med. 1995;19:150–9.
12. Noyes FRSR. The role of high tibial osteotomy in the anterior cruciate ligament-deficient knee with varus alignment. In: DeLee JCDD, editor. Orthopaedic sports medicine principles and practice. Philadelphia: WB Saunders; 1994. p. 1401–43.
13. Kosy JD, Mandalia VI. Plain radiographs can be used for routine assessment of ACL reconstruction tunnel position with three-dimensional imaging reserved for research and revision surgery. Knee Surg Sports Traumatol Arthrosc. 2018;26(2):534–49. https://doi.org/10.1007/s00167-017-4462-5. Epub 2017 Feb 21. PMID: 28224200.
14. Tapasvi S, Shekhar A. Revision ACL reconstruction: principles and practice. Indian J Orthop. 2021;55(2):263–75. https://doi.org/10.1007/s43465-020-00328-8. PMID: 33927805; PMCID: PMC8046893.
15. Ahn JH, Lee YS, Chang MJ, Yim HS. Analysis of revision anterior cruciate ligament reconstruction according to the combined injury, degenerative change, and MRI findings. Knee. 2011;18(6):382–6. https://doi.org/10.1016/j.knee.2010.11.001. Epub 2010 Dec 24. PMID: 21185729.
16. Parvizi J, Gehrke T. In: Parvizi J, Gehrke T, editors. Proceedings of the second international consensus on musculoskeletal infection. Brooklandville: Data Trace; 2018.
17. Gobbi A, Karnatzikos G, Chaurasia S, Abhishek M, Bulgherhoni E, Lane J. Postoperative infection after anterior cruciate ligament reconstruction. Sports Health. 2016;8(2):187–9. https://doi.org/10.1177/1941738115618638. PMID: 26603553; PMCID: PMC4789929.
18. Morgan JA, Dahm D, Levy B, Stuart MJ, MARS Study Group. Femoral tunnel malposition in ACL revision reconstruction. J Knee Surg. 2012;25(5):361–8.
19. Achtnich A, Ranuccio F, Willinger L, et al. High incidence of partially anatomic tunnel placement in primary single-bundle ACL reconstruction. Knee Surg Sports Traumatol Arthrosc. 2018;26(2):462–7.
20. Buelow J-U, Siebold R, Ellermann A. A prospective evaluation of tunnel enlargement in anterior cruciate ligament reconstruction with hamstrings: extracortical versus anatomical fixation. Knee Surg Sports Traumatol Arthrosc. 2002;10(2):80–5.
21. Mayr R, Rosenberger R, Agraharam D, Smekal V, El Attal R. Revision anterior cruciate ligament reconstruction: an update. Arch Orthop Trauma Surg. 2012;132(9):1299–313. https://doi.org/10.1007/s00402-012-1552-1. Epub 2012 Jun 6. PMID: 22669543.
22. MARS Group. Rehabilitation predictors of clinical outcome following revision ACL reconstruction in the MARS cohort. J Bone Joint Surg Am. 2019;101(9):779–86. https://doi.org/10.2106/JBJS.18.00397. PMID: 31045665; PMCID: PMC7292503.

23. Bhatia S, Korth K, Van Thiel GS, Frank RM, Gupta D, Cole BJ, Bach BR Jr, Verma NN. Effect of tibial tunnel diameter on femoral tunnel placement in transtibial single bundle ACL reconstruction. Knee Surg Sports Traumatol Arthrosc. 2016;24(1):51–7. https://doi.org/10.1007/s00167-014-3307-8. Epub 2014 Sep 19. PMID: 25236681.
24. Chahla J, Dean CS, Cram TR, Civitarese D, O'Brien L, Moulton SG, LaPrade RF. Two-stage revision anterior cruciate ligament reconstruction: bone grafting technique using an allograft bone matrix. Arthrosc Tech. 2016;5(1):e189–95. https://doi.org/10.1016/j.eats.2015.10.021. PMID: 27274452; PMCID: PMC4886209.
25. Schnetzke M, Vetter S, Linden PV, Grützner PA, von Recum J. Two-stage revision anterior cruciate ligament reconstruction using silicate-substituted calcium phosphate. Arthrosc Tech. 2019;8(10):e1239–46. https://doi.org/10.1016/j.eats.2019.06.015. PMID: 32042579; PMCID: PMC7000346.
26. Wolfson TS, Mannino B, Owens BD, Waterman BR, Alaia MJ. Tunnel management in revision anterior cruciate ligament reconstruction: current concepts. Am J Sports Med. 2023;51(2):545–56. https://doi.org/10.1177/03635465211045705.
27. Hofbauer M, Muller B, Murawski CD, Baraga M, van Eck CF, Fu FH. Strategies for revision surgery after primary double-bundle anterior cruciate ligament (ACL) reconstruction. Knee Surg Sports Traumatol Arthrosc. 2013;21(9):2072–80. https://doi.org/10.1007/s00167-013-2470-7. Epub 2013 Apr 12. PMID: 23579225.
28. Uchida R, Toritsuka Y, Mae T, Kusano M, Ohzono K. Healing of tibial bone tunnels after bone grafting for staged revision anterior cruciate ligament surgery: a prospective computed tomography analysis. Knee. 2016;23(5):830–6.
29. Rue J-P, Busam M, Detterline A, Bach B. Posterior wall blowout in anterior cruciate ligament reconstruction—avoidance, recognition, and salvage. J Knee Surg. 2008;21(3):235–40.
30. Temponi EF, Oliveira JNP, Soares LFM, de Carvalho LH Jr. A femoral tunnel view test during ACL reconstruction. Arthrosc Tech. 2017;6(4):e1177–81.
31. Mitchell JJ, Chahla J, Dean CS, et al. Posterior wall blowout during anterior cruciate ligament reconstruction: suspensory cortical fixation with a screw and washer post. Arthrosc Tech. 2016;5(3):e551–5.
32. Mitchell JJ, Dean CS, Chahla J, Menge TJ, Cram TR, LaPrade RF. Posterior wall blowout in anterior cruciate ligament reconstruction: a review of anatomic and surgical considerations. Orthop J Sports Med. 2016;4(6):2325967116652122.
33. Buda R, Ruffilli A, Di Caprio F, et al. Allograft salvage procedure in multiple-revision anterior cruciate ligament reconstruction. Am J Sports Med. 2013;41(2):402–10.
34. Usman MA, Kamei G, Adachi N, Deie M, Nakamae A, Ochi M. Revision single-bundle anterior cruciate ligament reconstruction with over-the-top route procedure. Orthop Traumatol Surg Res. 2015;101(1):71–5.
35. Grassi A, Nitri M, Moulton SG, et al. Does the type of graft affect the outcome of revision anterior cruciate ligament reconstruction? Bone Joint J. 2017;99(6):714–23.
36. Kaeding CC, Pedroza AD, Reinke EK, et al. Change in anterior cruciate ligament graft choice and outcomes over time. Arthroscopy. 2017;34(1):289–300.
37. Samuelsen BT, Webster KE, Johnson NR, Hewett TE, Krych AJ. Hamstring autograft versus patellar tendon autograft for ACL reconstruction: is there a difference in graft failure rate? A meta-analysis of 47,613 patients. Clin Orthop Relat Res. 2017;475(10):2459–68. https://doi.org/10.1007/s11999-017-5278-9. PMID: 28205075; PMCID: PMC5599382.
38. van Eck CF, Schkrohowsky JG, Working ZM, Irrgang JJ, Fu FH. Prospective analysis of failure rate and predictors of failure after anatomic anterior cruciate ligament reconstruction with allograft. Am J Sports Med. 2012;40(4):800–7. https://doi.org/10.1177/0363546511432545. Epub 2012 Jan 11. PMID: 22238055.
39. Dai W, Leng X, Wang J, Cheng J, Hu X, Ao Y. Quadriceps tendon autograft versus bone-Patellar Tendon-bone and Hamstring Tendon autografts for anterior cruciate ligament reconstruction: a systematic review and meta-analysis. Am J Sports Med. 2022;50(12):3425–39. https://doi.org/10.1177/03635465211030259. Epub 2021 Sep 8. PMID: 34494906.
40. Haner M, Bierke S, Petersen W. Anterior cruciate ligament revision surgery: ipsilateral quadriceps versus contralateral semitendinosusgracilis autografts. Arthroscopy. 2016;32(11):2308–17.
41. Brandon ML, Haynes PT, Bonamo JR, Flynn MI, Barrett GR, Sherman MF. The association between posterior-inferior tibial slope and anterior cruciate ligament insufficiency. Arthroscopy. 2006;22(8):894–9. https://doi.org/10.1016/j.arthro.2006.04.098. PMID: 16904590.
42. Meister K, Talley MC, Horodyski MB, Indelicato PA, Hartzel JS, Batts J. Caudal slope of the tibia and its relationship to noncontact injuries to the ACL. Am J Knee Surg. 1998;11(4):217–9. PMID: 9853999.
43. Beynnon BD, Hall JS, Sturnick DR, et al. Increased slope of the lateral tibial plateau subchondral bone is associated with greater risk of noncontact ACL injury in females but not in males: a prospective cohort study with a nested, matched case-control analysis. Am J Sports Med. 2014;42(5):1039–48.
44. Dæhlin L, Inderhaug E, Strand T, Parkar AP, Solheim E. The effect of posterior Tibial slope on the risk of revision surgery after anterior cruciate ligament reconstruction. Am J Sports Med. 2022;50(1):103–10.
45. Duerr RA, Ormseth B, DiBartola A, et al. Association of elevated posterior Tibial slope with revision anterior cruciate ligament graft failure in a matched cohort analysis. Am J Sports Med. 2023;51(1):38–48.

46. Guimarães TM, Giglio PN, Sobrado MF, et al. Knee hyperextension greater than 5° is a risk factor for failure in ACL reconstruction using hamstring graft. Orthop J Sport Med. 2021;9(11):4–9.
47. Dejour H, Bonnin M. Tibial translation after anterior cruciate ligament rupture. Two radiological tests compared. J Bone Joint Surg Br. 1994;76(5):745–9. PMID: 8083263.
48. Simon RA, Everhart JS, Nagaraja HN, Chaudhari AM. A case-control study of anterior cruciate ligament volume, tibial plateau slopes and intercondylar notch dimensions in ACL-injured knees. J Biomech. 2010;43(9):1702–7. https://doi.org/10.1016/j.jbiomech.2010.02.033. Epub 2010 Apr 10. PMID: 20385387; PMCID: PMC2882557.
49. Stijak L, Herzog RF, Schai P. Is there an influence of the tibial slope of the lateral condyle on the ACL lesion? A case-control study. Knee Surg Sports Traumatol Arthrosc. 2008;16(2):112–7. https://doi.org/10.1007/s00167-007-0438-1. Epub 2007 Nov 16. PMID: 18239948.
50. Webb JM, Salmon LJ, Leclerc E, Pinczewski LA, Roe JP. Posterior tibial slope and further anterior cruciate ligament injuries in the anterior cruciate ligament-reconstructed patient. Am J Sports Med. 2013;41(12):2800–4. https://doi.org/10.1177/0363546513503288. Epub 2013 Sep 13. PMID: 24036571.
51. Mariani PP. Posterior horn instability of the medial meniscus a sign of posterior meniscotibial ligament insufficiency. Knee Surg Sports Traumatol Arthrosc. 2011;19(7):1148–53. https://doi.org/10.1007/s00167-011-1424-1. Epub 2011 Feb 11. PMID: 21311863.
52. Dejour D, Saffarini M, Demey G, Baverel L. Tibial slope correction combined with second revision ACL produces good knee stability and prevents graft rupture. Knee Surg Sports Traumatol Arthrosc. 2015;23(10):2846–52. https://doi.org/10.1007/s00167-015-3758-6. Epub 2015 Aug 23. PMID: 26298711.
53. Dejour D, Khun A, Dejour H. Osteotomie tibiale de déflexion et laxité chronique antérieure à propos de 22 cas. Rev Chir Orthop. 1998:28–9.
54. Magnussen RA, Dahm DL, Neyret P. Osteotomy for slope correction following failed ACL reconstruction. In: Marx RG, editor. Revision ACL reconstruction: indications and technique. New York: Springer; 2014. p. 221–6. https://doi.org/10.1007/978-1-4614-0766-9.
55. Sonnery-Cottet B, Mogos S, Thaunat M, Archbold P, Fayard JM, Freychet B, Clechet J, Chambat P. Proximal tibial anterior closing wedge osteotomy in repeat revision of anterior cruciate ligament reconstruction. Am J Sports Med. 2014;42(8):1873–80.
56. Ganokroj P, Peebles AM, Mologne MS, Foster MJ, Provencher MT. Anterior closing-wedge high Tibial slope-correcting osteotomy using patient-specific preoperative planning software for failed anterior cruciate ligament reconstruction. Arthrosc Tech. 2022;11(11):e1989–95. https://doi.org/10.1016/j.eats.2022.07.015. PMID: 36457401; PMCID: PMC9705718.
57. Shekhar A, Tapasvi S, van Heerwaarden R. Anterior closing wedge osteotomy for failed anterior cruciate ligament reconstruction: state of the art. J Am Acad Orthop Surg Glob Res Rev. 2022;6(9):e22.00044. https://doi.org/10.5435/JAAOSGlobal-D-22-00044. PMID: 36121766; PMCID: PMC9484815
58. Dugdale TW, Noyes FR, Styer D. Preoperative planning for high tibial osteotomy. The effect of lateral tibiofemoral separation and tibiofemoral length. Clin Orthop Relat Res. 1992;274:248–64. PMID: 1729010.
59. Markolf KL, Burchfield DM, Shapiro MM, Shepard MF, Finerman GA, Slauterbeck JL. Combined knee loading states that generate high anterior cruciate ligament forces. J Orthop Res. 1995;13(6):930–5. https://doi.org/10.1002/jor.1100130618. PMID: 8544031.
60. Van de Pol GJ, Arnold MP, Verdonschot N, van Kampen A. Varus alignment leads to increased forces in the anterior cruciate ligament. Am J Sports Med. 2009;37(3):481–7. https://doi.org/10.1177/0363546508326715. Epub 2008 Dec 16. PMID: 19088054.
61. Sevim ÖF, Ergün S, Şahin Ediz S, Eceviz E, Karahan M. Comparison of side-to-side difference in posterior Tibial slope in knees with acute versus chronic anterior cruciate ligament deficiency. Orthop J Sports Med. 2024;12(5):23259671241247524. https://doi.org/10.1177/23259671241247524. PMID: 38726238; PMCID: PMC11080730.
62. Sonnery-Cottet B, Thaunat M, Freychet B, Pupim BH, Murphy CG, Claes S. Outcome of a combined anterior cruciate ligament and anterolateral ligament reconstruction technique with a minimum 2-year follow-up. Am J Sports Med. 2015;43(7):1598–605. https://doi.org/10.1177/0363546515571571. Epub 2015 Mar 4. PMID: 25740835.
63. Claes S, Bartholomeeusen S, Bellemans J. High prevalence of anterolateral ligament abnormalities in magnetic resonance images of anterior cruciate ligament-injured knees. Acta Orthop Belg. 2014;80(1):45–9. PMID: 24873084.
64. Gossner J. The anterolateral ligament of the knee - visibility on magnetic resonance imaging. Rev Bras Ortop. 2014;49(1):98–9. https://doi.org/10.1016/j.rboe.2014.01.004. PMID: 26229782; PMCID: PMC4511776.
65. Terry GC, Norwood LA, Hughston JC, Caldwell KM. How iliotibial tract injuries of the knee combine with acute anterior cruciate ligament tears to influence abnormal anterior tibial displacement. Am J Sports Med. 1993;21(1):55-60. https://doi.org/10.1177/036354659302100110. PMID: 8427369.
66. Claes S, Vereecke E, Maes M, Victor J, Verdonk P, Bellemans J. Anatomy of the anterolateral ligament of the knee. J Anat. 2013;223(4):321–8. https://doi.org/10.1111/joa.12087. Epub 2013 Aug 1. PMID: 23906341; PMCID: PMC3791125.

67. Borchers JR, Kaeding CC, Pedroza AD, Huston LJ, Spindler KP, Wright RW, MOON Consortium and the MARS Group. Intra-articular findings in primary and revision anterior cruciate ligament reconstruction surgery: a comparison of the MOON and MARS study groups. Am J Sports Med. 2011;39(9):1889–93. https://doi.org/10.1177/0363546511406871. Epub 2011 Jun 6. PMID: 21646434; PMCID: PMC3643140
68. Shoemaker SC, Markolf KL. The role of the meniscus in the anterior-posterior stability of the loaded anterior cruciate-deficient knee. Effects of partial versus total excision. J Bone Joint Surg Am. 1986;68(1):71–9. PMID: 3753605.
69. Lording T, Corbo G, Bryant D, Burkhart TA, Getgood A. Rotational laxity control by the anterolateral ligament and the lateral meniscus is dependent on knee flexion angle: a cadaveric biomechanical study. Clin Orthop Relat Res. 2017;475(10):2401–8. https://doi.org/10.1007/s11999-017-5364-z. Erratum in: Clin Orthop Relat Res. 2017 Oct;475(10):2613. PMID: 28536855; PMCID: PMC5599389.
70. Bollen SR. Posteromedial meniscocapsular injury associated with rupture of the anterior cruciate ligament: a previously unrecognised association. J Bone Joint Surg Br. 2010;92(2):222–3. https://doi.org/10.1302/0301-620X.92B2.22974. PMID: 20130312.
71. Liu X, Feng H, Zhang H, Hong L, Wang XS, Zhang J. Arthroscopic prevalence of ramp lesion in 868 patients with anterior cruciate ligament injury. Am J Sports Med. 2011;39(4):832–7. https://doi.org/10.1177/0363546510388933. Epub 2011 Jan 10. PMID: 21220541.
72. Robb C, Kempshall P, Getgood A, Standell H, Sprowson A, Thompson P, Spalding T. Meniscal integrity predicts laxity of anterior cruciate ligament reconstruction. Knee Surg Sports Traumatol Arthrosc. 2015;23(12):3683–90. https://doi.org/10.1007/s00167-014-3277-x. Epub 2014 Sep 13. PMID: 25217313.
73. Southam BR, Colosimo AJ, Grawe B. Underappreciated factors to consider in revision anterior cruciate ligament reconstruction: a current concepts review. Orthop J Sports Med. 2018;6(1):2325967117751689. https://doi.org/10.1177/2325967117751689. PMID: 29399591; PMCID: PMC5788104.
74. Svantesson E, Hamrin Senorski E, Alentorn-Geli E, Westin O, Sundemo D, Grassi A, Čustović S, Samuelsson K. Increased risk of ACL revision with non-surgical treatment of a concomitant medial collateral ligament injury: a study on 19,457 patients from the Swedish National Knee Ligament Registry. Knee Surg Sports Traumatol Arthrosc. 2019;27(8):2450–9. https://doi.org/10.1007/s00167-018-5237-3. Epub 2018 Oct 29. PMID: 30374568; PMCID: PMC6656795.
75. Pacheco RJ, Ayre CA, Bollen SR. Posterolateral corner injuries of the knee: a serious injury commonly missed. J Bone Joint Surg Br. 2011;93(2):194–7. https://doi.org/10.1302/0301-620X.93B2.25774. PMID: 21282758.
76. Duncan JB, Hunter R, Purnell M, Freeman J. Meniscal injuries associated with acute anterior cruciate ligament tears in alpine skiers. Am J Sports Med. 1995;23:170–2.
77. DeLee JC, Riley MB, Rockwood CA Jr. Acute posterolateral rotatory instability of the knee. Am J Sports Med. 1983;11:199–207.
78. Kannus P. Nonoperative treatment of grade II and III sprains of the lateral ligament compartment of the knee. Am J Sports Med. 1989;17:83–8.
79. Steadman JRBT. Principles of ACL revision surgery and rehabilitation. Sports Med Arthrosc Rev. 2005;13:53–8.
80. Fagelman M, Freedman KB. Revision reconstruction of the anterior cruciate ligament: evaluation and management. Am J Orthop. 2005;34:319–28.
81. Ellman MB. Chapter 1 rehabilitation of revision ACL reconstruction, 2011.

Part IV

Female ACL Injuries

Characteristics and Prevention of ACL Injuries in Females

20

Yiğit Umur Cırdı, Ali Eray Günay, and Umut Akgün

20.1 Characteristics of ACL Injuries in Females

Anterior cruciate ligament (ACL) injury is a prevalent and significant issue among athletes. The risk of ACL injury varies between 0.06 and 3.7 per 1000 hours of sports activity, including exercises [1–3]. After ACL injuries, athletes may not be able to return to their pre-injury level of performance, their sports career may be over, and they may encounter problems such as chronic pain and arthritis in the long term [4].

Female athletes are more likely to sustain ACL injuries, especially during competitive sports activities. The risk of ACL injury is 1.5 to 9 times higher than that of males of similar age groups [1–5]. This risk may vary according to the type of sports. Consistently, in a meta-analysis conducted by Prodromos et al. [6], the female/male injury risk ratio was found to be 1.18, 3.5, and 2.67 in the high-risk sports of lacrosse, basketball, and soccer, respectively, but this ratio was found to be 1.0 in professional alpine skiing. The diversity in injury risks across various sports indicates distinct factors that predispose women to ACL injuries. These reasons can be divided into two subgroups as unmodifiable (anatomical, genetic, etc.) and modifiable (behavioral) factors. Comprehending these attributes is crucial for devising efficient prevention tactics and reducing the incidence of such injuries.

20.1.1 Anatomical Features

Anatomical aspects that contribute to ACL injuries have been investigated widely. Due to the variances in anatomical features between genders, it is difficult to clarify the independent effects of these anatomical features. Although distinct sport-specific injury patterns were demonstrated via video analysis, the underlying cause is mostly the excessive valgus force generated secondary to anatomical features [1–3, 7]. Generally, the internal rotation of the trunk forces the knee towards valgus and causing ACL rupture [8, 9]. The anatomical factors causing the excessive valgus force and predisposing to ACL rupture should be identified and preventive strategies should be developed to minimize these forces [10, 11]. The anatomical factors are identified and detailed below.

Y. U. Cırdı · U. Akgün (✉)
ACIBADEM Mehmet Ali Aydınlar University, Department of Orthopedics and Traumatology, Ataşehir/Istanbul, Turkey

A. E. Günay
Kayseri City Education and Training Hospital, Department of Orthopedics and Traumatology, Kocasinan/Kayseri, Turkey

M. Karahan et al. (eds.), *Anterior Cruciate Ligament Injuries Throughout the Life Span*, https://doi.org/10.1007/978-3-031-84501-7_20

20.1.1.1 Quadriceps Angle

The quadriceps angle, known as the "Q-angle," is defined as the angle between the lines drawn parallel to the quadriceps muscle and the patellar tendon fibers (Fig. 20.1). While there is variation across different populations, generally, Q-angle ranges from 10 to 15 degrees in men and 14 to 20 degrees in women [12]. Elevated Q-angle levels have the potential to exert additional stress on the knee joint. Particularly in individuals with a Q-angle exceeding 19 degrees, there is a notable occurrence of excessive dynamic knee valgus and rotational moment during the post-jump landing phase, potentially elevating the susceptibility to ACL injuries [3, 4, 13]. Furthermore, higher hip internal rotation range and femoral anteversion in females contribute to an increase in valgus moment force as well [13].

20.1.1.2 Intercondylar Notch and Tibial Slope

The correlation between the morphology of the intercondylar notch (ICN) and ACL rupture has been well established over an extensive period. In 1988, Souryal and Freeman [14] defined the "notch width index (NWI)," which is delineated

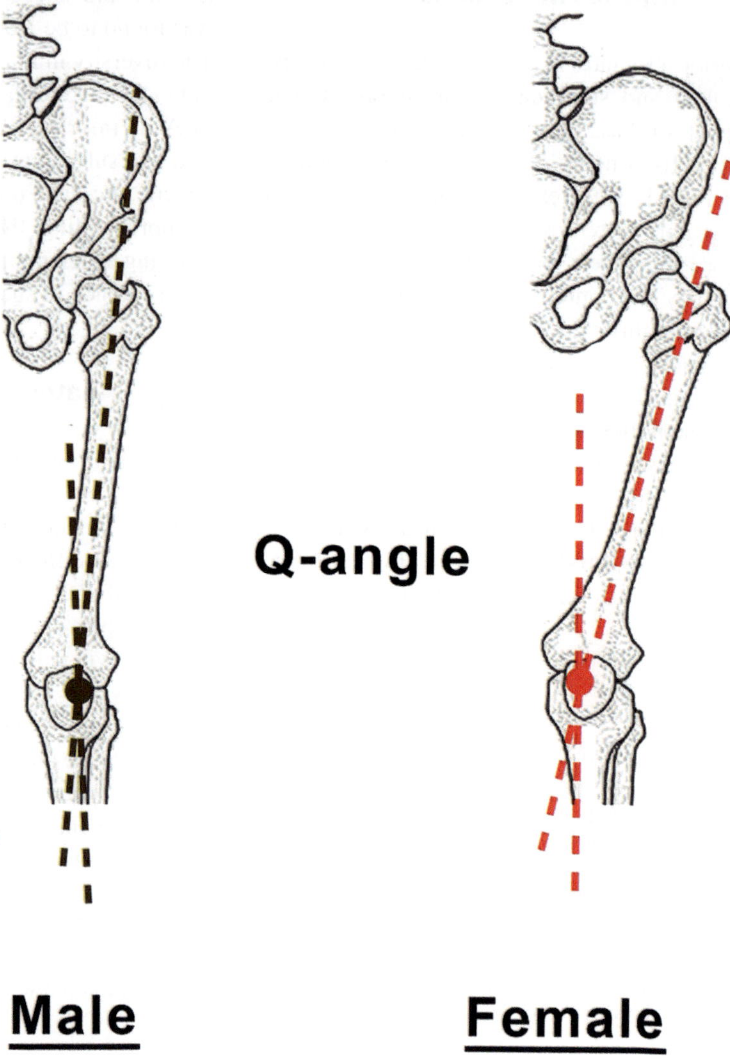

Fig. 20.1 The greater width of the pelvis in females results in higher Q angle. A higher Q angle results in the added loading of dynamic valgus-inducing forces on the knee

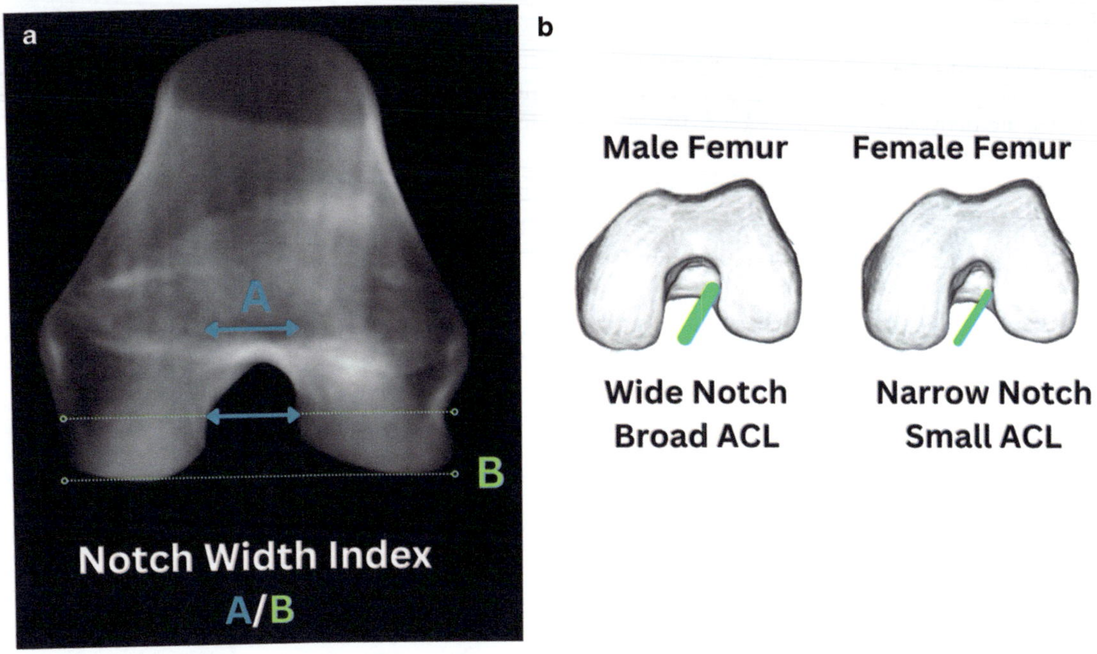

Fig. 20.2 Demonstration of notch width index. A: Distance of notch at the level of popliteal groove. B: Bony width at the level of popliteal groove. The male notch anatomy is broader, and the anterior cruciate ligament (ACL) is bigger; conversely, in females the notch is narrower, and the ACL is smaller

as the ratio of the width of the notch to the width of the distal femur at the level of popliteal groove. It was shown that narrow ICN and low NWI cause compression in the ACL and are also associated with smaller ACL volume [8] (Fig. 20.2), which is shown to be a potential risk factor for both ACL and PCL rupture [11, 15, 16]. In addition, the angle formed by the Blumensaat line and the femoral sagittal axis demonstrates the sagittal orientation of the ACL. Horizontally aligned ACLs have a higher risk of rupture; however, there is no definite difference between genders' ACL angular orientation [17] (Fig. 20.3).

The posterior tibial slope (PTS) is the angle between the tibial joint surface and the tibial axis; the normal value varies between 5 and 10 degrees, depending on the race and gender [18, 19]. The increase in the degree of the PTS transforms vertical forces into those that propel the distal femur posteriorly and the tibia anteriorly, which increase the likelihood of ACL graft failure by 5.7 times in individuals with a PTS value exceeding 10 degrees [19].

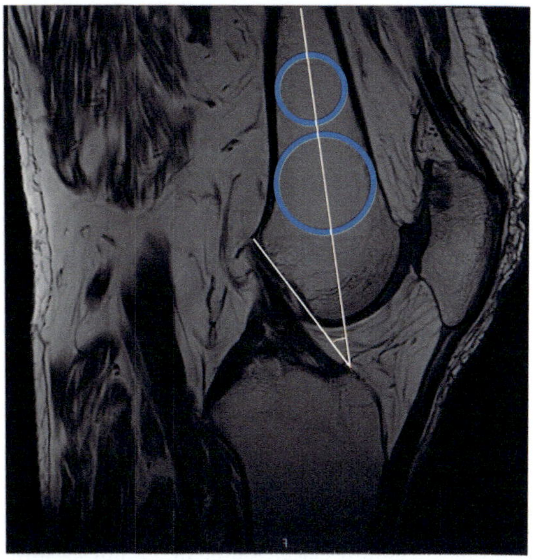

Fig. 20.3 Intercondylar Notch Angle (INA) is shown by drawing a line through the midline of the femur in the sagittal plane and using the Blumensaat line. A larger INA angle indicates that the anterior cruciate ligament is more horizontally oriented, which increases the risk of injury

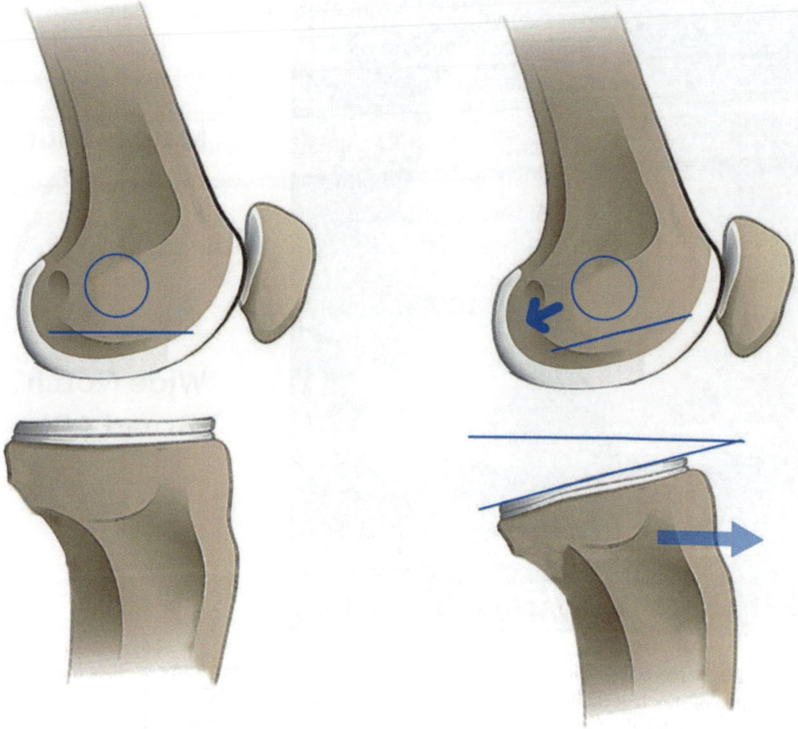

Fig. 20.4 An increase in the posterior tibial slope angle pushes the distal femur posteriorly, the tibia anteriorly, and increases the native ACL tension

Three different slopes can be measured for the medial, lateral, and middle column of the tibia. Gender disparities in the medial and lateral PTS might also be a potential predisposing factor for ACL injury. The PTS angle value of female tibias was found to be higher than that of male tibias [18–21]. Lipps et al. [22] compared the native tension of the ACL in male and female cadavers under equivalent loads, observing a higher tension in female cadavers, attributed to low NWI and high PTS (Fig. 20.4).

20.1.1.3 ACL Structural Features

The ACL is composed of two main bundles (anteromedial and posterolateral) that possess individual biomechanical roles along the range of motion. Its length varies between 27 and 38 mm, its width is 10–12 mm, and its cross-sectional area is between 30 and 53 mm^2. Histologically the ACL contains type I (90%) and type III (%10) collagen [23, 24].

Extensive research has been conducted to examine the correlation between the diameter of the ACL and the risk of injury. Shorter and thinner ACLs have been shown to generate less resilience against displacing forces [10, 11]. Since adolescence, the length, cross-sectional area, and volume of the ACL are smaller in women [23]. The disparity between genders is particularly noticeable in the distal one-third of the ACL [25]. In the same vein, Sturnick et al. [11] demonstrated that a 0.1 cm^3 decrease in the total volume of the ACL increases the risk of ACL rupture by 43%. Consistently, much of the available literature stated that smaller size of the ACL in women increases the risk of rupture [10, 11, 23, 24].

20.1.2 Biomechanical Factors

20.1.2.1 Movement Mechanics

According to the consensus announced in 1999 by the American Orthopaedic Society for Sports Medicine (AOSSM), National Athletic Trainers' Association (NATA), National Collegiate Athletic Association (NCAA), and Orthopaedic Research and Education Foundation (OREF), the risky movement phases for non-contact ACL rupture were reported as deceleration, stopping, change of direction, and landing [26].

In the video analyses, it was seen that the majority of the injuries were indirect-contact or non-contact traumas [8, 9]. The most common position at the moment of injury is the position where the trunk is turned toward the uninjured extremity, the hip is in abduction and internal rotation, and the knee is slightly flexed [8, 27].

A common feature in all phases is the dynamic valgus of the knee. Especially in the landing phase, women exhibit different landing mechanics compared to men. Women usually experience stiffer landing (less hip and knee flexion) [28, 29]. This landing pattern neutralizes the effect of the hamstrings and causes excessive stress on the ACL to provide translational stability of the knee.

20.1.2.2 Muscle Activation Patterns

The muscles surrounding the knee play a role in maintaining knee stability, but they can also exert compelling forces during certain maneuvers. The hamstring muscle group acts as an agonist for the ACL by preventing the forward translation of the tibia and hyperextension of the knee (Fig. 20.5).

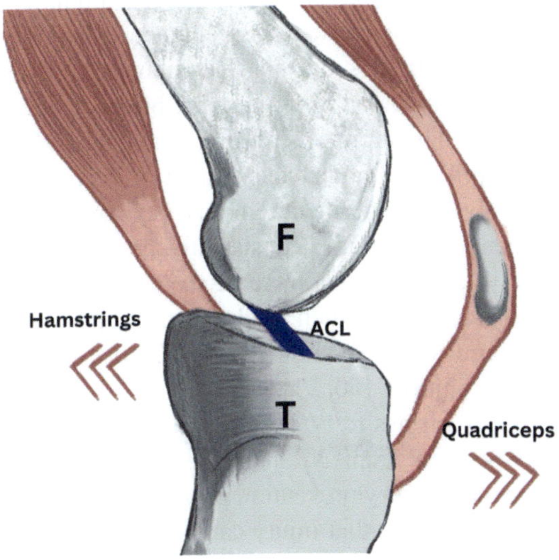

Fig. 20.5 Demonstration of the agonist and antagonist muscle effects on the anterior cruciate ligament (ACL). Please keep in mind that hamstring muscle contractions reduce the tension of the ACL, while quadriceps muscle contractions increase the tension

In contrast, quadriceps muscle contraction pulls the tibia forward and increases tension load on the ACL. The stabilizing musculature around the knee is weaker in women, and it takes longer time for the hamstrings to reach maximum contraction power [25, 30]. In addition, women tend to rely more on the quadriceps muscles rather than the hamstrings during athletic activities. This muscle activation pattern can overload the ACL, especially during sudden stops or directional changes [6, 10, 30] Nevertheless, the core stability generated by spinal and abdominal muscles is shown to be weaker in females, which exhibits significantly impaired knee biomechanics and strength and therefore responsible for increased ACL injury risk [25, 30, 31].

20.1.3 Hormonal, Genetic, and Molecular Factors

Female ligaments exhibit more laxity compared to males due to the hormonal status. The blood levels of sex hormones vary with menstrual cycle in females. High levels of estrogen and relaxin hormones increase the laxity of the ligaments. The ACL fibers contain receptors for estrogen, progesterone, and relaxin [31–33]. Consistently, recent studies demonstrated that the ACL injury prevalence is significantly higher during the pre-ovulatory and ovulatory periods when estrogen levels are at their peak levels [33, 34].

Furthermore, genetic variations are regarded as an additional predisposing factor for ACL injury [35]. The risk of injury is more than doubled in individuals with a family history [36, 37]. This situation is related to the genetic transmission of the anatomical structures that increases the risk, as well as the molecular structure of the ACL. Genetic background effecting collagen metabolism shows significant impact on the durability of the fiber bundles. Genomic variations affecting collagen synthesis at COL1A1, COL3A1, COL5A1, and COL12A1 genes have been demonstrated to be linked to widespread joint laxity. This is because ACL fibers consist of a significant number of helical collagen fibers, which increase the vulnerability to ACL injury [38, 39].

20.2 Current Prevention Strategies

Contemporary prevention strategies aim to augment landing biomechanics and enhance neuromuscular control around the knee. Three major components of the preventing programs include plyometrics, neuromuscular training, and strength training [40]. There is no single best exercise to fully prevent ACL injury. Therefore, optimal combination of injury preventive exercises is still under investigation.

Zazulak et al. showed that decreased neuromuscular control of the body core is related to increased risk of knee injury, postulating that the core strengthening may help to prevent knee injuries [41]. Koga et al. demonstrated that ACL ruptures mostly occur in the first 40 milliseconds after the initial ground contact, highlighting the importance of landing biomechanics in mechanisms of injury [42]. Therefore, research focused on enhanced landing alignment and core strengthening in high-risk athletes.

The initial step of the prevention is to be aware of the risk factors and successfully address the individuals who are at a higher risk. Following the identification of the cohort, sport-specific injury programs are developed. These injury programs aim to educate and train individuals on proper techniques and movements to reduce the likelihood of injury. Constant modifications should be implemented depending on the data gathered from the field. The ultimate goal is to prevent premature career endings and loss of sport participation.

The first publications on injury prevention appeared in the 1980s and there were no further articles published focused on injury prevention for a long time [43]. Interest in this subject has increased as injury prevention programs have been shown to reduce injury rates and prevent loss of time. Engebretsen et al. pioneered the use of prevention exercises on the professional field [44]. Then, Heidt et al. demonstrated an approximately 20% reduction in ACL injury events among soccer players with a specific preseason injury prevention program; however, the reduction rate was not significant when compared to the control group [45]. Subsequently, the Federation Internationale de Football Association (FIFA) initiated a program for the prevention of injuries, which was referred to as the "FIFA 11+" program. The objective of this program was to enhance the strength of the players and decrease the number of injuries that were sustained. Exercises that have been shown to be effective in preventing injuries were included in the program [46, 47]. These exercises focused on core stability, balance, dynamic stabilization, and eccentric hamstring strength. As a result of the introduction of the injury prevention program in Switzerland, the frequency of injuries sustained during matches decreased by 11.5%, and the incidence of injuries sustained while training decreased by 25.3% [48]. This observation was made in comparison to other teams that do not engage in regular injury prevention exercises. Additionally, prophylactic prevention efforts were shown to increase as the duration and consistency of the exercises increased [49].

When designing an injury prevention program, it is necessary to take into consideration a variety of perspectives, including biomechanical, physiological, socioeconomic, psychological, and ease of implementation [28]. The improper landing and lower extremity valgus alignment was the leading cause of non-contact ACL injury in female athletes and many components of the programs focus on improving the biomechanical balance. Therefore, increasing maximum knee and hip flexion angles and ultimately decreasing knee abduction moment remain a keystone of the prevention programs. Consistently, programs that involve exercises that increase the peak knee and hip flexion forces such as plyometrics and jump-landing tasks were shown to be effective in preventing injuries [50, 51].

Lastly, it has to be kept in mind that a coach and personal trainer remain the keystones of a successful injury prevention program. The coach has to be aware that injury can be prevented and motivate the players to perform preventive workout drills on a regular basis [52]. After all, players should be convinced to invest in prevention programs. The prevention rate was shown to increase as the players consistency improved [53].

Furthermore, the application of the injury prevention protocols was shown to be cost-effective and resulted in a cost reduction of $100 compared to the absence of training, and a $25 cost reduction compared to the screening and training approach for each individual case [54].

Current preferred preventive methods are as follows.

20.2.1 Neuromuscular Training

The females showed higher valgus force concomitant with internal rotation and therefore are at a higher risk for ACL injury compared to males. It is believed that the injury happens as a consequence of insufficient contraction of the muscles during the incident, which results in the muscle groups being unable to adequately resist rotational forces. In contrast to anatomical disparities, which cannot be altered, the harmony of muscle groups may be augmented, and their capabilities can be improved. Ultimately, the concept of training the peri-genicular muscles to enhance knee stability is proposed, which is expected to result in a reduction in the relative injury incidence among female athletes. This exercise package includes balance exercises, proprioceptive exercises, and stability exercises [55].

20.2.2 Plyometric Exercises

A plyometric workout executes exercises that are comparable to jumping with one's own bodyweight. This is accomplished through the utilization of the active stretch-shortening muscle cycle. The goal of this cycle is to create a shortcut between the neural and musculotendinous units, which will have the effect of boosting the force activation time. Consequently, it enhances agility by stimulating the strength and quickness of the individual and supplies power to requirements of sport demand [56]. These workouts improve landing and jumping biomechanics as well as inducing muscle hypertrophy (especially in fast-twitch fibers) [57]. In 1996, the first study demonstrating the effectiveness of plyometric exercises was published and showed a 72% decrease in non-contact ACL injuries in female athletes [48]. Furthermore, Gilchrist et al. conducted a study to compare the effectiveness of plyometric exercise programs in reducing ACL injuries in female athletes and demonstrated a 41 and 70% decrease in ACL injuries in overall and non-contact athletes, respectively [58]. Recent literature showed that neuromuscular training programs that include plyometrics can lead to decreased rates of ACL injuries in athletes and can be implemented on warm-up programs of different levels of sport participation [51, 59, 60]. Lack of compliance, improper skill level adjustment, and total exposure time still remain a major drawback of the injury prevention programs [44, 61].

20.2.3 Bracing and Taping Techniques

The mechanism of ACL rupture is primarily based on anatomical risk factors. Nonetheless, prophylactic knee bracing is intended to shape non-modifiable risk factors. It aims to provide extra stability on knee biomechanics or to reduce the magnitude of forces that can sprain the ACL. Ewing et al. demonstrated that the utilization of bracing during double leg landing results in increased flexion and extension torques, as well as higher peak muscle forces. Therefore, it enhances stability against ACL damage via hamstring and vastus femoris muscles [62]. A recent systematic analysis demonstrated that prophylactic knee braces (PKB) provided additional stabilization in the coronal and transverse planes, but not in the sagittal plane. Furthermore, they reduced ACL load and strain without significantly affecting ground reaction force [63]. Interestingly, Rishiraj et al. highlighted the challenge of performing a study to precisely assess the level of protection in healthy athletes. This difficulty arises from concerns that the use of preventive braces may hinder athletic performance [64]. There is still no consensus on the use of PKB in female athletes. Consequently, there is huge variety in literature in terms of the type of brace used, level of activity, and injury risk of the sports performed.

In recent years, preventive taping has become extensively utilized in sports. Like PKBs, taping is employed to offer some reinforcement and/or stability to the knee joint during physical activities. It is mostly favored as a preventative measure because it has fewer limitations on performance compared to PKBs. Huang et al. showed that taping resulted in a reduced knee abduction angle, an increased knee external rotation angle, and a decreased knee extensor moment during landing, which align the lower extremity with the desired natural landing mechanics [65]. Sheikhi et al. demonstrated that knee abduction and knee valgus during various landing activities were improved when taping was used for a duration of 72 hours [66]. However, there is scarce amount of evidence that evaluates the effectiveness of prophylactic taping especially in female athletes. This lack of research makes it difficult to draw definitive conclusions about the overall effectiveness of knee taping in preventing ACL injuries in female athletes.

20.3 Effectiveness of Prevention Strategies

PEP, Sportsmetrics, and Knäkontroll injury prevention programs have been shown to be effective in reducing ACL injury risk in women by various meta-analysis in the current literature [67–69]. Walden et al. demonstrated a 64% drop in ACL injury rate in female soccer players with 15 minutes of a neuromuscular warm-up program [70]. Myklebust et al. implemented neuromuscular control and planting/landing activities in female handball players for three consecutive seasons. Their study demonstrated a noteworthy decrease in the rate of ACL injuries in teams that followed the drills, as compared to those that did not [44]. Recently, three meta-analyses indicated that, in women, exercise-based injury prevention programs were effective in reducing the risk of all ACL injuries, with pooled odds ratios ranging from 0.40 to 0.64 [28]. Consequently, the most effective subgroup of exercises within neuromuscular training programs remains uncertain; however, programs that incorporate plyometric exercises have demonstrated overall effectiveness in reducing ACL injuries. Recent clinical practice guidelines from the Academy of Orthopedic Physical Therapy and the American Academy of Sports Physical Therapy offer recommendations for implementing neuromuscular training programs to prevent ACL injuries in female athletes [71].

20.4 Recommendations for ACL Injury Prevention in Females

- Neuromuscular training exercises should be implemented on the warm-up programs of athletes and should contain plyometric exercises.
- The effectiveness of injury prevention programs is closely correlated with the level of exposure and the athlete's compliance with the program. The application session should be conducted on a weekly basis, with multiple repetitions.
- Balance exercises are distinct from injury prevention regimens, although they are essentially a component of them. Make sure to use specialized training regimens tailored to the specific requirements of each sport.

20.5 Future Directions

Technologically oriented biomechanical measurements are still necessary to improve the understanding of ACL injuries in female athletes.

Sport-specific injury prevention programs should be improved and continuous evaluation must be encouraged based on the data obtained from high-level studies that are conducted.

References

1. Chia L, De Oliveira SD, Whalan M, McKay MJ, Sullivan J, Fuller CW, et al. Non-contact anterior cruciate ligament injury epidemiology in team-ball sports: a systematic review with meta-analysis by

sex, age, sport, participation level, and exposure type. Sports Med. 2022;52(10):2447–67.
2. Gornitzky AL, Lott A, Yellin JL, Fabricant PD, Lawrence JT, Ganley TJ. Sport-specific yearly risk and incidence of anterior cruciate ligament tears in high school athletes: a systematic review and meta-analysis. Am J Sports Med. 2016;44(10):2716–23.
3. Kaeding CC, Léger-St-Jean B, Magnussen RA. Epidemiology and diagnosis of anterior cruciate ligament injuries. Clin Sports Med. 2017;36(1):1–8.
4. Khan T, Alvand A, Prieto-Alhambra D, Culliford DJ, Judge A, Jackson WF, et al. ACL and meniscal injuries increase the risk of primary total knee replacement for osteoarthritis: a matched case-control study using the Clinical Practice Research Datalink (CPRD). Br J Sports Med. 2019;53(15):965–8.
5. Montalvo AM, Schneider DK, Silva PL, Yut L, Webster KE, Riley MA, et al. What's my risk of sustaining an ACL injury while playing football (soccer)?' A systematic review with meta-analysis. Br J Sports Med. 2019;53(21):1333–40.
6. Prodromos CC, Han Y, Rogowski J, Joyce B, Shi K. A meta-analysis of the incidence of anterior cruciate ligament tears as a function of gender, sport, and a knee injury-reduction regimen. Arthroscopy. 2007;23(12):1320–5.e6.
7. Kuntz A, Peters O, Bello A, Perkins R, Monti R, Murray L. An Anterior Cruciate Ligament (ACL) injury risk screening and reduction program for high school female athletes: a pilot study. Int J Sports Phys Ther. 2022;17(7):1318–29.
8. Boden BP, Torg JS, Knowles SB, Hewett TE. Video analysis of anterior cruciate ligament injury: abnormalities in hip and ankle kinematics. Am J Sports Med. 2009;37(2):252–9.
9. Della Villa F, Buckthorpe M, Grassi A, Nabiuzzi A, Tosarelli F, Zaffagnini S, et al. Systematic video analysis of ACL injuries in professional male football (soccer): injury mechanisms, situational patterns and biomechanics study on 134 consecutive cases. Br J Sports Med. 2020;54(23):1423–32.
10. Bayer S, Meredith SJ, Wilson KW, de Sa D, Pauyo T, Byrne K, et al. Knee morphological risk factors for anterior cruciate ligament injury: a systematic review. J Bone Joint Surg Am. 2020;102(8):703–18.
11. Sturnick DR, Vacek PM, DeSarno MJ, Gardner-Morse MG, Tourville TW, Slauterbeck JR, et al. Combined anatomic factors predicting risk of anterior cruciate ligament injury for males and females. Am J Sports Med. 2015;43(4):839–47.
12. Skouras AZ, Kanellopoulos AK, Stasi S, Triantafyllou A, Koulouvaris P, Papagiannis G, et al. Clinical significance of the static and dynamic Q-angle. Cureus. 2022;14(5):e24911.
13. Amraee D, Alizadeh MH, Minoonejhad H, Razi M, Amraee GH. Predictor factors for lower extremity malalignment and non-contact anterior cruciate ligament injuries in male athletes. Knee Surg Sports Traumatol Arthrosc. 2017;25(5):1625–31.
14. Souryal TO, Moore HA, Evans JP. Bilaterality in anterior cruciate ligament injuries: associated intercondylar notch stenosis. Am J Sports Med. 1988;16(5):449–54.
15. Sonnery-Cottet B, Archbold P, Cucurulo T, Fayard JM, Bortolletto J, Thaunat M, et al. The influence of the tibial slope and the size of the intercondylar notch on rupture of the anterior cruciate ligament. J Bone Joint Surg Br. 2011;93(11):1475–8.
16. van Kuijk KSR, Reijman M, Bierma-Zeinstra SMA, Meuffels DE. Smaller intercondylar notch size and smaller ACL volume increase posterior cruciate ligament rupture risk. Knee Surg Sports Traumatol Arthrosc. 2023;31(2):449–54.
17. Fernández-Jaén T, López-Alcorocho JM, Rodriguez-Iñigo E, Castellán F, Hernández JC, Guillén-García P. The importance of the Intercondylar Notch in anterior cruciate ligament tears. Orthop J Sports Med. 2015;3(8):2325967115597882.
18. Clinger BN, Plaster S, Passarelli T, Marshall J, Wascher DC. Differentiation in posterior Tibial slope by sex, age, and race: a cadaveric study utilizing 3-dimensional computerized tomography. Am J Sports Med. 2022;50(10):2698–704.
19. Fares A, Horteur C, Abou Al Ezz M, Hardy A, Rubens-Duval B, Karam K, et al. Posterior tibial slope (PTS) ≥ 10 degrees is a risk factor for further anterior cruciate ligament (ACL) injury; BMI is not. Eur J Orthop Surg Traumatol. 2023;33(5):2091–9.
20. Beynnon BD, Hall JS, Sturnick DR, Desarno MJ, Gardner-Morse M, Tourville TW, et al. Increased slope of the lateral tibial plateau subchondral bone is associated with greater risk of noncontact ACL injury in females but not in males: a prospective cohort study with a nested, matched case-control analysis. Am J Sports Med. 2014;42(5):1039–48.
21. Weinberg DS, Williamson DF, Gebhart JJ, Knapik DM, Voos JE. Differences in medial and lateral posterior Tibial slope: an osteological review of 1090 Tibiae comparing age, sex, and race. Am J Sports Med. 2017;45(1):106–13.
22. Lipps DB, Oh YK, Ashton-Miller JA, Wojtys EM. Morphologic characteristics help explain the gender difference in peak anterior cruciate ligament strain during a simulated pivot landing. Am J Sports Med. 2012;40(1):32–40.
23. Cone SG, Howe D, Fisher MB. Size and shape of the human anterior cruciate ligament and the impact of sex and skeletal growth: a systematic review. JBJS Rev. 2019;7(6):e8.
24. Morales-Avalos R, Torres-González EM, Padilla-Medina JR, Monllau JC. ACL anatomy: is there still something to learn? Rev Esp Cir Ortop Traumatol. 2024;68(4):422–7.
25. Menghini D, Kaushal SG, Flannery SW, Ecklund K, Murray MM, Fleming BC, et al. Three-dimensional magnetic resonance imaging analysis shows sex-specific patterns in changes in anterior cruciate ligament cross-sectional area along its length. J Orthop Res. 2023;41(4):771–8.

26. The female ACL. Why is it more prone to injury? J Orthop. 2016;13(2):A1–4.
27. Larwa J, Stoy C, Chafetz RS, Boniello M, Franklin C. Stiff landings, core stability, and dynamic knee valgus: a systematic review on documented anterior cruciate ligament ruptures in male and female athletes. Int J Environ Res Public Health. 2021;18(7):3826.
28. Arundale AJH, Silvers-Granelli HJ, Myklebust G. ACL injury prevention: where have we come from and where are we going? J Orthop Res. 2022;40(1):43–54.
29. Leppänen M, Pasanen K, Kujala UM, Vasankari T, Kannus P, Äyrämö S, et al. Stiff landings are associated with increased ACL injury risk in young female basketball and floorball players. Am J Sports Med. 2017;45(2):386–93.
30. Mancino F, Kayani B, Gabr A, Fontalis A, Plastow R, Haddad FS. Anterior cruciate ligament injuries in female athletes: risk factors and strategies for prevention. Bone Jt Open. 2024;5(2):94–100.
31. Marieswaran M, Jain I, Garg B, Sharma V, Kalyanasundaram D. A review on biomechanics of anterior cruciate ligament and materials for reconstruction. Appl Bionics Biomech. 2018;2018:4657824.
32. Bingzheng Z, Xinzhuo Z, Zhuo J, Xing Y, Bin L, Lunhao B. The effects of sex hormones during the menstrual cycle on knee kinematics. Front Bioeng Biotechnol. 2023;11:1209652.
33. Herzberg SD, Motu'apuaka ML, Lambert W, Fu R, Brady J, Guise JM. The effect of menstrual cycle and contraceptives on ACL injuries and laxity: a systematic review and meta-analysis. Orthop J Sports Med. 2017;5(7):2325967117718781.
34. Chidi-Ogbolu N, Baar K. Effect of estrogen on musculoskeletal performance and injury risk. Front Physiol. 2018;9:1834.
35. Magnusson K, Turkiewicz A, Hughes V, Frobell R, Englund M. High genetic contribution to anterior cruciate ligament rupture: Heritability ~69. Br J Sports Med. 2020;55:385.
36. Flynn RK, Pedersen CL, Birmingham TB, Kirkley A, Jackowski D, Fowler PJ. The familial predisposition toward tearing the anterior cruciate ligament: a case control study. Am J Sports Med. 2005;33(1):23–8.
37. Hasani S, Feller JA, Webster KE. Familial predisposition to anterior cruciate ligament injury: a systematic review with meta-analysis. Sports Med. 2022;52(11):2657–68.
38. Astur DC, Novaretti JV, Cohen M. Genetic and molecular factors and anterior cruciate ligament injuries: current concepts. J ISAKOS. 2017;2(3):123–6.
39. Bell RD, Shultz SJ, Wideman L, Henrich VC. Collagen gene variants previously associated with anterior cruciate ligament injury risk are also associated with joint laxity. Sports Health. 2012;4(4):312–8.
40. Nessler T, Denney L, Sampley J. ACL injury prevention: what does research tell us? Curr Rev Musculoskelet Med. 2017;10(3):281–8.
41. Zazulak BT, Hewett TE, Reeves NP, Goldberg B, Cholewicki J. Deficits in neuromuscular control of the trunk predict knee injury risk: a prospective biomechanical-epidemiologic study. Am J Sports Med. 2007;35(7):1123–30.
42. Koga H, Nakamae A, Shima Y, Iwasa J, Myklebust G, Engebretsen L, et al. Mechanisms for noncontact anterior cruciate ligament injuries: knee joint kinematics in 10 injury situations from female team handball and basketball. Am J Sports Med. 2010;38(11):2218–25.
43. Bizzini M, Junge A, Dvorak J. Implementation of the FIFA 11+ football warm up program: how to approach and convince the Football associations to invest in prevention. Br J Sports Med. 2013;47(12):803–6.
44. Myklebust G, Engebretsen L, Brækken IH, Skjølberg A, Olsen O-E, Bahr R. Prevention of anterior cruciate ligament injuries in female team handball players: a prospective intervention study over three seasons. Clin J Sport Med. 2003;13(2):71–8.
45. Heidt RS Jr, Sweeterman LM, Carlonas RL, Traub JA, Tekulve FX. Avoidance of soccer injuries with preseason conditioning. Am J Sports Med. 2000;28(5):659–62.
46. Mattu AT, Ghali B, Linton V, Zheng A, Pike I. Prevention of non-contact anterior cruciate ligament injuries among youth female athletes: an umbrella review. Int J Environ Res Public Health. 2022;19(8):4648.
47. Thompson JA, Tran AA, Gatewood CT, Shultz R, Silder A, Delp SL, et al. Biomechanical effects of an injury prevention program in preadolescent female Soccer athletes. Am J Sports Med. 2017;45(2):294–301.
48. Junge A, Lamprecht M, Stamm H, Hasler H, Bizzini M, Tschopp M, et al. Countrywide campaign to prevent Soccer injuries in Swiss Amateur players. Am J Sports Med. 2011;39(1):57–63.
49. Sugimoto D, Myer GD, Foss KD, Hewett TE. Dosage effects of neuromuscular training intervention to reduce anterior cruciate ligament injuries in female athletes: meta- and sub-group analyses. Sports Med. 2014;44(4):551–62.
50. Lopes TJA, Simic M, Myer GD, Ford KR, Hewett TE, Pappas E. The effects of injury prevention programs on the biomechanics of landing tasks: a systematic review with meta-analysis. Am J Sports Med. 2018;46(6):1492–9.
51. Al Attar WSA, Bakhsh JM, Khaledi EH, Ghulam H, Sanders RH. Injury prevention programs that include plyometric exercises reduce the incidence of anterior cruciate ligament injury: a systematic review of cluster randomised trials. J Physiother. 2022;68(4):255–61.
52. Ling DI, Boyle C, Schneider B, Janosky J, Kinderknecht J, Marx RG. Coach education improves adherence to anterior cruciate ligament injury prevention programs: a cluster-randomized controlled trial. Clin J Sport Med. 2022;32(4):348–54.
53. Silvers-Granelli H, Mandelbaum B, Adeniji O, Insler S, Bizzini M, Pohlig R, et al. Efficacy of the FIFA 11+ injury prevention program in the collegiate male Soccer player. Am J Sports Med. 2015;43(11):2628–37.

54. Swart E, Redler L, Fabricant PD, Mandelbaum BR, Ahmad CS, Wang YC. Prevention and screening programs for anterior cruciate ligament injuries in young athletes: a cost-effectiveness analysis. J Bone Joint Surg Am. 2014;96(9):705–11.
55. Söderman K, Werner S, Pietilä T, Engström B, Alfredson H. Balance board training: prevention of traumatic injuries of the lower extremities in female soccer players? A prospective randomized intervention study. Knee Surg Sports Traumatol Arthrosc. 2000;8(6):356–63.
56. Markovic G, Mikulic P. Neuro-musculoskeletal and performance adaptations to lower-extremity plyometric training. Sports Med. 2010;40(10):859–95.
57. Ramirez-Campillo R, García-Hermoso A, Moran J, Chaabene H, Negra Y, Scanlan AT. The effects of plyometric jump training on physical fitness attributes in basketball players: a meta-analysis. J Sport Health Sci. 2022;11(6):656–70.
58. Gilchrist J, Mandelbaum BR, Melancon H, Ryan GW, Silvers HJ, Griffin LY, et al. A randomized controlled trial to prevent noncontact anterior cruciate ligament injury in female collegiate Soccer players. Am J Sports Med. 2008;36(8):1476–83.
59. LaBella CR, Huxford MR, Grissom J, Kim K-Y, Peng J, Christoffel KK. Effect of neuromuscular warm-up on injuries in female Soccer and Basketball athletes in urban public high schools: cluster randomized controlled trial. Arch Pediatr Adolesc Med. 2011;165(11):1033–40.
60. Stevenson JH, Beattie CS, Schwartz JB, Busconi BD. Assessing the effectiveness of neuromuscular training programs in reducing the incidence of anterior cruciate ligament injuries in female athletes: a systematic review. Am J Sports Med. 2015;43(2):482–90.
61. Haddara R, Harandi VJ, Lee PVS. Effect of prophylactic knee bracing on anterior cruciate ligament agonist and antagonist muscle forces during perturbed walking. Orthop J Sports Med. 2021;9(2):2325967120981641.
62. Ewing KA, Fernandez JW, Begg RK, Galea MP, Lee PVS. Prophylactic knee bracing alters lower-limb muscle forces during a double-leg drop landing. J Biomech. 2016;49(14):3347–54.
63. Tuang BHH, Ng ZQ, Li JZ, Sirisena D. Biomechanical effects of prophylactic knee bracing on anterior cruciate ligament injury risk: a systematic review. Clin J Sport Med. 2023;33(1):78–89.
64. Rishiraj N, Taunton JE, Lloyd-Smith R, Woollard R, Regan W, Clement DB. The potential role of prophylactic/ functional knee bracing in preventing knee ligament injury. Sports Med. 2009;39(11):937–60.
65. Huang YL, Lin KW, Chou LW, Chang E. Immediate effect of anterior cruciate ligament protective knee taping on knee landing mechanics and muscle activations during side hops. Int J Environ Res Public Health. 2021;18(19):10110.
66. Sheikhi B, Letafatkar A, Hogg J, Naseri-Mobaraki E. The influence of kinesio taping on trunk and lower extremity motions during different landing tasks: implications for anterior cruciate ligament injury. J Exp Orthop. 2021;8(1):25.
67. Olsen O-E, Myklebust G, Engebretsen L, Holme I, Bahr R. Exercises to prevent lower limb injuries in youth sports: cluster randomised controlled trial. BMJ. 2005;330(7489):449.
68. Petersen W, Braun C, Bock W, Schmidt K, Weimann A, Drescher W, et al. A controlled prospective case control study of a prevention training program in female team handball players: the German experience. Arch Orthop Trauma Surg. 2005;125(9):614–21.
69. Arundale AJH, Bizzini M, Giordano A, Hewett TE, Logerstedt DS, Mandelbaum B, et al. Exercise-based knee and anterior cruciate ligament injury prevention. J Orthop Sports Phys Ther. 2018;48(9):A1–a42.
70. Waldén M, Atroshi I, Magnusson H, Wagner P, Hägglund M. Prevention of acute knee injuries in adolescent female football players: cluster randomised controlled trial. BMJ. 2012;344:e3042.
71. Arundale AJH, Bizzini M, Dix C, Giordano A, Kelly R, Logerstedt DS, et al. Exercise-based knee and anterior cruciate ligament injury prevention. J Orthop Sports Phys Ther. 2023;53(1):CPG1–CPG34.

Part V

High Level Athletes ACL Injuries

ACL Injuries in Female Athletes: The Ongoing Pandemic

21

Rafael Serrano, Celeste Dix, Jose Maria Oliva Lozano, Sanjeev Malik, Nate Jones, and George Chiampas

Abbreviations

ACL	Anterior cruciate ligament
ACLR	Anterior cruciate ligament reconstruction
BPTB	Bone patella tendon bone
IPPs	Injury prevention programs
OCPs	Oral contraceptive pills

21.1 Introduction

The anterior cruciate ligament (ACL) is the most common ligament injured in the knee. A large percentage of those injuries are a result of a non-contact mechanism [1, 2]. Athletes involved in pivoting and cutting sports are at increased risk of ACL injury [3–5]. Females experience a higher rate of ACL injuries compared to males [1–5]. The data show an increasingly alarming rate of ACL injuries in females worldwide [4].

Over the last decade, there has been an increased interest in understanding the risk for ACL injuries in female athletes. This chapter aims to review the risk factors that specifically affect females, analyze prevention programs, and discuss current surgical techniques and postoperative outcomes as well as psychological factors that may potentially affect outcomes.

R. Serrano (✉) · S. Malik · N. Jones · G. Chiampas
United States Soccer Federation, Chicago, IL, USA

C. Dix
Youth National Teams, United States Soccer Federation, Chicago, IL, USA

J. M. O. Lozano
Research and Innovation, United States Soccer Federation, Chicago, IL, USA

21.2 Specific Risk Factors for ACL Injuries in Females

Female sports participation has increased over the last decades [1]. As a result, the number of sports injuries in female athletes continues to rise. Data on ACL injuries show that ACL tears are three to six times more common in females [1].

Several reasons are behind the increased rate of ACL injuries in female athletes. Five large groups of risk factors have been identified; anatomic, hormonal, genetic, environmental, and biomechanical risk factors have been recognized as potential reasons that make females more prone to sustain an ACL injury [6].

21.2.1 Anatomy

The morphology of the knee in females differs slightly from males. Females have a narrower femoral intercondylar notch [7–9]. A stenotic femoral notch leaves less room for the ACL

M. Karahan et al. (eds.), *Anterior Cruciate Ligament Injuries Throughout the Life Span*,
https://doi.org/10.1007/978-3-031-84501-7_21

which increases the risk for impingement, placing excessive stress to the ligament and potentially increasing the risk for rupture [7–9].

The cross-sectional area of the ACL has been studied in great detail. On average, the female ACL is 17–39% smaller compared to males [10–12]. That might be a compensatory mechanism for a narrower notch. However, a thinner ligament will reach the yield point faster and rupture at a lower strain.

Tibial slope has been a topic of debate as a risk factor for ACL injury. It was proposed that females had steeper posterior tibial slopes than males [13]. Recent data have identified posterior tibial slope as a risk factor for ACL tears but have been inconclusive regarding whether females have steeper posterior tibial slopes compared to their male counterparts [14–16].

21.2.2 Biomechanical

Gender differences have been observed in neuromuscular activation. Female athletes spend significantly more time recruiting hamstring fibers allowing the quadriceps activation to take control. Quadriceps activation contributes to anterior tibial translation affecting knee stabilization [17].

The delay in neuromuscular activation along with poor landing biomechanics increases the risk for an ACL injury [17, 18]. Female athletes are at risk of landing in the so-termed "provocative position." The provocative position consists of landing in a single leg fashion, with the foot positioned in a heel strike or flat foot fashion during the initial contact, with minimal knee flexion and increased hip flexion combined with neutral or valgus moment of the knee [17–20]. Biomechanical analysis demonstrates that females are more prone to establish contact with the ground in the "provocative position." The provocative position leads to a multiplanar loading mechanism of the ACL. An anterior tibial force, along with internal tibial rotation and a knee abduction moment, translates into a force that reaches the ultimate stress capacity of the ACL prompting its rupture [20].

21.2.3 Genetics

It was suggested that certain genotypes involved in collagen expression were linked to an increased risk for ACL injuries especially in females [21]. Recent systematic reviews have not found a correlation [22–24]. However, certain ethnic groups may be at risk due to a genetic variant that regulates COL5A1. This variant increases the laxity of collagen fibers making those individuals more prone to ACL injuries due to increased knee laxity [25].

21.2.4 Hormones

Fluctuations in the hormone levels have shown to affect joint laxity [26]. Certain hormonal cycles or states can increase knee laxity ultimately increasing the risk for an ACL injury [27].

The influence of estradiol and progesterone on ACL laxity has been studied in the menstrual cycle. Both hormones are at their lowest in the first week of the menstrual cycle. Subsequently, estradiol starts to rise, reaching its peak at the time of ovulation. Conversely, progesterone rises late in the follicular phase just before the ovulation and reaches its highest level in the mid-luteal phase. The oscillations in hormone levels seem to affect ligament laxity [28]. Basic science research has shown that estradiol potentially reduces the synthesis of fibroblasts and collagen, which can potentially have a detrimental effect on tendon and ligament quality [28]. The effect is minimized by the addition of progesterone. Therefore, ACL injuries are more likely to happen during the follicular and ovulatory phases. Conversely, the luteal phase yields a lower rate of ACL injuries [28].

Oral contraceptive pills (OCPs) can regulate hormone fluctuations. Limited evidence shows that OCPs perhaps have the potential to decrease the rate of ACL injuries by avoiding fluctuations in estrogen and progesterone levels. The steady state of these hormones may limit knee laxity and decrease ACL rates by 20% [28–30].

Other hormones have been studied. Relaxin is a hormone produced by the placenta and ovary that regulates collagen expression. Relaxin may increase knee laxity. In a small cohort of college athletes, females with levels of Relaxin above 6.0 µg/mL had four times more risk of sustaining an ACL injury [31].

21.2.5 Environmental

National registry data on high school athletes show that female soccer players experience the highest rate of ACL injuries. In addition, weekend games yield the highest risk for ACL injuries compared to practice sessions [32].

For the last decade, synthetic turf has become a very popular surface in recreational and professional sports. The data does not show an increased risk for males. However, there is evidence that female soccer players are at a higher risk of experiencing an ACL injury when playing soccer on turf [33].

21.3 Prevention Programs

Given the high rates of ACL tears in athletes, injury prevention programs (IPPs) have been designed to lower the risk of ACL injuries [34]. IPPs can target a single component or multiple components of training such as agility, flexibility, and balance, among others. Multicomponent IPPs appear to be more effective in reducing injury rates than single component programs [35–38]. Multicomponent IPPs can help decrease contact and non-contact injuries. In addition, multicomponent IPPs improve lower extremity biomechanics and functional outcomes [34–38].

Multicomponent IPPs have a greater impact if they are implemented all year long including preseason, in-season, and off-season rather than specific parts of the season [39–41]. Additionally, multicomponent IPPs are more effective when they are performed two to three times a week at high intensity [41]. Moreover, data suggest that multicomponent IPPs can improve the overall performance of the team [42, 43].

Several meta-analyses have assessed the efficacy of multicomponent IPPs in female athletes. A meta-analysis of multicomponent IPPs in female athletes demonstrated that regular participation can reduce the risk of ACL injury by approximately 50 and 65% of the overall risk for non-contact injuries among female athletes [44]. Several multicomponent IPPs that have been studied showing positive results: The Prevent Injury and Enhance Performance program, HarmoKnee, and Knakontroll [42–44]. Additionally, the 11+ program, a general injury prevention program, has shown to reduce severe leg injuries in females by nearly half [45].

Although the data are promising, the continued high rates of ACL injuries in the literature as well as more casually in the news may not reflect the efficacy of multicomponent IPPs. A meta-analysis examining the incidence rate of ACL injuries in men and women demonstrated that despite the advent of multicomponent IPPs, the rate of ACL injuries among women soccer players has not decreased [46]. In a controlled research environment, multicomponent IPPs have been demonstrated to reduce injury rates. However, barriers including lack of time, lack of player engagement, and maintenance of the original multicomponent IPPs have limited successful real-world implementation [47]. Thus, current and future research should continue to address barriers to implementation and maintenance of multicomponent IPPs especially in female athletes.

21.4 Surgical Outcomes

Over the last decade, increased interest in anatomic ACL reconstruction (ACLR) has led surgeons to improve postoperative outcomes by restoring knee kinematics and lowering failure rates [48, 49]. Despite the efforts, females continue to experience worse postoperative outcomes, higher revision rates, and lower return to sports [50].

Female sex is among significant predictors for worse outcomes 10 years after ACLR. Females report lower International Knee Documentation Committee (IKDC), Knee injury and Osteoarthritis Outcome Score (KOOS), and Marx activity scale scores [51]. In addition, females experience more knee laxity and limb asymmetry following ACLR [52]. As a result, proper graft selection is paramount to minimize knee laxity. Moreover, postoperative rehabilitation is critical to optimize neuromuscular control and strength to decrease the risk of re-injury.

Graft selection has been a topic of debate since the inception of ACLR. Autografts are the preferred option [53, 54]. Hamstrings autografts are a very popular option across the globe [55]. However, hamstrings autografts yield a higher rate of instrumented knee laxity and revision compared to bone patella tendon bone grafts (BPTB) in females [56]. BPTB and quadriceps autografts have shown superiority in terms of knee laxity, failure rates, and return to sports [57, 58]. Conversely, BPTB yields a higher rate of anterior knee pain and quadriceps autografts have a higher rate of knee stiffness in female athletes especially when there is meniscus pathology associated [59, 60].

Optimal quadriceps strength is essential after ACLR [61]. Females experience significantly more quadriceps weakness following ACLR compared to males [62]. Muscle readiness may explain why females experience a higher risk for reinjury of the ipsilateral limb and a higher rate of contralateral ACL injury within 24 months of the surgery [63].

Knee laxity and muscle readiness may help understand altogether the reason why females have a lower rate of return to sport. Males return to sport on average at a rate 75 versus 65% in females [64].

Over the last decade, there has been a renewed interest in the anterolateral structures of the knee and their role in rotatory knee instability. Preliminary data show that lateral extra-articular tenodesis may decrease knee laxity and reduce ACL retear rates [65, 66]. Lateral extra-articular tenodesis could offer additional protection to female athletes, potentially improving their overall outcomes and return to sports [67].

Finally, there are some unanswered questions regarding ACL injuries in female athletes. Rehabilitation prior to surgery termed as prehab has been shown to help prior to ACLR [68]. However, there is no specific data on female athletes. Similarly, blood flow restriction is showing some early promising results [69]. Further research is necessary to learn whether blood flow restriction could improve quadriceps strengthening and muscle readiness in females.

21.5 Mental Health and ACL Injuries

Athletes who sustain an ACL tear typically face a season-ending injury. The mental and psychological stress of recovery plays a significant role in return to sport. Overall, 20–40% of athletes do not return to their pre-injury level [63]. The role of mental health and psychological readiness that contribute to an athlete's inability to return to sports is multifactorial and individual specific, noting that physical readiness and psychological readiness do not always coincide [70].

Multiple psychological factors contribute to an inability to return to pre-injury function and performance, including fear of reinjury, athlete expectation, feelings of loss of control, fear of letting team or coaches down, anxiety, and pressure to return to sport [71]. After an ACL injury, athletes may experience anger, depression, anxiety, decreased self-esteem, and, in more severe situations, post-traumatic stress disorder [72–74]. The severity of these symptoms is higher the first 6 weeks after ACLR. Additionally, elite athletes experience more severe symptoms compared to recreational athletes [75]. As patients progress through rehabilitation, they tend to experience fewer negative emotions and feel more positive about returning to sport [70].

Recognizing those athletes at risk of poor coping after injury and subsequent mental health struggles should be of utmost importance. This is true, not only postoperatively after ACLR but preoperatively as well. When looking at gender differences, women tend to exhibit greater anxiety and have more concern on how injury will affect their lives [76]. Women also demonstrate more emotionally oriented coping strategies with a greater propensity to report the severity of pain [77, 78]. In addition, when comparing female to

Table 21.1 Survey questionnaire to assess athlete's emotional vulnerability

(1) How have you been dealing with your injury/ recovery from an emotional standpoint?
(2) Do you feel sad or depressed about your injury/ recovery?
(3) Do you feel that you are able to control how well you will do with your rehabilitation?
(4) Do you think you have the mental and physical skills needed to get better?
(5) How are you feeling about yourself at this point in your recovery?
(6) Have you been able to stay involved with your team and teammates?

male soccer athletes, female soccer athletes return to sport at a lower rate than their male counterparts [64]. Identifying female athletes with low self-esteem, pessimism, anxiety, depression, unhealthy fear of reinjury, or those suffering from a threatened athletic identity can be one way to mitigate an athlete's inability to return to sport following ACLR [73].

There are no established protocols for identifying at-risk athletes, but questions, such as those in Table 21.1, could be a springboard for such screening or, at the minimum, initiate a conversation. In addition, validated depression and anxiety screens can be used. Tools such as QIDS (Quick Inventory of Depressive Symptomatology), PHQ2 (Patient Health Questionnaire), and HADS (Hospital Anxiety and Depression Scale) can prove invaluable and provide some objective data [77]. Once identified, treatment via a multidisciplinary approach can be initiated and should include an evaluation by a sports psychologist [79, 80].

There is a paucity in the medical literature regarding therapeutic interventions for the management of anxiety and depression following ACLR [80, 81]. Interventions such as positive self-talk or encouragement, guided imagery, relaxation techniques, goal-setting, and, most importantly, counseling could be part of a treatment strategy when dealing with patients recovering from ACLR who may be struggling with mental health [80–82]. In fact, such interventions are likely beneficial to all athletes after ACLR even those who do not appear to be struggling with their mental health. In addition to aiding in the physical recovery process, the athlete's mental health must be taken into account, as it has implications for overall self-worth and an individual's quality of life [77].

21.6 Conclusion

Female athletes are at an increased risk of sustaining an ACL injury compared to male counterparts. Several risk factors have been identified. Most of the risk factors are non-modifiable. However, synthetic turf may be a modifiable one. Synthetic turf yields a higher rate of ACL injuries when compared to other surfaces. Prevention programs have been designed to lower ACL injuries. Compliance to prevention programs throughout the season and off-season has shown to have a positive effect and lower injury rates. When facing an ACL tear, anatomic ACLR is paramount to restoring knee kinematics. ACLR with BPTB or quadriceps autograft could minimize knee laxity compared to other graft options. The addition of lateral extra-articular tenodesis may decrease failure rates. Supervised physical therapy is critical to restore quadriceps strength and muscle readiness to safely return to sports and prevent reinjury. Mental health has been recognized as an important topic with the potential to affect clinical outcomes. Recognizing athletes at risk is key to ensure they progress through the different stages of the recovery.

References

1. Gornitzky AL, Lott A, Yellin JL, Fabricant PD, Lawrence JT, Ganley TJ. Sport-specific yearly risk and incidence of anterior cruciate ligament tears in high school athletes: a systematic review and meta-analysis. Am J Sports Med. 2016;44(10):2716–23. https://doi.org/10.1177/0363546515617742. Epub 2015 Dec 11. PMID: 26657853.
2. Bram JT, Magee LC, Mehta NN, Patel NM, Ganley TJ. Anterior cruciate ligament injury incidence in adolescent athletes: a systematic review and meta-analysis. Am J Sports Med. 2021;49(7):1962–72. https://doi.org/10.1177/0363546520959619. Epub 2020 Oct 22. PMID: 33090889.

3. Paudel YR, Sommerfeldt M, Voaklander D. Increasing incidence of anterior cruciate ligament reconstruction: a 17-year population-based study. Knee Surg Sports Traumatol Arthrosc. 2023;31(1):248–55. https://doi.org/10.1007/s00167-022-07093-1. Epub 2022 Aug 10. PMID: 35948850.
4. Parsons JL, Coen SE, Bekker S. Anterior cruciate ligament injury: towards a gendered environmental approach. Br J Sports Med. 2021;55:984–90.
5. Hewett TE, Myer GD, Ford KR. Anterior cruciate ligament injuries in female athletes: part 1, mechanisms and risk factors. Am J Sports Med. 2006;34(2):299–311. https://doi.org/10.1177/0363546505284183. PMID: 16423913.
6. Griffin LY, Agel J, Albohm MJ, Arendt EA, Dick RW, Garrett WE, Garrick JG, Hewett TE, Huston L, Ireland ML, Johnson RJ, Kibler WB, Lephart S, Lewis JL, Lindenfeld TN, Mandelbaum BR, Marchak P, Teitz CC, Wojtys EM. Noncontact anterior cruciate ligament injuries: risk factors and prevention strategies. J Am Acad Orthop Surg. 2000;8(3):141–50. https://doi.org/10.5435/00124635-200005000-00001. PMID: 10874221.
7. Whitney DC, Sturnick DR, Vacek PM, DeSarno MJ, Gardner-Morse M, Tourville TW, Smith HC, Slauterbeck JR, Johnson RJ, Shultz SJ, Hashemi J, Beynnon BD. Relationship between the risk of suffering a first-time noncontact ACL injury and geometry of the femoral notch and ACL: a prospective cohort study with a nested case-control analysis. Am J Sports Med. 2014;42(8):1796–805. https://doi.org/10.1177/0363546514534182. Epub 2014 May 27. PMID: 24866891; PMCID: PMC6604056.
8. Bouras T, Fennema P, Burke S, Bosman H. Stenotic intercondylar notch type is correlated with anterior cruciate ligament injury in female patients using magnetic resonance imaging. Knee Surg Sports Traumatol Arthrosc. 2018;26(4):1252–7. https://doi.org/10.1007/s00167-017-4625-4. Epub 2017 Jun 23. PMID: 28646381.
9. Charlton WPH, John TAST, Ciccotti MG, Harrison N, Schweitzer M. Differences in femoral notch anatomy between men and women: a magnetic resonance imaging study. Am J Sports Med. 2002;30(3):329–33. https://doi.org/10.1177/03635465020300030501.
10. Cone SG, Howe D, Fisher MB. Size and shape of the human anterior cruciate ligament and the impact of sex and skeletal growth: a systematic review. JBJS Rev. 2019;7(6):e8. https://doi.org/10.2106/JBJS.RVW.18.00145. PMID: 31246862; PMCID: PMC7771555.
11. Chandrashekar N, Slauterbeck J, Hashemi J. Sex-based differences in the anthropometric characteristics of the anterior cruciate ligament and its relation to intercondylar notch geometry: a cadaveric study. Am J Sports Med. 2005;33(10):1492–8. https://doi.org/10.1177/0363546504274149. Epub 2005 Jul 11. PMID: 16009992.
12. Hashemi J, Mansouri H, Chandrashekar N, Slauterbeck JR, Hardy DM, Beynnon BD. Age, sex, body anthropometry, and ACL size predict the structural properties of the human anterior cruciate ligament. J Orthop Res. 2011;29(7):993–1001. https://doi.org/10.1002/jor.21245. Epub 2011 Jan 18. PMID: 21246609.
13. Lipps DB, Oh YK, Ashton-Miller JA, Wojtys EM. Morphologic characteristics help explain the gender difference in peak anterior cruciate ligament strain during a simulated pivot landing. Am J Sports Med. 2012;40(1):32–40. https://doi.org/10.1177/0363546511422325.
14. Aiwaiole A, Gurbani A, Motamedi K, et al. Relationship of ACL injury and posterior Tibial slope with patient age, sex, and race. Orthop J Sports Med. 2016;4(11) https://doi.org/10.1177/2325967116672852.
15. Hohmann E, Tetsworth K, Glatt V, Ngcelwane M, Keough N. Medial and lateral posterior Tibial slope are independent risk factors for noncontact ACL injury in both men and women. Orthop J Sports Med. 2021;9(8) https://doi.org/10.1177/23259671211015940.
16. Bayer S, Meredith SJ, Wilson KW, de Sa D, Pauyo T, Byrne K, McDonough CM, Musahl V. Knee morphological risk factors for anterior cruciate ligament injury: a systematic review. J Bone Joint Surg Am. 2020;102(8):703–18. https://doi.org/10.2106/JBJS.19.00535. Erratum in: J Bone Joint Surg Am. 2020;102(14):e85. PMID: 31977822.
17. Lucarno S, Zago M, Buckthorpe M, Grassi A, Tosarelli F, Smith R, Della VF. Systematic video analysis of anterior cruciate ligament injuries in professional female soccer players. Am J Sports Med. 2021;49(7):1794–802. https://doi.org/10.1177/03635465211008169. Epub 2021 May 14. PMID: 33989090.
18. Beaulieu ML, Ashton-Miller JA, Wojtys EM. Loading mechanisms of the anterior cruciate ligament. Sports Biomech. 2023;22(1):1–29. https://doi.org/10.1080/14763141.2021.1916578. Epub 2021 May 7. PMID: 33957846; PMCID: PMC9097243.
19. Levine JW, Kiapour AM, Quatman CE, Wordeman SC, Goel VK, Hewett TE, Demetropoulos CK. Clinically relevant injury patterns after an anterior cruciate ligament injury provide insight into injury mechanisms. Am J Sports Med. 2013;41(2):385–95. https://doi.org/10.1177/0363546512465167. Epub 2012 Nov 9. PMID: 23144366; PMCID: PMC3935824.
20. Bates NA, Schilaty ND, Nagelli CV, Krych AJ, Hewett TE. Multiplanar loading of the knee and its influence on anterior cruciate ligament and medial collateral ligament strain during simulated landings and noncontact tears. Am J Sports Med. 2019;47(8):1844–53. https://doi.org/10.1177/0363546519850165. Epub 2019 May 31. PMID: 31150273; PMCID: PMC6988507.
21. Posthumus M, September AV, O'Cuinneagain D, van der Merwe W, Schwellnus MP, Collins M. The COL5A1 gene is associated with increased risk of anterior cruciate ligament ruptures in female participants. Am J Sports Med. 2009;37(11):2234–40. https://doi.org/10.1177/0363546509338266.

22. Sun Z, Cięszczyk P, Humińska-Lisowska K, Michałowska-Sawczyn M, Yue S. Genetic determinants of the anterior cruciate ligament rupture in sport: an up-to-date systematic review. J Hum Kinet. 2023;87:105–17. https://doi.org/10.5114/jhk/163073. PMID: 37559763; PMCID: PMC10407318.
23. Perini JA, Lopes LR, Guimarães JAM, Goes RA, Pereira LFA, Pereira CG, Mandarino M, Villardi AM, de Sousa EB, Cossich VRA. Influence of type I collagen polymorphisms and risk of anterior cruciate ligament rupture in athletes: a case-control study. BMC Musculoskelet Disord. 2022;23(1):154. https://doi.org/10.1186/s12891-022-05105-2. PMID: 35172811; PMCID: PMC8848903.
24. Sivertsen EA, Haug KBF, Kristianslund EK, et al. No association between risk of anterior cruciate ligament rupture and selected candidate collagen gene variants in female elite athletes from high-risk team sports. Am J Sports Med. 2019;47(1):52–8. https://doi.org/10.1177/0363546518808467.
25. Alvarez-Romero J, Laguette MN, Seale K, Jacques M, Voisin S, Hiam D, Feller JA, Tirosh O, Miyamoto-Mikami E, Kumagai H, Kikuchi N, Kamiya N, Fuku N, Collins M, September AV, Eynon N. Genetic variants within the *COL5A1*gene are associated with ligament injuries in physically active populations from Australia, South Africa, and Japan. Eur J Sport Sci. 2023;23(2):284–93. https://doi.org/10.1080/17461391.2021.2011426. Epub 2021 Dec 30. PMID: 34821541.
26. Yu WD, Liu SH, Hatch JD, Panossian V, Finerman GA. Effect of estrogen on cellular metabolism of the human anterior cruciate ligament. Clin Orthop Relat Res. 1999;366:229–38. https://doi.org/10.1097/00003086-199909000-00030. PMID: 10627740.
27. Yu WD, Panossian V, Hatch JD, Liu SH, Finerman GA. Combined effects of estrogen and progesterone on the anterior cruciate ligament. Clin Orthop Relat Res. 2001;383:268–81. https://doi.org/10.1097/00003086-200102000-00031. PMID: 11210964.
28. Herzberg SD, Motu'apuaka ML, Lambert W, Fu R, Brady J, Guise JM. The effect of menstrual cycle and contraceptives on ACL injuries and laxity: a systematic review and meta-analysis. Orthop J Sports Med. 2017;5(7):2325967117718781. https://doi.org/10.1177/2325967117718781. PMID: 28795075; PMCID: PMC5524267.
29. Rahr-Wagner L, Thillemann TM, Mehnert F, Pedersen AB, Lind M. Is the use of oral contraceptives associated with operatively treated anterior cruciate ligament injury? A case-control study from the Danish Knee Ligament Reconstruction Registry. Am J Sports Med. 2014;42(12):2897–905. https://doi.org/10.1177/0363546514557240. PMID: 25428957.
30. Samuelson K, Balk EM, Sevetson EL, Fleming BC. Limited evidence suggests a protective association between oral contraceptive pill use and anterior cruciate ligament injuries in females: a systematic review. Sports Health. 2017;9(6):498–510. https://doi.org/10.1177/1941738117734164. Epub 2017 Oct 10. PMID: 29016234; PMCID: PMC5665118.
31. Dragoo JL, Castillo TN, Braun HJ, Ridley BA, Kennedy AC, Golish SR. Prospective correlation between serum relaxin concentration and anterior cruciate ligament tears among elite collegiate female athletes. Am J Sports Med. 2011;39(10):2175–80. https://doi.org/10.1177/0363546511413378. Epub 2011 Jul 7. PMID: 21737831.
32. Joseph AM, Collins CL, Henke NM, Yard EE, Fields SK, Comstock RD. A multisport epidemiologic comparison of anterior cruciate ligament injuries in high school athletics. J Athl Train. 2013;48(6):810–7. https://doi.org/10.4085/1062-6050-48.6.03. Epub 2013 Oct 23. PMID: 24143905; PMCID: PMC3867093.
33. Xiao M, Lemos JL, Hwang CE, Sherman SL, Safran MR, Abrams GD. Increased risk of ACL injury for female but not male soccer players on artificial turf versus natural grass: a systematic review and meta-analysis. Orthop J Sports Med. 2022;10(8) https://doi.org/10.1177/23259671221114353.
34. Crossley KM, Patterson BE, Culvenor AG, Bruder AM, Mosler AB, Mentiplay BF. Making football safer for women : a systematic review analysis of injury prevention programmes in 11 773 female football (soccer) players. Br J Sports Med. 2020;2:1–12. https://doi.org/10.1136/bjsports-2019-101587.
35. Sugimoto D, Myer GD, Foss KDB, Hewett TE. Specific exercise effects of preventive neuromuscular training intervention on anterior cruciate ligament injury risk reduction in young females: meta-analysis and subgroup analysis. Br J Sports Med. 2015;49(5):282–9. https://doi.org/10.1136/bjsports-2014-093461.
36. Sugimoto D, Myer GD, McKeon JM, Hewett TE. Evaluation of the effectiveness of neuromuscular training to reduce ACL injury in female athletes. Br J Sport Med. 2014;46(14):979–88. https://doi.org/10.1136/bjsports-2011-090895.
37. Myer GD, Sugimoto D, Thomas S, Hewett TE. The influence of age on the effectiveness of neuromuscular training to reduce anterior cruciate ligament injury in female athletes: a meta-analysis. Am J Sports Med. 2013;41(1):203–15. https://doi.org/10.1177/0363546512460637.
38. Waldén M, Atroshi I, Magnusson H, Wagner P, Hägglund M. Prevention of acute knee injuries in adolescent female football players: cluster randomised controlled trial. BMJ. 2012;344(7858):1–12. https://doi.org/10.1136/bmj.e3042.
39. Soligard T, Nilstad A, Steffen K, et al. Compliance with a comprehensive warm-up programme to prevent injuries in youth football. Br J Sports Med. 2010;44(11):787–93. https://doi.org/10.1136/bjsm.2009.070672.

40. Halvorsen KC, Marx RG, Wolfe I, et al. Higher adherence to anterior cruciate ligament injury prevention programs is associated with lower injury rates: a meta-analysis and, meta-gegression. HSS J. 2023;19(2):154–62. https://doi.org/10.1177/15563316221140860.
41. Donnell-Fink LA, Klara K, Collins JE, et al. Effectiveness of knee injury and anterior cruciate ligament tear prevention programs: a meta-analysis. PLoS One. 2015;10(12):1–17. https://doi.org/10.1371/journal.pone.0144063.
42. Hägglund M, Waldén M, Magnusson H, Kristenson K, Bengtsson H, Ekstrand J. Injuries affect team performance negatively in professional football: an 11-year follow-up of the UEFA Champions League injury study. Br J Sports Med. 2013;47(12):738–42. https://doi.org/10.1136/bjsports-2013-092215.
43. Silvers-granelli H. High compliance with the 11 + injury prevention program results in better win-loss records. Int J Sports Phys Ther. 2023;18(5):1065–75. https://doi.org/10.26603/001c.87502.
44. Gilchrist J, Mandelbaum BR, Melancon H, et al. A randomized controlled trial to prevent noncontact anterior cruciate ligament injury in female collegiate soccer players. Am J Sports Med. 2008;36(8):1476–83. https://doi.org/10.1177/0363546508318188.
45. Silvers-Granelli HJ, Bizzini M, Arundale A, Mandelbaum BR, Snyder-Mackler L. Does the FIFA 11+ injury prevention program reduce the incidence of ACL injury in male soccer players? Clin Orthop Relat Res. 2017;475(10):2447–55. https://doi.org/10.1007/s11999-017-5342-5.
46. Montalvo AM, Schneider DK, Silva PL, et al. "What's my risk of sustaining an ACL injury while playing football (soccer)?" a systematic review with meta-analysis. Br J Sports Med. 2019;53(21):1333–40. https://doi.org/10.1136/bjsports-2016-097261.
47. Lindblom H, Waldén M, Hägglund M. Adherence to injury prevention exercise programmes in amateur adolescent and adult football: a detailed description of programme use from a randomised study. Sport Med Open. 2023;9(1):57. https://doi.org/10.1186/s40798-023-00608-1.
48. Musahl V, Nazzal EM, Lucidi GA, Serrano R, Hughes JD, Margheritini F, Zaffagnini S, Fu FH, Karlsson J. Current trends in the anterior cruciate ligament part 1: biology and biomechanics. Knee Surg Sports Traumatol Arthrosc. 2022;30(1):20–33. https://doi.org/10.1007/s00167-021-06826-y. Epub 2021 Dec 20. PMID: 34927221.
49. Musahl V, Engler ID, Nazzal EM, Dalton JF, Lucidi GA, Hughes JD, Zaffagnini S, Della Villa F, Irrgang JJ, Fu FH, Karlsson J. Current trends in the anterior cruciate ligament part II: evaluation, surgical technique, prevention, and rehabilitation. Knee Surg Sports Traumatol Arthrosc. 2022;30(1):34–51. https://doi.org/10.1007/s00167-021-06825-z. Epub 2021 Dec 5. PMID: 34865182.
50. Tan SH, Lau BP, Khin LW, Lingaraj K. The importance of patient sex in the outcomes of anterior cruciate ligament reconstructions: a systematic review and meta-analysis. Am J Sports Med. 2016;44(1):242–54. https://doi.org/10.1177/0363546515573008. Epub 2015 Mar 23. PMID: 25802119.
51. MOON Knee Group, Spindler KP, Huston LJ, Chagin KM, Kattan MW, Reinke EK, Amendola A, Andrish JT, Brophy RH, Cox CL, Dunn WR, Flanigan DC, Jones MH, Kaeding CC, Magnussen RA, Marx RG, Matava MJ, McCarty EC, Parker RD, Pedroza AD, Vidal AF, Wolcott ML, Wolf BR, Wright RW. Ten-year outcomes and risk factors after anterior cruciate ligament reconstruction: a MOON longitudinal prospective cohort study. Am J Sports Med. 2018;46(4):815–25. https://doi.org/10.1177/0363546517749850. PMID: 29543512; PMCID: PMC6036619.
52. Jackson GR, Lee J, Tuthill T, Chan J, Sugrañes J, Mowers CC, Batra A, Khan ZA, Mameri ES, Brusalis CM, Chahla J, Verma NN. Higher rates of residual postoperative instability after anterior cruciate ligament reconstruction in female patients: a systematic review of level II studies. Arthrosc Sports Med Rehabil. 2023;5(5):100772.
53. Kaeding CC, Aros B, Pedroza A, Pifel E, Amendola A, Andrish JT, Dunn WR, Marx RG, McCarty EC, Parker RD, Wright RW, Spindler KP. Allograft versus autograft anterior cruciate ligament reconstruction: predictors of failure from a MOON prospective longitudinal cohort. Sports Health. 2011;3(1):73–81. https://doi.org/10.1177/1941738110386185. PMID: 23015994; PMCID: PMC3445196.
54. MOON Knee Group, Spindler KP, Huston LJ, Zajichek A, Reinke EK, Amendola A, Andrish JT, Brophy RH, Dunn WR, Flanigan DC, Jones MH, Kaeding CC, Marx RG, Matava MJ, McCarty EC, Parker RD, Vidal AF, Wolcott ML, Wolf BR, Wright RW. Anterior cruciate ligament reconstruction in high school and college-aged athletes: does autograft choice influence anterior cruciate ligament revision rates? Am J Sports Med. 2020;48(2):298–309. https://doi.org/10.1177/0363546519892991. Epub 2020 Jan 9. PMID: 31917613; PMCID: PMC7319140.
55. Arnold MP, Calcei JG, Vogel N, Magnussen RA, Clatworthy M, Spalding T, Campbell JD, Bergfeld JA, Sherman SL, ACL Study Group. ACL Study Group survey reveals the evolution of anterior cruciate ligament reconstruction graft choice over the past three decades. Knee Surg Sports Traumatol Arthrosc. 2021;29(11):3871–6. https://doi.org/10.1007/s00167-021-06443-9. Epub 2021 Jan 24. PMID: 33486558.
56. Cristiani R, Sarakatsianos V, Engström B, Samuelsson K, Forssblad M, Stålman A. Increased knee laxity with hamstring tendon autograft compared to patellar tendon autograft: a cohort study of 5462 patients with primary anterior cruciate ligament reconstruction. Knee Surg Sports Traumatol Arthrosc.

2019;27(2):381–8. https://doi.org/10.1007/s00167-018-5029-9. Epub 2018 Jun 28. PMID: 29955930; PMCID: PMC6394544.

57. DeFazio MW, Curry EJ, Gustin MJ, et al. Return to sport after ACL reconstruction with a BTB versus hamstring tendon autograft: a systematic review and meta-analysis. Orthop J Sports Med. 2020;8(12) https://doi.org/10.1177/2325967120964919.
58. Hurley ET, Mojica ES, Kanakamedala AC, Meislin RJ, Strauss EJ, Campbell KA, Alaia MJ. Quadriceps tendon has a lower re-rupture rate than hamstring tendon autograft for anterior cruciate ligament reconstruction - a meta-analysis. J ISAKOS. 2022;7(2):87–93. https://doi.org/10.1016/j.jisako.2021.10.001. Epub 2021 Nov 17. PMID: 35543668.
59. Salem HS, Varzhapetyan V, Patel N, Dodson CC, Tjoumakaris FP, Freedman KB. Anterior cruciate ligament reconstruction in young female athletes: patellar versus hamstring tendon autografts. Am J Sports Med. 2019;47(9):2086–92. https://doi.org/10.1177/0363546519854762. Epub 2019 Jun 24. PMID: 31233335.
60. Xerogeanes JW, Godfrey W, Gebrelul A, Premkumar A, Mignemi D, Gottschalk MB, Dalwadi P, Fellow R, Slone H. Clinical outcomes of all soft tissue quadriceps tendon autograft in ACL reconstruction. Orthop J Sports Med. 2017;5(7 suppl6):2325967117S00310. https://doi.org/10.1177/2325967117S00310. PMCID: PMC5542337.
61. Brown C, Marinko L, LaValley MP, Kumar D. Quadriceps strength after anterior cruciate ligament reconstruction compared with uninjured matched controls: a systematic review and meta-analysis. Orthop J Sports Med. 2021;9(4) https://doi.org/10.1177/2325967121991534.
62. Kuenze C, Pietrosimone B, Lisee C, Rutherford M, Birchmeier T, Lepley A, Hart JM. SEX DIFFERENCES IN QUADRICEPS STRENGTH AMONG ADOLESCENTS WITHIN 1 YEAR OF ACL RECONSTRUCTION. Orthop. J Sports Med. 2019;7(3 Suppl):2325967119S00089. https://doi.org/10.1177/2325967119S00089. PMCID: PMC6700908.
63. Paterno MV, Rauh MJ, Schmitt LC, Ford KR, Hewett TE. Incidence of second ACL injuries 2 years after primary ACL reconstruction and return to sport. Am J Sports Med. 2014;42(7):1567–73. https://doi.org/10.1177/0363546514530088. Epub 2014 Apr 21. PMID: 24753238; PMCID: PMC4205204.
64. Klemm HJ, Feller JA, Webster KE. Comparison of return-to-sports rates between male and female Australian athletes after ACL reconstruction. Orthop J Sports Med. 2023;11(6):23259671231169199. https://doi.org/10.1177/23259671231169199. PMID: 37347025; PMCID: PMC10280513
65. Marom N, Ouanezar H, Jahandar H, Zayyad ZA, Fraychineaud T, Hurwit D, Imhauser CW, Wickiewicz TL, Pearle AD, Nawabi DH. Lateral extra-articular Tenodesis reduces anterior cruciate ligament graft force and anterior Tibial translation in response to applied pivoting and anterior drawer loads. Am J Sports Med. 2020;48(13):3183–93. https://doi.org/10.1177/0363546520959322. Epub 2020 Oct 5. PMID: 33017168.
66. Lai S, Zhang Z, Li J, Fu WL. Comparison of anterior cruciate ligament reconstruction with versus without anterolateral augmentation: a systematic review and meta-analysis of randomized controlled trials. Orthop J Sports Med. 2023;11(3):23259671221149403. https://doi.org/10.1177/23259671221149403. PMID: 37025126; PMCID: PMC10071203.
67. Borque KA, Jones M, Laughlin MS, et al. Effect of lateral extra-articular Tenodesis on the rate of revision anterior cruciate ligament reconstruction in elite athletes. Am J Sports Med. 2022;50(13):3487–92. https://doi.org/10.1177/03635465221128828.
68. Cunha J, Solomon DJ. ACL Prehabilitation improves postoperative strength and motion and return to sport in athletes. Arthrosc Sports Med Rehabil. 2022;4(1):e65–9. https://doi.org/10.1016/j.asmr.2021.11.001. PMID: 35141537; PMCID: PMC8811524.
69. Charles D, White R, Reyes C, Palmer D. A systematic review of the effects of blood flow restriction training on quadriceps muscle atrophy and circumference post ACL reconstruction. Int J Sports Phys Ther. 2020;15(6):882–91. https://doi.org/10.26603/ijspt20200882. PMID: 33344004; PMCID: PMC7727417.
70. Langford JL, Webster KE, Feller JA. A prospective longitudinal study to assess psychological changes following anterior cruciate ligament reconstruction surgery. Br J Sports Med. 2009;43(5):377–81. https://doi.org/10.1136/bjsm.2007.044818. Epub 2008 Nov 19. PMID: 19019910.
71. Podlog L, Dimmock J, Miller J. A review of return to sport concerns following injury rehabilitation: practitioner strategies for enhancing recovery outcomes. Phys Ther Sport. 2011;12(1):36–42.
72. Brewer BW, Cornelius AE, Stephan Y, Van Raalte J. Self-protective changes in athletic identity following anterior cruciate ligament reconstruction. Psychol Sport Exerc. 2010;11(1):1–5.
73. Morrey MA, Stuart MJ, Smith AM, Wiese- Bjornstal DM. A longitudinal examination of athletes 'emotional and cognitive responses to anterior cruciate ligament injury. Clin J Sport Med. 1999;9(2):63–9.
74. Newcomer RR, Perna FM. Features of posttraumatic distress among adolescent athletes. J Athl Train. 2003;38(2):163–6.
75. Piussi R, Berghdal T, Sundemo D, Grassi A, Zaffagnini S, Sansone M, Samuelsson K, Senorski EH. Self-reported symptoms of depression and anxiety after ACL injury: a systematic review. Orthop J Sports Med. 2022;10(1):23259671211066493. https://doi.org/10.1177/23259671211066493. PMID: 35071657; PMCID: PMC8777351.

76. Prodromos CC, Han Y, Rogowski J, Joyce B, Shi K. A meta-analysis of the incidence of anterior cruciate ligament tears as a function of gender, sport, and a knee injury-reduction regimen. Arthroscopy. 2007;23(12):1320–1325.e6.
77. Sims M, Mulcahey MK. Sex-specific differences in psychological response to injury and return to sport following ACL reconstruction. JBJS Rev. 2018;6(7):e9. https://doi.org/10.2106/JBJS.RVW.17.00170. PMID: 30063501.
78. Udry E, Donald Shelbourne K, Gray T. Psychological readiness for anterior cruciate ligament surgery: describing and comparing the adolescent and adult experience. J Athl Train. 2003;38(2):167–71.
79. Ashton ML, Kraeutler MJ, Brown SM, Mulcahey MK. Psychological readiness to return to sport following anterior cruciate ligament reconstruction. JBJS Rev. 2020;8(3):e0110. https://doi.org/10.2106/JBJS.RVW.19.00110.
80. Christino MA, Fantry AJ, Vopat BG. Psychological aspects of recovery following anterior cruciate ligament reconstruction. J Am Acad Orthop Surg. 2015;23(8):501–9. https://doi.org/10.5435/JAAOS-D-14-00173. PMID: 26209145.
81. Myers CA, Peyton DD, Jensen BJ. Treatment acceptability in NCAA division I football athletes: rehabilitation intervention strategies. J Sport Behav. 2004;27:205–17.
82. Webster KE, Feller JA, Lambros C. Development and preliminary validation of a scale to measure the psychological impact of returning to sport following anterior cruciate ligament reconstruction surgery. Phys Ther Sport. 2008;9(1):9–15. https://doi.org/10.1016/j.ptsp.2007.09.003. Epub 2007 Nov 5. PMID: 19083699.

Management of ACL Injury for Different Sports Types

22

Woo Chin Yee, Benedict Tan,
and Yee Han Dave Lee

22.1 Introduction

Knee injuries occur at a higher rate in professional athletes, with the anterior cruciate ligament (ACL) as the most frequently injured knee ligament in both contact and non-contact sports [1–3]. Such injuries jeopardise an athlete's career and performance, with some unable to return to pre-injury participation levels [4].

22.2 Epidemiology of Sports-Related ACL Injuries

The registry data from Scandinavia, the United States of America (USA) and the United Kingdom (UK) revealed that soccer was the most commonly reported causative activity at the time of ACL injury, followed by winter sports and other sporting activities (e.g. handball, rugby, basketball, netball) [5] (Table 22.1). This is aligned with the popular sports within the countries surveyed. In all cohorts, males were more frequently observed, and the most common age group at the time of ACL reconstruction surgery was 15–19 years in all countries except the UK where 25–29 years was more common [5].

A meta-analysis of the incidence of ACL injuries also reported the following sports in descending order based on the number of exposures for each sport from studies available: basketball, soccer, alpine skiing, lacrosse, American football, handball, rugby, volleyball and wrestling [6].

In professional soccer players, a high-level men's team can anticipate 0.4 ACL injuries per season and women's team 0.7 ACL injuries per season [7]. The rates of ACL injuries differed in player positions, in the following descending order: midfielders, defenders, strikers and goalkeepers [8, 9]. Midfielders had the most ACL injuries—43% of ACL injuries were in midfielders and most of the injuries also occurred in the first 30 minutes of the game [10]. More ACL injuries in soccer occurred on natural grass as opposed to AstroTurf [11].

Among competitive skiers, one can expect 8.5 ACL injuries per 100 skier-seasons [12]. In basketball, one can expect one ACL injury for each regular season [13]. The incidence of ACL injury for male and female handball athletes are 0.24 and 0.86 per 1000 player game hours, respectively [14]. For ice hockey, a reported

W. C. Yee
Department of Orthopaedic Surgery, KK Women's and Children's Hospital, Singapore, Singapore
e-mail: woo.chin.yee@singhealth.com.sg

B. Tan
Singhealth Duke-NUS Sport & Exercise Medicine Centre, Singapore, Singapore
e-mail: benedict.tan.c.l@singhealth.com.sg

Y. H. D. Lee (✉)
Division of Sports Shoulder & Elbow Surgery, National University Hospital Singapore, Singapore, Singapore
e-mail: dave_yh_lee@nuhs.edu.sg

M. Karahan et al. (eds.), *Anterior Cruciate Ligament Injuries Throughout the Life Span*,
https://doi.org/10.1007/978-3-031-84501-7_22

Table 22.1 Common causative sports in ACL injuries

Denmark, Norway	Sweden	UK	USA
Soccer	Soccer	Soccer	Soccer
Winter sports	Winter sports	Winter sports	Basketball, Netball, American football
Handball	Floorball	Rugby	Winter sports

UK United Kingdom, *USA* United States of America

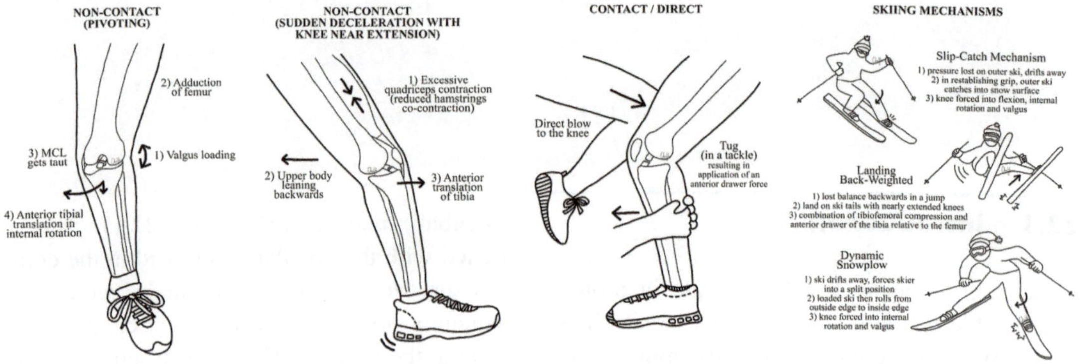

Fig. 22.1 Mechanisms of ACL injury in sports; ACL anterior cruciate ligament

incidence rate over a 10-year period for hockey league players was 0.42 per 1000 player game hours [15].

Female athletes have a greater susceptibility and are at four to six times greater risk of suffering an ACL injury from jumping and pivoting compared to their male counterparts [7, 14, 16, 17]. Studies have shown that apart from anatomical and hormonal risk factors which are non-modifiable, neuromuscular risk factors which are modifiable also predispose female athletes to ACL injuries [18]. Female athletes often land with a higher knee abduction moment, that is, valgus torque [19]. Prevention programmes and neuromuscular training can help to reduce the risk of ACL injuries in female athletes [18].

In a study of US college athletes [1], the authors revealed that the highest rates of recurrent ACL injuries occurred in male American football players. This was followed by female gymnasts and female soccer players. Among all sports, men had a significantly higher rate of recurrent ACL injuries, likely due to increased exposure. However, women soccer players had a significantly higher rate of recurrent ACL injuries than male soccer players.

22.3 Differences in ACL Injuries in Each Sport

Non-contact mechanisms (e.g. cutting, pivoting, jumping) account for the majority of ACL injuries [7, 9, 10, 17, 20–22], while contact mechanisms (e.g. direct blow or tackle) are less common. Non-contact injuries happen during deceleration and acceleration movements with excessive quadriceps contraction and reduced hamstrings co-contraction at or near full knee extension (applying an anterior drawer force). ACL loading is also increased when there is a combined knee valgus load with internal rotation (Fig. 22.1).

Additionally, use of equipment (e.g. in skiing) contributes to injury via unique mechanisms [23] (Fig. 22.1). The *slip-catch mechanism* occurs when the skier turns and loses his balance backward, causing the outer ski to drift away from the body, then catching the snow surface and forcing the extended knee into flexion, internal rotation and valgus. The *landing back-weighted mechanism* occurs when a skier lands from a jump out of balance backward, generating a combination of tibiofemoral compression and anterior drawer of the tibia with respect to the femur as the athlete lands on the ski tails. The *dynamic snowplow mechanism* occurs when the skier loses balance backward with more weight on one ski than the other, causing the unweighted ski to drift away and forcing the skier into a split position; when the skier attempts to correct his position, the weighted ski then engages the snow surface and forces the knee into internal rotation and/or valgus.

Skiing and American football injuries were more likely than soccer to involve the posterior cruciate ligament (PCL), medial collateral ligament (MCL) and/or multi-ligamentous injuries than isolated ACL injuries [3]. Basketball injuries had a higher likelihood for cartilage and lateral meniscus injuries compared to soccer injuries [3].

22.4 Management of ACL Injuries in Each Sport

The goals of management for ACL injuries include restoring knee function, preventing further injury, reducing the risk of osteoarthritis and enabling the athlete to resume activity participation. This is often achieved with an ACL reconstruction. A single-incision, arthroscopically assisted, single-bundle ACL reconstruction was the most common technique utilised [24].

22.4.1 The Evidence for Non-operative Management in an Athlete

Not all athletes who tear their ACL should undergo surgery. There is research to suggest that it is possible for an individual to return to sports following an ACL injury without surgery [25, 26]. However, there is currently a paucity of high-quality trials that compare rehabilitation alone versus ACL reconstruction in the management of ACL injuries [27]. There is only one randomised controlled trial in the literature—the KANON trial [28].

The KANON trial compared outcomes following ACL injury with rehabilitation plus optional delayed ACL reconstruction versus ACL reconstruction and post-operative rehabilitation [28]. The two groups had similar 2-year and 5-year self-reported physical activity levels, rates of meniscus surgery, reported symptoms, pain, range of motion and radiographic changes. The authors reported that there were no differences in the Knee Injury and Osteoarthritis Outcome Score (KOOS) for those who had rehabilitation initially followed by a delayed ACL reconstruction versus those that had an early ACL reconstruction.

Likewise, a more recent cohort study comparing outcomes following ACL injury managed with strength and neuromuscular training versus ACL reconstruction showed that both treatment groups had similar 5-year functional and radiographic outcomes, but the ACL reconstruction group had better self-reported knee function scores and lower self-reported fear [29].

If an athlete is involved in a sport that does not require pivoting actions, the athlete may be able to return to play without any surgical intervention. Ice hockey is one of the sports where it is believed that athletes with an isolated ACL injury and minimal knee instability can attempt to return to play through rehabilitation alone without ACL reconstruction [30–32]. It has been reported that ice hockey players are able to complete their season with a torn ACL and postpone ACL reconstruction to their off season [30–32]. This is postulated to be due to the low friction activity of on-ice skating and little to no pivoting required for the game. ACL injuries in ice hockey occur commonly due to on-ice collisions with another player or the goalpost. However, ACL reconstruction is still recommended for such athletes if there is knee instability and to prevent further intra-articular damage [30–33].

A review of non-operative management of ACL injuries and return to sporting activities recommended the use of a set of criteria to identify potential copers [34]. This includes using a series of hop tests where limb symmetry has to be >80%, a self-reported functional survey (Knee Outcome Survey Activities of Daily Living Scale, KOS ADLS) score of >80, a self-reported global knee function rating of >60 and subjective report of knee giving way of 1 episode or less. These identified copers (who would do well with non-operative management) are then placed on a structured rehabilitation programme comprising of the following:

- Acute phase—haemarthrosis management, improving and maintaining knee range of motion, training normal quadriceps contraction
- Neuromuscular training phase—must achieve no episodes of giving way
- Return to sport phase

The choice of surgical or non-operative management should be individualised after a careful and open discussion between the athlete and the surgeon.

22.4.2 Timing of ACL Surgery

From the literature, most athletes adopt the option of ACL reconstruction for an earlier and more predictable return to sports [11, 24]. In a 2002 survey of National Football League (NFL) physicians, all physicians recommended ACL reconstruction for NFL players after an ACL injury [11]. The majority surveyed recommended ACL reconstruction to be performed as early as possible especially in quarterbacks [11].

In that survey, most NFL physicians recommended that ACL reconstruction be performed once swelling has subsided, range of motion has improved and quadriceps tone returned. Those physicians who recommended a wait and see plan would commence with functional bracing to allow the athlete to continue participation in competition and then perform ACL reconstruction once the season has ended [11].

When combined with a medial collateral ligament (MCL) injury, approximately half of the surveyed physicians (45%) waited 4–6 weeks to allow the MCL to heal before proceeding with ACL reconstruction. Thirteen per cent would repair the MCL along with ACL reconstruction acutely once knee swelling had subsided [11].

In another survey of Major League Soccer (MLS) physicians, most surveyed recommended ACL reconstruction to be performed within 4 weeks from injury [24]. The optimal time to perform ACL reconstruction surgery is when the knee range of motion has returned to normal and knee effusion is no longer present [35]. This is usually at 3–6 weeks post-injury. The odds of articular cartilage injury increase by 1% for every month the surgery is delayed, and the odds of meniscus injury are twice as large if there is an articular cartilage injury [35].

For competitive athletes that engage in pivoting sports, the literature supports early ACL reconstruction. This is recommended once there is minimal knee effusion and full knee extension has been achieved. This will allow these athletes to return back to their competitive sports at the earliest time with rehabilitation after ACL reconstruction.

22.4.3 ACL Graft Options

Autografts are favoured over allografts in ACL reconstruction, as allografts had a higher rate of ACL graft rupture than autografts [36]. The odds of ACL graft rupture with allograft ACL reconstruction are four times higher than those with autograft ACL reconstructions [37].

The bone-patella-tendon-bone (BPTB) autograft has historically been considered the 'gold standard' for ACL reconstruction in athletes as bone-to-bone healing is faster and stronger, and studies have shown excellent long-term results [38]. The reviews reported a higher rate of donor site morbidity (anterior knee pain and kneeling pain) in patients who underwent ACL reconstruction with BPTB autograft compared with hamstring autograft [39–41]. There was no difference in ACL graft rupture rates between BPTB and hamstring autografts [39–41]. There was also no difference in time to return to play after ACL reconstruction between hamstring and BPTB autografts [7, 42].

Despite the resultant hamstring weakness, the hamstring autograft used in ACL reconstruction is reported to achieve better knee range of motion post-operatively and avoids anterior knee pain and pain with kneeling [39–41]. This hamstring weakness is pertinent in female athletes as shown in a study by Briem et al. [43]. Female athletes who had a history of hamstring autograft ACL reconstruction demonstrated inter-limb difference in relative medial versus lateral hamstring muscle activity not found in uninjured controls. This is an important factor to consider and work on in their post-surgical rehabilitation to reduce the risk of an ACL re-injury after surgery [43].

In a study, the cases in which quadriceps tendon autograft were used for ACL reconstruction showed better isokinetic hamstring to quadriceps ratio (H/Q) compared to when hamstring autografts were used [44]. This higher H/Q could be an advantage in reducing ACL graft tear as hamstring activation reduces the load on the ACL graft after surgery. A low H/Q ratio has been associated with ACL injuries, especially in female athletes [45, 46].

In the literature mainly from North American professional sports, surgeons prefer to use the BPTB autograft for ACL reconstruction in elite athletes [11, 24, 42, 47, 48] as compared to hamstrings and quadriceps tendon autograft as their primary ACL graft choice. The current graft of choice for ACL reconstruction among surgeons looking after MLS teams shows a preference for the use of a single bundle BPTB autograft (68%) [24]. In this study, 86% of surgeons selected autograft as the preferred choice of graft for ACL reconstruction [24].

Most sports surgeons looking after NFL and National Collegiate Athletic Association (NCAA) American football running backs would choose to perform a single bundle ACL reconstruction (99%) with a BPTB autograft (86.1%), using an accessory medial portal (67%) [47]. However, among the same surgeons, when asked for their preferred graft choice in recreational athletes between 25 and 35 years of age, less than 15% chose BPTB autograft as their graft of choice [47].

In baseball players, the most common graft type for catchers was a hamstring autograft, while the most common graft type for all other positions was the BPTB autograft. This is because of the demands on the knees of catchers and the need to pop up from a squatting position during the game [42]. This shows that age, level of sports participation and even position played affect the graft choice by sports surgeons in ACL reconstruction for competitive athletes.

Table 22.2 ACL graft choices for athletes; ACL: anterior cruciate ligament

Bone-patella-tendon bone (BPTB) autograft	Sports that do not do well with hamstring weakness: Soccer Rugby Sprinters
Hamstrings autograft	Sports that entail a risk for anterior knee pain (jumping, kneeling): Basketball Volleyball Alpine skiing Wrestling/martial arts
Quadriceps autograft	Young males in contact sports (alternative to BPTB autograft)

It is important to note that graft choice is also dependent on factors such as surgeon preference and region of practice. A survey of surgeons treating professional soccer players in Brazil revealed that 50% of the patients underwent a transportal hamstring autograft ACL reconstruction [10]. An international survey of 261 orthopaedic surgeons within the ACL Study Group in 2012 demonstrated that the hamstring tendon graft was the favoured graft in primary ACL reconstruction in 63% of surgeons, followed by BPTB in 26% of surgeons [49].

The authors have provided their recommendations for high-level athletes in Table 22.2.

22.4.4 Timing of Return to Sport After ACL Surgery

Currently, return to sport guidelines is not consistent. In a systematic review on return to sport after ACL reconstruction, it was found that 65% of studies did not report whether any criteria were used to allow a patient to return to sport. In the remaining studies, requirements to allow return to sport differed. These included objective strength testing, subjective assessment of thigh atrophy, examination of knee range of motion, examination of knee stability, assessment for the absence or presence of effusion, proprioception and functional testing [50].

A survey of NFL and NCAA team orthopaedic surgeons showed that 55% of surgeons waited a minimum of 6 months before allowing the athlete to return to sport [47]. In another systematic review, it was reported that 39% of surgeons permitted running at 3 months, 45% permitted return to cutting/pivoting sports at 6 months and 51% permitted return to sport without restrictions at 6 months [50].

In a study of European Football Champions League soccer players, the players required 7 months to return to first training, 10 months to return to regular training and 12 months to return to match play [7]. Other surveys from North American and South American soccer leagues reported that most surgeons permitted return to sport without restrictions at 6–8 months following ACL reconstruction [24, 51].

Current literature from surveys in North American professional sports leagues suggests a return to sports time frame of between 9 and 12 months [42]. This is consistent with reports that reinjury rates are significantly reduced by 51% for each month return to sport was delayed until 9 months after the surgery, after which no further risk reduction was observed [52].

More recently, authors have reported a higher secondary injury rate for patients who return to sport within 2 years of ACL reconstruction [53, 54]. The decrease in risk of ACL graft rupture with time has been postulated to be biological and related to graft healing and recovering joint homeostasis [55]. The native properties of the graft will require cell proliferation and revascularisation to confer it the rigidity and compliance necessary of a functional ACL, and this process continues to take place after 1 year post-operatively [56–58].

In addition, quadriceps and hamstring strengthening are a crucial step in working towards return to sport. The time it takes to eliminate strength deficits in the operative limb is another important factor to consider when evaluating return to sport. When athletes were assessed for knee extensor and flexor strength after ACL reconstruction, near symmetrical knee extensor and flexor strength were reported only at 2 years post-operatively [55].

This has led some experts to caution early return to full sports for fear of re-injury. This also accounts for reports that professional NFL players gradually return to baseline participation in terms of games per season and minutes per game only at two and three seasons post-ACL reconstruction [59].

22.4.5 Return to Sport Rates After ACL Surgery

Lindanger et al. [59] in 2019 reported that 83% of patients who underwent ACL reconstruction returned to pivoting sports, but only 53% returned to their preinjury level of sports participation. They reported that 73% of elite athletes returned to elite level after ACL reconstruction [59]. A systematic review of studies on return to sports following ACL reconstruction found that 63% of patients returned to pre-injury sports participation rate. The authors also reported that less than half of patients who underwent ACL reconstruction returned to competitive sport [60].

Mai et al. reported that there were differences in return to sports levels when comparing among different professional sports cohorts [61]. The return to sport rates for the professional athletes were as follows:

- National Football League (NFL) 85.5%
- National Hockey League (NHL) 95.8%
- Major League Baseball (MLB) 80%
- National Basketball Association (NBA) 82.4%

In a survey across four North American professional athlete groups (the NFL, NBA, NHL and MLS), Erickson et al. [62] reported that NFL athletes fared worse after ACL reconstruction with the lowest survival rate, shortest post-operative career length and sustained decreases in performance. NHL athletes fared best with highest rates of return to play, highest survival rates, longest post-operative career lengths and no significant changes in performances. NBA and NFL athletes showed decreased performance at season 1 after ACL reconstruction. NFL players also continued to have lower performance at seasons 2 and 3, while NBA players had recovered to their pre-surgery performance level.

American Football Mai et al. reported that 82.4% of NFL athletes returned to play after ACL reconstruction [63]. ACL reconstruction led to decreases in performance in post-operative season 1 compared with preinjury values. The decreased performance in seasons 2 and 3 continued for players in the ACL reconstruction cohort.

Soccer The Multicentre Orthopaedic Outcomes Network (MOON) group published outcomes of 100 male and female soccer players undergoing ACL reconstruction at a mean of 7-year follow-up. They found a return to sport rate of 72% [64]. Erickson et al. reported a return to play of 77% for MLS professional soccer league players. They did not report a difference in performance post-ACL reconstruction compared to pre-ACL reconstruction. The mean career length in the soccer league after ACL reconstruction was reported to be 4.0 +/− 2.8 years [65].

In their survey of professional European soccer players, Zaffagnini et al. reported a 95% return to sport rate to pre-injury level for 21 professional male soccer players after ACL reconstruction [66]. Niederer et al. performed a media-based collection and analysis of return to play and quality of play data after ACL rupture in top level football in Europe. They reported that the return to play rate after ACL reconstruction was 98.2%; however, the return to play at same pre-injury level rate was 59.4% [67]. The return to real game performance at a previous level may take up to 2 years. Career duration was shorter and game performance was of a lower quality after an ACL tear [67]. Similarly, at three seasons post-ACL reconstruction, Barth et al. reported that athletes were still starting fewer games per season and scoring fewer goals in a 90-minute game [8].

In addition, it was reported that soccer goalkeepers and centre backs needed significantly shorter time for return to play than forwards; this suggested that the more offensive a player was positioned, the longer the return to sport process was required [67]. Five years after return to play, 69.9% were still engaged in football, with only 40.9% of the players playing at their same level [67].

Australian Rules Football In a 15-year survey of Australian Football League (AFL) players [68], the rate of return to play after an ACL reconstruction was 77%. The rate of a subsequent ACL injury to either knee was 30% and this was especially high in players less than 21 years of age (50%). After a subsequent ACL injury, 71% of players were able to return to play.

In another study, 104 Australian Football League players who underwent ACL reconstruction between 1999 and 2013 were reviewed for return to play and return to pre-injury performance. The authors reported that 68% returned to their pre-injury levels of performance and the remaining players who did not return to their pre-injury levels still performed comparably to average levels of play in the AFL [69]. Those patients who were younger and those who weighed less than 90 kg were more likely to return to their pre-injury levels [69].

Basketball In their study of 58 NBA players that underwent ACL reconstruction [13], Harris et al. reported that 98% of players returned to sports in the NBA season following ACL reconstruction (11.6 ± 4.1 months from injury). The players' performance upon return declined in terms of games per season, minutes, points and rebounds per game and field goal percentage.

Ice Hockey Sikka et al. reported on 47 NHL players who underwent ACL reconstruction [70]. In this cohort of NHL players, 89.4% of players were able to return to play in the NHL after ACL reconstruction. Their career length and performance may be significantly decreased compared to controls. The presence of meniscal injury was associated with decreased length of career compared to the control group [70].

Baseball Erickson et al. reported that the rate of return to sports for professional baseball players after ACL reconstruction was 80% [42]. In this study, 73% of the players returned at the same level or higher at a mean of 10.2 ± 3.6 months. There was no difference in performance with regard to the type of graft used or the technique of femoral tunnel drilling.

X-Games Skiers/Snowboarders The rate of return to sport in X-Games sports practitioners following ACL reconstruction was 80%: 87% in skiers and 70% in snowboarders [71]. The skiers earned a similar number of medals after ACL reconstruction, as compared to prior to their ACL tear. The snowboarders that returned to competition after ACL reconstruction earned more medals than prior to their ACL tear.

22.4.6 Concept of Kinesiophobia

Kinesiophobia, a fear of activity due to feeling of vulnerability to injury or re-injury, is a psychological factor that is increasingly recognised for its contribution to success or failure in an athlete's return to sport [60, 72–76]. A meta-analysis and systematic review of studies on return to sports following ACL reconstruction revealed that despite successful surgical outcomes based on objective scores and measures of knee function, only 44% of patients returned to competitive sport [60].

Lee et al. studied a cohort of patients who underwent ACL reconstruction with 5 years of follow-up. In the analysis, having an unstable knee (17.8%) and fear of re-injury (20%) were the two main reasons for athletes not returning to their previous sporting levels [76]. Similarly, Flanigan et al.'s survey of patients who had undergone ACL reconstruction also reported that apart from persistent knee symptoms, the second most common reason for lack of return to sport was kinesiophobia [75]. Athletes who had returned to their pre-injury level of participation demonstrated significantly less fear of re-injury than those who had not returned to their pre-injury level [73]. This may suggest that fear of re-injury is a significant barrier that hinders athletes' return to their pre-injury level of sports participation. An athlete's pre-operative prospective judgement of their ability to return to sport predicted his likelihood of returning to their pre-injury level [74]. These important psychological barriers should be addressed in pre-operative counselling and during their rehabilitation process after ACL reconstruction.

☑ Rehabilitation or Surgery
☑ Timing of Surgery
☑ Graft Choice
☑ Timing of return to training
☑ Psychological considerations
☑ Timing of return to sports
☑ Timing of return to previous performance
☑ Prevention of reinjury

Fig. 22.2 Checklist for the management of ACL injury in an athlete; ACL anterior cruciate ligament

Finally, it is important to think of return to sport after ACL reconstruction to be a continuum that begins with return to participation, progresses to return to sport and finally a return to performance. On average, males tend to have higher return to sport rates by about 10% than females, and the return to sport rates for younger athletes are 30% higher than the older athletes [77]. The motivation to return to sports in competitive athletes may be the factor that sees a higher percentage of them returning to competitive sports [76, 78, 79]. It is important to be aware that the risk for ACL graft rupture is highest during the first 2 post-operative years and the risk for contralateral ACL injury occurs later [77].

The following is a checklist of considerations in the management of ACL injury in an athlete (Fig. 22.2).

22.5 Conclusion

- The most common causative sport for ACL injuries is soccer; other sports include skiing, basketball, netball, rugby, handball, floorball.
- Female athletes have a greater susceptibility for ACL injuries compared to their male counterparts.
- Non-contact mechanisms account for the majority of ACL tears.
- There may be a role for non-operative management for ACL tears in athletes; the key is to consider their sports participation level,

identify potential copers with screening tests, then placing the athlete on a structured rehabilitation programme.

- ACL reconstruction should be performed only when knee effusion has resolved and knee range of motion has returned to normal to prevent arthrofibrosis.
- BPTB grafts are the gold standard, but graft selection is dependent on surgeon preference/ expertise and the athlete's exact sporting requirement.
- As a guide, return to sport should be between 9 and 12 months post-surgery, with the athlete having near symmetrical quadriceps and hamstrings strength recovery and based on a return to sport criteria.
- Return to sport should be a continuum: return to participation, return to sport and finally return to performance.
- It must be recognised that any return to sport carries the risk of further injury to the affected knee and risk of ACL injury in the contralateral knee.
- The psychological aspect, the fear of reinjury during the time of return to sport, needs to be managed.

References

1. Gans I, Retzky JS, Jones LC, Tanaka MJ. Epidemiology of recurrent anterior cruciate ligament injuries in National Collegiate Athletic Association Sports: the injury surveillance program, 2004-2014. Orthop J Sports Med. 2018;6(6):2325967118777823. https://doi.org/10.1177/2325967118777823.
2. Gianotti SM, Marshall SW, Hume PA, Bunt L. Incidence of anterior cruciate ligament injury and other knee ligament injuries: a national population-based study. J Sci Med Sport. 2009;12(6):622–7. https://doi.org/10.1016/j.jsams.2008.07.005.
3. Granan LP, Inacio MC, Maletis GB, Funahashi TT, Engebretsen L. Sport-specific injury pattern recorded during anterior cruciate ligament reconstruction. Am J Sports Med. 2013;41(12):2814–8. https://doi.org/10.1177/0363546513501791.
4. Lai CCH, Ardern CL, Feller JA, Webster KE. Eighty-three per cent of elite athletes return to preinjury sport after anterior cruciate ligament reconstruction: a systematic review with meta-analysis of return to sport rates, graft rupture rates and performance outcomes. Br J Sports Med. 2018;52(2):128–38. https://doi.org/10.1136/bjsports-2016-096836.
5. Prentice HA, Lind M, Mouton C, Persson A, Magnusson H, Gabr A, et al. Patient demographic and surgical characteristics in anterior cruciate ligament reconstruction: a description of registries from six countries. Br J Sports Med. 2018;52(11):716–22. https://doi.org/10.1136/bjsports-2017-098674.
6. Prodromos CC, Han Y, Rogowski J, Joyce B, Shi K. A meta-analysis of the incidence of anterior cruciate ligament tears as a function of gender, sport, and a knee injury-reduction regimen. Arthroscop J Arthroscop Relat Surg. 2007;23(12):1320–5.e6. https://doi.org/10.1016/j.arthro.2007.07.003.
7. Walden M, Hagglund M, Magnusson H, Ekstrand J. Anterior cruciate ligament injury in elite football: a prospective three-cohort study. Knee Surg Sports Traumatol Arthrosc. 2011;19(1):11–9. https://doi.org/10.1007/s00167-010-1170-9.
8. Barth KA, Lawton CD, Touhey DC, Selley RS, Li DD, Balderama ES, et al. The negative impact of anterior cruciate ligament reconstruction in professional male footballers. Knee. 2019;26(1):142–8. https://doi.org/10.1016/j.knee.2018.10.004.
9. Rekik RN, Tabben M, Eirale C, Landreau P, Bouras R, Wilson MG, et al. ACL injury incidence, severity and patterns in professional male soccer players in a Middle Eastern league. BMJ Open Sport Exerc Med. 2018;4(1):e000461. https://doi.org/10.1136/bmjsem-2018-000461.
10. Arliani GG, Lara PHS, Astur DC, Pedrinelli A, Pagura JR, Cohen M. Prospective evaluation of injuries ocurred during a professional soccer championship in 2016 in Sao Paulo, Brazil. Acta Ortopedica Brasileira. 2017;25(5):212–5. https://doi.org/10.1590/1413-785220172505167238.
11. Bradley JP, Klimkiewicz JJ, Rytel MJ, Powell JW. Anterior cruciate ligament injuries in the National Football League: epidemiology and current treatment trends among team physicians. Arthroscop J Arthroscop Relat Surg. 2002;18(5):502–9. https://doi.org/10.1053/jars.2002.30649.
12. Pujol N, Blanchi MP, Chambat P. The incidence of anterior cruciate ligament injuries among competitive Alpine skiers: a 25-year investigation. Am J Sports Med. 2007;35(7):1070–4. https://doi.org/10.1177/0363546507301083.
13. Harris JD, Erickson BJ, Bach BR Jr, Abrams GD, Cvetanovich GL, Forsythe B, et al. Return-to-sport and performance after anterior cruciate ligament reconstruction in National Basketball Association Players. Sports Health. 2013;5(6):562–8. https://doi.org/10.1177/1941738113495788.
14. Hewett TE, Ford KR, Hoogenboom BJ, Myer GD. Understanding and preventing acl injuries: current biomechanical and epidemiologic considerations - update 2010. N Am J Sports Phys Ther. 2010;5(4):234–51. https://pmc-ncbi-nlm-nih-gov.libproxy1.nus.edu.sg/articles/PMC3096145/
15. Longstaffe R, Leiter J, MacDonald P. Anterior cruciate ligament injuries in the National Hockey League: epidemiology and performance impact. Clin

J Sport Med. 2018;30:224. https://doi.org/10.1097/JSM.0000000000000584.
16. Arendt E, Dick R. Knee injury patterns among men and women in collegiate basketball and soccer. NCAA data and review of literature. Am J Sports Med. 1995;23(6):694–701. https://doi.org/10.1177/036354659502300611.
17. Delfico AJ, Garrett WE Jr. Mechanisms of injury of the anterior cruciate ligament in soccer players. Clin Sports Med. 1998;17(4):779–85., vii. https://doi.org/10.1016/s0278-5919(05)70118-6.
18. Laible C, Sherman OH. Risk factors and prevention strategies of non-contact anterior cruciate ligament injuries. Bullet Hosp Joint Dis (2013). 2014;72(1):70–5. https://pubmed-ncbi-nlm-nih-gov.libproxy1.nus.edu.sg/25150329/
19. Hewett TE, Myer GD, Ford KR, Heidt RS Jr, Colosimo AJ, McLean SG, et al. Biomechanical measures of neuromuscular control and valgus loading of the knee predict anterior cruciate ligament injury risk in female athletes: a prospective study. Am J Sports Med. 2005;33(4):492–501. https://doi.org/10.1177/0363546504269591.
20. Schiffner E, Latz D, Grassmann JP, Schek A, Thelen S, Windolf J, et al. Anterior cruciate ligament ruptures in German elite soccer players: epidemiology, mechanisms, and return to play. Knee. 2018;25(2):219–25. https://doi.org/10.1016/j.knee.2018.01.010.
21. Shimokochi Y, Shultz SJ. Mechanisms of noncontact anterior cruciate ligament injury. J Athl Train. 2008;43(4):396–408. https://pmc-ncbi-nlm-nih-gov.libproxy1.nus.edu.sg/articles/PMC2474820/
22. Boden BP, Dean GS, Feagin JA Jr, Garrett WE Jr. Mechanisms of anterior cruciate ligament injury. Orthopedics. 2000;23(6):573–8. https://doi.org/10.3928/0147-7447-20000601-15.
23. Bere T, Florenes TW, Krosshaug T, Koga H, Nordsletten L, Irving C, et al. Mechanisms of anterior cruciate ligament injury in World Cup alpine skiing: a systematic video analysis of 20 cases. Am J Sports Med. 2011;39(7):1421–9. https://doi.org/10.1177/0363546511405147.
24. Farber J, Harris JD, Kolstad K, McCulloch PC. Treatment of anterior cruciate ligament injuries by Major League Soccer team physicians. Orthop J Sports Med. 2014;2(11):2325967114559892. https://doi.org/10.1177/2325967114559892.
25. Frobell RB, Roos HP, Roos EM, Roemer FW, Ranstam J, Lohmander LS. Treatment for acute anterior cruciate ligament tear: five year outcome of randomised trial. BMJ (Clinical research ed). 2013;346:f232. https://doi.org/10.1136/bmj.f232.
26. Grindem H, Eitzen I, Moksnes H, Snyder-Mackler L, Risberg MA. A pair-matched comparison of return to pivoting sports at 1 year in anterior cruciate ligament-injured patients after a nonoperative versus an operative treatment course. Am J Sports Med. 2012;40(11):2509–16. https://doi.org/10.1177/0363546512458424.
27. Kay J, Memon M, Sa D, Simunovic N, Musahl V, Fu FH, et al. A historical analysis of randomized controlled trials in anterior cruciate ligament surgery. J Bone Joint Surg Am. 2017;99(24):2062–8. https://doi.org/10.2106/JBJS.16.01408.
28. Frobell RB, Roos EM, Roos HP, Ranstam J, Lohmander LS. A randomized trial of treatment for acute anterior cruciate ligament tears. N Engl J Med. 2010;363(4):331–42. https://doi.org/10.1056/NEJMoa0907797.
29. Wellsandt E, Failla MJ, Axe MJ, Snyder-Mackler L. Does anterior cruciate ligament reconstruction improve functional and radiographic outcomes over nonoperative management 5 years after injury? Am J Sports Med. 2018;46(9):2103–12. https://doi.org/10.1177/0363546518782698.
30. Laprade RF, Surowiec RK, Sochanska AN, Hentkowski BS, Martin BM, Engebretsen L, et al. Epidemiology, identification, treatment and return to play of musculoskeletal-based ice hockey injuries. Br J Sports Med. 2014;48(1):4–10. https://doi.org/10.1136/bjsports-2013-093020.
31. Moslener MD, Wadsworth LT. Ice hockey: a team physician's perspective. Curr Sports Med Rep. 2010;9(3):134–8. https://doi.org/10.1249/JSR.0b013e3181dfccd9.
32. LaPrade RF, Wijdicks CA, Griffith CJ. Division I intercollegiate ice hockey team coverage. Br J Sports Med. 2009;43(13):1000–5. https://doi.org/10.1136/bjsm.2009.067884.
33. Mosenthal W, Kim M, Holzshu R, Hanypsiak B, Athiviraham A. Common ice hockey injuries and treatment: a current concepts review. Curr Sports Med Rep. 2017;16(5):357–62. https://doi.org/10.1249/JSR.0000000000000402.
34. Paterno MV. Non-operative Care of the Patient with an ACL-deficient knee. Curr Rev Musculoskelet Med. 2017;10(3):322–7. https://doi.org/10.1007/s12178-017-9431-6.
35. Renstrom PA. Eight clinical conundrums relating to anterior cruciate ligament (ACL) injury in sport: recent evidence and a personal reflection. Br J Sports Med. 2013;47(6):367–72. https://doi.org/10.1136/bjsports-2012-091623.
36. Krych AJ, Jackson JD, Hoskin TL, Dahm DL. A meta-analysis of patellar tendon autograft versus patellar tendon allograft in anterior cruciate ligament reconstruction. Arthroscop J Arthroscop Relat Surg. 2008;24(3):292–8. https://doi.org/10.1016/j.arthro.2007.08.029.
37. Kaeding CC, Aros B, Pedroza A, Pifel E, Amendola A, Andrish JT, et al. Allograft versus autograft anterior cruciate ligament reconstruction: predictors of failure from a MOON prospective longitudinal cohort. Sports Health. 2011;3(1):73–81. https://doi.org/10.1177/1941738110386185.
38. Koh HS, In Y, Kong CG, Won HY, Kim KH, Lee JH. Factors affecting patients' graft choice in anterior cruciate ligament reconstruction. Clin Orthop

Surg. 2010;2(2):69–75. https://doi.org/10.4055/cios.2010.2.2.69.

39. Poolman RW, Abouali JA, Conter HJ, Bhandari M. Overlapping systematic reviews of anterior cruciate ligament reconstruction comparing hamstring autograft with bone-patellar tendon-bone autograft: why are they different? J Bone Joint Surg Am. 2007;89(7):1542–52. https://doi.org/10.2106/JBJS.F.01292.
40. Spindler KP, Kuhn JE, Freedman KB, Matthews CE, Dittus RS, Harrell FE Jr. Anterior cruciate ligament reconstruction autograft choice: bone-tendon-bone versus hamstring: does it really matter? A systematic review. Am J Sports Med. 2004;32(8):1986–95. https://doi.org/10.1177/0363546504271211.
41. Samuelsson K, Andersson D, Karlsson J. Treatment of anterior cruciate ligament injuries with special reference to graft type and surgical technique: an assessment of randomized controlled trials. Arthroscop J Arthroscop Relat Surg. 2009;25(10):1139–74. https://doi.org/10.1016/j.arthro.2009.07.021.
42. Erickson BJ, Chalmers PN, D'Angelo J, Ma K, Dahm DL, Romeo AA, et al. Performance and return to sport after anterior cruciate ligament reconstruction in professional baseball players. Orthop J Sports Med. 2019;7(10):2325967119878431. https://doi.org/10.1177/2325967119878431.
43. Briem K, Ragnarsdottir AM, Arnason SI, Sveinsson T. Altered medial versus lateral hamstring muscle activity during hop testing in female athletes 1-6 years after anterior cruciate ligament reconstruction. Knee Surg Sports Traumatol Arthrosc. 2016;24(1):12–7. https://doi.org/10.1007/s00167-014-3333-6.
44. Martin-Alguacil JL, Arroyo-Morales M, Martin-Gomez JL, Monje-Cabrera IM, Abellan-Guillen JF, Esparza-Ros F, et al. Strength recovery after anterior cruciate ligament reconstruction with quadriceps tendon versus hamstring tendon autografts in soccer players: a randomized controlled trial. Knee. 2018;25(4):704–14. https://doi.org/10.1016/j.knee.2018.03.011.
45. Myer GD, Ford KR, Barber Foss KD, Liu C, Nick TG, Hewett TE. The relationship of hamstrings and quadriceps strength to anterior cruciate ligament injury in female athletes. Clin J Sport Med. 2009;19(1):3–8. https://doi.org/10.1097/JSM.0b013e318190bddb.
46. Withrow TJ, Huston LJ, Wojtys EM, Ashton-Miller JA. Effect of varying hamstring tension on anterior cruciate ligament strain during in vitro impulsive knee flexion and compression loading. J Bone Joint Surg Am. 2008;90(4):815–23. https://doi.org/10.2106/JBJS.F.01352.
47. Erickson BJ, Harris JD, Fillingham YA, Frank RM, Bush-Joseph CA, Bach BR Jr, et al. Anterior cruciate ligament reconstruction practice patterns by NFL and NCAA football team physicians. Arthroscop J Arthroscop Relat Surg. 2014;30(6):731–8. https://doi.org/10.1016/j.arthro.2014.02.034.
48. Porter A, Yang S, Chauhan A, Early S, Challa S, D'Angelo J, et al. Professional advancement and performance of amateur baseball players selected in the Major League Baseball draft with previous anterior cruciate ligament reconstruction. Am J Sports Med. 2020;48(3):581–7. https://doi.org/10.1177/0363546519898194.
49. Chechik O, Amar E, Khashan M, Lador R, Eyal G, Gold A. An international survey on anterior cruciate ligament reconstruction practices. Int Orthop. 2013;37(2):201–6. https://doi.org/10.1007/s00264-012-1611-9.
50. Harris JD, Abrams GD, Bach BR, Williams D, Heidloff D, Bush-Joseph CA, et al. Return to sport after ACL reconstruction. Orthopedics. 2014;37(2):e103–8. https://doi.org/10.3928/01477447-20140124-10.
51. Arliani GG, Pereira VL, Leao RG, Lara PS, Ejnisman B, Cohen M. Treatment of anterior cruciate ligament injuries in professional soccer players by orthopedic surgeons. Rev Bras Ortop. 2019;54(6):703–8. https://doi.org/10.1055/s-0039-1697017.
52. Grindem H, Snyder-Mackler L, Moksnes H, Engebretsen L, Risberg MA. Simple decision rules can reduce reinjury risk by 84% after ACL reconstruction: the Delaware-Oslo ACL cohort study. Br J Sports Med. 2016;50(13):804–8. https://doi.org/10.1136/bjsports-2016-096031.
53. Roos EM, Roos HP, Lohmander LS, Ekdahl C, Beynnon BD. Knee Injury and Osteoarthritis Outcome Score (KOOS)--development of a self-administered outcome measure. J Orthop Sports Phys Ther. 1998;28(2):88–96. https://doi.org/10.2519/jospt.1998.28.2.88.
54. LaPrade CM, Dornan GJ, Granan LP, LaPrade RF, Engebretsen L. Outcomes after anterior cruciate ligament reconstruction using the Norwegian knee ligament registry of 4691 patients: how does meniscal repair or resection affect short-term outcomes? Am J Sports Med. 2015;43(7):1591–7. https://doi.org/10.1177/0363546515577364.
55. Paterno MV, Rauh MJ, Schmitt LC, Ford KR, Hewett TE. Incidence of second ACL injuries 2 years after primary ACL reconstruction and return to sport. Am J Sports Med. 2014;42(7):1567–73. https://doi.org/10.1177/0363546514530088.
56. Nagelli CV, Hewett TE. Should return to sport be delayed until 2 years after anterior cruciate ligament reconstruction? Biological and functional considerations. Sports Med (Auckland, NZ). 2017;47(2):221–32. https://doi.org/10.1007/s40279-016-0584-z.
57. Salmon L, Russell V, Musgrove T, Pinczewski L, Refshauge K. Incidence and risk factors for graft rupture and contralateral rupture after anterior cruciate ligament reconstruction. Arthroscop J Arthroscop Relat Surg. 2005;21(8):948–57. https://doi.org/10.1016/j.arthro.2005.04.110.
58. Scheffler SU, Unterhauser FN, Weiler A. Graft remodeling and ligamentization after cruciate ligament reconstruction. Knee Surg Sports Traumatol Arthroscop. 2008;16(9):834–42. https://doi.org/10.1007/s00167-008-0560-8.

59. Lindanger L, Strand T, Molster AO, Solheim E, Inderhaug E. Return to play and long-term participation in pivoting sports after anterior cruciate ligament reconstruction. Am J Sports Med. 2019;47(14):3339–46. https://doi.org/10.1177/0363546519878159.
60. Ardern CL, Webster KE, Taylor NF, Feller JA. Return to sport following anterior cruciate ligament reconstruction surgery: a systematic review and meta-analysis of the state of play. Br J Sports Med. 2011;45(7):596–606. https://doi.org/10.1136/bjsm.2010.076364.
61. Mai HT, Chun DS, Schneider AD, Erickson BJ, Freshman RD, Kester B, et al. Performance-based outcomes after anterior cruciate ligament reconstruction in professional athletes differ between sports. Am J Sports Med. 2017;45(10):2226–32. https://doi.org/10.1177/0363546517704834.
62. Erickson BJ, Harris JD, Fillingham YA, Cvetanovich GL, Bush-Joseph C, Cole BJ, et al. Orthopedic practice patterns relating to anterior cruciate ligament reconstruction in elite athletes. Am J Orthop (Belle Mead NJ). 2015;44(12):E480–5. https://pubmed--ncbi-nlm-nih-gov.libproxy1.nus.edu.sg/26665248/
63. Mai HT, Alvarez AP, Freshman RD, Chun DS, Minhas SV, Patel AA, et al. The NFL Orthopaedic Surgery Outcomes Database (NO-SOD): the effect of common orthopaedic procedures on football careers. Am J Sports Med. 2016;44(9):2255–62. https://doi.org/10.1177/0363546516651426.
64. Brophy RH, Schmitz L, Wright RW, Dunn WR, Parker RD, Andrish JT, et al. Return to play and future ACL injury risk after ACL reconstruction in soccer athletes from the Multicenter Orthopaedic Outcomes Network (MOON) group. Am J Sports Med. 2012;40(11):2517–22. https://doi.org/10.1177/0363546512459476.
65. Erickson BJ, Harris JD, Cvetanovich GL, Bach BR, Bush-Joseph CA, Abrams GD, et al. Performance and return to sport after anterior cruciate ligament reconstruction in male Major League Soccer players. Orthop J Sports Med. 2013;1(2):2325967113497189. https://doi.org/10.1177/2325967113497189.
66. Zaffagnini S, Grassi A, Marcheggiani Muccioli GM, Tsapralis K, Ricci M, Bragonzoni L, et al. Return to sport after anterior cruciate ligament reconstruction in professional soccer players. Knee. 2014;21(3):731–5. https://doi.org/10.1016/j.knee.2014.02.005.
67. Niederer D, Engeroff T, Wilke J, Vogt L, Banzer W. Return to play, performance, and career duration after anterior cruciate ligament rupture: a case-control study in the five biggest football nations in Europe. Scand J Med Sci Sports. 2018;28(10):2226–33. https://doi.org/10.1111/sms.13245.
68. Lai CCH, Feller JA, Webster KE. Fifteen-year audit of anterior cruciate ligament reconstructions in the Australian Football League from 1999 to 2013: return to play and subsequent ACL injury. Am J Sports Med. 2018;46(14):3353–60. https://doi.org/10.1177/0363546518803932.
69. Lai CCH, Feller JA, Webster KE. Playing performance after anterior cruciate ligament reconstruction among Australian football league players from 1999 to 2013. Am J Sports Med. 2019;47(7):1550–6. https://doi.org/10.1177/0363546519843908.
70. Sikka R, Kurtenbach C, Steubs JT, Boyd JL, Nelson BJ. Anterior cruciate ligament injuries in professional hockey players. Am J Sports Med. 2016;44(2):378–83. https://doi.org/10.1177/0363546515616802.
71. Erickson BJ, Harris JD, Fillingham YA, Cvetanovich GL, Bhatia S, Bach BR Jr, et al. Performance and return to sport after anterior cruciate ligament reconstruction in X-games skiers and snowboarders. Orthop J Sports Med. 2013;1(6):2325967113511196. https://doi.org/10.1177/2325967113511196.
72. Ardern CL, Osterberg A, Tagesson S, Gauffin H, Webster KE, Kvist J. The impact of psychological readiness to return to sport and recreational activities after anterior cruciate ligament reconstruction. Br J Sports Med. 2014;48(22):1613–9. https://doi.org/10.1136/bjsports-2014-093842.
73. Ardern CL, Taylor NF, Feller JA, Webster KE. Fear of re-injury in people who have returned to sport following anterior cruciate ligament reconstruction surgery. J Sci Med Sport. 2012;15(6):488–95. https://doi.org/10.1016/j.jsams.2012.03.015.
74. Ardern CL, Taylor NF, Feller JA, Whitehead TS, Webster KE. Psychological responses matter in returning to preinjury level of sport after anterior cruciate ligament reconstruction surgery. Am J Sports Med. 2013;41(7):1549–58. https://doi.org/10.1177/0363546513489284.
75. Flanigan DC, Everhart JS, Pedroza A, Smith T, Kaeding CC. Fear of reinjury (kinesiophobia) and persistent knee symptoms are common factors for lack of return to sport after anterior cruciate ligament reconstruction. Arthroscop J Arthroscop Relat Surg. 2013;29(8):1322–9. https://doi.org/10.1016/j.arthro.2013.05.015.
76. Lee DY, Karim SA, Chang HC. Return to sports after anterior cruciate ligament reconstruction - a review of patients with minimum 5-year follow-up. Ann Acad Med Singap. 2008;37(4):273–8. https://pubmed-ncbi-nlm-nih-gov.libproxy1.nus.edu.sg/18461210/
77. Webster KE, Feller JA. A research update on the state of play for return to sport after anterior cruciate ligament reconstruction. J Orthop Traumatol. 2019;20(1):10. https://doi.org/10.1186/s10195-018-0516-9.
78. Fabbriciani C, Milano G, Mulas PD, Ziranu F, Severini G. Anterior cruciate ligament reconstruction with doubled semitendinosus and gracilis tendon graft in rugby players. Knee Surg Sports Traumatol Arthroscop. 2005;13(1):2–7. https://doi.org/10.1007/s00167-004-0505-9.
79. Smith FW, Rosenlund EA, Aune AK, MacLean JA, Hillis SW. Subjective functional assessments and the return to competitive sport after anterior cruciate ligament reconstruction. Br J Sports Med. 2004;38(3):279–84. https://doi.org/10.1136/bjsm.2002.001982.

23 Return to Sports After ACL Reconstruction in High-Level Athletes

José Lito Mónico, Renato Andrade, Francesco Della Villa, Matthew Buckthorpe, Rogério Pereira, Fernando Fonseca, and João Espregueira-Mendes

23.1 Background

Anterior cruciate ligament (ACL) injuries are among the most common major injuries in sports. The annual ACL injury incidence rate is 0.15–3.67% in professional athletes which is substantially higher than in general population (median rate of 0.03%) [1]. An ACL injury can lead to knee instability, reduced performance, increased risk of developing knee osteoarthritis (OA), and long-term disability [2]. Besides the biomechanical instability, ACL injuries are also becoming recognized as neurophysiological injuries, with neurocognitive and sensorimotor control dysfunction [3]. Changes in the static and dynamic loading of the knee are believed to be

J. L. Mónico (✉)
Department of Orthopaedics, Centro Hospitalar e Universitário de Coimbra, Coimbra, Portugal

Faculty of Medicine, University of Porto, Porto, Portugal

R. Andrade
Clínica Espregueira - FIFA Medical Centre of Excellence, Porto, Portugal

Dom Henrique Research Centre, Porto, Portugal

Porto Biomechanics Laboratory (LABIOMEP), Faculty of Sports, University of Porto, Porto, Portugal

F. D. Villa
Education and Research Department, Isokinetic Medical Group, FIFA Medical Centre of Excellence, Bologna, Italy

M. Buckthorpe
Education and Research Department, Isokinetic Medical Group, FIFA Medical Centre of Excellence, Bologna, Italy

Isokinetic London, Isokinetic Medical Group FIFA Medical Centre of Excellence, London, UK

R. Pereira
Clínica Espregueira - FIFA Medical Centre of Excellence, Porto, Portugal

Dom Henrique Research Centre, Porto, Portugal

University Fernando Pessoa, Porto, Portugal

F. Fonseca
Department of Orthopaedics, Centro Hospitalar e Universitário de Coimbra, Coimbra, Portugal

Faculty of Medicine, University of Coimbra, Coimbra, Portugal

J. Espregueira-Mendes
Clínica Espregueira - FIFA Medical Centre of Excellence, Porto, Portugal

M. Karahan et al. (eds.), *Anterior Cruciate Ligament Injuries Throughout the Life Span*,
https://doi.org/10.1007/978-3-031-84501-7_23

the leading mechanism of early OA [4]. Nonoperative treatment and modifying sporting activities are an option for those trying to avoid surgery, but if there is ongoing knee instability and the athlete is a non-coper, ACL reconstruction may be considered because it seems to decrease the risk of secondary meniscal, chondral pathology, and onset of early OA [5]. However, the evidence that ACL reconstruction reduces the risk of developing early knee OA is still too weak to make definitive conclusions [6]. Aiming to increase their likelihood of returning to competitive sport [7] and potentially decrease the risk of developing early OA [8], many athletes opt for ACL reconstruction. Up to 80% of ACL-reconstructed patients return to some kind of sporting activities, but many athletes do not return to competitive sports (only 55%) [9], and the incidence of knee OA is still high after ACL reconstruction [10].

23.2 Importance of Rehabilitation

Optimal return to sport (RTS) implies optimization of every step of the rehabilitation process to allow the athlete to RTS as fast and as safe as possible to the preinjury level of performance [11]. The ACL injury and surgical trauma cause pain and impair joint function, including swelling, reduced strength, range of motion, and stability. Rehabilitation is crucial to reduce symptoms, improve knee function, and decrease the risk of reinjury and safely RTS [12]. Progressing in rehabilitation before milestones are met can be counterproductive and dangerous.

When a player RTS after suffering an ACL injury, the risk of developing a new knee injury is high. There is a pooled 23% rate of secondary ACL injury for those that are younger than 25 years old [13], and rates can go as high as 31% [14]. On the contralateral nonoperated knee, ACL injury risk is nearly double compared to graft retear [15]. There is a tendency to overload the uninjured knee to protect the injured side [16], loss of proprioception (exposing both knees to injury) [17], and unbalanced rehabilitation (with focus on the operative knee only) [18]. These are some possible explanations to the higher risk of contralateral injury that we should couple with the logical increased risk of injury due to proxy intrinsic nonmodifiable factors (e.g., genetic predisposition, bony morphology). Suffering a graft retear or an injury of the contralateral ACL can be devastating. Ranging from 1 to 11% [19, 20], graft retears may be caused by another traumatic injury, biologic failure of graft integration, technical surgical factors [21], or inadequate rehabilitation [22].

23.3 Rehabilitation

Following a progressive and stepwise rehabilitation, based on objective criteria, is crucial to ensure a safe RTS (Fig. 23.1) [11]. Rehabilitation should follow high-quality clinical practice guidelines to apply safe and effective evidence-based interventions [23].

Aiming to a fast and safe RTS, reconditioning is a crucial step and should start as soon as possible. Reconditioning can be introduced in a stepwise rehabilitation program in many ways and should aim to restore neuromuscular strength, proprioceptive sense, motor skills, cardiorespiratory, and metabolic fitness [11]. Besides muscle anatomic changes, neural factors play an important role in muscle strength gains. After a musculoskeletal injury, an athlete might lose proprioceptive afferents affecting muscle control, postural reflexes, spatial and temporal aspects of volitional movement [24]. Proprioceptive training, with both passive and active movements with and without visual feedback, can improve somatosensory and sensorimotor function in an impaired motor function [25]. Motor skills are very important for specific movements and sporting performance. Including sport-specific exercises in rehabilitation will increase an athlete's adherence and compliance [26].

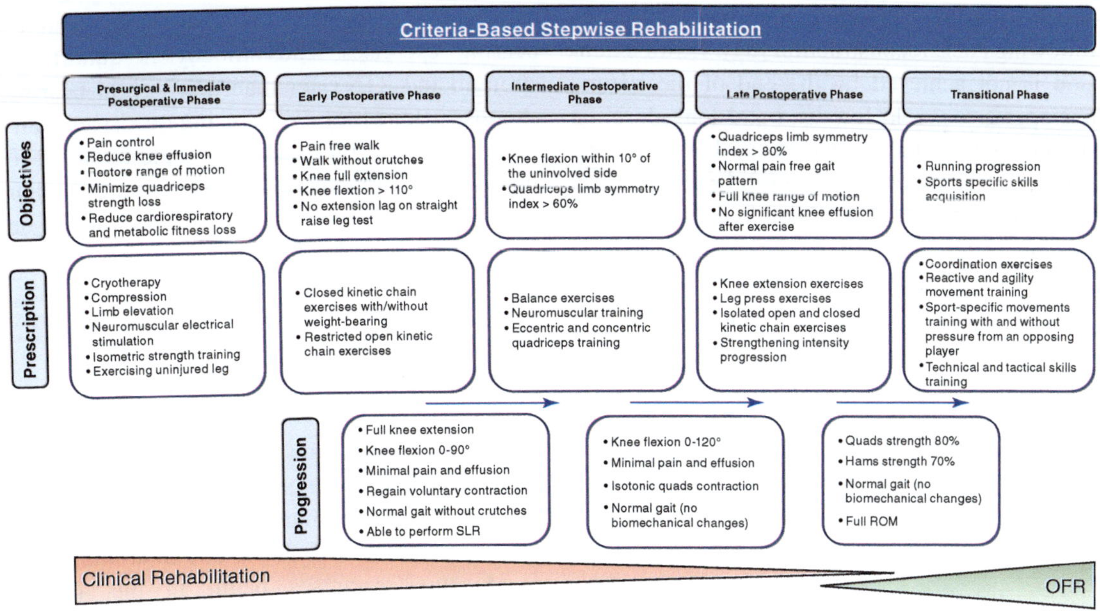

Fig. 23.1 Stepwise and goal-based rehabilitation. OFR On-field rehabilitation, SLR Straight leg raise, ROM Range of motion

23.4 Goal-Based and Stepwise Rehabilitation

23.4.1 Presurgical Rehabilitation

In the days after ACL injury, the goal for every patient should be pain control, reduce knee effusion, restore range of motion, minimize quadriceps strength loss, and reduce cardiorespiratory and metabolic fitness loss. Several studies pinpoint that extensive preoperative rehabilitation results in better functional outcomes and RTS rates [27, 28]. Preoperatively, achieving full knee extension, absent or minimal effusion, no knee extension lag during a straight leg raise, and good quadriceps strength [27, 29] are associated with better postsurgical outcomes, with improved knee function and higher levels of activity [27]. Indeed, quadriceps strength is a good predictor of clinical outcomes 2 years after ACL reconstruction. Achieving 80% symmetry of quadriceps strength before surgery is associated with improved 2 years clinical outcomes [30]. Cryotherapy, compression, and elevation are simple but essential measures to control the knee effusion. Physical exercises involving the uninjured leg and upper limbs can be very useful in maintaining cardiorespiratory and metabolic fitness, while exercises for the injured knee (active/passive range of motion and active quadriceps contraction) are crucial to improve knee function.

23.4.2 Immediate Postoperative Phase

ACL reconstruction causes further trauma to the knee. The milestones for the immediate postoperative phase are similar to the presurgical rehabilitation. The loss of knee range of motion is a common issue, usually associated with weaker quadriceps and postoperative complications [31]. Even small losses of knee extension (up to 5°)

adversely affect subjective and objective results following ACL reconstruction [31]. If full extension is not achieved by the end of the second postoperative week, low-load and long-duration stretching techniques, like prone hangs or bag hangs, can be used to restore full knee extension [32]. After ACL reconstruction, quadriceps muscles are often affected by arthrogenic muscle inhibition, limiting volitional contraction. Neuromuscular electrical stimulation (NMES) addition to isometric strength training during the first postoperative week can be used to re-educate voluntary contraction and augment quadriceps strengthening activities [33].

23.4.3 Early Postoperative Phase

The early postoperative phase begins as soon as pain and knee effusion are controlled and the athlete can engage in a more demanding rehabilitation exercises, usually in the second postoperative week. The aim is to achieve knee flexion greater than 110 degrees, walking pain free without crutches with the knee fully extended, perform a straight raise leg test without an extension lag, and use the cycle or stair climber without difficulty [29]. Once the athlete is able to walk without pain, crutches may be discontinued. Non-weight-bearing and weight-bearing exercises (closed-chain) can be safely incorporated into the rehabilitation process, as long as they are pain free and undue strain on the healing graft is avoided [34]. Early (within first week) full weight-bearing is encouraged but clinicians should progress as tolerated [33, 35]. Isometric quadriceps strengthening exercises can be initiated from the first postoperative week (if asymptomatic). Closed kinetic chain exercises should be prioritized in the first 4 weeks to decrease the risk of patellofemoral pain. Open kinetic chain exercises (45–90 degrees of knee flexion) can be added after 4 weeks. However, extra-caution should be taken in the first 12 weeks in order to avoid submitting the graft to great in situ forces, making it vulnerable to early graft failure [33]. Combining non-weight-bearing and weight-bearing exercises leads to higher quadriceps strength and RTS rates than non-weight-bearing exercises alone [36]. When using quadriceps tendon autografts, some caution should be noted to implement early intervention to promote patellar mobility, knee range-of-motion, and soft tissue compliance in order to reduce the risk of joint stiffness [37].

23.4.4 Intermediate Postoperative

Achieving knee flexion within 10 degrees of the uninvolved side and a quadriceps limb symmetry index (LSI) greater than 60% are the milestones for this phase [29]. Balance and neuromuscular re-education exercises begin in this phase. Eccentric (in closed kinetic chain) and concentric quadriceps training should be added at 3 weeks postoperative [35]. Neuromuscular alterations, such as muscle inhibition and impaired sensorimotor function, may contribute to clinical impairment, strength loss, and altered function. Combining neuromuscular re-education balance and proprioceptive exercises with standard rehabilitation methods improves clinical outcomes without deleterious effects [38].

23.4.5 Late Postoperative Phase

At the late postoperative phase, the treatment frequency depends on the remaining impairments. At this stage, the goals are to achieve quadriceps and hamstrings LSI greater than 80% in isokinetic assessment, a normal gait pattern, full knee range of motion, and no significant knee effusion or pain after exercise [29]. Those with large knee extensor strength deficits adopt movement patterns in which they utilize adjacent muscular groups, such as hip extensors, altering their walking gait pattern [18]. Besides standard strengthening exercises, open and closed chain knee extensor strengthening exercises should be included (knee extension or leg press) [39].

Isolated open and closed kinetic chain exercises will reduce athlete's adjacent muscular groups recruitment, enabling a targeted work of the desired muscle group. Although there are concerns about the safety of open kinetic chain exercises, they are unsupported by substantial published research [40]. It is important to avoid undue stress on the healing ACL graft, performing strengthening exercises at specific knee angles (45–90° open kinetic chain knee extension) and carefully incorporating other open kinetic chain strength exercises [29]. When using patella tendon and quadriceps tendon autografts, isolated progressive quadriceps overload training will be a crucial step of the rehabilitation program [37]. More specific to quadriceps tendon autografts, strengthening should target the superficial layer of the quadriceps tendon and rectus femoris with open kinetic chain exercises [41], and progress from active range-of-motion toward open kinetic chain quadriceps strengthening on a knee extension machine. Open kinetic chain exercises will be performed preferably at deeper levels of knee flexion to better load the quadriceps tendon, but some caution is advised as they may provoke irritability at the donor harvest site [41]. Restoring knee flexor strength is also crucial in the rehabilitation process. Hamstring strength deficits can persist for many years after ACL reconstruction [42], increasing nearly 10-fold the risk of ACL re-injury for every 10% decrease in knee flexor to extensor strength ratio of the injured limb in professional football players [43]. Regaining knee flexor strength is particularly important for athletes submitted to ACL reconstruction with hamstrings tendons graft, due to selective muscle atrophy [44]. In these athletes, knee flexor strengthening programs should be delayed for 6 weeks after surgery, allowing the donor site to heal [45]. Progression on intensity of strengthening can be based on the soreness rules, where the level exercise directly depends on the soreness felt by the athletes [46]. Rehabilitating other muscles, like the lumbo-pelvic and hip muscles, is also paramount. Suffering from weakness or reduced activation, lumbo-pelvic and hip muscles may contribute to an altered movement pattern, increase knee and ACL loading, and constitute an important risk factor for ACL injury [47].

Mimicking the situations in which muscles are commonly used and training athletes' functional strength is fundamental for a safe RTS. Closed kinetic chain strength (isometric or dynamic squat strength) correlates with athletic performance tasks, such as jumping, sprint running, and change of direction ability. It is fundamental to restore the necessary closed kinetic chain strength to support more demanding functional tasks, eliminate movement compensations, reduce chronic overload, and lower the odds of suffering new acute injuries [48]. Muscle imbalances/weakness, arthrokinetic dysfunction, altered reciprocal muscle inhibition, and synergistic dominance and altered proprioception [25] leads to impaired movement quality. Addressing these factors with corrective exercises is crucial to optimize movement quality. These exercise programs should include strengthening weak/inhibited muscles, muscle release techniques, and flexibility training of tight muscles [49].

23.4.6 Transitional Phase

Once a patient has met all the criteria of the late postoperative phase, a running progression may start. Although running is a relatively low demanding exercise to the knee, it marks the transition from early rehabilitation to a more advanced rehabilitation focused on sport-specific tasks [29]. Return to running is usually allowed around 12 weeks, but it can take from 5 to 39 weeks. The decision to return to running should be made on case-to-case basis and rely on meeting clinical (pain and range of motion) and physical performance-based (strength and hop testing) criteria and not only on the time elapsed. Progression to jogging to running can be based on knee effusion and reported pain [29].

Aiming to restore movement quality, improving sport-specific skills, physical re-conditioning and progressive chronic training load development, exercise prescription should be personalized and focused on the individual needs [50]. Neuromuscular training is not enough to directly restore movement quality during sport-specific movements. An athlete must practice the sport-specific movements to relearn and improve movement coordination [50], progressively incrementing the trained movement complexity. We start with a general coordination program, with pre-planned movements, to prepare for more progressive sport-specific movement retraining. Following this, re-active and agility movement training takes place, where an athlete moves and reacts to an external stimulus. At a later stage, athletes should retrain sport-specific movements with and without pressure from an opposing player [50].

Sport-specific physical re-conditioning is another key part of on-field rehabilitation. Appropriate cardiorespiratory and metabolic fitness should be trained as soon as possible because it is fundamental to perform relevant movement tasks and endure matches [50]. Specific training regimens for aerobic and anaerobic fitness are fundamental to on-field rehabilitation. Fatigue may alter movement quality and biomechanics [51], increasing the risk of new ACL injury [52].

Besides sport-specific movements, team-based sports (like football, basketball or volleyball) requires both technical and tactical skills. After a substantial time away from play, an athlete must practice specific tasks prior to RTS to achieve optimal coordination and performance. On-field rehabilitation is the ideal environment to improve both individual technical skills and tactical play. Gradual transition from individual training to small-group and then team-based training is mandatory [50]. Lastly, load progression is also a key part in every rehabilitation protocol. Inappropriate loading (chronic or acute) negatively affects sporting performance and increases injury risk [53]. On-field rehabilitation comprises the progression from returning to sport-specific rehabilitation through return to play (unrestricted team training) and return to competition (first competitive match) [54]. Buckthorpe et al. describe a five-stage soccer-specific on-field program that can serve as guidance for on-field rehabilitation [55].

23.5 Factors Affecting Return to Sports

A wide variety of factors can affect the clinical outcome of ACL reconstruction; thus, predicting if and when a player is ready to RTS is a difficult task. Acknowledging these factors is fundamental to optimizing rehabilitation and managing expectations.

23.5.1 Surgical Factors

Changes in loading of the knee after ACL injury can result in posterior medial meniscus and cartilage injuries; thus, studies suggest a potential benefit from early ACL reconstruction [56]. However, delayed surgery per se does seems to not impact the clinical outcomes, if there is no additional meniscal or chondral damage [57].

Using ACL autografts, bone-patellar tendon-bone (BPTB), hamstrings tendons or quadriceps tendon, might not differ in clinical and functional outcomes [58]. Still, autograft selection implies donor site morbidity and should be considered when rehabilitating patients. BPTB harvesting can cause anterior knee pain, especially during kneeling and quadriceps strengthening exercises [59]. If hamstrings autograft is used, some authors recommend to wait 12 weeks before engaging in hamstrings open kinetic chain rehabilitation exercises to allow soft tissue healing [35]. While we avoid strenuous activity on the hamstrings to allow for donor site recovery, typically in the first 6 postoperative weeks [45], we recommend to progressively implement strengthening exercises to restore the muscle strength as soon as possible. More recently, the quadriceps tendon has also been successfully used as an alternative option [60]. With decreased donor site morbidity [61], quadriceps autografts show similar results in knee stability tests [58]. An allograft selection does not imply donor site morbidity, nonirradiated allografts have similar strength to that of autografts, and patients have faster immediate postoperative recovery and less operative pain [62]. On the other hand, allografts have delayed incorporation and remodeling [63], combined with a premature attempt at RTS, might be the cause for higher retear rates using allografts [64]. Younger patients might benefit from an autograft [65], but there is still no evidence that delaying RTS decreases the odds of allograft rupture [29, 62] The choice of graft may influence the time to RTS, and the BPTB autograft may take longer to complete postoperative rehabilitation, recover quadriceps strength, and meet RTS criteria, as compared to hamstring autograft or allografts [62]. The addition of lateral extra-articular tenodesis (LEAT) has recently re-emerged and may be considered in revision cases or in younger athletes that have a high-risk profile including hyperlaxity, high rotatory knee laxity, and high-risk associated bone morphology. Adding a LEAT decreases the forces in the graft [66] (especially high rotatory torques due to its longer lever arm) and does not increase the risk of knee OA [67]. It can reduce the risk of clinical failure by 38% (ACL revision or persistent rotatory knee laxity) and the risk of graft failure by 67% [68].

Surgical technical errors are often cited as a leading cause of ACL reconstruction failure [69]. Several authors have argued that nonanatomical tunnel placement arises from errors in surgical technique [69]. Anatomic tunnel placement, which is the current trend in ACL reconstruction, is believed to result in improved outcomes [70] and should be the goal of ACL reconstruction regardless of the selected surgical technique [71]. In the transtibial technique, the femoral tunnel is created through the tibial tunnel; thus, the position of the femoral tunnel is determined by the tibial tunnel. On the other hand, using the anatomical technique, the tibial tunnel and the femoral tunnel are created independently and their positions are determined by the surgeon. By avoiding the constraints of the tibial tunnel, several authors argue that a more anatomic tunnel placement results in a better elimination of the pivot-shift phenomenon [72]. Tunnel malposition influences knee kinematics and may cause knee instability, reduced mobility, graft impingement, progressive graft loosening, and graft damage [73]. Infection is a rare surgical complication, with a 0.5% rate of occurrence [21], that can have catastrophic results. It may lead to further surgical interventions, delayed rehabilitation, additional chondral damage, graft tear, fibrous adhesions, and arthrofibrosis development [21]. Clinically, arthrofibrosis manifests as a stiff knee with reduced range of motion and a flexed gait, conditioning secondary patellofemoral overload, pain, and further chondral wear down [21]. Graft tension can determine the success of the surgery. Despite preconditioning protocol, after several cycles of knee flexion and extension, there can be loss of graft tension [74], resulting in residual knee instability and poor clinical outcome [75]. On the other hand, excessive graft tension can increase articular pressure, cause joint stiffness, secondary patellofemoral overload, and pain [21].

23.5.2 Concomitant Injuries

Concomitant injuries are frequent at the time of ACL reconstruction, with an incidence as high as 60% [76]. These lesions can be previous or contemporary to the ACL tear, or can occur while the patient is awaiting surgical ACL reconstruction. Meniscal injuries are the most frequent, but there could also be chondral damage or other ligament lesions [29]. Concomitant injuries are associated with premature OA [77], lower patient-reported outcomes measures (PROMs) [78] and, if untreated, can be responsible for more than 15% of ACL reconstruction failures [21].

23.5.2.1 Meniscal Lesions

Losing meniscal tissue may negatively affect the knee biomechanics and accelerate the degeneration process, with early onset of OA after ACL reconstruction [79]. Thus, conserving or repairing the menisci, whenever possible, is fundamental for a better clinical outcome [29]. When feasible, meniscal repairs are the gold standard treatment [80]. Several studies reported that meniscal repairs performed at the time of ACL reconstruction have superior healing rates and better outcomes than isolated meniscal repairs [80]. Rehabilitation protocols are, however, not consensual. After partial arthroscopic meniscectomy, no rehabilitation modifications are necessary [29]. For meniscal repair, some surgeons limit weight-bearing and knee flexion during the first 8 weeks (45 degrees in the first 4 weeks and progressive flexion during the remaining weeks) [81], while others suggest that early weight-bearing and unrestricted mobilization show similar success rates [82]. After the first 8 weeks, there are no additional restrictions to rehabilitation progressions for an ACL reconstruction with meniscal repair [29].

23.5.2.2 Other Ligaments

The MCL is sometimes injured during an ACL tear, especially in athletes. While ACL reconstruction is the gold standard treatment, the treatment of concomitant MCL lesions is controversially in the literature [83]. MCL injuries can be graded as I, II, or III. Conservative treatment of MCL injuries grade I is generally advised, but the treatment regime for grades II and III is still controversial [83]. In combined ACL–MCL injuries, anteromedial rotatory knee instability is a frequent complication, with increasing incidence according to the severity of the MCL lesion [83]. The evaluation of this instability can be a crucial decision factor between surgical or nonsurgical treatment of concomitant MCL lesions [84]. In the acute setting, surgical treatment of both the ACL and MCL can lead to higher rates of knee stiffness with decreased range of motion, requiring more aggressive physical therapy regimens; [85] initial conservative management (4–6 weeks) with delayed ACL reconstruction allows the patient to regain range of motion and strength faster [29]. The rehabilitation regimen for acute combined ACL–MCL injuries must be restricted to exercises performed in the sagittal plane. During these exercises, tibial internal rotation helps to eliminate valgus stress for a proper MCL healing process [29]. When initial conservative treatment fails and chronic MCL insufficiency develops, combined MCL and ACL reconstruction is indicated. In this setting, patients should use crutches for 5 weeks, an immobilizer with knee in full extension for 6 weeks, and weight-bear as tolerated [86]. After this period, patients should resume regular ACL reconstruction guidelines for isolated ACL reconstruction [29].

Multiligament injuries can involve the ACL, posterior cruciate ligament (PCL), and collateral ligaments rupture. These are devastating injuries with very variable clinical results. Around 6 to 8 weeks of non-weight-bearing with a knee immobilizer is recommended after multiligament reconstructive surgery [87]. PCL reconstruction techniques have less predictable outcomes when compared to ACL reconstruction surgeries. Thus, protecting the PCL graft is the key point of rehabilitation after combined ACL and PCL reconstruction surgery [88]. Therefore, these patients should follow a more conservative PCL rehabilitation. Preventing tibial sliding, by avoiding hyperextension and resisted knee flexion for 12 weeks, is essential for a proper healing process. With concomitant posterolateral corner injury, it is recommended to eliminate tibial external rotation and varus stress [88].

23.5.3 Intrinsic and Extrinsic Factors

Several factors can influence the clinical outcomes and RTS, but scientific knowledge on these factors is still limited. Intrinsic factors are individual-level characteristics such as age, gender, genome, and psychological, physiological, and pathological factors. Extrinsic factors are a broad category of outside the body elements that can yield their influence on the rehabilitation outcomes and include quality of rehabilitation, adherence, and compliance with the rehabilitation program, type, and level of sports activity, among others.

23.5.3.1 Playing Field

Understanding how field surface type affects the risk of lower extremity injuries is essential. Synthetic surfaces can be harder and lack the ability to release the cleat in a potentially injurious overload situation, generating higher shear forces and torque on the lower limb [89]. Meanwhile, grass turf can release the cleat easily from the surface, which serves as an inherent force-limiting mechanism. Synthetic turfs can record higher rates of lower extremity injuries caused by a twisting or shearing mechanism than natural turfs [90] and should thus be taken into account when releasing the athlete to sports.

23.5.3.2 Age and Level of Activity

The rate of ACL graft retear and contralateral ACL tear is higher in younger athletes (<25 years) that returned to sports [91] and younger age has been considered a critical factor for graft tear. More recent evidence shows that after adjusting some confounding factors (considering only those that pass the RTS criteria and return to level 9 or 10 sports according to the Tegner activity scale), there is no association between age and risk of a second ACL injury [92]. There is conflicting evidence regarding the effect of age on PROMs. While some studies associate older age (≥30 years) with increased odds of achieving better PROMs [57, 93], others correlate younger age (<30 years) with better PROMs [94] or did not found any correlation at all [95].

We cannot, however, separate the age factor from the activity level. Participating in sporting activities carries an injury risk, which depends on numerous factors including higher level of competition, game day, preseason, resting period, and type of sports (pivoting and collision or contact sports). Returning to level I sports has a fourfold increased risk of a new knee injury. Younger athletes (<25 years) are more active (higher baseline level of sports) with higher sporting demands and expectations, which could justify lower PROMs [57] and the often-reported higher retear rates. Younger athletes also experience higher rates of return to preinjury level of sports, which can also explain the higher retear rates [96].

23.5.3.3 Gender

There is conflicting evidence regarding gender and return to preinjury sports rates. While Ardern et al. [96] found no differences, Walden et al. [97] reported that males exceeded females in some age groups. There are, however, a number of studies pinpointing a higher risk of ipsilateral reinjuries in younger males [98, 99]. Some studies report residual knee laxity after ACL reconstruction in females using instrumented testing, but within the clinical setting these differences were not replicated [100]. Female gender has also been associated with inferior PROMs after ACL reconstruction [57, 95]. Differences in muscle function between gender are sometimes hypothesized but no differences were found in any objective evaluation [57].

23.5.3.4 Psychological Factors

Despite the often-reported successful outcomes, rates of athletes resuming to their preinjury levels are lower than expected, with only 55% returning to competitive sports [9]. Increasing evidence supports the role of cognitive and emotional factors in the RTS after ACL reconstruction [101]. Despite the recent reports on psychological factors, these are still largely unexplored.

Fear of reinjury, kinesiophobia, and other negative feelings are common reasons cited by athletes who do not RTS after ACL reconstruction [102]. Kinesiophobia and fear of re-injury can be assessed through the Tampa Scale [103]. Meanwhile, the Anterior Cruciate Ligament–Return to Sport after

Injury (ACL-RSI) scale was developed to assess the athlete's psychological readiness to RTS after ACL reconstruction. The ACL-RSI aims to identify psychological aspects for which patients may need counselling to improve their chances of returning to preinjury activity level [104]. The psychological response, assessed pre-operatively and during early rehabilitation, can predict sporting ability and increase the odds of return to preinjury level of sports [9]. Athletes who do not return to preinjury level of sports often display greater fear of re-injury [105] and might even attribute this fear to their inability or lack of confidence to return to preinjury levels. Greater fear of reinjury is also associated with an increased risk of a second ACL injury [105]. Psychological readiness can differ between gender, with females recording lower scores and lower RTS rates [102]. On the other hand, athletes with a more positive attitude toward RTS and higher ACL-RSI scores have greater odds of returning to preinjury activity levels [9]. Merging the psychosocial factors surveillance with the rehabilitation can create a more comprehensive approach with a more positive and successful rehabilitation environment. Thus, some authors suggest assessing ACL-RSI scores preoperatively; 3 months postoperative, when individuals can start RTS-specific drills; and 6 months postoperatively, when athletes can return to training [101, 104]

23.5.3.5 High-Quality Rehabilitation

An ACL reconstruction alone is not enough to guarantee a safe return to preinjury activity levels. After ACL reconstruction, appropriate physical and psychological recovery are regarded as key factors prior to RTS [106], and are associated with superior clinical and patient reported outcomes [57]. Muscular asymmetries may be a determinant risk factor for ACL graft tear or new knee injuries [106]. Inadequate rehabilitation protocols can injure the healing graft or other peripherical structures [106]. Following individualized and goal-based progressive rehabilitation programs is fundamental to achieve a safe RTS. The clinician should focus on meeting the RTS criteria, whilst reconditioning the athletes to their preinjury level of performance. This is especially important in elite professional athletes, where the stakeholders' incentives and dedicated rehabilitation facilities can enhance the quality of rehabilitation.

23.6 Criteria-Based Return to Sports

To achieve medical clearance to RTS, athletes must pass all the necessary criteria [29]. Time has been historically used as the main or sole criterion to allow the athlete to RTS; but, since the 1990s other criteria (although infrequently) have been used to allow a more objective and comprehensive assessment of the athlete readiness to RTS, including strength (41% of studies), clinical examination (26% of studies), physical performance-based criteria (20% of studies), and PROMs (12% of studies) [107].

Strict time and functional-based objective criteria have been proposed to decide on the readiness of the athlete to RTS. Those athletes who fail to meet the discharge criteria show a fourfold increased risk of knee reinjuries and early graft tears [106]. Poor function at 1 year is also associated with a 3.7 times increased risk of worsening patellofemoral chondral damage [108] Passing the criteria for a definitive decision to RTS is still controversial because the reduced risk for second ACL injuries (same knee or contralateral) is not statistically significant [109]. Recent studies [22] show that there is a low rate of meeting the RTS criteria at 6 and 9 months (3 and 11%, respectively) and that nearly half of athletes do not meet the discharge criteria when they RTS, suggesting insufficient rehabilitation and low compliance with criteria-based RTS. Despite the controversy, a criteria-based RTS decision indicates the level of readiness of the athlete and can identify important deficits that require intervention before allowing the athlete to RTS. Box 1 displays suggested criteria that can be used for RTS.

Box 1. Recommended Criteria for RTS

Clinical Examination

- Ligament stability, full ROM, no effusion or pain

Chronobiological

- Minimum 6 months, but 9 months are suggested

Strength Testing

- Quadriceps LSI > 90% (100% if elite pivot sport)

Functional Testing

- Single-Leg Hop Tests LSI > 90% (100% if elite pivot sport)

Movement quality

- No biomechanical deficits in functional and reactive decision tasks

Patient reported-outcome measures

- IKDC = 100

Psychological readiness

- ACL-RSI > 56 points

Agility Testing (changes of direction)

- Agility T Test < 11s

Physically ready without symptoms

- No pain / effusion during team training & reactive decisions

Tolerance to maximal sprints

- Sprints > 8 m/s [25 Km/h] - only for sports involving sprints

23.6.1 Chronobiological

Following ACL reconstruction, the graft goes through a remodeling and maturation process during the first 12 months [110], explaining graft weakness and higher risk of reinjury. Although some studies suggest that maturation can last up to 2 or 3 years [110], after a 12-month period, the graft may display similar risk of injury than the native contralateral ACL. Maturation of graft varies between auto and allograft, and between autograft options [110]. If available, the clinician may ask for a magnetic resonance imaging (MRI) to assess the graft healing [111]. An increased T2 signal intensity may indicate increased graft laxity [112]. The use of MRI for assessing readiness to RTS is however still not well supported by evidence and should not be used as sole criterion, as signal intensity correlates poorly with clinical outcomes [113].

Aiming to respect the healing process and protect the graft, time is widely used as a RTS criterion. Time is unfortunately still the major criterion used in the RTS decision but time as a criterion to RTS remains controversial. A total of 85% of studies use time as primary criterion for RTS and 44% of studies use time as the sole criterion to RTS [107]. The hypothesis is that waiting longer before RTS provides extra time for the graft to continue the healing process and the athlete to the improve strength, motor control (physical readiness), and psychological readiness to return to activity.

Looking at the available evidence, various studies explored the correlation between time to RTS time and second ACL injury risk. The two papers with the highest sample size found no difference in second ACL injury rate with various time to RTS [114, 115] On over 1400 patients available at follow-up, Shelbourne and colleagues [114] found no difference in secondary ACL risk (within 5 years follow-up) in players that returned before or after 6 months. King et al. [115] on a very similar cohort of 1432 high-level athletes (2-year follow-up), found no difference in the second ACL injury rate if RTS was between 6 and 9 months, 9 and12 months, or 13 and 16 months. Similarly, Zwolski et al. [116] on a cohort of 188 young athletes, found no difference in second ACL injury risk basing on RTS time. McPherson et al. [102] on a cohort of 329 patients, also reported no significant difference in RTS time between second ACL injured athletes and the remaining group.

On the other hand, other studies, with smaller sample sizes, show different findings. Beischer et al. [117] found a higher risk of second ACL injury if RTS happened before 9 months in a cohort of 159 young athletes. Grindem et al. [106] on a cohort of 106 patients, found an increased risk of second knee injury (all knee injuries, not only at the ACL) if RTS was before 9 months, with a 51% decreased risk of knee injury for each month the RTS was delayed, up to 9 months. All athletes that RTS before 5 months had a knee injury within 2 months of return. Finally, Dekker et al. [118] on a cohort of 85 young athletes (<18 years old), reported no significant difference in RTS time between the second ACL injury group and the remaining one, but there was a small protective effect for each month the RTS was delayed.

There is no clear evidence to date that RTS time is correlated with second ACL injury risk. However, clinical practice guidelines are now suggesting a prolonged rehabilitation (9–12 months), depending on the goals of the athlete [35] without a strong scientific support from the literature. On one side, one group proposed to wait up to 2 years to RTS, especially referring to the youngest patients [119]. On the other side, this is not practically feasible, especially when referring to high-level athletes, that expect to RTS as fast and as safe as possible. There is indeed a difficult balance between RTS at the preinjury level and second ACL injury risk that can be solved with high-quality rehabilitation, based on correct volume and frequency of sessions with a personalized load progression until the fulfilment of the RTS criteria. Buckthorpe et al. [120] proposed a model with a strong focus on the transition between rehabilitation and performance. We strongly advise on completing a period of on-field rehabilitation that can act as a bridge before the return to unrestricted team practice.

23.6.2 Strength Testing

Strength is a key factor to RTS, knee stabilization, and motor control [43, 106], and may directly influence the clinical and patient reported outcomes [57] and the reinjury rate [106]. Strength is among the most frequently used RTS criterion (more than 40% of studies) [107]. Achieving a LSI of 90% have previously been suggested as the milestone to determine normal limb symmetry in quadriceps strength and functional testing [121]. Other authors have suggested a LSI of 100% for both knee extensor and flexor strength in competitive contact and pivoting athletes [122]. Isokinetic testing provides an objective and popular method of strength evaluation. Although scientific papers report multiple evaluation protocols, knee-extensors strength evaluation at 60 degrees/s is the most used protocol [107]. In our practice, we also use the 90 degrees/s and 180 degrees/s that are more adjusted to the sports-specific skills that required faster production of strength (e.g., soccer shot to the goal). The knee-flexors strength should also be considered for the LSI, especially in those that the hamstrings graft was used. The quadriceps/hamstrings conventional and functional ratios show association with the risk of graft tear [43] and have been included by some authors as passing criterion [107] but require further scientific validation to obtain reliable cut-offs.

23.6.3 Functional Testing

Single-leg hop tests are popular functional performance tests that can assess and objectively quantify knee functional performance [123]. By using the uninjured limb as a control for comparisons between limbs, single-leg hop tests have also been used as an RTS criterion in 14% of studies [107], with better results associated with higher RTS rates [124]. Achieving a LSI of 90% or more is the suggested milestone, with some authors suggesting again the 100% goal [35, 122]. If available, the hop performance should achieve the preinjury or pre-season levels, because the contralateral limb also suffers a deconditioning after injury [125]. Single-leg hop tests usually may include the single hop for distance, triple hop for distance, crossover hop for distance, and the 6-m timed hop [126]. Recently, two new single-leg hop tests have been proposed—the medial side triple hop for distance and the 90° medial rotation hop for distance—that showed high reliability and are more likely to show limb asymmetries in ACL-reconstructed athletes compared to the usual forward hop tests [127]. Agility tests (T test, Y balance test or star balance test) are also described, but less frequently [107]. The agility tests involve changes of direction and sport-specific reactive decisions. For instance, ACL injuries in soccer occur mostly as noncontact mechanism during defensive actions [128], and we should thus test the athlete's functional competence during decision-making process while performing defensive actions.

Despite the functional tests available, interpreting the LSI is limited. Possible altered knee's biomechanics, the absence of an accurate baseline, or a preinjury measure can limit the applicability of LSI in RTS, requiring further investigation [129]. Hop testing only assesses the physical performance, but fails to evaluate the biomechanical competence (kinetic and kinematic deficits) [130]. In fact, although many athletes show an LSI > 90%, many display a stiff knee landing [131]. It is crucial to also assess the quality of movement and identify any biomechanical deficit before discharging the athlete to RTS [35]. There are available qualitative scoring systems (Jump Landing System and Landing Error Scoring System) [132] and quality of movement tests (tuck jump [133], drop vertical jump [134], side stepping [135] and pivoting maneuvers) that can be used when evaluating the ACL-reconstructed athlete. A Landing Error Scoring System below 5 points during a jump-landing task was suggested as an RTS criterion [136], but requires further validation.

23.6.4 Clinical Examination

Physical examination of ligament stability, range of motion, effusion, and pain is fundamental for clinical assessment after ACL reconstruction. As a RTS criterion, clinical examination is reported in 26% of the studies [107]. Achieving full knee range of motion, absence of effusion, and no pain (after one-on-one sport-specific exercises) are clinical milestones commonly used as RTS criteria [107]. Knee residual laxity is a major risk factor for reinjury, although it is not frequently used as an RTS criterion (only in 6% reported studies) [107]. Clinical manual tests to access knee stability (pivot-shift, Lachman or anterior drawer tests) are easy to perform, quick, and efficient. However, they are subjective tests that lack standardization or objective grading and are user-dependent, making them hard to interpret in the clinical setting [137]. Instrumented-assisted devices to access knee laxity might solve this problem, but these are poorly available in the clinical setting [137]. In our experience, we rely on a MRI-compatible stress arthrometer—the Porto Knee Testing Device [138]—to measure knee laxity after ACL reconstruction which helps us to objectively measure knee laxity before discharging the athlete to RTS and identify those with residual knee laxity.

23.6.5 Patient Report Outcome Measures (PROMs)

The PROMs are a clinical-related outcome reported by the patient that are essential for clinical evaluation of outcomes (in the patient perspective) but rarely used as a RTS criterion [107]. The interpretability of PROMs is not straightforward, where a successful surgery or rehabilitation for the clinician may not correspond to a patient's satisfactory knee function, limiting its use as an objective RTS criterion [139]. Clinical practice guidelines have suggested a variety of PROMs to be used during rehabilitation and at the RTS assessment, including the Knee and Osteoarthritis Outcome Score (KOOS), the International Knee Documentation Committee (IKDC) 2000 questionnaire, the Lysholm Knee Scale, the Marx Activity Level Scale, the Tegner Activity Level Scale, the Knee Self-Efficacy Scale (K-SES), the Tampa Scale-11 (TSK-11), and the ACL-RSI score [23]. However, their use as an objective RTS criterion is still poorly validated. The Multicenter Orthopaedic Outcomes Network (MOON) clinical practice guidelines recommend a scoring over 9 points in the IKDC Question #10 [33]. Other authors have suggested an IKDC score within 15th percentile of healthy subjects [136, 140] The ACL-RSI has been widely used during rehabilitation and RTS testing. The psychosocial readiness measures are important to assess the athlete's mental resiliency and should be included in the RTS decision process. Validated cut-offs are still missing, but some authors [136, 140] suggest an ACL-RSI score over 56 points as RTS criterion.

23.7 Secondary Prevention

Athletes who RTS have higher odds of sustaining a reinjury [141], which are usually more severe. A stepwise rehabilitation [142] and meeting all discharge criteria before RTS is the very first step to lower the odds of sustaining reinjuries and graft tears [43, 106]. It is also important to identify the underlaying modifiable risk factors and implement neuromuscular preventive strategies in late phase rehabilitation [143].

Secondary prevention should always be part of a RTS program [144] and recent studies show that secondary prevention programs can decrease the reinjury rate by more than 40% [144]. Following RTS, specific training programs with agility and stabilization exercises, rather than regular strengthening and stretching exercises, can decrease the risk for reinjury (7.7 vs. 70%) [142]. A recent study has shown that 10 sessions of a secondary prevention program resulted in a small reinjury rate of 2.5% [145]. Video feedback, virtual reality, and verbally directed optimization techniques can help to enhance the ACL injury prevention programs [146] and help the athlete to recover a safe and effective sport-specific movement.

23.8 Take-Home Messages

Deciding when the athlete is ready to RTS is a complex task. A multifactorial approach, considering intrinsic and extrinsic factors, an individualized high-quality rehabilitation, and strict and objective RTS criteria, is needed to increase the odds and ensuring a safe RTS (Fig. 23.2).

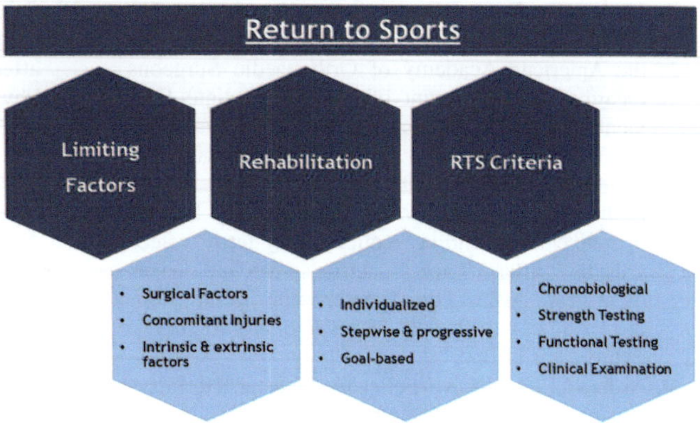

Fig. 23.2 Factors to be considered during rehabilitation and on the RTS decision. RTS Return to sport

23.9 Top Five Evidence-Based References

1. Adams D, Logerstedt D, Hunter-Giordano A, Axe MJ, Snyder-Mackler L. Current concepts for anterior cruciate ligament reconstruction: A criterion-based rehabilitation progression. *J Orthop Sports Phys Ther.* 2012;42(7):601–614.
2. Buckthorpe M. Optimising the Late-Stage Rehabilitation and Return-to-Sport Training and Testing Process After ACL Reconstruction. *Sport Med.* 2019;49(7):1043–1058.
3. Cristiani R, Mikkelsen C, Edman G, Forssblad M, Engström B, Stålman A. Age, gender, quadriceps strength and hop test performance are the most important factors affecting the achievement of a patient-acceptable symptom state after ACL reconstruction. *Knee Surgery, Sport Traumatol Arthrosc.* 2019;28(2):369–380.
4. Burgi CR, Peters S, Ardern CL, et al. Which criteria are used to clear patients to return to sport after primary ACL reconstruction? A scoping review. *Br J Sports Med.* 2019;53(18), 1154–1161.
5. King E, Richter C, Jackson M., et al. Factors Influencing Return to Play and Second Anterior Cruciate Ligament Injury Rates in Level 1 Athletes After Primary Anterior Cruciate Ligament Reconstruction: 2-Year Follow-up on 1432 Reconstructions at a Single Center. Am J Sports Med. 2020;48(4):812–824.

References

1. Moses B, Orchard J, Orchard J. Systematic review: annual incidence of ACL injury and surgery in various populations. Res Sports Med. 2012;20(3–4):157–79. https://doi.org/10.1080/15438627.2012.680633.
2. Montalvo AM, et al. 'what's my risk of sustaining an ACL injury while playing sports?' A systematic review with meta-analysis. Br J Sports Med. 2019;53(16):1003–12. https://doi.org/10.1136/bjsports-2016-096274.
3. Dingenen B, Gokeler A. Optimization of the return-to-sport paradigm after anterior cruciate ligament reconstruction: a critical step back to move forward. Sports Med. 2017;47(8):1487–500. https://doi.org/10.1007/s40279-017-0674-6.
4. Buckwalter JA. Articular cartilage injuries. Clin Orthop Relat Res. 2002;402:21–37. https://doi.org/10.1097/00003086-200209000-00004.
5. Weiler R. Unknown unknowns and lessons from non-operative rehabilitation and return to play of a complete anterior cruciate ligament injury in English Premier League football. Br J Sports Med. 2016;50(5):261–2. https://doi.org/10.1136/bjsports-2015-095141.
6. Ekås GR, Ardern CL, Grindem H, Engebretsen L. Evidence too weak to guide surgical treatment decisions for anterior cruciate ligament injury: a systematic review of the risk of new meniscal tears after anterior cruciate ligament injury. Br J Sports Med. 2020;54(9):520–7. https://doi.org/10.1136/bjsports-2019-100956.

7. Marx RG, Jones EC, Angel M, Wickiewicz TL, Warren RF. Beliefs and attitudes of members of the American Academy of Orthopaedic Surgeons regarding the treatment of anterior cruciate ligament injury. Arthroscop J Arthroscop Relat Surg. 2003;19(7):762–70. https://doi.org/10.1016/S0749-8063(03)00398-0.
8. Richmond JC, Lubowitz JH, Poehling GG. Prompt operative intervention reduces long-term osteoarthritis after knee anterior cruciate ligament tear. Arthroscop J Arthroscop Relat Surg. 2011;27(2):149–52. https://doi.org/10.1016/j.arthro.2010.11.060.
9. Ardern CL, Taylor NF, Feller JA, Webster KE. Fifty-five per cent return to competitive sport following anterior cruciate ligament reconstruction surgery: an updated systematic review and meta-analysis including aspects of physical functioning and contextual factors. Br J Sports Med. 2014;48(21):1543–52. https://doi.org/10.1136/bjsports-2013-093398.
10. Lie MM, Risberg MA, Storheim K, Engebretsen L, Øiestad BE. What's the rate of knee osteoarthritis 10 years after anterior cruciate ligament injury? An updated systematic review. Br J Sports Med. 2019;53(18):1162–7. https://doi.org/10.1136/bjsports-2018-099751.
11. Rebelo-Marques A, Andrade R, Pereira R, João E-M. Return to Play (RTP). In: The sports medicine physician. Berlin Heidelberg: Springer; 2019. p. 149–69.
12. Thomeé R, et al. Variability in leg muscle power and hop performance after anterior cruciate ligament reconstruction. Knee Surg Sports Traumatol Arthrosc. 2012;20(6):1143–51. https://doi.org/10.1007/s00167-012-1912-y.
13. McCormack RG, Hutchinson MR. Time to be honest regarding outcomes of ACL reconstructions: should we be quoting 55-65% success rates for high-level athletes? Br J Sports Med. 2016;50(19):1167–8. https://doi.org/10.1136/bjsports-2016-096324.
14. Webster KE, Feller JA, Leigh WB, Richmond AK. Younger patients are at increased risk for graft rupture and contralateral injury after anterior cruciate ligament reconstruction. Am J Sports Med. 2014;42(3):641–7. https://doi.org/10.1177/0363546513517540.
15. Magnussen RA, Meschbach NT, Kaeding CC, Wright RW, Spindler KP. ACL graft and contralateral ACL tear risk within ten years following reconstruction. J Bone Joint Surg. 2015;3(1):1–8.
16. Ernst GP, Saliba E. Lower-extremity compensations following anterior cruciate ligament reconstruction. Phys Ther. 2000;80(3):251–60. https://doi.org/10.1093/ptj/80.3.251.
17. Wojtys EM, Huston LJ. Longitudinal effects of anterior cruciate ligament injury and patellar tendon autograft reconstruction on neuromuscular performance. Am J Sports Med. 2000;28(3):336–44. https://doi.org/10.1177/03635465000280030901.
18. Salem GJ, Salinas R, Harding FV. Bilateral kinematic and kinetic analysis of the squat exercise after anterior cruciate ligament reconstruction. Arch Phys Med Rehabil. 2003;84(8):1211–6. https://doi.org/10.1016/S0003-9993(03)00034-0.
19. Allen MM, et al. Are female soccer players at an increased risk of second anterior cruciate ligament injury compared with their athletic peers? Am J Sports Med. 2016;44(10):2492–8. https://doi.org/10.1177/0363546516648439.
20. Lind M, Menhert F, Pedersen AB. Incidence and outcome after revision anterior cruciate ligament reconstruction: results from the Danish registry for knee ligament reconstructions. Am J Sports Med. 2012;40(7):1551–7. https://doi.org/10.1177/0363546512446000.
21. Alentorn-geli E, Cugat R, Farmer KW, Moser MW. Failure of ACL reconstruction. Arch Bone Joint Surg. 2015;3(4):220–40. https://doi.org/10.22038/abjs.2015.4584.
22. Ebert JR, et al. Strength and functional symmetry is associated with post-operative rehabilitation in patients following anterior cruciate ligament reconstruction. Knee Surg Sports Traumatol Arthrosc. 2018;26(8):2353–61. https://doi.org/10.1007/s00167-017-4712-6.
23. Andrade R, Pereira R, Van Cingel R, Staal JB, Espregueira-Mendes J. How should clinicians rehabilitate patients after ACL reconstruction? A systematic review of clinical practice guidelines (CPGs) with a focus on quality appraisal (AGREE II). Br J Sports Med. 2019:1–9. https://doi.org/10.1136/bjsports-2018-100310.
24. Allum JHJ, Bloem BR, Carpenter MG, Hulliger M, Hadders-Algra M. Proprioceptive control of posture: a review of new concepts. Gait Posture. 1998;8(3):214–42. https://doi.org/10.1016/S0966-6362(98)00027-7.
25. Aman JE, Elangovan N, Yeh IL, Konczak J. The effectiveness of proprioceptive training for improving motor function: a systematic review. Front Hum Neurosci. 2015;8(JAN):1–18. https://doi.org/10.3389/fnhum.2014.01075.
26. Della Villa S, et al. Clinical outcomes and return-to-sports participation of 50 soccer players after anterior cruciate ligament reconstruction through a sport-specific rehabilitation protocol. Sports Health. 2012;4(1):17–24. https://doi.org/10.1177/1941738111417564.
27. Failla MJ, et al. Does extended preoperative rehabilitation influence outcomes 2 years after ACL reconstruction? Am J Sports Med. 2016;44(10):2608–14. https://doi.org/10.1177/0363546516652594.
28. Grindem H, Granan LP, Risberg MA, Engebretsen L, Snyder-Mackler L, Eitzen I. How does a combined preoperative and postoperative rehabilitation programme influence the outcome of ACL reconstruction 2 years after surgery? A comparison between patients in the Delaware-Oslo ACL Cohort and

the Norwegian National Knee Ligament Registry. Br J Sports Med. 2015;49(6):385–9. https://doi.org/10.1136/bjsports-2014-093891.

29. Adams D, Logerstedt D, Hunter-Giordano A, Axe MJ, Snyder-Mackler L. Current concepts for anterior cruciate ligament reconstruction: a criterion-based rehabilitation progression. J Orthop Sports Phys Ther. 2012;42(7):601–14. https://doi.org/10.2519/jospt.2012.3871.
30. Eitzen I, Holm I, Risberg MA. Preoperative quadriceps strength is a significant predictor of knee function two years after anterior cruciate ligament reconstruction. Br J Sports Med. 2009;43(5):371–6. https://doi.org/10.1136/bjsm.2008.057059.
31. Shelbourne KD, Gray T. Minimum 10-year results after anterior cruciate ligament reconstruction: how the loss of normal knee motion compounds other factors related to the development of osteoarthritis after surgery. Am J Sports Med. 2009;37(3):471–80. https://doi.org/10.1177/0363546508326709.
32. Logerstedt DS, et al. Knee pain and mobility impairments: meniscal and articular cartilage lesions. J Orthop Sports Phys Ther. 2010;40(6) https://doi.org/10.2519/jospt.2010.0304.
33. Wright RW, et al. Anterior cruciate ligament reconstruction rehabilitation: MOON guidelines. Sports Health. 2015;7(3):239–43. https://doi.org/10.1177/1941738113517855.
34. Bynum EB, Barrack RL, Alexander AH. Open versus closed chain kinetic exercises after anterior cruciate ligament reconstruction: a prospective randomized study. Am J Sports Med. 1995;23(4):401–6. https://doi.org/10.1177/036354659502300405.
35. Van Melick N, et al. Evidence-based clinical practice update: practice guidelines for anterior cruciate ligament rehabilitation based on a systematic review and multidisciplinary consensus. Br J Sports Med. 2016;50(24):1506–15. https://doi.org/10.1136/bjsports-2015-095898.
36. Mikkelsen C, Werner S, Eriksson E. Closed kinetic chain alone compared to combined open and closed kinetic chain exercises for quadriceps strengthening after anterior cruciate ligament reconstruction with respect to return to sports: a prospective matched follow-up study. Knee Surg Sports Traumatol Arthrosc. 2000;8(6):337–42. https://doi.org/10.1007/s001670000143.
37. Solie B, Monson J, Larson C. Graft-specific surgical and rehabilitation considerations for anterior cruciate ligament reconstruction with the quadriceps tendon autograft. Int J Sports Phys Ther. 2023;18(2):493.
38. Cooper RL, Taylor NF, Feller JA. A systematic review of the effect of proprioceptive and balance exercises on people with an injured or reconstructed anterior cruciate ligament. Res Sports Med. 2005;13(2):163–78. https://doi.org/10.1080/15438620590956197.
39. Buckthorpe M, La Rosa G, Della Villa F. Restoring knee extensor strength after anterior cruciate ligament reconstruction: a clinical commentary. Int J Sports Phys Ther. 2019;14(1):159–72. https://doi.org/10.26603/ijspt20190159.
40. Jewiss D, Ostman C, Smart N. Open versus closed kinetic chain exercises following an anterior cruciate ligament reconstruction: a systematic review and meta-analysis. J Sports Med. 2017:1–10. https://doi.org/10.1155/2017/4721548.
41. Maeo S, Saito A, Otsuka S, Shan X, Kanehisa H, Kawakami Y. Localization of muscle damage within the quadriceps femoris induced by different types of eccentric exercises. Scand J Med Sci Sports. 2018;28(1):95–106.
42. Ardern CL, Webster KE, Taylor NF, Feller JA. Hamstring strength recovery after hamstring tendon harvest for anterior cruciate ligament reconstruction: a comparison between graft types. Arthrosc J Arthroscop Relat Surg. 2010;26(4):462–9. https://doi.org/10.1016/j.arthro.2009.08.018.
43. Kyritsis P, Bahr R, Landreau P, Miladi R, Witvrouw E. Likelihood of ACL graft rupture: not meeting six clinical discharge criteria before return to sport is associated with a four times greater risk of rupture. Br J Sports Med. 2016;50(15):946–51. https://doi.org/10.1136/bjsports-2015-095908.
44. Bourne MN, Bruder AM, Mentiplay BF, Carey DL, Patterson BE, Crossley KM. Eccentric knee flexor weakness in elite female footballers 1–10 years following anterior cruciate ligament reconstruction. Phys Ther Sport. 2019;37:144–9. https://doi.org/10.1016/j.ptsp.2019.03.010.
45. Escamilla RF, MacLeod TD, Wilk KE, Paulos L, Andrews JR. Anterior cruciate ligament strain and tensile forces for weight-bearing and non-weight-bearing exercises: a guide to exercise selection. J Orthop Sports Phys Ther. 2012;42(3):208–20. https://doi.org/10.2519/jospt.2012.3768.
46. Fees M, Decker T, Snyder-Mackler L, Axe MJ. Upper extremity weight-training modifications for the injured athlete: a clinical perspective. Am J Sports Med. 1998;26(5):732–42. https://doi.org/10.1177/03635465980260052301.
47. Khayambashi K, Ghoddosi N, Straub RK, Powers CM. Hip muscle strength predicts noncontact anterior cruciate ligament injury in male and female athletes: a prospective study. Am J Sports Med. 2016;44(2):355–61. https://doi.org/10.1177/0363546515616237.
48. Buckthorpe M. Optimising the late-stage rehabilitation and return-to-sport training and testing process after ACL reconstruction. Sports Med. 2019;49(7):1043–58. https://doi.org/10.1007/s40279-019-01102-z.
49. Buckthorpe M, Della Villa F. Optimising the 'Mid-Stage' training and testing process after ACL reconstruction. Sports Med. 2019;0123456789 https://doi.org/10.1007/s40279-019-01222-6.
50. Buckthorpe M, Della Villa F, Della Villa S, Roi GS. On-field rehabilitation part 1: 4 pillars of high-quality on-field rehabilitation are restoring movement quality, physical conditioning, restor-

ing sport-specific skills, and progressively developing chronic training load. J Orthop Sports Phys Ther. 2019;49(8):565–9. https://doi.org/10.2519/jospt.2019.8954.
51. Gokeler A, et al. Effect of fatigue on landing performance assessed with the landing error scoring system (less) in patients after ACL reconstruction. A pilot study. Int J Sports Phys Ther. 2014;9(3):302–11. [Online]. Available: http://www.ncbi.nlm.nih.gov/pubmed/24944848%0A. http://www.pubmedcentral.nih.gov/articlerender.fcgi?artid=PMC4060307.
52. Chappell JD, Herman DC, Knight BS, Kirkendall DT, Garrett WE, Yu B. Effect of fatigue on knee kinetics and kinematics in stop-jump tasks. Am J Sports Med. 2005;33(7):1022–9. https://doi.org/10.1177/0363546504273047.
53. Gabbett TJ. How much? How fast? How soon? Three simple concepts for progressing training loads to minimize injury risk and enhance performance. J Orthop Sports Phys Ther. 2019:1–9. https://doi.org/10.2519/jospt.2020.9256.
54. Bloch RH, Klein C, Luig P. Development and implementation of a modular return-to-play test battery after ACL reconstruction. In: Return to play in football: an evidence-based approach. Berlin: Springer International Publishing; 2018. p. 217–36.
55. Buckthorpe M, Della Villa F, Della Villa S, Roi GS. On-field rehabilitation part 2: a 5-stage program for the soccer player focused on linear movements, multidirectional movements, soccer-specific skills, soccer-specific movements, and modified practice. J Orthop Sports Phys Ther. 2019;49(8):570–5. https://doi.org/10.2519/jospt.2019.8952.
56. Thompson WO, Thaete FL, Fu FH, Dye SF. Tibial meniscal dynamics using three-dimensional reconstruction of magnetic resonance images. Am J Sports Med. 1991;19(3):210–6. https://doi.org/10.1177/036354659101900302.
57. Cristiani R, Mikkelsen C, Edman G, Forssblad M, Engström B, Stålman A. Age, gender, quadriceps strength and hop test performance are the most important factors affecting the achievement of a patient-acceptable symptom state after ACL reconstruction. Knee Surg Sports Traumatol Arthrosc. 2019;28(2):369–80. https://doi.org/10.1007/s00167-019-05576-2.
58. Lee JK, Lee S, Lee MC. Outcomes of anatomic anterior cruciate ligament reconstruction: bone–quadriceps tendon graft versus double-bundle hamstring tendon graft. Am J Sports Med. 2016;44(9):2323–9. https://doi.org/10.1177/0363546516650666.
59. Kartus J, Movin T, Karlsson J. Donor-site morbidity and anterior knee problems after anterior cruciate ligament reconstruction using autografts. Arthroscopy. 2001;17(9):971–80. https://doi.org/10.1053/jars.2001.28979.
60. Slone HS, Romine SE, Premkumar A, Xerogeanes JW. Quadriceps tendon autograft for anterior cruciate ligament reconstruction: a comprehensive review of current literature and systematic review of clinical results. Arthrosc J Arthroscop Relat Surg. 2015;31(3):541–54. https://doi.org/10.1016/j.arthro.2014.11.010.
61. Hurley ET, Calvo-Gurry M, Withers D, Farrington SK, Moran R, Moran CJ. Quadriceps tendon autograft in anterior cruciate ligament reconstruction: a systematic review. Arthrosc J Arthroscop Relat Surg. 2018;34(5):1690–8. https://doi.org/10.1016/j.arthro.2018.01.046.
62. Smith AH, Capin J, Zarzycki R, Snyder-Mackler L. Athletes with bone-patellar tendon bone autograft for ACL reconstruction were months slower to meet rehabilitation milestones and return to sport criteria than athletes with hamstring tendon autograft or soft tissue allograft: secondary analysis from the. J Orthop Sports Phys Ther. 2019:1–28. https://doi.org/10.2519/jospt.2020.9111.
63. Samuelsson K, Andersson D, Karlsson J. Treatment of anterior cruciate ligament injuries with special reference to graft type and surgical technique: an assessment of randomized controlled trials. Arthroscop J Arthroscop Relat Surg. 2009;25(10):1139–74. https://doi.org/10.1016/j.arthro.2009.07.021.
64. Borchers JR, Pedroza A, Kaeding C. Activity level and graft type as risk factors for anterior cruciate ligament graft failure: a case-control study. Am J Sports Med. 2009;37(12):2362–7. https://doi.org/10.1177/0363546509340633.
65. Spindler KP, et al. The prognosis and predictors of sports function and activity at minimum 6 years after anterior cruciate ligament reconstruction: a population cohort study. Am J Sports Med. 2011;39(2):348–59. https://doi.org/10.1177/0363546510383481.
66. Engebretsen L, Lew WD, Lewis JL, Hunter RE. The effect of an iliotibial tenodesis on intraarticular graft forces and knee joint motion. Am J Sports Med. 1990;18(2):169–76. https://doi.org/10.1177/036354659001800210.
67. Devitt BM, Bouguennec N, Barfod KW, Porter T, Webster KE, Feller JA. Combined anterior cruciate ligament reconstruction and lateral extra-articular tenodesis does not result in an increased rate of osteoarthritis: a systematic review and best evidence synthesis. Knee Surg Sports Traumatol Arthrosc. 2017;25(4):1149–60. https://doi.org/10.1007/s00167-017-4510-1.
68. Getgood AMJ, et al. Lateral extra-articular Tenodesis reduces failure of hamstring tendon autograft anterior cruciate ligament reconstruction: 2-year outcomes from the STABILITY study randomized clinical trial. Am J Sports Med. 2020;48(2):285–97. https://doi.org/10.1177/0363546519896333.
69. Scopp JM, Jasper LE, Belkoff SM, Moorman CT. The effect of oblique femoral tunnel placement on rotational constraint of the knee reconstructed using patellar tendon autografts. Arthroscop J Arthroscop Relat Surg. 2004;20(3):294–9. https://doi.org/10.1016/j.arthro.2004.01.001.

70. Tantuway V, Mustafa Johar SA, Patel V, Nagla A, Gupta R, Bhambani P. Assessment of foot print of femoral tunnel placement with commercially available off set guide in arthroscopic ACL reconstruction. Int J Res Orthop. 2016;3(1):43. https://doi.org/10.18203/issn.2455-4510.intjresorthop20164627.
71. Pinczewski LA, Lyman J, Salmon LJ, Russell VJ, Roe J, Linklater J. A 10-year comparison of anterior cruciate ligament reconstructions with hamstring tendon and patellar tendon autograft: a controlled, prospective trial. Am J Sports Med. 2007;35(4):564–74. https://doi.org/10.1177/0363546506296042.
72. Marchant BG, Noyes FR, Barber-Westin SD, Fleckenstein C. Prevalence of nonanatomical graft placement in a series of failed anterior cruciate ligament reconstructions. Am J Sports Med. 2010;38(10):1987–96. https://doi.org/10.1177/0363546510372797.
73. Xu H, et al. A systematic review of anterior cruciate ligament femoral footprint location evaluated by quadrant method for single-bundle and double-bundle anatomic reconstruction. Arthroscop J Arthroscop Relat Surg. 2016;32(8):1724–34. https://doi.org/10.1016/j.arthro.2016.01.065.
74. Arnold MR, Lie DTT, Verdonschot N, De Graaf R, Amis AA, Van Kampen A. The remains of anterior cruciate ligament graft tension after cyclic knee motion. Am J Sports Med. 2005;33(4):536–42. https://doi.org/10.1177/0363546504269938.
75. Getelman MH, Friedman MJ. Revision anterior cruciate ligament reconstruction surgery. J Am Acad Orthop Surg. 1999;7(3):189–98. https://doi.org/10.5435/00124635-199905000-00005.
76. dos Santos Figueiredo FW, et al. Anterior Cruciate Ligament rupture and Concomitant Injuries. Int Arch Med. 2015;55(11):1–6. https://doi.org/10.3823/1835.
77. Røtterud JH, Sivertsen EA, Forssblad M, Engebretsen L, Årøen A. Effect of gender and sports on the risk of full-thickness articular cartilage lesions in anterior cruciate ligament-injured knees: a nationwide cohort study from Sweden and Norway of 15 783 patients. Am J Sports Med. 2011;39(7):1387–94. https://doi.org/10.1177/0363546510397813.
78. Cox CL, et al. Are articular cartilage lesions and meniscus tears predictive of IKDC, KOOS, and Marx activity level outcomes after anterior cruciate ligament reconstruction?: a 6-year multicenter cohort study. Am J Sports Med. 2014;42(5):1058–67. https://doi.org/10.1177/0363546514525910.
79. Barenius B, Ponzer S, Shalabi A, Bujak R, Norlén L, Eriksson K. Increased risk of osteoarthritis after anterior cruciate ligament reconstruction: a 14-year follow-up study of a randomized controlled trial. Am J Sports Med. 2014;42(5):1049–57. https://doi.org/10.1177/0363546514526139.
80. Toman CV, et al. Success of meniscal repair at anterior cruciate ligament reconstruction. Am J Sports Med. 2009;37(6):1111–5. https://doi.org/10.1177/0363546509337010.
81. Stärke C, Kopf S, Petersen W, Becker R. Meniscal Repair. Arthroscop J Arthroscop Relat Surg. 2009;25(9):1033–44. https://doi.org/10.1016/j.arthro.2008.12.010.
82. Barber FA, Click SD. Meniscus repair rehabilitation with concurrent anterior cruciate reconstruction. Arthroscopy. 1997;13(4):433–7. https://doi.org/10.1016/S0749-8063(97)90120-1.
83. Grant JA, Tannenbaum E, Miller BS, Bedi A. Treatment of combined complete tears of the anterior cruciate and medial collateral ligaments. Arthroscop J Arthroscop Relat Surg. 2012;28(1):110–22. https://doi.org/10.1016/j.arthro.2011.08.293.
84. Blanke F, et al. Results of a new treatment concept for concomitant lesion of medial collateral ligament in patients with rupture of anterior cruciate ligament. J Knee Surg. 2017;30(7):652–8. https://doi.org/10.1055/s-0036-1593873.
85. Halinen J, Lindahl J, Hirvensalo E. Range of motion and quadriceps muscle power after early surgical treatment of acute combined anterior cruciate and grade-III medial collateral ligament injuries: a prospective randomized study. J Bone Joint Surg Ser A. 2009;91(6):1305–12. https://doi.org/10.2106/JBJS.G.01571.
86. Osti L, Papalia R, Del Buono A, Merlo F, Denaro V, Maffulli N. Simultaneous surgical management of chronic grade-2 valgus instability of the knee and anterior cruciate ligament deficiency in athletes. Knee Surg Sports Traumatol Arthrosc. 2010;18(3):312–6. https://doi.org/10.1007/s00167-009-0966-y.
87. Edson CJ. Conservative and postoperative rehabilitation of isolated and combined injuries of the medial collateral ligament. Sports Med Arthrosc Rev. 2006;14(2):105–10. https://doi.org/10.1097/01.jsa.0000212308.32076.f2.
88. Medvecky MJ, Zazulak BT, Hewett TE. A multidisciplinary approach to the evaluation, reconstruction and rehabilitation of the multi-ligament injured athlete. Sports Med. 2007;37(2):169–87. https://doi.org/10.2165/00007256-200737020-00005.
89. Kent R, Forman JL, Crandall J, Lessley D. The mechanical interactions between an American football cleat and playing surfaces in-situ at loads and rates generated by elite athletes: a comparison of playing surfaces. Sports Biomech. 2015;14(1):1–17. https://doi.org/10.1080/14763141.2015.1024277.
90. Balazs GC, Pavey GJ, Brelin AM, Pickett A, Keblish DJ, Rue JPH. Risk of anterior cruciate ligament injury in athletes on synthetic playing surfaces: a systematic review. Am J Sports Med. 2015;43(7):1798–804. https://doi.org/10.1177/0363546514545864.
91. Zbrojkiewicz D, Vertullo C, Grayson JE. Increasing rates of anterior cruciate ligament reconstruction in young Australians, 2000-2015. Med J Aust. 2018;208(8):354–8. https://doi.org/10.5694/mja17.00974.

92. Grindem H, Engebretsen L, Axe M, Snyder-Mackler L, Risberg MA. Activity and functional readiness, not age, are the critical factors for second anterior cruciate ligament injury - the Delaware-Oslo ACL cohort study. Br J Sports Med. 2020:1–5. https://doi.org/10.1136/bjsports-2019-100623.
93. Desai N, Björnsson H, Samuelsson K, Karlsson J, Forssblad M. Outcomes after ACL reconstruction with focus on older patients: results from the Swedish National Anterior Cruciate Ligament Register. Knee Surg Sports Traumatol Arthrosc. 2014;22(2):379–86. https://doi.org/10.1007/s00167-013-2803-6.
94. Hamrin Senorski E, et al. Factors affecting the achievement of a patient-acceptable symptom state 1 year after anterior cruciate ligament reconstruction: a cohort study of 343 patients from 2 registries. Orthop J Sports Med. 2018;6(4):1–16. https://doi.org/10.1177/2325967118764317.
95. Ageberg E, Forssblad M, Herbertsson P, Roos EM. Sex differences in patient-reported outcomes after anterior cruciate ligament reconstruction: data from the swedish knee ligament register. Am J Sports Med. 2010;38(7):1334–42. https://doi.org/10.1177/0363546510361218.
96. Ardern CL, Taylor NF, Feller JA, Webster KE. Return-to-sport outcomes at 2 to 7 years after anterior cruciate ligament reconstruction surgery. Am J Sports Med. 2012;40(1):41–8. https://doi.org/10.1177/0363546511422999.
97. Waldén M, Hägglund M, Magnusson H, Ekstrand J. Anterior cruciate ligament injury in elite football: a prospective three-cohort study. Knee Surg Sports Traumatol Arthrosc. 2011;19(1):11–9. https://doi.org/10.1007/s00167-010-1170-9.
98. Thompson SM, Salmon LJ, Waller A, Linklater J, Roe JP, Pinczewski LA. Twenty-year outcome of a longitudinal prospective evaluation of isolated endoscopic anterior cruciate ligament reconstruction with Patellar Tendon or hamstring autograft. Am J Sports Med. 2016;44(12):3083–94. https://doi.org/10.1177/0363546516658041.
99. Schlumberger M, et al. Traumatic graft rupture after primary and revision anterior cruciate ligament reconstruction: retrospective analysis of incidence and risk factors in 2915 cases. Knee Surg Sports Traumatol Arthrosc. 2017;25(5):1535–41. https://doi.org/10.1007/s00167-015-3699-0.
100. Tan SHS, Lau BPH, Khin LW, Lingaraj K. The importance of patient sex in the outcomes of anterior cruciate ligament reconstructions. Am J Sports Med. 2016;44(1):242–54. https://doi.org/10.1177/0363546515573008.
101. Burland JP, Toonstra JL, Howard JS. Psychosocial barriers after anterior cruciate ligament reconstruction: a clinical review of factors influencing postoperative success. Sports Health. 2019;11(6):528–34. https://doi.org/10.1177/1941738119869333.
102. McPherson AL, Feller JA, Hewett TE, Webster KE. Psychological readiness to return to sport is associated with second anterior cruciate ligament injuries. Am J Sports Med. 2019;47(4):857–62. https://doi.org/10.1177/0363546518825258.
103. Hart HF, Culvenor AG, Guermazi A, Crossley KM. Worse knee confidence, fear of movement, psychological readiness to return-to-sport and pain are associated with worse function after ACL reconstruction. Phys Ther Sport. 2020;41:1–8. https://doi.org/10.1016/j.ptsp.2019.10.006.
104. Webster KE, Feller JA, Lambros C. Development and preliminary validation of a scale to measure the psychological impact of returning to sport following anterior cruciate ligament reconstruction surgery. Phys Ther Sport. 2008;9(1):9–15. https://doi.org/10.1016/j.ptsp.2007.09.003.
105. Kvist J, Ek A, Sporrstedt K, Good L. Fear of re-injury: a hindrance for returning to sports after anterior cruciate ligament reconstruction. Knee Surg Sports Traumatol Arthrosc. 2005;13(5):393–7. https://doi.org/10.1007/s00167-004-0591-8.
106. Grindem H, Snyder-Mackler H, Lynn Moksnes L, Engebretsen, Risberg MA. Simple decision rules reduce reinjury risk after anterior cruciate ligament reconstruction: the Delaware-Oslo ACL cohort study. Br J Sports Med. 2016;50(13):804–8. https://doi.org/10.1136/bjsports-2016-096031.
107. Burgi CR, et al. Which criteria are used to clear patients to return to sport after primary ACL reconstruction? A scoping review. Br J Sports Med. 2019:1–10. https://doi.org/10.1136/bjsports-2018-099982.
108. Patterson B, et al. Poor functional performance 1 year after ACL reconstruction increases the risk of early osteoarthritis progression. Br J Sports Med. 2020;54(9):246–53. https://doi.org/10.1136/bjsports-2019-101503.
109. Webster KE, Hewett TE. What is the evidence for and validity of return-to-sport testing after anterior cruciate ligament reconstruction surgery? A systematic review and meta-analysis. Sports Med. 2019;49:917–29. https://doi.org/10.1007/s40279-019-01093-x.
110. Claes S, Verdonk P, Forsyth R, Bellemans J. The 'ligamentization' process in anterior cruciate ligament reconstruction: what happens to the human graft? A systematic review of the literature. Am J Sports Med. 2011;39(11):2476–83. https://doi.org/10.1177/0363546511402662.
111. Grassi A, et al. Magnetic resonance imaging after anterior cruciate ligament reconstruction: a practical guide. World J Orthop. 2016;7(10):638–49. https://doi.org/10.5312/wjo.v7.i10.638.
112. Ma Y, Murawski CD, Rahnemai-Azar AA, Maldjian C, Lynch AD, Fu FH. Graft maturity of the reconstructed anterior cruciate ligament 6 months postoperatively: a magnetic resonance imaging evaluation of quadriceps tendon with bone block and hamstring tendon autografts. Knee Surg Sports Traumatol Arthrosc. 2015;23(3):661–8. https://doi.org/10.1007/s00167-014-3302-0.

113. Van Dyck P, Zazulia K, Smekens C, Heusdens CHW, Janssens T, Sijbers J. Assessment of anterior cruciate ligament graft maturity with conventional magnetic resonance imaging: a systematic literature review. Orthop J Sports Med. 2019;7(6):1–9. https://doi.org/10.1177/2325967119849012.
114. Shelbourne KD, Gray T, Haro M. Incidence of subsequent injury to either knee within 5 years after anterior cruciate ligament reconstruction with patellar tendon autograft. Am J Sports Med. 2009;37(2):246–51. https://doi.org/10.1177/0363546508325665.
115. King E, et al. Factors influencing return to play and second anterior cruciate ligament injury rates in level 1 athletes after primary anterior cruciate ligament reconstruction: 2-year follow-up on 1432 reconstructions at a single center. Am J Sports Med. 2020;48(4):812–24. https://doi.org/10.1177/0363546519900170.
116. Zwolski CM, Schmitt LC, Thomas S, Paterno MV. The incidence of second anterior cruciate ligament injury in young athletes is not influenced by time to return to sport. Orthop J Sports Med. 2020;8(4_suppl3):2325967120S0018. https://doi.org/10.1177/2325967120s00188.
117. Beischer S, et al. Young athletes who return to sport before 9 months after anterior cruciate ligament reconstruction have a rate of new injury 7 times that of those who delay return. J Orthop Sports Phys Ther. 2020;50(2):83–90. https://doi.org/10.2519/jospt.2020.9071.
118. Dekker TJ, Godin JA, Dale KM, Garrett WE, Taylor DC, Riboh JC. Return to sport after pediatric anterior cruciate ligament reconstruction and its effect on subsequent anterior cruciate ligament injury. J Bone Joint Surg Am. 2017;99(11):897–904. https://doi.org/10.2106/JBJS.16.00758.
119. Nagelli CV, Hewett TE. Should return to sport be delayed until two years after anterior cruciate ligament reconstruction? Biological and functional considerations. Sports Med. 2017;47(2):221–32. https://doi.org/10.1016/j.physbeh.2017.03.040.
120. Buckthorpe M, Frizziero A, Roi GS. Update on functional recovery process for the injured athlete: return to sport continuum redefined. Br J Sports Med. 2019;53(5):265–7. https://doi.org/10.1136/bjsports-2018-099341.
121. Hartigan EH, Axe MJ, Snyder-Mackler L. Time line for noncopers to pass return-to-sports criteria after anterior cruciate ligament reconstruction. J Orthop Sports Phys Ther. 2010;40(3):141–54. https://doi.org/10.2519/jospt.2010.3168.
122. Thomeé R, et al. Muscle strength and hop performance criteria prior to return to sports after ACL reconstruction. Knee Surg Sports Traumatol Arthrosc. 2011;19(11):1798–805. https://doi.org/10.1007/s00167-011-1669-8.
123. Logerstedt D, et al. Single-legged hop tests as predictors of self-reported knee function after anterior cruciate ligament reconstruction: the Delaware-Oslo ACL cohort study. Am J Sports Med. 2012;40(10):2348–56. https://doi.org/10.1177/0363546512457551.
124. Ardern CL, Taylor NF, Feller JA, Whitehead TS, Webster KE. Sports participation 2 years after anterior cruciate ligament reconstruction in athletes who had not returned to sport at 1 year: a prospective follow-up of physical function and psychological factors in 122 athletes. Am J Sports Med. 2015;43(4):848–56. https://doi.org/10.1177/0363546514563282.
125. Patterson BE, et al. Limb symmetry index on a functional test battery improves between one and five years after anterior cruciate ligament reconstruction, primarily due to worsening contralateral limb function. Phys Ther Sport. 2020;44:67–74. https://doi.org/10.1016/j.ptsp.2020.04.031.
126. Reid A, Birmingham TB, Stratford PW, Alcock GK, Giffin JR. Hop testing provides a reliable and valid outcome measure during rehabilitation after anterior cruciate ligament reconstruction. Phys Ther. 2007;87(3):337–49. https://doi.org/10.2522/ptj.20060143.
127. Dingenen B, Truijen J, Bellemans J, Gokeler A. Test-retest reliability and discriminative ability of forward, medial and rotational single-leg hop tests. Knee. 2019;xxxx:1–10. https://doi.org/10.1016/j.knee.2019.06.010.
128. Waldén M, Krosshaug T, Bjørneboe J, Andersen TE, Faul O, Hägglund M. Three distinct mechanisms predominate in noncontact anterior cruciate ligament injuries in male professional football players: a systematic video analysis of 39 cases. Br J Sports Med. 2015;49(22):1452–60. https://doi.org/10.1136/bjsports-2014-094573.
129. Gokeler A, Welling W, Benjaminse A, Lemmink K, Seil R, Zaffagnini S. A critical analysis of limb symmetry indices of hop tests in athletes after anterior cruciate ligament reconstruction: a case control study. Orthop Traumatol Surg Res. 2017;103(6):947–51. https://doi.org/10.1016/j.otsr.2017.02.015.
130. Kotsifaki A, Korakakis V, Whiteley R, Van Rossom S, Jonkers I. Measuring only hop distance during single leg hop testing is insufficient to detect deficits in knee function after ACL reconstruction: a systematic review and meta-analysis. Br J Sports Med. 2020;54(3):139–53. https://doi.org/10.1136/bjsports-2018-099918.
131. Welling W, Benjaminse A, Seil R, Lemmink K, Gokeler A. Altered movement during single leg hop test after ACL reconstruction: implications to incorporate 2-D video movement analysis for hop tests. Knee Surg Sports Traumatol Arthrosc. 2018;26(10):3012–9. https://doi.org/10.1007/s00167-018-4893-7.
132. Bell DR, Smith MD, Pennuto AP, Stiffler MR, Olson ME. Jump-landing mechanics after anterior cruciate ligament reconstruction: a landing error scoring system study. J Athl Train. 2014;49(4):435–41. https://doi.org/10.4085/1062-6050-49.3.21.
133. Myer GD, Ford KR, Hewett TE. Tuck jump assessment for reducing anterior cruciate ligament injury

risk. Athlet Ther Today. 2008;13(5):39–44. https://doi.org/10.1123/att.13.5.39.
134. Redler LH, Watling JP, Dennis ER, Swart E, Ahmad CS. Reliability of a field-based drop vertical jump screening test for ACL injury risk assessment. Phys Sports Med. 2016;44(1):46–52. https://doi.org/10.1080/00913847.2016.1131107.
135. Pollard CD, Stearns KM, Hayes AT, Heiderscheit BC. Altered lower extremity movement variability in female soccer players during side-step cutting after anterior cruciate ligament reconstruction. Am J Sports Med. 2015;43(2):460–5. https://doi.org/10.1177/0363546514560153.
136. Gokeler A, Welling W, Zaffagnini S, Seil R, Padua D. Development of a test battery to enhance safe return to sports after anterior cruciate ligament reconstruction. Knee Surg Sports Traumatol Arthrosc. 2017;25(1):192–9. https://doi.org/10.1007/s00167-016-4246-3.
137. Branch TP, Mayr HO, Browne JE, Campbell JC, Stoehr A, Jacobs CA. Instrumented examination of anterior cruciate ligament injuries: minimizing flaws of the manual clinical examination. Arthroscop J Arthroscop Relat Surg. 2010;26(7):997–1004. https://doi.org/10.1016/j.arthro.2010.01.019.
138. Espregueira-Mendes J, et al. Global rotation has high sensitivity in ACL lesions within stress MRI. Knee Surg Sports Traumatol Arthrosc. 2017;25(10):2993–3003. https://doi.org/10.1007/s00167-016-4281-0.
139. Muller B, et al. Defining thresholds for the patient acceptable symptom state for the IKDC subjective knee form and KOOS for patients who underwent ACL reconstruction. Am J Sports Med. 2016;44(11):2820–6. https://doi.org/10.1177/0363546516652888.
140. Welling W, Benjaminse A, Seil R, Lemmink K, Zaffagnini S, Gokeler A. Low rates of patients meeting return to sport criteria 9 months after anterior cruciate ligament reconstruction: a prospective longitudinal study. Knee Surg Sports Traumatol Arthrosc. 2018;26(12):3636–44. https://doi.org/10.1007/s00167-018-4916-4.
141. Wiggins AJ, Grandhi RK, Schneider DK, Stanfield D, Webster KE, Myer GD. Risk of secondary injury in younger athletes after anterior cruciate ligament reconstruction. Am J Sports Med. 2016;44(7):1861–76. https://doi.org/10.1177/0363546515621554.
142. Wangensteen A, et al. Hamstring Reinjuries occur at the same location and early after return to sport. Am J Sports Med. 2016;44(8):2112–21. https://doi.org/10.1177/0363546516646086.
143. Wordeman SC, Hewett TE. Research-based and clinical considerations for effective neuromuscular training to prevent second anterior cruciate ligament injury. Oper Tech Sports Med. 2016;24(1):7–11. https://doi.org/10.1053/j.otsm.2015.09.008.
144. Pereira R, Andrade R, Rebelo-Marques A, Espregueira-Mendes J. Sport injury primary and secondary prevention. In: The sports medicine physician. Springer; 2019. p. 121–47.
145. Arundale AJH, Capin JJ, Zarzycki R, Smith AH, Snyder-Mackler L. Two year ACL Reinjury rate of 2.5%: outcomes report of the men in a secondary ACL injury prevention program (ACL-sports). Int J Sports Phys Ther. 2018;13(3):422–31. https://doi.org/10.26603/ijspt20180422.
146. Gokeler A, Benjaminse A, Seil R, Kerkhoffs G, Verhagen E. Using principles of motor learning to enhance ACL injury prevention programs. Sports Orthop Traumatol. 2018;34(1):23–30. https://doi.org/10.1016/j.orthtr.2017.12.006.

24 Prevention of Anterior Cruciate Injuries in High-Level Athletes

Stephen E. Marcaccio, Joseph D. Giusto, and Jonathan D. Hughes

24.1 Introduction

Anterior cruciate ligament (ACL) injuries are common in young and active individuals, with over 120,000 injuries occurring annually in the United States [1]. Further research has shown that this incidence is slowly increasing, and the majority of these injuries occur during the high school and collegiate years, particularly in females [1]. In high-level athletes, including collegiate, professional, and select high school athletes, an ACL injury often requires surgical intervention in the form of ACL reconstruction, which requires significant time for rehabilitation before returning to sport, often resulting in missing at least one competitive season.

Due to the significant consequences of ACL injuries in high-level athletes, many studies have been conducted on this specific population in an attempt to prevent these injuries from occurring, maximizing competitive availability and career longevity. As a result, several sport-specific, anatomic, biologic, biomechanical, and neuromuscular risk factors have been identified as targets for prevention initiatives [2].

The purpose of this chapter is to review the epidemiology of ACL injuries in high-level athletes and to identify several anatomic, biologic, biomechanical, and neuromuscular risk factors for injury. Further, this chapter will review several different ACL prevention initiatives, including neuromuscular training programs and screening programs for youth and adolescent athletes.

24.2 Epidemiology

The increasing incidence of ACL tears is an important consideration when discussing ACL injury prevention strategies, which occurs at an estimated rate of 68.6 per 100,000 person-years [3]. Both ACL injury and ACL reconstruction rates have continued to increase since the 1990s, particularly in young adolescent athletes [3–6]. An increased incidence has been described in elite athletes with one study of professional soccer players reporting a 6% annual increase [6].

Several epidemiologic studies have further characterized the types and levels of sports participation that are most commonly associated with ACL injury. Among adolescents and high

S. E. Marcaccio · J. D. Giusto
Department of Orthopaedic Surgery, UPMC Freddie Fu Center for Sports Medicine, University of Pittsburgh Medical Center, Pittsburgh, PA, USA

J. D. Hughes (✉)
University of Pittsburgh/UPMC Freddie Fu Sports Medicine Center, Pittsburgh, PA, USA
e-mail: hughesjd3@upmc.edu

M. Karahan et al. (eds.), *Anterior Cruciate Ligament Injuries Throughout the Life Span*,
https://doi.org/10.1007/978-3-031-84501-7_24

school athletes, the incidence of ACL injury ranges from 0.61 to 0.69 per 10,000 athlete exposures [7–9]. A higher incidence is observed in collegiate-level athletes, occurring at a rate of 0.86–1.50 per 10,000 athlete exposures [9, 10]. When comparing the risk of noncontact ACL injuries in high school versus collegiate athletes, collegiate athletes have twice the risk of injury (relative risk 2.38) after controlling for sex and type of sport participation [9].

Elite athletes experience a high rate of ACL injury, with one study of the top eight professional European soccer leagues reporting an annual incidence of 1.42% [11]. In the National Football League (NFL), 219 ACL injuries were recorded over a 3-year period from 2010 to 2013 [12]. Several ACL injury prevention programs have subsequently been developed in collaboration with professional leagues, including the Fédération Internationale de Football Association (FIFA) 11+ prevention program, due to the impact of ACL injuries [13, 14].

Sports requiring pivoting and/or cutting maneuvers, including soccer and American football, consistently demonstrate a higher incidence of ACL injuries compared to straight plane activities like running. In high school athletes, a relative risk of 3.80 and 3.67 was found for participation in basketball and soccer, respectively [8]. Among National Collegiate Athletic Association (NCAA) athletes, American football demonstrated the highest incidence among male athletes (1.44 injuries per 10,000 athlete exposures), whereas soccer was the highest for female athletes (2.60 injuries per 10,000 athlete exposures) [10]. Player position in each sport has an influence on injury incidence. For example, a review of ACL injuries occurring in the NFL found that receivers, linebackers, and halfbacks had a higher rate of injury than lineman, kickers, and quarterbacks [12]. Injuries occur more frequently in the setting of a game/match than routine practice across all levels of play, likely due to the increased effort and intensity of competitive play [7, 15] (Table 24.1).

Table 24.1 Epidemiology and risk factors associated with ACL injury

Population	*ACL injury incidence*
Level of sport	
High school	0.61–0.69
Collegiate	0.86–1.50
Professional	
European soccer	1.42% annual incidence
NFL	2% of all injuries
Type of sport	
Pivoting and/or cutting sports[a]	
Adolescents (male)	0.27–1.01
Adolescents (female)	0.35–1.66
NCAA (male)	0.18–1.44
NCAA (female)	0.14–2.60
Non-pivoting sports[b]	
Adolescents (male)	0.00–0.34
Adolescents (female)	0.14–1.14
NCAA (male)	0.02–1.00
NCAA (female)	0.00–1.77
Competitive setting	
Routine practice	0.05–0.41
Competitive game/match	0.10–4.67
Variable	*Increased risk of ACL injury*
Anatomic	
Intercondylar notch size	Narrow (<17 mm)
Notch shape	"A" shaped
Lateral femoral condylar offset ratio	Increased offset (>63%)
Posterior tibial slope	Increased slope (>12°)
Biologic	
Sex	Female sex (2–8× higher risk)
Genetic	Family history of ACL injury
Biomechanical and neuromuscular	
Knee kinematics	Dynamic valgus, decreased knee flexion
Quadriceps/hamstrings balance	Excessive quadriceps strength combined with hamstring weakness

ACL anterior cruciate ligament, *NFL* National Football League, *NCAA* National Collegiate Athletic Association

Incidence reported as number of injuries per 10,000 athletic exposures unless otherwise specified

[a]Includes basketball, American football, soccer, lacrosse, field hockey, tennis

[b]Includes softball/baseball, track and field, gymnastics, volleyball, wrestling, cheerleading, ice hockey, cross country, swimming and diving

24.3 Risk Factors for ACL Injury in Athletes

24.3.1 Anatomic and Biologic Risk Factors

Anatomic risk factors for ACL injury have been well-described and include a narrow femoral intercondylar notch (<17 mm), increased lateral femoral condylar offset ratio (>63%), and increased posterior tibial slope (>12 degrees) [16–18]. Increased femoral anteversion has been reported among patients with ACL injuries and may be related to the excessive valgus mechanism of ACL injury [19]. Recent attention has been placed on the posterior tibial slope as this is one of the few anatomic risk factors that can be surgically modified. Recent observational studies have shown that a slope-reducing anterior-closing-wedge high-tibial-osteotomy can effectively decrease the posterior tibial slope and lead to improvements in International Knee Documentation Committee (IKDC) scores in patients undergoing revision ACL reconstruction [20, 21].

Patient sex is a key biologic factor that influences the risk of ACL injury, with female athletes having a two- to eightfold higher risk of injury compared to males regardless of sport type or level [5, 7–10, 22, 23]. Explanations for the sex-based differences in ACL injury risk are multifactorial. Anatomic differences between males and females, particularly with respect to the quadriceps angle (Q-angle), as females have been shown to have a Q-angle that is 3.1–3.4 degrees larger than males [22, 24]. However, hereditary factors cannot be overlooked. The inheritance of ACL injury risk was previously described in an identical twin study, which reported the genetic contribution to be 69% (relative risk in identical vs. fraternal twins 8.6 vs. 1.9, respectively) [25] (Table 24.1).

24.3.2 Biomechanical and Neuromuscular Risk Factors

Several in vivo biomechanical studies have analyzed kinematic and neuromuscular patterns associated with ACL injuries in an effort to identify high risk athletes who may benefit from injury prevention training. A prospective study of 205 female athletes evaluated the preseason kinematic patterns of athletes during a 13-month observation period and recorded those who subsequently experienced an ACL injury. ACL injured athletes demonstrated significantly decreased knee flexion and increased knee abduction moments on initial presentation compared to their uninjured counterparts, corresponding to more dynamic valgus and a 20% higher vertical ground reaction force [26].

Subsequent comparisons of movement patterns between males and females have found that healthy, uninjured female athletes have more dynamic valgus than males [27, 28]. Sex-based differences also exist with regard to the timing of maximum dynamic valgus when measured using a drop-jump landing test. In female athletes, dynamic valgus peaked during the deceleration phase of landing, whereas the peak was delayed in male athletes and occurred during the subsequent acceleration phase [29]. Females showed a greater angular velocity of knee valgus than males (mean difference 25.53 deg/sec) [29].

Dynamic stabilization of the ACL by hamstring and quadriceps muscle co-contraction has been well described. Hamstring muscles agonize the ACL and function to limit anterior tibial translation, whereas the quadriceps antagonize the ACL and increase anterior translation [30, 31]. In a cadaveric model, a quadriceps load increased ACL forces from 27.8 N at full knee extension to 44.9 N at 15 degrees of flexion, which was significantly reduced by 30% with the addition of a hamstring load [31]. A “ligament-muscle reflex arc” has been described a cat model in which loading of the ACL resulted in increased electromyography (EMG) activity of the hamstrings and decreased EMG activity of the quadriceps [30]. Clinically, patients who injure their ACL have been shown to have an imbalance of quadriceps/hamstrings activity, which may increase ACL forces [32]. Excessive quadriceps activity combined with hamstring muscle weakness has also been found among healthy female athletes, which may further explain the increased risk of ACL injuries among females in addition to the dynamic valgus kinematics [27, 28] (Table 24.1).

24.4 ACL Injury Prevention Strategies

24.4.1 Neuromuscular Training Programs

With the incidence of ACL injuries increasing across adolescent, collegiate, and professional athletes, there has been a similar increase in efforts tailored to preventing these injuries from a neuromuscular training perspective. While complete prevention is not possible, the majority of ACL injuries occur via noncontact or indirect contact mechanisms, especially in American football and soccer players; therefore, many of these injuries may be prevented with proper training [2]. Neuromuscular training programs focus on improving several biomechanical risk factors associated with ACL injuries, including hamstring and quadriceps strength imbalances, increased dynamic knee valgus loading, decreased knee flexion loading, and increased knee abduction moment [31]. Multiple neuromuscular training programs have been initiated across a variety of levels of sport with varying success. Two examples of such training programs are the FIFA 11 and FIFA 11+ programs.

The FIFA 11 program was developed with the goal of reducing the rates of primary ACL injuries in soccer players. This program includes several different types of exercises that are meant to be performed several times weekly to specifically target key muscle groups in the lower extremities to increase strength and decrease imbalance between muscle groups [33]. The exercises in the FIFA 11 program include those focusing on core stability, balance, plyometrics, and strength. Once familiarized, this program requires roughly 15 minutes to perform and is meant to be performed prior to initiating soccer activity. There are several studies that have evaluated the effectiveness of the FIFA 11 program in reducing primary ACL injuries. A cluster-randomized controlled trial evaluated over 2000 female youth Scandinavian soccer players, randomized to an “intervention” group and a control group, where the intervention group implemented the FIFA 11 program over the course of their 8-month season [34]. The authors found no difference in injury rates nor in the incidence of any type of lower extremity injury between the two groups, but noted that compliance was extremely low, with only 14 of the 58 included teams (24.1%) completing the minimal suggested training sessions [34]. While the authors attributed this lack of difference between the groups to poor overall compliance, a more recent randomized control trial conducted a similar evaluation of over 450 adult male amateur soccer players [35]. Similar to the previously mentioned study, the authors found no difference between the two groups in terms of total injury incidence (9.6 and 9.7 injuries per 1000 sports hours for intervention and control groups, respectively) and injury severity. However, the authors found athletes in the intervention group sustained significantly fewer knee injuries when compared to the control group, suggesting some efficacy to the program when specifically targeting knee injuries [35]. Further, in contrast to the previously mentioned study, the overall compliance in this study was high, with 73% of teams and 71% of individual players implementing the program over the course of the season [35]. Finally, a recent systematic review was unable to find a clear preventative effect of the FIFA 11 program, even when controlling for compliance within the included studies [33]. It should be noted that these studies evaluated injuries on a broad scale, focusing on various types of knee injuries rather than isolated ACL injuries.

In 2006, the FIFA 11+ program was developed in response to early results of the FIFA 11 program and feedback from its member organizations. Similar to the FIFA 11 program, the FIFA 11+ program includes several different exercises that are meant to be performed as a warmup before competitive activity, and includes several exercises that focus on core stability, balance, strength, and plyometrics [33]. However, the FIFA 11+ program is a more strenuous and time intensive program, taking at least 20 minutes to complete and involving higher loads during exercise, more repetitions, and longer active muscle contractions [33]. Further, this program introduces multiple progression levels for each exercise, therefore the number of exercises and muscular time under tension is significantly increased when compared to the FIFA 11 program [33].

The FIFA 11+ program has demonstrated success in preventing knee injuries when applied to athletes of various levels. In one of the first studies to evaluate the efficacy of the FIFA 11+ program, a randomized control trial compared over 1800 female youth soccer players who implemented the program and found a significantly reduced risk of knee injuries overall, overuse injuries, and severe injuries when compared to those who did not implement the program [36]. Further, a more recent randomized control trial evaluated over 200 Canadian youth female soccer players and found a significant reduction in knee injuries (as high as 50%) in those who implemented the program consistently [37]. Further, both of these studies found that players with higher adherence to the program demonstrated lower injury risks, suggesting that compliance with the program is essential for successful outcomes [36, 37]. The FIFA 11+ program has also demonstrated similar success in youth and collegiate male soccer players [14, 38].

Although there is growing evidence that neuromuscular training programs, such as the FIFA 11+ program, is effective in preventing knee injuries (including ACL injuries) in athletes of various levels, the implementation of these programs has remained a significant challenge. While FIFA has implemented training for the program into all of its official coaching courses, the program is only implemented by 10% of all affiliated member organizations [39]. Successful implementation of these programs requires the commitment of league leadership, coaches, and players on the international level.

24.4.2 Preventive Screening Tools

With the documented success of several injury reduction training programs, there have been attempts to implement screening tools at the youth and adolescent levels to identify patients with specific biomechanical factors that place them at risk for sustaining ACL injuries, including those with low knee abduction moments with cutting and jumping activities [2]. A recent study utilized video-analysis to evaluate 34 competitive soccer players and developed a scoring system to identify athletes with elevated knee abduction moments, demonstrating high intra-rater reliability in the system [40]. The authors concluded this scoring system could be utilized to identify at-risk players at an earlier age to best maximize the preventive effect of neuromuscular training programs [40]. While these screening strategies are intriguing, consistent effectiveness has yet to be demonstrated [2].

24.4.3 Preventive Bracing

While there is some belief that functional knee braces, used preoperatively or in the postoperative setting following ACL reconstruction, can help protect the internal structures of the knee and minimize translational and rotatory forces on the knee, there is currently limited evidence to support the universal use of such braces for the purposes of ACL injury prevention [41]. Further, use of functional bracing following ACL reconstruction has been associated with increased risk of additional injury with improper brace wear, increased muscle atrophy and fatigability with strength testing, and increased cost [41]. Preventative bracing has demonstrated largely conflicting results and is not currently recommended in routine practice.

24.4.4 Role of Playing Surface in ACL Injury

The role of environmental factors, including playing surface and weather in ACL injury is becoming increasingly popular, particularly the impact of artificial turf on non-contact ACL injuries. To date, the literature evaluating this topic remains controversial. A recent systematic review evaluated 10 studies and over 960 ACL injuries in soccer and American football players and found a significantly increased rate of ACL injury on artificial playing surfaces in American football, but no differences in injury rates among soccer players [42]. A more recent systematic review evaluated the role of artificial surface in ACL injuries among soccer players of various levels, and found a significantly elevated risk of ACL injury in females when playing on artificial surfaces, but no difference in risk profile for male soccer players [43]. Regarding weather, a recent study found that high evaporation and low rainfall conditions before matches are associated with non-contact ACL injury in Australian football [44]. More research is required to evaluate the role of these environmental factors in ACL injuries.

24.5 Conclusion

High-level athletes are at an increased risk for ACL injury, particularly when participating in competitive pivoting sports at a young age. Prevention of ACL injuries in this population has been widely researched with promising results, although adequate patient adherence is a notable barrier. Programs should be designed to emphasize the correction of kinematic patterns associated with a higher risk of ACL injury, including dynamic valgus and quadriceps muscle dominance. Preventative bracing has demonstrated largely conflicting results and is not currently recommended in routine practice. Discussions surrounding the influence of environmental factors on ACL injury risk, including playing on artificial surfaces, are becoming increasingly popular. However, more studies are needed to determine the efficacy of changing the competitive athletic environment given the strong influence of other anatomic and biomechanical risk factors.

References

1. Kaeding CC, Léger-St-Jean B, Magnussen RA. Epidemiology and diagnosis of anterior cruciate ligament injuries. Clin Sports Med. 2017;36:1.
2. Musahl V, Engler ID, Nazzal EM, Dalton JF, Lucidi GA, Hughes JD, et al. Current trends in the anterior cruciate ligament part II: evaluation, surgical technique, prevention, and rehabilitation. Knee Surg Sports Traumatol Arthrosc. 2022;30:34.
3. Sanders TL, Maradit Kremers H, Bryan AJ, Larson DR, Dahm DL, Levy BA, et al. Incidence of anterior cruciate ligament tears and reconstruction: a 21-year population-based study. Am J Sports Med. 2016;44(6):1502.
4. Herzog MM, Marshall SW, Lund JL, Pate V, Mack CD, Spang JT. Trends in incidence of ACL reconstruction and concomitant procedures among commercially insured individuals in the United States, 2002-2014. Sports Health. 2018;10(6):523.
5. Hootman JM, Dick R, Agel J. Epidemiology of collegiate injuries for 15 sports: summary and recommendations for injury prevention initiatives. J Athl Train. 2007;42:311.
6. Waldén M, Hägglund M, Magnusson H, Ekstrand J. ACL injuries in men's professional football: a 15-year prospective study on time trends and return-to-play rates reveals only 65% of players still play at the top level 3 years after ACL rupture. Br J Sports Med. 2016;50(12):744.
7. Bram JT, Magee LC, Mehta NN, Patel NM, Ganley TJ. Anterior cruciate ligament injury incidence in adolescent athletes: a systematic review and meta-analysis. Am J Sports Med. 2021;49:1962.
8. Gornitzky AL, Lott A, Yellin JL, Fabricant PD, Lawrence JT, Ganley TJ. Sport-specific yearly risk and incidence of anterior cruciate ligament tears in high school athletes: a systematic review and meta-analysis. Am J Sports Med. 2016;44:2716.

9. Beynnon BD, Vacek PM, Newell MK, Tourville TW, Smith HC, Shultz SJ, et al. The effects of level of competition, sport, and sex on the incidence of first-time noncontact anterior cruciate ligament injury. Am J Sports Med. 2014;42(8):1806.
10. Dewig DR, Boltz AJ, Moffit RE, Rao N, Collins CL, Chandran A. Epidemiology of anterior cruciate ligament tears in National Collegiate Athletic Association Athletes: 2014/2015-2018/2019. Med Sci Sports Exerc. 2024;56(1):29.
11. Mazza D, Viglietta E, Monaco E, Iorio R, Marzilli F, Princi G, et al. Impact of anterior cruciate ligament injury on European professional soccer players. Orthop J Sport Med. 2022;10(2):23259671221076865.
12. Dodson CC, Secrist ES, Bhat SB, Woods DP, Deluca PF. Anterior cruciate ligament injuries in National Football League Athletes from 2010 to 2013: a descriptive epidemiology study. Orthop J Sport Med. 2016;4(3):2325967116631949.
13. Buerba RA, Zaffagnini S, Kuroda R, Musahl V. ACL reconstruction in the professional or elite athlete: state of the art. J ISAKOS. 2021;6:226.
14. Silvers-Granelli H, Mandelbaum B, Adeniji O, Insler S, Bizzini M, Pohlig R, et al. Efficacy of the FIFA 11+ injury prevention program in the collegiate male soccer player. Am J Sports Med. 2015;43(11):2628.
15. Bradley JP, Klimkiewicz JJ, Rytel MJ, Powell JW. Anterior cruciate ligament injuries in the National Football League: epidemiology and current treatment trends among team physicians. Arthroscopy. 2002;18(5):502.
16. Bayer S, Meredith SJ, Wilson K, De Sa D, Pauyo T, Byrne K, et al. Knee morphological risk factors for anterior cruciate ligament injury: a systematic review. J Bone Joint Surg (Am Vol). 2020;102:703.
17. Todd MS, Lalliss S, Garcia E, DeBerardino TM, Cameron KL. The relationship between posterior tibial slope and anterior cruciate ligament injuries. Am J Sports Med. 2010;38(1):63.
18. Pfeiffer TR, Burnham JM, Hughes JD, Kanakamedala AC, Herbst E, Popchak A, et al. An increased lateral femoral condyle ratio is a risk factor for anterior cruciate ligament injury. J Bone Jt Surg - Am Vol. 2018;100(10):857.
19. Alpay Y, Ezici A, Kurk MB, Ozyalvac ON, Akpinar E, Bayhan AI. Increased femoral anteversion related to infratrochanteric femoral torsion is associated with ACL rupture. Knee Surg Sport Traumatol Arthrosc. 2020;28(8):2567.
20. Vivacqua T, Thomassen S, Winkler PW, Lucidi GA, Rousseau-Saine A, Firth AD, et al. Closing-wedge posterior Tibial slope–reducing osteotomy in complex revision ACL reconstruction. Orthop J Sport Med. 2023;11(1):23259671221144786.
21. Mabrouk A, Kley K, Jacquet C, Fayard JM, An JS, Ollivier M. Outcomes of slope-reducing proximal Tibial osteotomy combined with a third anterior cruciate ligament reconstruction procedure with a focus on return to impact sports. Am J Sports Med. 2023;51(13):3454.
22. Griffin LY, Agel J, Albohm MJ, Arendt EA, Dick RW, Garrett WE, et al. Noncontact anterior cruciate ligament injuries: risk factors and prevention strategies. J Am Acad Orthop Surg. 2000;8:141.
23. Sutton KM, Bullock JM. Anterior cruciate ligament rupture: differences between males and females. J Am Acad Orthop Surg. 2013;21:41.
24. Woodland LH, Francis RS. Parameters and comparisons of the quadriceps angle of college-aged men and women in the supine and standing positions. Am J Sports Med. 1992;20(2):208.
25. Magnusson K, Turkiewicz A, Hughes V, Frobell R, Englund M. High genetic contribution to anterior cruciate ligament rupture: heritability ~69%. Br J Sports Med. 2021;55:385.
26. Hewett TE, Myer GD, Ford KR, Heidt RS, Colosimo AJ, McLean SG, et al. Biomechanical measures of neuromuscular control and valgus loading of the knee predict anterior cruciate ligament injury risk in female athletes: a prospective study. Am J Sports Med. 2005;33(4):492.
27. Chappell JD, Creighton RA, Giuliani C, Yu B, Garrett WE. Kinematics and electromyography of landing preparation in vertical stop-jump: risks for noncontact anterior cruciate ligament injury. Am J Sports Med. 2007;35(2):e3.
28. Orishimo KF, Liederbach M, Kremenic IJ, Hagins M, Pappas E. Comparison of landing biomechanics between male and female dancers and athletes, part 1: influence of sex on risk of anterior cruciate ligament injury. Am J Sports Med. 2014;42(5):1082.
29. Joseph MF, Rahl M, Sheehan J, MacDougall B, Horn E, Denegar CR, et al. Timing of lower extremity frontal plane motion differs between female and male athletes during a landing task. Am J Sports Med. 2011;39(7):1517.
30. Solomonow M, Baratta R, Zhou BH, Shoji H, Bose W, Beck C, et al. The synergistic action of the anterior cruciate ligament and thigh muscles in maintaining joint stability. Am J Sports Med. 1987;15(3):207.
31. Li G, Rudy TW, Sakane M, Kanamori A, Ma CB, Woo SLY. The importance of quadriceps and hamstring muscle loading on knee kinematics and in-situ forces in the ACL. J Biomech. 1999;32(4):395.
32. Myer GD, Ford KR, Barber Foss KD, Liu C, Nick TG, Hewett TE. The relationship of hamstrings and quadriceps strength to anterior cruciate ligament injury in female athletes. Clin J Sport Med. 2009;19(1):3.
33. Thorborg K, Krommes KK, Esteve E, Clausen MB, Bartels EM, Rathleff MS. Effect of specific exercise-based football injury prevention programmes on the overall injury rate in football: a systematic review and meta-analysis of the FIFA 11 and 11+ programmes. Br J Sports Med. 2017;51:562.
34. Steffen K, Myklebust G, Olsen OE, Holme I, Bahr R. Preventing injuries in female youth football - a cluster-randomized controlled trial. Scand J Med Sci Sport. 2008;18(5):605.
35. Van Beijsterveldt AMC, Van De Port IGL, Krist MR, Schmikli SL, Stubbe JH, Frederiks JE, et al.

Effectiveness of an injury prevention programme for adult male amateur soccer players: a cluster-randomised controlled trial. Br J Sports Med. 2012;46(16):1114.
36. Soligard T, Myklebust G, Steffen K, Holme I, Silvers H, Bizzini M, et al. Comprehensive warm-up programme to prevent injuries in young female footballers: cluster randomised controlled trial. BMJ. 2009;338(7686):a2469.
37. Steffen K, Emery CA, Romiti M, Kang J, Bizzini M, Dvorak J, et al. High adherence to a neuromuscular injury prevention programme (FIFA 11+) improves functional balance and reduces injury risk in Canadian youth female football players: a cluster randomised trial. Br J Sports Med. 2013;47(12):794.
38. Owoeye OBA, Akinbo SRA, Tella BA, Olawale OA. Efficacy of the FIFA 11+ warm-up programme in male youth football: a cluster randomised controlled trial. J Sport Sci Med. 2014;13(2):321.
39. Bizzini M, Dvorak J. FIFA 11+: An effective programme to prevent football injuries in various player groups worldwide - a narrative review. Br J Sports Med. 2015;49:577.
40. Della Villa F, Di Paolo S, Santagati D, Della Croc E, Lopomo NF, Grassi A, et al. A 2D video-analysi scoring system of 90° change of direction techniqu identifies football players with high knee abduc tion moment. Knee Surg Sport Traumatol Arthrosc 2022;30(11):3616.
41. Lowe WR, Warth RJ, Davis EP, Bailey L. Functiona bracing after anterior cruciate ligament reconstruc tion: a systematic review. J Am Acad Orthop Surg 2017;25(3):239.
42. Balazs GC, Pavey GJ, Brelin AM, Pickett A, Keblish DJ, Rue JPH. Risk of anterior cruciate ligament injury in athletes on synthetic playing surfaces: a systematic review. Am J Sports Med. 2015;43:1798.
43. Xiao M, Lemos JL, Hwang CE, Sherman SL, Safran MR, Abrams GD. Increased risk of ACL injury for female but not male soccer players on artificial turf versus natural grass: a systematic review and meta-analysis. Orthop J Sports Med. 2022;10:23259671221114353.
44. Webster KE, Hewett TE, Feller JA. Anterior cruciate ligament injuries in Australian rules football: incidence, prevention and return to play outcomes. Open Access J Sports Med. 2021;12:33.

Risks and Complications Related to Anterior Cruciate Ligament Reconstruction in High-Level Athletes

25

Nicola Blucher and Brian M. Devitt

25.1 Introduction

Elite athletes are individuals who have achieved distinction within the context of their sport. The commitment, dedication and motivation they display in pursuit of excellence are quite extraordinary, whether participating at an amateur or professional level. There are a number of common characteristics found in elite athletes, such as ambition, focus, confidence and single-mindedness, which undoubtedly contribute to their success in the sporting arena. However, the environment in which they perform is one of intense pressure with extremes of rapturous joy and crushing disappointment. As such, more negative traits such as obsessiveness, compulsion, worry and anxiety are also prevalent. There is a fine line between these contrasting qualities in normal circumstances but when injury strikes the consequences can be quite devastating and overwhelming for the athlete. Therefore, while dealing with elite athlete as patients can be highly rewarding, it can also be incredibly challenging.

Mens sana in corpore sano

The above Latin phrase is translated as '*a healthy mind in a healthy body*'. The phrase is widely used in sporting context to express the theory that physical exercise is an important or essential part of mental and psychological well-being. But, what happens to the mind when the body is not healthy?

When dealing with high-level athletes, it is essential to strip down the prestige, celebrity, and cachet that often accompany them and consider them first as people who are vulnerable and exposed. It is not always easy, because the traditional doctor–patient relationship is often distorted or absent. A high-level athlete rarely arrives to the first consultation alone, but is typically accompanied by support staff, medical personnel, agents, managers and family all with a vested interest in the outcome of the meeting [1, 2]. A doctor patient-team triad frequently exists, which creates a scenario in which the team's priority can conflict with or even supplant the doctor's primary obligation to player well-being [2]. As such, the customary ethical norms that govern most forms of clinical practice, such as patient confidentiality and autonomy, are not easily translated into sports medicine.

Athletes may excel physically and perform superhuman feats on the field of play, yet they remain human. Therefore, following anterior cruciate ligament (ACL) injury they are exposed to the same risks and complications as the normal population and, in some cases, perhaps even more [3]. Caring for this unique group of patients as a surgeon can represent a double-edged sword; on the one hand, it brings with it a certain notoriety

N. Blucher · B. M. Devitt (✉)
OrthoSport Victoria, Melbourne, VIC, Australia

UPMC Sports Surgery Clinic, Dublin and Dublin City University, Dublin, Ireland

M. Karahan et al. (eds.), *Anterior Cruciate Ligament Injuries Throughout the Life Span*,
https://doi.org/10.1007/978-3-031-84501-7_25

and status, yet it can also be an immense burden, particularly when complications occur. The following chapter outlines the specific risks and complications associated with ACL reconstruction (ACLR) in high-level athletes. In doing so, we have presented a number of questions frequently asked of us, as surgeons, by our patients. We have provided answers supported by the latest evidence and made suggestions and recommendations on how to mitigate the risks and avoid complications.

25.2 Preoperative Considerations

When can I play again, Doc?

This is one of the early questions asked of an orthopaedic surgeon dealing with a high-level athlete following an ACL injury [3]. There is often pressure to carry out surgery urgently in order to return the athlete to the field of play as quickly as possible, but early surgery may not always be in the athlete's best interest. Therefore, before embarking on surgery it is important to take time to consider the readiness of the athlete, now patient, both psychologically and physically.

25.2.1 Psychological Preparedness

Fail to prepare, prepare to fail!

It is well established that athletes, especially high-level athletes, can suffer a period of depression following ACL injury [4]. As with any traumatic event, the patient should be allowed to go through the normal phases of grief; anger, denial, bargaining, depression and finally acceptance of the injury. Patients who embark on surgery with a negative mindset are known to have greater issues with pain, reduced motivation and poorer outcomes [5]. Therefore, a preoperative delay can allow the patient to be in full psychological control at the time of surgery, facilitating a more positive approach to rehabilitation which can lead to a smoother postoperative recovery [4, 6]. It is advisable to encourage the athlete to consult with a sports psychologist or talk to a teammate who has gone through a similar surgery. This ensures the athlete has realistic expectations for what is to come.

It is also important to establish a doctor–patient relationship, which can be difficult with the entourage that typically follows high-level athletes. The patient needs to be aware that his/her best interests are paramount, not simply the team's. It is worthwhile to try to create some time alone with the player, away from the team doctor, management or medical personnel to not only explain in detail the risks of surgery but also to answer any of the patient's questions. The decision to undergo surgery and timing of this can have big implications for an athlete's contract or career aspirations and it is important to tease these out prior to any wielding of a knife.

25.2.2 Physical Readiness and Timing of Surgery

Although the concept of early surgery is appealing, one of the big concerns is the development of arthrofibrosis, which is the most common postoperative complication of ACLR with a reported incidence of 2–35% [7–10]. When arthrofibrosis affects the knee, the condition is frequently more debilitating than the original injury [11]. Although a stiff athletic knee is stable, patients are not satisfied with the function of their knee on return to competition [6]. For a high-level athlete, these symptoms can therefore be career ending and will negatively impact on both their physical and emotional well-being [12].

When is the best time for surgery?

Unfortunately, there is no consensus as to what is the ideal time for a patient to undergo ACLR [13]. In a landmark study from the early 1990s, Shelbourne et al. sought to determine the optimal time to perform acute ACLR and explore the effects of accelerated versus conventional rehabilitation [13]. They conducted a retrospective review of 169 young athletes (average age, 22 years) undergoing acute ACLR and revealed that patients who had surgery delayed for 21 days had earlier return of strength, greater knee flexion and less loss of knee extension at 13 weeks postoperatively compared with patients who had surgery prior to 21 days from initial injury. Even with an accelerated rehabilitation programme, 7% of patients who underwent surgery within 7 days of injury developed arthrofibrosis compared to none in the group who had surgery after 21 days. It is important to consider that the technique for ACLR in this series was open, using a patellar tendon graft, which may not reflect modern-day techniques.

However, in a more recent randomized controlled trial comparing acute ACLR within 8 days of injury with delayed ACLR after normalized range of movement (ROM) (6–10 weeks after injury), Eriksson et al. determined that at 3 months there was no significant difference in flexion, extension or total ROM between the groups [14]. These results are supported by three other randomized control trials which conclude that early surgery does not appreciably affect postoperative ROM nor confer a poorer outcome compared to delayed surgery [15–17].

It would appear that the time from injury to surgery is not as important as the extent of inflammation or '*irritation*' of the knee to determine the readiness for ACLR [12, 13]. The response to injury may vary between patients; some present with significant swelling, effusion, hyperthermia, tissue induration and reduced ROM, while others have a relatively '*quiet* knee' with minimal inflammatory response and near normal knee motion. It has been shown that preoperative extension equivalent to the contralateral uninjured side is a key factor in determining suitable candidates for ACLR within the first two weeks following injury [6]. Considering that impaired postoperative ROM is often related to limited preoperative ROM [12, 18–20], delaying surgery until the knee swelling has subsided and full ROM has been restored should be considered. To achieve hyperextension compared to the normal contralateral side and regaining quadriceps control is vital and has been shown to predict a quicker return of postoperative quadriceps strength [13, 21].

Table 25.1 Checklist for readiness for ACLR

Mental preparation of the patient
Choosing an optimal time for the rehabilitation of both patient and his/her team
Minimal knee swelling
Full ROM achieved compared with the contralateral knee, including hyperextension
Good quadriceps activation and leg control
Associated intra-articular injuries preventing full extension
Associated ligamentous injuries requiring urgent repair

ACLR Anterior cruciate ligament reconstruction, *ROM* Range of motion

There are clearly many factors to consider with regard to timing of ACLR [13, 19, 22], which vary for each patient (Table 25.1). Rather than using a specific time period from injury to surgery, it is probably more useful to confirm that various preoperative targets have been met to both optimize recovery and reduce complications. This should ensure a faster and safer return to full activity with reduced risk of need for further surgical intervention. Athletes can be reassured that if surgery is delayed, it has not been shown to lead to a delay in return to sports, and it is a time when preoperative therapy can be optimized [13].

25.2.3 Indications for Acute Surgery

There are certain situations when urgent surgery is required in the setting of reduced ROM. One such scenario is in the setting of meniscal tears which occur frequently in association with acute ACL injuries [23]. Although uncommon in the setting of acute primary ACL injury, a displaced bucket handle tear, which may block full extension, is often considered an exemption to any surgical delay due to diminished ROM [13, 19]. The key decision in this setting is whether to perform a combined meniscal repair and ACLR or stage the surgery, repairing the meniscus initially and waiting to reconstruct the ACL once full ROM has been restored.

Shelbourne et al. postulated that if a meniscal tear does occur with acute ACL injury, it is typically peripheral and does not displace during the index trauma. It usually takes a subsequent episode of instability and further tearing of the meniscus to cause displacement leading to a block to full extension [13]. As such, they advise that immediate arthroscopic intervention for the management of the meniscal lesions in a patient with an acute ACL tear is not necessary and they propose that meniscal tears can be addressed at the time of the delayed ACLR if further instability episodes are prevented prior to surgery. However, it has also been shown that in the presence of meniscal tears delay in ACLR for 6 weeks or longer has been associated with an increased risk of meniscal injuries requiring resection [16, 24].

Therefore, when dealing with an elite athlete with all things considered, it is reasonable to perform a meniscal repair and ACLR acutely in the one setting. Of course, there are multiple other scenarios where urgent surgery is required in the setting of multi-ligamentous injury. Management of these complex injuries is beyond the scope of this chapter but has been dealt with in this book [25, 26].

25.3 Perioperative and Early Postoperative Considerations

Blood, sweat and tears

The perioperative period is probably the most difficult time for the athlete physically and emotionally. After surgery there is often a certain euphoria that patients feel but this is quickly replaced by the staunch realization of the mammoth task ahead. The above idiom is quite appropriate in describing the effort required but also the complications that may arise.

25.3.1 Septic Arthritis

Septic arthritis (SA) following ACLR is a rare but serious complication. The consequences can be catastrophic, including destruction of articular cartilage and the need to remove the graft and associated implants [27–29]. Sonnery-Cottet et al. concluded that the prevalence of SA in the overall population was 0.61% [30], which is similar to other published reports [31–34]. However, this compares to a rate of up to 5.7% in professional athletes [30, 34, 35]. It appears that apart from the possibility of an increased operative time due to the addition of a lateral extra-articular tenodesis (LET), which is commonly performed in this high-risk group, the only other reason for the increased risk of postoperative infection is being a professional athlete [32, 35]. But, why is this?

The environments in which high-level athletes compete, practice, receive therapy for injuries and travel provide varied opportunities for the transmission of infectious organisms [36]. Close contact in locker rooms, benches and during practice and competition make athletes extremely likely to spread infections among themselves [37]. The sweaty skin of an athlete, especially under tight-fitting clothes or equipment, is an excellent reservoir for bacterial growth [37]. Higher skin infection rates are reported in professional athletes involved in frequent skin-to-skin direct contact sports, such as soccer, rugby or wrestling [36, 38, 39]. The combination of physical contact and frequent skin abrasions (turf burns, lacerations) are potential risk factors for postoperative infection [36, 38–40].

Skin and soft tissue infections organisms such as *Staphylococcus aureus*, *Streptococcus pyogenes*, and Herpes simplex virus commonly affect athletes [30]. The most common bacteria found in SA post-ACLR are coagulase negative *Staphylococcus aureus* and epidermidis [30, 35]; other organisms have included *Streptococcus pyogenes* and Propionibacterium. It has been postulated that professional athletes are predisposed to develop a specific skin flora that increases their risk of postoperative infection and that these patients may benefit from a specific pre-operative protocol to reduce the risk of infection [30].

Based on this information, it is imperative that steps are taken to mitigate SA. Surgery should be delayed if a patient reports or shows any signs of current infection or skin lesions. Good hygiene should be encouraged, including a thorough wash/scrub of the knee in the shower or bath with soap the day before surgery. Inform patients not to shave pre-operatively; shaving is associated with a significantly higher surgical site Infection (SSI) risk than either the use of depilatory agents or no hair removal [41]. Shaving immediately before the operation compared to shaving within 24 hours preoperatively is also associated with decreased SSI rates (3.1 vs. 7.1%); shaving performed >24 hours prior to surgery has shown a SSI rate of over 20% [41].

A single dose of IV prophylactic antibiotics should be given prior to surgery at the same time as shaving of hair [42]. Chlorhexidine or Betadine and alcohol preparation is recommended and this can be used to paint the knee prior to draping [42]. While doing this, it is best to position the knee in 80–90 degrees flexion to maximize coverage of the solution and open up skin creases in the knee to ensure all areas have been optimally treated. Intra-operatively pre-soaking of the autograft has been shown to reduce the post-surgical infection risk significantly [43]. Using a 5-mg/mL vancomycin solution to soak the graft following harvest and prior to implantation in a soaked gauze, Perez-Prieto et al. did not observe any bacterial growth on the graft prior to implantation [44]. Postoperatively, it is advisable to cover all wounds and advise the athlete not to return to team practice/rehabilitation until the wounds have healed to prevent breaches of the skin barrier during recovery. Reinforce the importance of good hygiene, including showering and washing with soap after all practices and competitions and encourage them to seek medical help early if there are any concerns with the wound or skin lesions.

25.3.2 Early Rehabilitation

Rehabilitation is critical for the successful return of high-level athletes to pre-injury level activities. They are usually highly motivated individuals and keen to get going as quickly as possible; however, there is a risk of being over-aggressive with early physiotherapy. There needs to be balance; loading the knee excessively too early can cause increased irritation of the joint [45]. Early ROM exercises are crucial and reduce the risk of developing arthrofibrosis and its associated problems [7, 46, 47]. In the early period following ACLR, the graft is undergoing biological bone incorporation, vascularization and ligamentization and is most at risk of injury [48, 49]. Therefore, overly aggressive rehabilitation can possibly compromise this process leading to attenuation of the graft and ultimately failure of the ACLR [48, 49].

It is useful to share the various components of postoperative care and rehabilitation with athletes. Athletes are normally goal-oriented and setting achievable weekly targets tends to keep them focused on the task at hand and avoids despondency regarding the magnitude of mission ahead. Although tempting to use the latest rehabilitation technology, it is important not to lose sight of the tried and trusted rehabilitation methods in this group. Studies have demonstrated no advantage in the use of continuous passive motion postoperatively nor have they found any significant added benefit in using a brace following an uncomplicated ACLR [50–54].

Accelerated rehabilitation protocols are widely accepted nowadays and have been shown to be safe and effective [21, 47, 55, 56]. As part of an accelerated rehabilitation programme, early weight-bearing and motion are beneficial and may decrease patellofemoral pain [57]. Early eccentric exercises have been associated with increased quadriceps muscle strength, and isokinetic hamstring muscle strengthening can be safely incorporated into rehabilitation protocols as early as 3 weeks after surgery, pain allowing [58, 59]. Closed chain exercises are safe and effective and are believed to offer the advantages of less stress on the healing graft and less patellofemoral pain [60]. Variety in training is useful to avoid boredom and burnout. Water-based therapy, slide board and cycling programmes are safe and are popular with athletes [61–63]. Neuromuscular electrical stimulation has been shown to be effective, but if used as part of rehabilitation, then it should be instituted early in the postoperative timeframe and should be of high intensity to achieve meaningful results [53, 63].

25.3.3 Pain Management

Good pain management is critical following surgery and improves early mobilization and rehabilitation. *'No pain, no gain'* is a common mantra among athletes in describing their acceptance of pain as a normal consequence of athletic activity [64]. However, widespread use of both prescription and nonprescription medication is also common in athletes. Use of pain medications has therefore got to be taken into consideration when treating high-level athletes. A dependence on analgesics has the potential to exacerbate post-recovery because of the inhibition of the pain response and/or the inflammatory cascade and increase analgesic requirements [64].

Multimodal analgesia is most appropriate and combination of different methods is used including; oral analgesics, nerve blocks and intra-articular injections [65]. Certain medications are also prohibited *'In-Competition'* and every efforts should be taken to prevent the use of these substances in high-level athletes to avert any potential conduct breaches on return to sport [66]. It is important to communicate with the medical team about therapeutic use exemption forms.

An instructional video concerning the psychological aspects of ACLR—anxiety, pain and functional outcomes—has been developed and could be helpful in managing expectations of pain and athlete's approach to rehabilitation following surgery (75, 76).

25.3.4 Haemarthrosis Formation

Haemarthrosis is a cause of pain and can disrupt early rehabilitation, which can lead to functional limitations and poor long-term outcomes, especially in high-level athletes [67–70]. It is usually associated with immediate morbidity, as it typically develops in the first 24 hours but it can increase in the first few days. Although hard to avoid, there are certain measures that can be taken to reduce the risk of haemarthrosis formation. The use of a tourniquet is useful in controlling blood loss, improving arthroscopic visibility and reducing surgical time [71, 72]. Tranexamic acid (TXA) has been shown in ACLR to produce a clinical difference in the development of postoperative haemarthrosis with intra-articular infiltration or intravenous administration [67, 73–76]. Although the use of drains has been advocated both intra-articularly and at the graft harvest site, several studies have reported that drains are not necessary and do not affect functional outcomes [73, 77–79]. What is perhaps most important is to recognise excessive haemarthrosis formation early and address it in a timely manner to avoid potential long-term complications, including arthrogenic muscle inhibition and the development of arthrofibrosis [80]. The treatment options include aspiration of the haemarthrosis, if the haematoma has liquified sufficiently; arthroscopic washout, if there is excessive pain and limitation in joint motion; or, 'watchful neglect', if muscle activation and range of motion are maintained sufficiently to not hinder rehabilitation.

25.3.5 Recurrent Effusions

Joint effusions following both ACL injuries and ACLR are common findings and are thought to be part of the process of joint cartilage degradation after joint trauma and surgery [81, 82]. Resolution of the effusion is one of the primary goals of preoperative rehabilitation [47]. Patients who have persistent effusions often complain of pain and swelling that cause problems with various activities, which impede rehabilitation and return to sport [83]. Lindstrom et al. reported an association between excessive joint effusion, the presence of a partial meniscus injury at surgery, female sex and a higher pre-injury Tegner activity level, which included the elite athlete group [84]. It has been hypothesized that a joint effusion and knee pain causes arthrogenic muscle inhibition (AMI), a reflex inhibition of voluntary muscle contraction in the extensor muscles of the thigh [85], which is thought to effect quadriceps strength and reduce quadriceps functional performance [86, 87]. This can severely hamper rehabilitation, causing gait asymmetry as a result of reduced quadriceps muscle activation [88–90].

High-intensity activities may also increase the amount of effusion in the knee joint if the activities are an irritant to the knee joint, allowing AMI to persist. Relative inactivity can decrease the amount of effusion by not irritating the joint; therefore, it is important to advise rest in an athlete when an effusion is noted [45]. Other inflammatory pathways, such as pain and chemical markers of inflammation increase the activity of receptors in the joint, increasing AMI [45]. If inflammation is a cause, then nonsteroidal anti-inflammatory drugs have a potential benefit. The use of an ice pack is an effective method for improving pain and knee ROM after ACLR [91]. Some studies demonstrate that after ACLR, cryotherapy can reduce pain, drug intake, length of hospital stay, improvement knee ROM and the patient's quality of life [92, 93].

25.3.6 Early Graft-Related Complications

During the first year after ACLR, the strength and resistance of the graft are only 30% and 50% of the original ACL, respectively [48, 94]. Excessive loads in this early period may induce plastic deformity and a lengthening/stretching of the graft [95]. With accelerated rehabilitation, especially in high-level athletes, the graft undergoes more cycles of flexion in the early postoperative stages; this can contribute to early elongation and failure of the graft fibres [96]. Graft loosening can ultimately lead to graft injury. In high-level athletes the surgery may be considered to have failed even before complete rupture as even subtle laxity/loosening of the graft leaves the knee with an abnormal anterior laxity and symptoms of instability that impede rehabilitation and return to sport.

25.3.7 Cyclops Lesions/Syndrome

Cyclops lesions are fibrovascular nodules that frequently form and occur after ACLR [97]. Cyclops syndrome describes partial knee extension loss, attributed to impingement of a cyclops lesion within the intercondylar notch [98]. Development of cyclops syndrome can be devastating to high-level athletes and will often require further surgery [97]. In a recent study, Delaloye et al. investigated different potential risk factors that can lead to the development of cyclops syndrome in a group of 3633 patients [98]. They concluded that regaining full extension in the early postoperative period was the only significant risk factor and importantly preservation of a large ACL remnant did not predispose to the development of cyclops syndrome [98]. The authors suggest that lack of extension could be related to quadriceps weakness and activation failure, secondary to AMI. Therapeutic interventions that have been shown to be effective for AMI are cryotherapy and physical exercise therapy [80]. Rehabilitation programmes should therefore focus on regaining early extension so as to minimize the risk of cyclops syndrome in high-level athletes.

25.3.8 Graft Harvest Site Morbidity

Each graft has specific harvest related complications that can cause issues in the postoperative period. There are reports of patella fractures after ACLR using patella tendon grafts, an incidence between 0.23 and 2.3% [99–101]. A famous example in a high-level athlete is Jerry Rice, National Football League (NFL) wide receiver and superstar athlete. He returned to play after only 3 months following ACLR, far ahead of his original schedule. On his first game back, he made a victorious touchdown but sadly fell directly onto his knee, which resulted in a season ending fractured patella. It is thought that reversing the graft by taking the larger trapezoidal bone plug from the tibia and the triangular plug from the patella may reduce the amount of patella excised and risk of fracture [99]. Filling in the defect may preserve the anatomic integrity of the donor site as well as enhance the early stages of healing [102]. Disruption of the extensor mechanism is another potential but rare complication after patellar tendon graft harvest [103] and can lead to further surgeries, delay recovery and worse outcomes [104, 105].

Hamstring tendon harvest also causes morbidity; symptoms of early pain and tearing are common and typically occur when the hamstring muscles are subjected to increased load shortly following surgery or with commencement of sprinting in later rehabilitation [106]. Although the tendon of semitendinosus may regenerate, the muscle loses its volume and function to a large extent which increases the reliance on the remaining hamstring muscles for strength [106, 107]. Harvesting of the hamstring and patellar tendons may also damage branches of the saphenous and infrapatellar nerves, respectively, and result in comparable disturbances of sensitivity in the knee region [108, 109] . Studies have reported that hamstring harvests cause fewer kneeling and knee-walking difficulties compared with patella tendon grafts [109, 110]. The quadriceps tendon has also gained popularity and studies have shown that harvesting this graft was safer, causing with the smallest area of hypesthesia, compared with patella and hamstring grafts [111]. It also concluded that the scar from quadriceps harvest was more cosmetic and shorter. It is therefore important to consider these issues when deciding on graft type in high-level athletes, but it will often depend on the surgeon's comfort level, the sport played, not to mention the athlete's preference.

25.4 Postoperative Considerations

25.4.1 Return to Sport

Will I ever play again?

Surgical success after ACLR is often judged by whether a player returns quickly to their pre-injury level of play. This is of particular importance to high-level athletes, not only because it impacts on the individual's professional earning capacity; it is also an important concern to the team, coaching staff, managers and supporters [112]. It has been reported that between 33 and 92% of patients returned to some level of sporting activity by 12 months after ACLR [113–117]. Lai et al., in a systematic review of 24 studies reporting on 1272 elite level athletes, revealed that 83% returned to their pre-injury sport [118]. The same authors, in case series of elite Australian Rules Football (AFL) players, reported that 77% returned to play at the highest level after ACLR [119]. AFL is a unique and demanding sport that involves high-risk pivoting and, therefore, represents a sport with a particularly high risk of ACL injury [112].

25.4.2 Timing of Return to Sport

When will I be back?

This refrain from an athlete following an ACLR can sometimes sound like the lamentations of a bored child on a long journey: *'Are we there yet? Are we there YET?'* In many ways, the orthopaedic surgeon needs to assume the role of a patient parent with the response: *'No, not yet, but soon. I'll let you know when we're close!'*. There is a lack of current evidence and consensus regarding objective assessment criteria for patients to return to sport after ACLR. Time periods for return can also vary widely from 12 weeks to well beyond 12 months [120, 121]. There are huge variations in functional outcomes that teams use to ascertain when players should return to sport [122–130] . In a systematic review, Barber-Westin et al. concluded that factors related to muscle strength, stability, neuromuscular control and lower limb function should be measured before release to unrestricted activities [131]. In a recent study, Webster and Feller demonstrated that younger athletes (≤25 years of age) who had returned to level 1 sport were more likely to have met criteria that are used to indicate readiness to return compared to an older patient cohort [132]. In a later review, Webster et al. also concluded that there is some utility in return to sport criteria, especially to provide the patient with feedback with regard to their rehabilitation progress but meeting these criteria does not necessarily prevent re-injury to the same knee or a new injury to contralateral side [133].

Barriers to successful return to sport and timing are clearly multifactorial. In high-level athletes, these factors can be influenced not only by the athlete's confidence level in the stability of the knee but also by a desire to restore athletic identity and monetary considerations of returning to sport [114, 134]. In elite sports, where other players are readily available on lists and commercial interests are at stake, delays in restoration of player confidence are not tolerated [112]. Therefore, the team surrounding the player and those involved in rehabilitation have a profound effect on their psychological attitudes. It may also be of significant benefit to the patient to utilize questionnaires or other measures to gauge psychological preparedness prior to return to sport [135, 136].

25.4.3 Return to Elite Performance

Will I be just as good?

This is a difficult question to answer, but it is important to be honest with athletes, so that both they and we, as surgeons, have realistic expectations for their future performance following ACLR [137]. Although return to sport is the initial objective of most athletes following ACLR, return to an elite level of performance is really the ultimate goal. While for many the former is possible, unfortunately, for a significant number the latter remains elusive. Lai et al., in a study of 104 AFL players following ACLR, reported that only 53 (51%) returned to play at least 10 matches in two seasons after surgery; of these 53, 36 (68%) returned to their preinjury levels of performance [138]. Interestingly, younger (<25 years) and lighter players (<90 kg) were more likely to return to their preinjury level of performance which possibly relates to decisions on keeping older and previously injured players on the playing list. Liptak et al., in a similar study of AFL players, suggested that of the athletes who did return to preinjury form in some cases it took up to 3 years [112]; this possibly also explains the increased chances of delisting.

Looking at evidence from elite athletes in the NFL, another high-risk sport, Carey et al. showed that 21% of players did not return to play at the elite level [139]. The authors also found a statistically significant reduction in player form, defined as running back and wide receiver performance, on return to competition. Secrist et al. reported that on average, ACL-injured players earned $2,070,521 less than salary-matched controls over the 4 years after injury. Players initially earning less than $2 million per year have lower mean salaries and are less likely to remain in the league than uninjured controls [140]. The results are not as positive in the National Basketball Association (NBA), where Busfield et al. found that only 22% of athletes returned to the NBA after ACLR but 44% suffered a decrease in performance using a designated scale [141]. An ACL injury is perhaps even more catastrophic for the paediatric/adolescent population with aspirations of elite sport. Hamrin Senorski et al. concluded in their review that only the occasional patient will succeed in this goal [142]. Unfortunately, the details regarding specific numbers of paediatric/adolescent patients returning to sport are deficient and there is an urgent future need for more detailed data in this high-risk younger population.

So, what happens to those players on the wrong side of the above statistics? Although the majority of high-level athletes are initially able to resume their sports following ACL injury, studies have shown that the retirement rate may be higher amongst these athletes [3]. Roos et al. reported on elite soccer players 3–7 years after the ACL injury and found that 30% were active in soccer after 3 years compared with 80% in an uninjured control population [143]. In addition, they showed that, after 7 years, none of the injured elite players were active at the same level regardless of the type of treatment.

25.4.4 Reinjury and Graft Re-injury

Will it happen again?

Sport by its very nature is a game of risk and reward, but it gets riskier following ACL injury. By returning to sport, athletes are putting them-

Table 25.2 Factors associated with graft failure in high-level athletes

Early return to sport [144, 155]
High-risk sports that involve cutting, jumping and pivoting [146, 156]
Inadequate rehabilitation [21, 123, 131, 157–160]
Young age [156, 159, 161, 162]
Postoperative infection [163]

selves in harm's way which involves damage of the ACL graft; the rates of ACL graft injury vary from 4 to 30% [114, 121, 124, 144–146]. The highest incidence is within 3 years of surgery; however, there is still a significant risk of reinjury to the graft after this with continued sports participation [3, 121, 146]. Athletes who return to pivoting sports have a higher risk of ACL graft failure, as shown in both AFL and NFL players [119, 147, 148]. Bjordal et al. found a correlation between the risk of re-injury, the number of hours of sport and the competitive level [149]. Therefore, returning to a high activity level after ACLR is a stronger risk factor for ACL graft failure than many other variables and high-level athletes should be counselled about this pre-operatively [150].

There are many other factors that are thought to contribute to graft failure and lead to re-injury rates (Table 25.2). To minimize the risk, in this specific patient population, the addition of a concomitant lateral extra-articular tenodesis (LET) or anterolateral ligament (ALL) reconstruction has been shown to improve clinical outcomes and most importantly improve knee stability and lower re-injury rates [151–153]. Getgood et al. have showed that the addition of LET to an ACLR in young patients at high risk of failure results in a statistically significant reduction in graft injury 2 years after surgery. Although the indications for LET are as yet unproven, it would appear that many surgeons would seriously consider the addition of this procedure to an ACLR in elite athletes [154].

25.4.5 Contralateral ACL Injury and Other Injuries

What about my other knee?

As with an ACL graft injury, a contralateral ACL injury or an injury to other structures of the knee (menisci, cartilage or other ligaments) remains another devastating complication for high-level athletes. Some studies have documented a higher rate of contralateral injury than a re-injury of the ACL-reconstructed knee, with rates of contralateral ACL injury ranging from 4 to 30% [121, 123, 127, 145, 146, 157, 159, 164]. Liptak et al. found that 28% of players who returned at the elite level had a further ACL injury on either the ipsilateral or contralateral knee [112]. As with graft re-injury, return to sports after ACLR increases exposure to activities that put the ACL at risk, particularly in those who return to competitive sports that require jumping, pivoting and side-stepping [146]. Brophy et al. concluded that soccer athletes who injure their non-dominant limb are at higher risk of injuring the ACL in their contralateral dominant limb in the future [117]. It is postulated that the reason for this is that injury to the non-dominant limb may lead to even greater dependence on the dominant limb during play, potentially increasing the risk for future injury [117]. A similar finding was reported in AFL players by Orchard et al. who found that there was a higher risk of graft injury in the first 12 months after ACLR than injury to the contralateral knee [147]. After 12 months, the risk of sustaining a graft injury or a contralateral ACL injury was similar.

High-level athletes need to be aware of these risks and to try, with the support of their rehabilitation team, to reduce them. This may include activity modification, rehabilitation protocol modification and return to sports training modification that puts emphasis on also training and rehabilitating the contralateral knee [121].

25.5 Long-Term Considerations

25.5.1 Osteoarthritis

It has long been recognized that ACL injuries have been associated with development of osteoarthritis (OA), in the young or middle-aged adult; '*the young patient with an old knee*' [82]. The primary risk factor frequently reported for the development of OA is meniscectomy combined with ACLR at the time of ACLR [165–169]. It has been shown that participation in and return to competitive high-level sports activities involving cutting, pivoting, and twisting following ACLR are risk factors for the development of OA [165, 170].

Smith et al. reported that in a group of elite American footballer players previous knee surgery, specifically previous meniscectomy and/or prior ACLR and a body mass index >30, was associated with a higher prevalence of knee OA [171]. Deacon et al. found that a group of AFL footballers had a significant risk of both functional and radiological OA, with a history of intra-articular ligament or meniscal injury increasing this risk [172]. It has been established that a delay in surgical reconstruction of the ACL increases the risk of OA in high-level athletes [173, 174]. Early surgery is recommended because it has the potential to reduce the risk of sustaining additional secondary injuries and risk of long-term degenerative changes in athletes who continue participation in pivoting sports with ongoing instability due to ACL deficiency [134, 167].

25.6 Conclusion

High-level athletes who participate in sport do so at their own risk. An ACL injury represents a significant injury that can potentially derail a successful sporting career. The insult sustained to the knee initiates a cascade of events which ultimately can lead to the development of OA, no matter what treatment, other risk factors involved or how the knee is loaded during subsequent years [3]. Thus, the prevention of knee injuries is key to help protect high-level athletes from these injuries [175]. It is incumbent on orthopaedic surgeons to advocate for injury reduction programmes in elite sport. As key stakeholders in this ever-increasing problem, we must endeavour to reduce the risk of ACL injury and not simply stand by waiting at the stretcher with knife and arthroscope in hand.

References

1. Bernstein J, Perlis C, Bartolozzi AR. Ethics in sports medicine. Clin Orthop Relat Res. 2000;378:50–60.
2. Devitt BM, McCarthy C. 'I am in blood Stepp'd in so far...': ethical dilemmas and the sports team doctor. Br J Sports Med. 2010;44(3):175–8.
3. Myklebust G, Bahr R. Return to play guidelines after anterior cruciate ligament surgery. Br J Sports Med. 2005;39(3):127–31.
4. Smith AM, Scott SG, O'Fallon WM, Young ML. Emotional responses of athletes to injury. Mayo Clin Proc. 1990;65(1):38–50.
5. Ghoneim MM, O'Hara MW. Depression and postoperative complications: an overview. BMC Surg. 2016;16:5.
6. Shelbourne KD, Wilckens JH, Mollabashy A, DeCarlo M. Arthrofibrosis in acute anterior cruciate ligament reconstruction. The effect of timing of reconstruction and rehabilitation. Am J Sports Med. 1991;19(4):332–6.
7. Ekhtiari S, Horner NS, de Sa D, Simunovic N, Hirschmann MT, Ogilvie R, et al. Arthrofibrosis after ACL reconstruction is best treated in a step-wise approach with early recognition and intervention: a systematic review. Knee Surg Sports Traumatol Arthrosc. 2017;25(12):3929–37.
8. Noll S, Garrison JC, Bothwell J, Conway JE. Knee extension range of motion at 4 weeks is related to knee extension loss at 12 weeks after anterior cruciate ligament reconstruction. Orthop J Sports Med. 2015;3(5):2325967115583632.
9. Sanders TL, Kremers HM, Bryan AJ, Kremers WK, Stuart MJ, Krych AJ. Procedural intervention for arthrofibrosis after ACL reconstruction: trends over two decades. Knee Surg Sports Traumatol Arthrosc. 2017;25(2):532–7.
10. Sachs RA, Daniel DM, Stone ML, Garfein RF. Patellofemoral problems after anterior cruciate ligament reconstruction. Am J Sports Med. 1989;17(6):760–5.
11. Werner BC, Cancienne JM, Miller MD, Gwathmey FW. Incidence of manipulation under anesthesia or lysis of adhesions after arthroscopic knee surgery. Am J Sports Med. 2015;43(7):1656–61.
12. Mayr HO, Weig TG, Plitz W. Arthrofibrosis following ACL reconstruction--reasons and outcome. Arch Orthop Trauma Surg. 2004;124(8):518–22.
13. Shelbourne KD, Patel DV. Timing of surgery in anterior cruciate ligament-injured knees. Knee Surg Sports Traumatol Arthrosc. 1995;3(3):148–56.
14. Eriksson K, von Essen C, Jonhagen S, Barenius B. No risk of arthrofibrosis after acute anterior cruciate ligament reconstruction. Knee Surg Sports Traumatol Arthrosc. 2018;26(10):2875–82.
15. Bottoni CR, Liddell TR, Trainor TJ, Freccero DM, Lindell KK. Postoperative range of motion following anterior cruciate ligament reconstruction using autograft hamstrings: a prospective, randomized clinical trial of early versus delayed reconstructions. Am J Sports Med. 2008;36(4):656–62.
16. Frobell RB, Roos EM, Roos HP, Ranstam J, Lohmander LS. A randomized trial of treatment for acute anterior cruciate ligament tears. N Engl J Med. 2010;363(4):331–42.
17. Meighan AA, Keating JF, Will E. Outcome after reconstruction of the anterior cruciate ligament in

athletic patients. A comparison of early versus delayed surgery. J Bone Joint Surg Br. 2003;85(4):521–4.
18. Quelard B, Sonnery-Cottet B, Zayni R, Ogassawara R, Prost T, Chambat P. Preoperative factors correlating with prolonged range of motion deficit after anterior cruciate ligament reconstruction. Am J Sports Med. 2010;38(10):2034–9.
19. Mauro CS, Irrgang JJ, Williams BA, Harner CD. Loss of extension following anterior cruciate ligament reconstruction: analysis of incidence and etiology using IKDC criteria. Arthroscopy. 2008;24(2):146–53.
20. Wasilewski SA, Covall DJ, Cohen S. Effect of surgical timing on recovery and associated injuries after anterior cruciate ligament reconstruction. Am J Sports Med. 1993;21(3):338–42.
21. Shelbourne KD, Klotz C. What I have learned about the ACL: utilizing a progressive rehabilitation scheme to achieve total knee symmetry after anterior cruciate ligament reconstruction. J Orthop Sci. 2006;11(3):318–25.
22. Kwok CS, Harrison T, Servant C. The optimal timing for anterior cruciate ligament reconstruction with respect to the risk of postoperative stiffness. Arthroscopy. 2013;29(3):556–65.
23. Borchers JR, Kaeding CC, Pedroza AD, Huston LJ, Spindler KP, Wright RW, et al. Intra-articular findings in primary and revision anterior cruciate ligament reconstruction surgery: a comparison of the MOON and MARS study groups. Am J Sports Med. 2011;39(9):1889–93.
24. Church S, Keating JF. Reconstruction of the anterior cruciate ligament: timing of surgery and the incidence of meniscal tears and degenerative change. J Bone Joint Surg Br. 2005;87(12):1639–42.
25. Clancy WG Jr, Sutherland TB. Combined posterior cruciate ligament injuries. Clin Sports Med. 1994;13(3):629–47.
26. Chahla J, Moatshe G, Dean CS, LaPrade RF. Posterolateral corner of the knee: current concepts. Arch Bone Jt Surg. 2016;4(2):97–103.
27. Schub DL, Schmitz LM, Sakamoto FA, Winalski CS, Parker RD. Long-term outcomes of postoperative septic arthritis after anterior cruciate ligament reconstruction. Am J Sports Med. 2012;40(12):2764–70.
28. Wang C, Ao Y, Wang J, Hu Y, Cui G, Yu J. Septic arthritis after arthroscopic anterior cruciate ligament reconstruction: a retrospective analysis of incidence, presentation, treatment, and cause. Arthroscopy. 2009;25(3):243–9.
29. Wang C, Lee YH, Siebold R. Recommendations for the management of septic arthritis after ACL reconstruction. Knee Surg Sports Traumatol Arthrosc. 2014;22(9):2136–44.
30. Sonnery-Cottet B, Saithna A, Abreu FG, Franck F, de Abreu GV, Vieira TD, et al. Professional athletes are at higher risk of septic arthritis after anterior cruciate ligament reconstruction: an analysis of 4421 consecutive patients including 265 elite athlete from the SANTI Study Group. Am J Sports Med 2019;47(12):2910–8.
31. Indelli PF, Dillingham M, Fanton G, Schurman DJ. Septic arthritis in postoperative anterior cruciate ligament reconstruction. Clin Orthop Relat Res 2002;398:182–8.
32. Williams RJ 3rd, Laurencin CT, Warren RF, Speciale AC, Brause BD, O'Brien S. Septic arthritis after arthroscopic anterior cruciate ligament reconstruction. Diagnosis and management. Am J Sports Med. 1997;25(2):261–7.
33. Van Tongel A, Stuyck J, Bellemans J, Vandenneucker H. Septic arthritis after arthroscopic anterior cruciate ligament reconstruction: a retrospective analysis of incidence, management and outcome. Am J Sports Med. 2007;35(7):1059–63.
34. Ristic V, Maljanovic M, Harhaji V, Milankov M. Infections after reconstructions of anterior cruciate ligament. Med Pregl. 2014;67(1–2):11–5.
35. Sonnery-Cottet B, Archbold P, Zayni R, Bortolletto J, Thaunat M, Prost T, et al. Prevalence of septic arthritis after anterior cruciate ligament reconstruction among professional athletes. Am J Sports Med. 2011;39(11):2371–6.
36. Turbeville SD, Cowan LD, Greenfield RA. Infectious disease outbreaks in competitive sports: a review of the literature. Am J Sports Med. 2006;34(11):1860–5.
37. Pecci M, Comeau D, Chawla V. Skin conditions in the athlete. Am J Sports Med. 2009;37(2):406–18.
38. Bowers AL, Huffman GR, Sennett BJ. Methicillin-resistant Staphylococcus aureus infections in collegiate football players. Med Sci Sports Exerc. 2008;40(8):1362–7.
39. Rihn JA, Michaels MG, Harner CD. Community-acquired methicillin-resistant staphylococcus aureus: an emerging problem in the athletic population. Am J Sports Med. 2005;33(12):1924–9.
40. Grosset-Janin A, Nicolas X, Saraux A. Sport and infectious risk: a systematic review of the literature over 20 years. Med Mal Infect. 2012;42(11):533–44.
41. Seropian R, Reynolds BM. Wound infections after preoperative depilatory versus razor preparation. Am J Surg. 1971;121(3):251–4.
42. Darouiche RO, Wall MJ Jr, Itani KM, Otterson MF, Webb AL, Carrick MM, et al. Chlorhexidine-alcohol versus povidone-iodine for surgical-site antisepsis. N Engl J Med. 2010;362(1):18–26.
43. Phegan M, Grayson JE, Vertullo CJ. No infections in 1300 anterior cruciate ligament reconstructions with vancomycin pre-soaking of hamstring grafts. Knee Surg Sports Traumatol Arthrosc. 2016;24(9):2729–35.
44. Perez-Prieto D, Torres-Claramunt R, Gelber PE, Shehata TMA, Pelfort X, Monllau JC. Autograft soaking in vancomycin reduces the risk of infection after anterior cruciate ligament reconstruction. Knee Surg Sports Traumatol Arthrosc. 2016;24(9):2724–8.

45. Lynch AD, Logerstedt DS, Axe MJ, Snyder-Mackler L. Quadriceps activation failure after anterior cruciate ligament rupture is not mediated by knee joint effusion. J Orthop Sports Phys Ther. 2012;42(6):502–10.
46. Noyes FR, Berrios-Torres S, Barber-Westin SD, Heckmann TP. Prevention of permanent arthrofibrosis after anterior cruciate ligament reconstruction alone or combined with associated procedures: a prospective study in 443 knees. Knee Surg Sports Traumatol Arthrosc. 2000;8(4):196–206.
47. Shelbourne KD, Nitz P. Accelerated rehabilitation after anterior cruciate ligament reconstruction. J Orthop Sports Phys Ther. 1992;15(6):256–64.
48. Clancy WG Jr, Narechania RG, Rosenberg TD, Gmeiner JG, Wisnefske DD, Lange TA. Anterior and posterior cruciate ligament reconstruction in rhesus monkeys. J Bone Joint Surg Am. 1981;63(8):1270–84.
49. Corsetti JR, Jackson DW. Failure of anterior cruciate ligament reconstruction: the biologic basis. Clin Orthop Relat Res. 1996;325:42–9.
50. Rosen MA, Jackson DW, Atwell EA. The efficacy of continuous passive motion in the rehabilitation of anterior cruciate ligament reconstructions. Am J Sports Med. 1992;20(2):122–7.
51. Engstrom B, Sperber A, Wredmark T. Continuous passive motion in rehabilitation after anterior cruciate ligament reconstruction. Knee Surg Sports Traumatol Arthrosc. 1995;3(1):18–20.
52. Wright RW, Preston E, Fleming BC, Amendola A, Andrish JT, Bergfeld JA, et al. A systematic review of anterior cruciate ligament reconstruction rehabilitation: part I: continuous passive motion, early weight bearing, postoperative bracing, and home-based rehabilitation. J Knee Surg. 2008;21(3):217–24.
53. Kruse LM, Gray B, Wright RW. Rehabilitation after anterior cruciate ligament reconstruction: a systematic review. J Bone Joint Surg Am. 2012;94(19):1737–48.
54. McCarthy MR, Yates CK, Anderson MA, Yates-McCarthy JL. The effects of immediate continuous passive motion on pain during the inflammatory phase of soft tissue healing following anterior cruciate ligament reconstruction. J Orthop Sports Phys Ther. 1993;17(2):96–101.
55. Beynnon BD, Uh BS, Johnson RJ, Abate JA, Nichols CE, Fleming BC, et al. Rehabilitation after anterior cruciate ligament reconstruction: a prospective, randomized, double-blind comparison of programs administered over 2 different time intervals. Am J Sports Med. 2005;33(3):347–59.
56. Shelbourne KD, Nitz P. Accelerated rehabilitation after anterior cruciate ligament reconstruction. Am J Sports Med. 1990;18(3):292–9.
57. Tyler TF, McHugh MP, Gleim GW, Nicholas SJ. The effect of immediate weightbearing after anterior cruciate ligament reconstruction. Clin Orthop Relat Res. 1998;357:141–8.
58. Gerber JP, Marcus RL, Dibble LE, Greis PE, Burks RT, LaStayo PC. Effects of early progressive eccentric exercise on muscle size and function after anterior cruciate ligament reconstruction: a 1-year follow-up study of a randomized clinical trial. Phys Ther. 2009;89(1):51–9.
59. Sekir U, Gur H, Akova B. Early versus late start of isokinetic hamstring-strengthening exercise after anterior cruciate ligament reconstruction with patellar tendon graft. Am J Sports Med. 2010;38(3):492–500.
60. Bynum EB, Barrack RL, Alexander AH. Open versus closed chain kinetic exercises after anterior cruciate ligament reconstruction. A prospective randomized study. Am J Sports Med. 1995;23(4):401–6.
61. Tovin BJ, Wolf SL, Greenfield BH, Crouse J, Woodfin BA. Comparison of the effects of exercise in water and on land on the rehabilitation of patients with intra-articular anterior cruciate ligament reconstructions. Phys Ther. 1994;74(8):710–9.
62. Blanpied P, Carroll R, Douglas T, Lyons M, Macalisang R, Pires L. Effectiveness of lateral slide exercise in an anterior cruciate ligament reconstruction rehabilitation home exercise program. J Orthop Sports Phys Ther. 2000;30(10):602–8; discussion 9-11.
63. Wright RW, Preston E, Fleming BC, Amendola A, Andrish JT, Bergfeld JA, et al. A systematic review of anterior cruciate ligament reconstruction rehabilitation: part II: open versus closed kinetic chain exercises, neuromuscular electrical stimulation, accelerated rehabilitation, and miscellaneous topics. J Knee Surg. 2008;21(3):225–34.
64. Matava MJ. Ethical considerations for analgesic use in sports medicine. Clin Sports Med. 2016;35(2):227–43.
65. Baverel L, Cucurulo T, Lutz C, Colombet CJ, Dalmay F, et al. Anesthesia and analgesia methods for outpatient anterior cruciate ligament reconstruction. Orthop Traumatol Surg Res. 2016;102(8S):S251–S5.
66. Vernec A, Pipe A, Slack A. A painful dilemma? Analgesic use in sport and the role of anti-doping. Br J Sports Med. 2017;51(17):1243–4.
67. Karaaslan F, Karaoglu S, Yurdakul E. Reducing intra-articular Hemarthrosis after arthroscopic anterior cruciate ligament reconstruction by the Administration of Intravenous Tranexamic Acid: a prospective, randomized controlled trial. Am J Sports Med. 2015;43(11):2720–6.
68. Bahl V, Goyal A, Jain V, Joshi D, Chaudhary D. Effect of haemarthrosis on the rehabilitation of anterior cruciate ligament reconstruction--single bundle versus double bundle. J Orthop Surg Res. 2013;8:5.
69. Simpson LA, Barrett JP Jr. Factors associated with poor results following arthroscopic subcutaneous lateral retinacular release. Clin Orthop Relat Res. 1984;186:165–71.

70. Coupens SD, Yates CK. The effect of tourniquet use and hemovac drainage on postoperative hemarthrosis. Arthroscopy. 1991;7(3):278–82.
71. Wu H, Chen LX, Li YL, Wu Q, Wu QL, Ning GZ, et al. Tourniquet used in anterior cruciate ligament reconstruction: a system review. Eur J Orthop Surg Traumatol. 2014;24(6):999–1003.
72. Nakayama H, Yoshiya S. The effect of tourniquet use on operative performance and early postoperative results of anatomic double-bundle anterior cruciate ligament reconstruction. J Orthop Sci. 2013;18(4):586–91.
73. McCormack RG, Greenhow RJ, Fogagnolo F, Shrier I. Intra-articular drain versus no drain after arthroscopic anterior cruciate ligament reconstruction: a randomized, prospective clinical trial. Arthroscopy. 2006;22(8):889–93.
74. Seo JG, Moon YW, Park SH, Kim SM, Ko KR. The comparative efficacies of intra-articular and IV tranexamic acid for reducing blood loss during total knee arthroplasty. Knee Surg Sports Traumatol Arthrosc. 2013;21(8):1869–74.
75. Camarasa MA, Olle G, Serra-Prat M, Martin A, Sanchez M, Ricos P, et al. Efficacy of aminocaproic, tranexamic acids in the control of bleeding during total knee replacement: a randomized clinical trial. Br J Anaesth. 2006;96(5):576–82.
76. Chiang ER, Chen KH, Wang ST, Ma HL, Chang MC, Liu CL, et al. Intra-articular injection of Tranexamic acid reduced postoperative Hemarthrosis in arthroscopic anterior cruciate ligament reconstruction: a prospective randomized study. Arthroscopy. 2019;35(7):2127–32.
77. Karahan M, Erol B, Bekiroglu N, Uyan D. Effect of drain placed in the donor site in the early postoperative period after arthroscopically assisted anterior cruciate ligament reconstruction with quadrupled hamstring tendons. Am J Sports Med. 2005;33(6):900–6.
78. Clifton R, Haleem S, McKee A, Parker MJ. Closed suction surgical wound drainage after anterior cruciate ligament reconstruction: a systematic review of randomised controlled trials. Knee. 2007;14(5):348–51.
79. Dhawan A, Doukas WC, Papazis JA, Scoville CR. Effect of drain use in the early postoperative period after arthroscopically assisted anterior cruciate ligament reconstruction with bone-patellar tendon-bone graft. Am J Sports Med. 2003;31(3):419–24.
80. Sonnery-Cottet B, Saithna A, Quelard B, Daggett M, Borade A, Ouanezar H, et al. Arthrogenic muscle inhibition after ACL reconstruction: a scoping review of the efficacy of interventions. Br J Sports Med. 2019;53(5):289–98.
81. Sward P, Frobell R, Englund M, Roos H, Struglics A. Cartilage and bone markers and inflammatory cytokines are increased in synovial fluid in the acute phase of knee injury (hemarthrosis)—a cross-sectional analysis. Osteoarthr Cartil. 2012;20(11):1302–8.
82. Lohmander LS, Englund PM, Dahl LL, Roos EM. The long-term consequence of anterior cruciate ligament and meniscus injuries: osteoarthritis. Am J Sports Med. 2007;35(10):1756–69.
83. Yamamoto H, Ishibashi T, Muneta T, Furuya K, Mizuta T. Effusions after anterior cruciate ligament reconstruction using the ligament augmentation device. Arthroscopy. 1992;8(3):305–10.
84. Lindstrom M, Wredmark T, Wretling ML, Henriksson M, Fellander-Tsai L. Post-operative bracing after ACL reconstruction has no effect on knee joint effusion. A prospective, randomized study. Knee. 2015;22(6):559–64.
85. Palmieri-Smith RM, Thomas AC, Wojtys EM. Maximizing quadriceps strength after ACL reconstruction. Clin Sports Med. 2008;27(3):405–24, vii-ix.
86. Fahrer H, Rentsch HU, Gerber NJ, Beyeler C, Hess CW, Grunig B. Knee effusion and reflex inhibition of the quadriceps. A bar to effective retraining. J Bone Joint Surg Br. 1988;70(4):635–8.
87. Stevens JE, Mizner RL, Snyder-Mackler L. Quadriceps strength and volitional activation before and after total knee arthroplasty for osteoarthritis. J Orthop Res. 2003;21(5):775–9.
88. Palmieri-Smith RM, Kreinbrink J, Ashton-Miller JA, Wojtys EM. Quadriceps inhibition induced by an experimental knee joint effusion affects knee joint mechanics during a single-legged drop landing. Am J Sports Med. 2007;35(8):1269–75.
89. Hart JM, Pietrosimone B, Hertel J, Ingersoll CD. Quadriceps activation following knee injuries: a systematic review. J Athl Train. 2010;45(1):87–97.
90. Palmieri RM, Ingersoll CD, Hoffman MA, Cordova ML, Porter DA, Edwards JE, et al. Arthrogenic muscle response to a simulated ankle joint effusion. Br J Sports Med. 2004;38(1):26–30.
91. Bleakley C, McDonough S, MacAuley D. The use of ice in the treatment of acute soft-tissue injury: a systematic review of randomized controlled trials. Am J Sports Med. 2004;32(1):251–61.
92. Airaksinen OV, Kyrklund N, Latvala K, Kouri JP, Gronblad M, Kolari P. Efficacy of cold gel for soft tissue injuries: a prospective randomized double-blinded trial. Am J Sports Med. 2003;31(5):680–4.
93. Martin SS, Spindler KP, Tarter JW, Detwiler K, Petersen HA. Cryotherapy: an effective modality for decreasing intraarticular temperature after knee arthroscopy. Am J Sports Med. 2001;29(3):288–91.
94. Drez DJ Jr, DeLee J, Holden JP, Arnoczky S, Noyes FR, Roberts TS. Anterior cruciate ligament reconstruction using bone-patellar tendon-bone allografts. A biological and biomechanical evaluation in goats. Am J Sports Med. 1991;19(3):256–63.
95. Kamath GV, Murphy T, Creighton RA, Viradia N, Taft TN, Spang JT. Anterior cruciate ligament injury, return to play, and reinjury in the Elite Collegiate Athlete: analysis of an NCAA division I Cohort. Am J Sports Med. 2014;42(7):1638–43.

96. Samitier G, Marcano AI, Alentorn-Geli E, Cugat R, Farmer KW, Moser MW. Failure of anterior cruciate ligament reconstruction. Arch Bone Jt Surg. 2015;3(4):220–40.
97. Wang J, Ao Y. Analysis of different kinds of cyclops lesions with or without extension loss. Arthroscopy. 2009;25(6):626–31.
98. Delaloye JR, Murar J, Vieira TD, Franck F, Pioger C, Helfer L, et al. Knee extension deficit in the early postoperative period predisposes to cyclops syndrome after anterior cruciate ligament reconstruction: a risk factor analysis in 3633 patients from the SANTI study group database. Am J Sports Med. 2020;48(3):565–72.
99. Christen B, Jakob RP. Fractures associated with patellar ligament grafts in cruciate ligament surgery. J Bone Joint Surg Br. 1992;74(4):617–9.
100. Viola R, Vianello R. Three cases of patella fracture in 1,320 anterior cruciate ligament reconstructions with bone-patellar tendon-bone autograft. Arthroscopy. 1999;15(1):93–7.
101. Berg EE. Management of patella fractures associated with central third bone-patella tendon-bone autograft ACL reconstructions. Arthroscopy. 1996;12(6):756–9.
102. Stein DA, Hunt SA, Rosen JE, Sherman OH. The incidence and outcome of patella fractures after anterior cruciate ligament reconstruction. Arthroscopy. 2002;18(6):578–83.
103. Benner RW, Shelbourne KD, Urch SE, Lazarus D. Tear patterns, surgical repair, and clinical outcomes of patellar tendon ruptures after anterior cruciate ligament reconstruction with a bone-patellar tendon-bone autograft. Am J Sports Med. 2012;40(8):1834–41.
104. Lee GH, McCulloch P, Cole BJ, Bush-Joseph CA, Bach BR Jr. The incidence of acute patellar tendon harvest complications for anterior cruciate ligament reconstruction. Arthroscopy. 2008;24(2):162–6.
105. Milankov Ziva M, Semnic R, Miljkovic N, Harhaji V. Reconstruction of patellar tendon rupture after anterior cruciate ligament reconstruction: a case report. Knee. 2008;15(5):419–22.
106. Nakamae A, Ochi M, Deie M, Adachi N. Unsuccessful regeneration of the semitendinosus tendon harvested for anterior cruciate ligament reconstruction: report of two cases. Orthop Traumatol Surg Res. 2012;98(8):932–5.
107. Nishino A, Sanada A, Kanehisa H, Fukubayashi T. Knee-flexion torque and morphology of the semitendinosus after ACL reconstruction. Med Sci Sports Exerc. 2006;38(11):1895–900.
108. Brown CH Jr, Steiner ME, Carson EW. The use of hamstring tendons for anterior cruciate ligament reconstruction. Technique and results. Clin Sports Med. 1993;12(4):723–56.
109. Eriksson E. Does harvesting the hamstrings for ACL reconstruction impair the muscular function of the knee joint? Knee Surg Sports Traumatol Arthrosc. 2008;16(1):1.
110. Corry IS, Webb JM, Clingeleffer AJ, Pinczewski LA. Arthroscopic reconstruction of the anterior cruciate ligament. A comparison of patellar tendon autograft and four-strand hamstring tendon autograft. Am J Sports Med. 1999;27(4):444–54.
111. Mouarbes D, Dagneaux L, Olivier M, Lavoue V, Peque E, Berard E, et al. Lower donor-site morbidity using QT autografts for ACL reconstruction. Knee Surg Sports Traumatol Arthrosc. 2020;28:2558.
112. Liptak MG, Angel KR. Return to play and player performance after anterior cruciate ligament injury in elite Australian rules football players. Orthop J Sports Med. 2017;5(6):2325967117711885.
113. Ardern CL, Webster KE, Taylor NF, Feller JA. Return to sport following anterior cruciate ligament reconstruction surgery: a systematic review and meta-analysis of the state of play. Br J Sports Med. 2011;45(7):596–606.
114. Czuppon S, Racette BA, Klein SE, Harris-Hayes M. Variables associated with return to sport following anterior cruciate ligament reconstruction: a systematic review. Br J Sports Med. 2014;48(5):356–64.
115. Ardern CL, Taylor NF, Feller JA, Webster KE. Return-to-sport outcomes at 2 to 7 years after anterior cruciate ligament reconstruction surgery. Am J Sports Med. 2012;40(1):41–8.
116. Lee DY, Karim SA, Chang HC. Return to sports after anterior cruciate ligament reconstruction - a review of patients with minimum 5-year follow-up. Ann Acad Med Singap. 2008;37(4):273–8.
117. Brophy RH, Schmitz L, Wright RW, Dunn WR, Parker RD, Andrish JT, et al. Return to play and future ACL injury risk after ACL reconstruction in soccer athletes from the Multicenter Orthopaedic Outcomes Network (MOON) group. Am J Sports Med. 2012;40(11):2517–22.
118. Lai CCH, Ardern CL, Feller JA, Webster KE. Eighty-three per cent of elite athletes return to preinjury sport after anterior cruciate ligament reconstruction: a systematic review with meta-analysis of return to sport rates, graft rupture rates and performance outcomes. Br J Sports Med. 2018;52(2):128–38.
119. Lai CCH, Feller JA, Webster KE. Fifteen-year audit of anterior cruciate ligament reconstructions in the Australian Football League from 1999 to 2013: return to play and subsequent ACL injury. Am J Sports Med. 2018;46(14):3353–60.
120. Barber-Westin SD, Noyes FR. Factors used to determine return to unrestricted sports activities after anterior cruciate ligament reconstruction. Arthroscopy. 2011;27(12):1697–705.
121. Wright RW, Magnussen RA, Dunn WR, Spindler KP. Ipsilateral graft and contralateral ACL rupture at five years or more following ACL reconstruction: a systematic review. J Bone Joint Surg Am. 2011;93(12):1159–65.
122. Myer GD, Schmitt LC, Brent JL, Ford KR, Barber Foss KD, Scherer BJ, et al. Utilization of modified NFL combine testing to identify functional deficits

in athletes following ACL reconstruction. J Orthop Sports Phys Ther. 2011;41(6):377–87.
123. Myer GD, Paterno MV, Ford KR, Quatman CE, Hewett TE. Rehabilitation after anterior cruciate ligament reconstruction: criteria-based progression through the return-to-sport phase. J Orthop Sports Phys Ther. 2006;36(6):385–402.
124. Myer GD, Martin L Jr, Ford KR, Paterno MV, Schmitt LC, Heidt RS Jr, et al. No association of time from surgery with functional deficits in athletes after anterior cruciate ligament reconstruction: evidence for objective return-to-sport criteria. Am J Sports Med. 2012;40(10):2256–63.
125. Myer GD, Paterno MV, Ford KR, Hewett TE. Neuromuscular training techniques to target deficits before return to sport after anterior cruciate ligament reconstruction. J Strength Cond Res. 2008;22(3):987–1014.
126. Paterno MV, Ford KR, Myer GD, Heyl R, Hewett TE. Limb asymmetries in landing and jumping 2 years following anterior cruciate ligament reconstruction. Clin J Sport Med. 2007;17(4):258–62.
127. Paterno MV, Schmitt LC, Ford KR, Rauh MJ, Myer GD, Huang B, et al. Biomechanical measures during landing and postural stability predict second anterior cruciate ligament injury after anterior cruciate ligament reconstruction and return to sport. Am J Sports Med. 2010;38(10):1968–78.
128. Kvist J. Rehabilitation following anterior cruciate ligament injury: current recommendations for sports participation. Sports Med. 2004;34(4):269–80.
129. Hartigan EH, Axe MJ, Snyder-Mackler L. Time line for noncopers to pass return-to-sports criteria after anterior cruciate ligament reconstruction. J Orthop Sports Phys Ther. 2010;40(3):141–54.
130. Mascarenhas R, Tranovich M, Karpie JC, Irrgang JJ, Fu FH, Harner CD. Patellar tendon anterior cruciate ligament reconstruction in the high-demand patient: evaluation of autograft versus allograft reconstruction. Arthroscopy. 2010;26(9 Suppl):S58–66.
131. Barber-Westin S, Noyes FR. One in 5 athletes sustain reinjury upon return to high-risk sports after ACL reconstruction: a systematic review in 1239 athletes younger than 20 years. Sports Health. 2020;12:587.
132. Webster KE, Feller JA. Return to level I sports after anterior cruciate ligament reconstruction: evaluation of age, sex, and readiness to return criteria. Orthop J Sports Med. 2018;6(8):2325967118788045.
133. Webster KE, Feller JA. A research update on the state of play for return to sport after anterior cruciate ligament reconstruction. J Orthop Traumatol. 2019;20(1):10.
134. Hoffelner T, Resch H, Moroder P, Atzwanger J, Wiplinger M, Hitzl W, et al. No increased occurrence of osteoarthritis after anterior cruciate ligament reconstruction after isolated anterior cruciate ligament injury in athletes. Arthroscopy. 2012;28(4):517–25.
135. Ardern CL, Osterberg A, Tagesson S, Gauffin H, Webster KE, Kvist J. The impact of psychological readiness to return to sport and recreational activities after anterior cruciate ligament reconstruction. Br J Sports Med. 2014;48(22):1613–9.
136. Lentz TA, Zeppieri G Jr, George SZ, Tillman SM, Moser MW, Farmer KW, et al. Comparison of physical impairment, functional, and psychosocial measures based on fear of reinjury/lack of confidence and return-to-sport status after ACL reconstruction. Am J Sports Med. 2015;43(2):345–53.
137. McCullough KA, Phelps KD, Spindler KP, Matava MJ, Dunn WR, Parker RD, et al. Return to high school- and college-level football after anterior cruciate ligament reconstruction: a Multicenter Orthopaedic Outcomes Network (MOON) cohort study. Am J Sports Med. 2012;40(11):2523–9.
138. Lai CCH, Feller JA, Webster KE. Playing performance after anterior cruciate ligament reconstruction among Australian football league players from 1999 to 2013. Am J Sports Med. 2019;47(7):1550–6.
139. Carey JL, Huffman GR, Parekh SG, Sennett BJ. Outcomes of anterior cruciate ligament injuries to running backs and wide receivers in the National Football League. Am J Sports Med. 2006;34(12):1911–7.
140. Secrist ES, Bhat SB, Dodson CC. The financial and professional impact of anterior cruciate ligament injuries in National Football League Athletes. Orthop J Sports Med. 2016;4(8):2325967116663921.
141. Busfield BT, Kharrazi FD, Starkey C, Lombardo SJ, Seegmiller J. Performance outcomes of anterior cruciate ligament reconstruction in the National Basketball Association. Arthroscopy. 2009;25(8):825–30.
142. Hamrin Senorski E, Seil R, Svantesson E, Feller JA, Webster KE, Engebretsen L, et al. "I never made it to the pros..." return to sport and becoming an elite athlete after pediatric and adolescent anterior cruciate ligament injury-current evidence and future directions. Knee Surg Sports Traumatol Arthrosc. 2018;26(4):1011–8.
143. Roos H, Ornell M, Gardsell P, Lohmander LS, Lindstrand A. Soccer after anterior cruciate ligament injury--an incompatible combination? A national survey of incidence and risk factors and a 7-year follow-up of 310 players. Acta Orthop Scand. 1995;66(2):107–12.
144. Paterno MV, Rauh MJ, Schmitt LC, Ford KR, Hewett TE. Incidence of second ACL injuries 2 years after primary ACL reconstruction and return to sport. Am J Sports Med. 2014;42(7):1567–73.
145. Pinczewski LA, Lyman J, Salmon LJ, Russell VJ, Roe J, Linklater J. A 10-year comparison of anterior cruciate ligament reconstructions with hamstring tendon and patellar tendon autograft: a controlled, prospective trial. Am J Sports Med. 2007;35(4):564–74.

146. Salmon L, Russell V, Musgrove T, Pinczewski L, Refshauge K. Incidence and risk factors for graft rupture and contralateral rupture after anterior cruciate ligament reconstruction. Arthroscopy. 2005;21(8):948–57.
147. Orchard J, Seward H, McGivern J, Hood S. Intrinsic and extrinsic risk factors for anterior cruciate ligament injury in Australian footballers. Am J Sports Med. 2001;29(2):196–200.
148. Bradley JP, Klimkiewicz JJ, Rytel MJ, Powell JW. Anterior cruciate ligament injuries in the National Football League: epidemiology and current treatment trends among team physicians. Arthroscopy. 2002;18(5):502–9.
149. Bjordal JM, Arnly F, Hannestad B, Strand T. Epidemiology of anterior cruciate ligament injuries in soccer. Am J Sports Med. 1997;25(3):341–5.
150. Borchers JR, Pedroza A, Kaeding C. Activity level and graft type as risk factors for anterior cruciate ligament graft failure: a case-control study. Am J Sports Med. 2009;37(12):2362–7.
151. Hurley ET, Fried JW, Kingery MT, Strauss EJ, Alaia MJ. Antero-lateral ligament reconstruction improves knee stability alongside anterior cruciate ligament reconstruction. Knee Surg Sports Traumatol Arthrosc. 2020;29:764.
152. Getgood A, Hewison C, Bryant D, Litchfield R, Heard M, Buchko G, et al. No difference in functional outcomes when lateral extra-articular Tenodesis is added to anterior cruciate ligament reconstruction in young active patients: the stability study. Arthroscopy. 2020;36(6):1690–701.
153. Feller JA, Devitt BM, Webster KE, Klemm HJ. Augmentation of primary ACL reconstruction with a modified Ellison lateral extra-articular Tenodesis in high-risk patients: a pilot study. Orthop J Sports Med. 2021;9(8):23259671211021351.
154. Sonnery-Cottet B, Vieira TD, Ouanezar H. Anterolateral ligament of the knee: diagnosis, indications, technique. Outcomes Arthroscop. 2019;35(2):302–3.
155. Laboute E, Savalli L, Puig P, Trouve P, Sabot G, Monnier G, et al. Analysis of return to competition and repeat rupture for 298 anterior cruciate ligament reconstructions with patellar or hamstring tendon autograft in sportspeople. Ann Phys Rehabil Med. 2010;53(10):598–614.
156. Webster KE, Feller JA. Exploring the high reinjury rate in younger patients undergoing anterior cruciate ligament reconstruction. Am J Sports Med. 2016;44(11):2827–32.
157. Paterno MV, Rauh MJ, Schmitt LC, Ford KR, Hewett TE. Incidence of contralateral and ipsilateral anterior cruciate ligament (ACL) injury after primary ACL reconstruction and return to sport. Clin J Sport Med. 2012;22(2):116–21.
158. Bien DP, Dubuque TJ. Considerations for late stage acl rehabilitation and return to sport to limit re-injury risk and maximize athletic performance. Int J Sports Phys Ther. 2015;10(2):256–71.
159. Wiggins AJ, Grandhi RK, Schneider DK, Stanfield D, Webster KE, Myer GD. Risk of secondary injury in younger athletes after anterior cruciate ligament reconstruction: a systematic review and meta-analysis. Am J Sports Med. 2016;44(7):1861–76.
160. Hewett TE, Di Stasi SL, Myer GD. Current concepts for injury prevention in athletes after anterior cruciate ligament reconstruction. Am J Sports Med. 2013;41(1):216–24.
161. Kamath GV, Redfern JC, Greis PE, Burks RT. Revision anterior cruciate ligament reconstruction. Am J Sports Med. 2011;39(1):199–217.
162. Magnussen RA, Lawrence JT, West RL, Toth AP, Taylor DC, Garrett WE. Graft size and patient age are predictors of early revision after anterior cruciate ligament reconstruction with hamstring autograft. Arthroscopy. 2012;28(4):526–31.
163. Gobbi A, Karnatzikos G, Chaurasia S, Abhishek M, Bulgherhoni E, Lane J. Postoperative infection after anterior cruciate ligament reconstruction. Sports Health. 2016;8(2):187–9.
164. Hui C, Salmon LJ, Kok A, Maeno S, Linklater J, Pinczewski LA. Fifteen-year outcome of endoscopic anterior cruciate ligament reconstruction with patellar tendon autograft for "isolated" anterior cruciate ligament tear. Am J Sports Med. 2011;39(1):89–98.
165. Oiestad BE, Engebretsen L, Storheim K, Risberg MA. Knee osteoarthritis after anterior cruciate ligament injury: a systematic review. Am J Sports Med. 2009;37(7):1434–43.
166. Ait Si Selmi T, Fithian D, Neyret P. The evolution of osteoarthritis in 103 patients with ACL reconstruction at 17 years follow-up. Knee. 2006;13(5):353–8.
167. Barenius B, Ponzer S, Shalabi A, Bujak R, Norlen L, Eriksson K. Increased risk of osteoarthritis after anterior cruciate ligament reconstruction: a 14-year follow-up study of a randomized controlled trial. Am J Sports Med. 2014;42(5):1049–57.
168. Claes S, Hermie L, Verdonk R, Bellemans J, Verdonk P. Is osteoarthritis an inevitable consequence of anterior cruciate ligament reconstruction? A meta-analysis. Knee Surg Sports Traumatol Arthrosc. 2013;21(9):1967–76.
169. Englund M, Lohmander LS. Risk factors for symptomatic knee osteoarthritis fifteen to twenty-two years after meniscectomy. Arthritis Rheum. 2004;50(9):2811–9.
170. Kessler MA, Behrend H, Henz S, Stutz G, Rukavina A, Kuster MS. Function, osteoarthritis and activity after ACL-rupture: 11 years follow-up results of conservative versus reconstructive treatment. Knee Surg Sports Traumatol Arthrosc. 2008;16(5):442–8.
171. Smith MV, Nepple JJ, Wright RW, Matava MJ, Brophy RH. Knee osteoarthritis is associated with previous meniscus and anterior cruciate ligament surgery among elite college American football athletes. Sports Health. 2017;9(3):247–51.

172. Deacon A, Bennell K, Kiss ZS, Crossley K, Brukner P. Osteoarthritis of the knee in retired, elite Australian rules footballers. Med J Aust. 1997;166(4):187–90.
173. Liden M, Sernert N, Rostgard-Christensen L, Kartus C, Ejerhed L. Osteoarthritic changes after anterior cruciate ligament reconstruction using bone-patellar tendon-bone or hamstring tendon autografts: a retrospective, 7-year radiographic and clinical follow-up study. Arthroscopy. 2008;24(8):899–908.
174. Louboutin H, Debarge R, Richou J, Selmi TA, Donell ST, Neyret P, et al. Osteoarthritis in patients with anterior cruciate ligament rupture: a review of risk factors. Knee. 2009;16(4):239–44.
175. von Porat A, Roos EM, Roos H. High prevalence of osteoarthritis 14 years after an anterior cruciate ligament tear in male soccer players: a study of radiographic and patient relevant outcomes. Ann Rheum Dis. 2004;63(3):269–73.

Part VI

Young at Heart

Management of ACL Injuries in the Young at Heart (Elderly) Patients

26

Facundo Gigante, Manuel Perez Zabala, and Miguel Lapera

26.1 Introduction

Aging population will be one the most relevant social changes of twenty-first century. According to the study on World Populations Prospects from 2022 by the Department of Economic and Social Affairs of the United Nations, globally, life expectancy reached 72.8 years in 2019, an increase of almost 9 years since 1990. Further reductions in mortality are projected to result in an average life expectancy of around 77.2 years globally in 2050 [1].

The US Census Bureau claims that, for the year 2050, people over 65 years of age will amount to 16% of the world's population, meaning 1 in every 6 people will be over 65 years old [2, 3].

It is likely that this increase in life expectancy is directly related to individuals adopting ever healthier lifestyles added to advances in the field of medicine and treatment of diseases.

Evidently the need to change certain harmful or dangerous habits for healthier ones and increasing both recreational and competitive sporting activities is a fundamental part of many peoples' lives, regardless of age or gender.

This has generated an increase in the number of adults engaged in recreational sports activities with higher intensity and frequency. The same applies to high performance or professional athletes, with longer careers and levels comparable to younger athletes. Today, it is not unusual to find professional athletes over 40 years of age with high performance levels and individuals over 50 or 60 years of age engaged in recreational sports. For example, the Institut National de la Statistique et des Études Économiques (INSEE) of France reports that 40% of its population are 50 years or older and 50% of that age group are engaged in some physical activity or sport [4].

As a result, the population between 40 and 60 years has a similar physical demand to that of a younger population and, as such, the number of injuries in this age group has increased proportionally, including anterior cruciate ligament (ACL) injuries.

Although the average age for ACL reconstruction remains the same (29 years old), in the last few years the incidence of ACL reconstruction in patients over 40 years of age has increased by 200%, similar to the increased incidence in minors [5].

In the USA, according to Stone, between 1994 and 2006, the number of ACL reconstructions for this age group increased an average of 4.5 times, which is most likely related to the increase in the number of older people engaged in physical activity and active lifestyles [6].

These changes, which have come about in the last 25 to 30 years, set up both a philosophical

F. Gigante (✉) · M. Perez Zabala · M. Lapera
Arthroscopy Surgery Service, Hospital Militar Central, Buenos Aires, Argentina

M. Karahan et al. (eds.), *Anterior Cruciate Ligament Injuries Throughout the Life Span*,
https://doi.org/10.1007/978-3-031-84501-7_26

and medical challenge. The increasing number of elderly patients who wish to continue their active lifestyles forces us to change our views on the dogmas set two or three decades ago. Although there is no firm consensus for the ideal treatment of ACL injuries in this age group, many studies and research papers show good outcomes for ACL reconstruction in elderly patients [7–9].

To force lifestyle changes after an ACL injury would greatly impact quality of life and many, in this population, would find it unacceptable to reduce, or altogether stop, their physical activities.

In this chapter, we will try to answer questions about management and treatment of ACL injuries in elderly patients, considering the pros and cons of each treatment and assessing final outcomes.

26.2 Relationship Between Age and ACL Injury

As mentioned before, there is no age limit for participating in sports or other physical activities. It is important to determine what age group we are actually studying when referring to 'elderly patients'. In this case, it seems more fitting to say patients over 50 years of age. Although, in this age group, physical activity is usually recreational and not high performance, physical demand is ever higher and, therefore, medical consults about orthopaedic injuries are increasingly frequent.

Is there a relationship between age and ACL injury? There are no doubts that aging causes degenerative changes in joints. But how does aging specifically affect the ACL?

On the one hand, age plays an important part in the biological changes of ligament ultrastructure through the years. These changes show a decrease in the number of mesenchymal stem cells, growth factor receptors, cellular metabolic activity and production of collagen, which result in a lower regenerative potential of the ACL [5]. On the other hand, aging also brings changes in ligament-to-bone ACL insertion sites, with structural modification, in particular at the femoral attachment. In addition to this, changes in morphology are significant and a decrease in the number of mechanoreceptors has also been observed [5].

Table 26.1 Relationship of age and changes in the anterior cruciate ligament (ACL)

Relationship—Age and ACL
Mesenchymal stem cells decrease
Growth factors receptors decrease
Cellular metabolic activity decrease
Collagen production decrease
Mechanoreceptors decrease
Healing potential decrease

With these findings in mind, we can see that age may very well have an impact on the number of inflammatory processes, mucoid degeneration and the reorganisation and amount of collagen fibres of the ACL which considerably alter its intrinsic biological repairing capabilities (Table 26.1).

According to these findings, the ACL is affected by age, the likelihood of ACL injuries in the over 40 years of age group is greater, and the healing capacity in partial injuries is lower, regardless of the physical activity engaged in. Clearly, though, those engaged in pivoting sports, included in this age group, have a higher probability of ACL injury than patients under 40 years old (5).

26.3 Initial Management of ACL Injuries

Initial management of ACL injuries in older patients is no different from management in younger patients. We prescribe analgesic and anti-inflammatory measures such as ice and rest, just as with any other patient. Since we address isolated injuries, we never prescribe immobilisation, on the contrary, we focus on restoring mobility and preserving muscle strength, especially in older patients, whose muscle strength is lower than in younger patients, and the likelihood and risk of stiffness due to immobilisation is greater.

We start kinesiology and rehabilitation as soon as possible to improve mobility, reduce inflammation and bone oedema of the lateral compartment. This is commonly associated to ACL injury and often causes range of movement limitations which, in the event of an ACL reconstruction, may become a surgical complication. Regarding thc strategy chosen, there is no specific program recommended and often depends on the experience and decision of the medical team and physical therapists.

Although the treatment does not differ from what we should do with a younger patient, there are some important points to assess in order to further establish the definitive treatment [10].

Stability tests help assess both subjective and objective stability and are the main tool to establish a diagnosis and prognosis. It should be taken into consideration that some patients already show signs of advanced osteoarthritis and these changes in joint structure may reduce instability, whether subjective or objective (Figs. 26.1 and 26.2).

Assessing rotatory knee stability is important, but we must also evaluate the state of the knee joint, which depends not only on age but previous injuries, surgeries and the time since the ACL injury. It is not unusual to find patients with previous knee trauma, where an MRI shows an ACL injury but with no subjective knee instability.

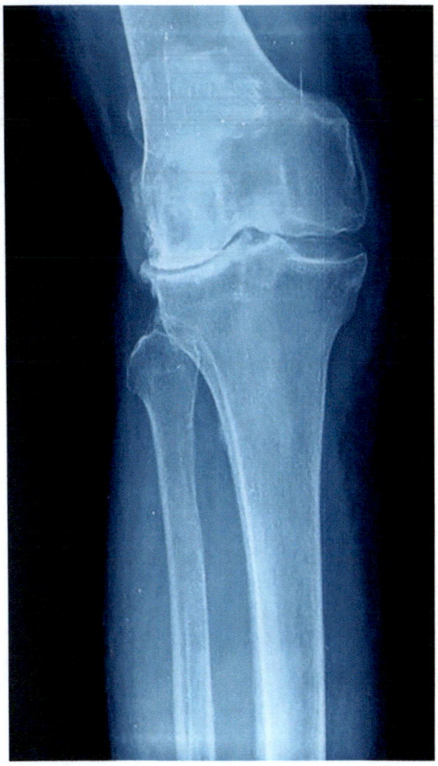

Fig. 26.2 Anteroposterior view of right knee radiograph of chronic ACL rupture in a 53-year-old man with an arthroscopic lateral meniscectomy 7 years before. Note the significant narrowing of the lateral compartment and the development of tricompartmental osteoarthritis. ACL anterior cruciate ligament

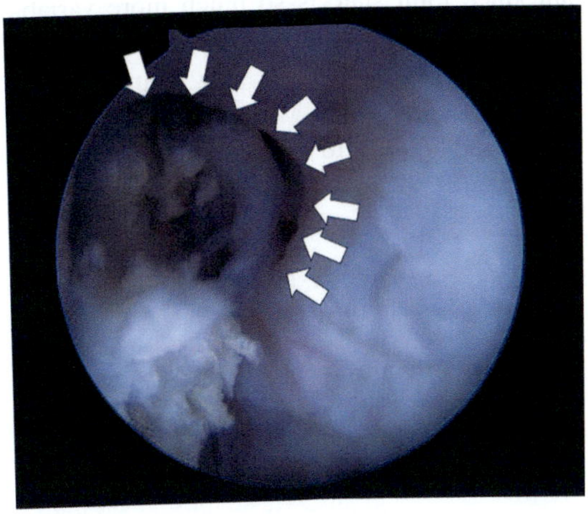

Fig. 26.1 Anterolateral arthroscopic view of chronic ACL rerupture in a 55-year-old man with more than 5 years without treatment. Observe the notch stenosis which can block the instability. Chronic rupture of the neoligament can be seen probably due to a tunnel in an anterior position. ACL anterior cruciate ligament

This chronic ACL insufficiency produces, in time, significant joint damage. These patients rarely show instability and only, eventually, seek medical advice for their painful knee arthritis.

Age, occupation, level of physical activity, motivation, expectation and even body mass index (BMI), all play a role and must be taken into consideration. This information becomes essential in planning the ideal treatment for, and with, the patient who will eventually have the final say.

The most important imaging tests, in this case, are knee radiographs to properly gauge the joint's condition and alignment (Figs. 26.3 and 26.4).

Magnetic resonance imaging (MRI) is the other routine imaging test which confirms ACL injury and helps evaluate the state of the articular cartilage and any injury to the meniscus.

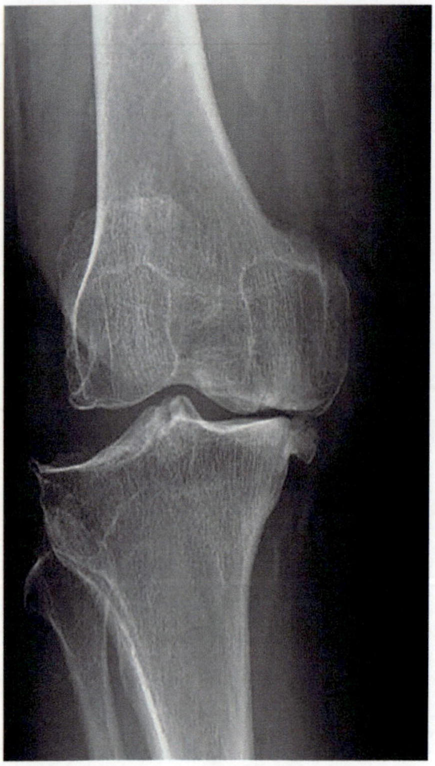

Fig. 26.3 Right knee. Genu varus. Anteroposterior radiograph control. Although the image is not a full-length radiograph to evaluate varus the irregular axis and the severe narrowing in the medial joint space can be observed

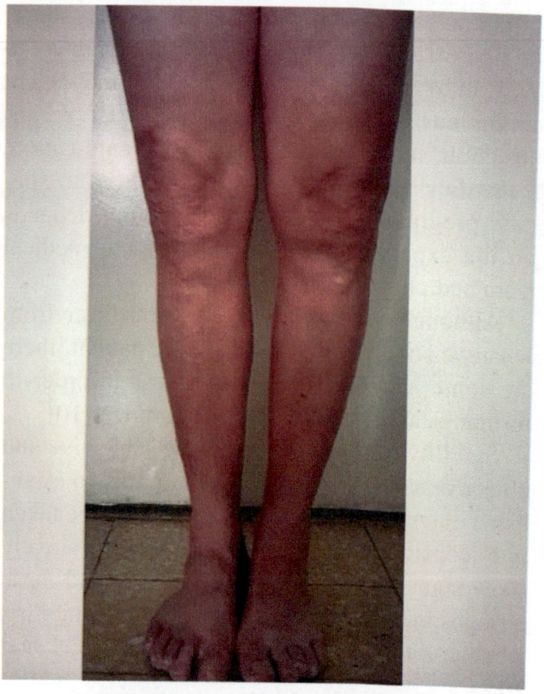

Fig. 26.4 Genu varus acquired. Predictive factor of poor outcomes and formal contraindication for anterior cruciate ligament (ACL) reconstruction in elderly patients

Articular alignment, the state of the menisci and articular cartilage will be key factors in deciding treatment plans and determining prognosis should the choice be ACL reconstruction [7].

In short, proper assessment of the knee joint, especially in this age group, is extremely important since joint state is so much more variable than in younger patients but they might still expect similar outcomes. Using a simple initial evaluation form eases a first impression of the patient and helps us to know where we stand with each one (Fig. 26.5).

ACL - HMC

Name:
Age:
Phone number:
Physician:
Knee Side:

ID:
M/F:
email:

For the Patient		
Occupation:		
Sport / Activity:		Times per week:
When was the trauma?:		
Previous surgerys in the same knee:		
Do you fell the knee fails?:	Yes / No	When it happens: 1. When I'm working
		2. When I do sports
		3. Always or at any time

For the Physician	Name:
BMI:	
Rx Ahlback:	
Rx Axis:	
Lachman:	
Pivot Shift:	

Fig. 26.5 Initial Evaluation Form

26.4 Non-Operative Management of ACL Injury

Thirty years ago, an ACL injury meant instant retirement for many athletes over the age of 40 years, especially in any sport requiring pivoting movements of the knee. Quite probably at the time, we would have suggested such a patient should abandon that sport, completely modify their physical activity and focus on non-operative treatment of joint instability without the need for an ACL reconstruction.

The belief was that this type of surgery might greatly increase the chances of knee stiffness and reduce knee motion due to reduced healing potential and graft integration. Thromboembolic complications and apparent poor postoperative results seemed to justify this kind of treatment.

On the other hand, physical activity in people over 40 years old was not as it is today. Quite possibly this was another reason that tipped the scales in favour of non-operative treatment, since many might already have had a lower physical demand and were more receptive to changing their lifestyles. The 'rule of thirds' was proposed over 25 years ago with the idea of classifying patients into groups that might help in choosing the optimal treatment course [5]. This 'rule' separates patients into 3 groups: the '*Copers*', those patients that go back to their usual activities after ACL injury without surgery since they do not show instability symptoms; the '*Adapters*', patients with ACL injury who modify their activity to avoid symptoms of instability and surgery, and are satisfied with their new reality; and the '*Noncopers'*, who refuse to modify their lifestyles and, because of knee instability, must therefore undergo ACL reconstruction [11].

This simple, and quite popular, rule has no real scientific basis and was introduced in a time when a 40-year-old was considered too old for high physical demand activities. Applying this rule, most patients would have to be treated non-operatively either because they had no symptoms or because they were willing to change their physical activity. These days, this is unacceptable for patients over 40 years old, and an increasing amount of patients over 50 years old.

Today, knee stiffness and mobility loss are not linked to age, therefore, our question should be whether non-operative treatment or ACL reconstruction is more beneficial in patients over 50 years old.

Most surgeons agree that ACL reconstruction is the ideal treatment of younger patients, but there is less of a consensus for older patients. As mentioned before, these patients are more complex and thus the strategy and planning of treatment must be more finely tuned to each patient. We have already mentioned the importance of information such as age, occupation, physical activity, functional demand and expectations of the patient [12, 13]. Clinical data such as joint alignment, knee rotatory instability, lower limb axis, previous surgeries or trauma, associated ligament and meniscal injury, articular cartilage and overall degree of osteoarthritis are equally important. For example, some authors, like Fayard, note the presence of osteoarthritic degeneration of the medial femorotibial compartment as a predictive sign of poor results for ACL reconstruction [14].

Based on this initial assessment, it is important to clarify to the patient the options and prognosis of each treatment. This group of patients might be more open and accepting of lifestyle changes and in more sedentary patients, it might be easier to convince against choosing a surgical procedure, but it is always important to inform of the risks of residual rotatory instability and the benefits of ACL reconstruction (Table 26.2).

The main objective of non-operative treatment of ACL injuries is to reduce pain, improve mobility and, above all, minimise the effects of the rotatory knee instability. Kinesiological rehabilitation is essential for this, and it does not differ from other protocols used on younger patients. Regardless, the protocol applied depends on the experiences and planning of each medical team.

As previously stated, it is essential to explain to patients the risks posed by instability with an ACL injury. In addition to altering quality of life, it increases the risk of meniscal and chondral damage, and potentially risks osteoarthritis in patients that already have degenerative changes due to age. Therefore, the most efficient way to reduce these risks caused by instability is by reducing activity, both in sedentary and active patients. The issue is convincing the group of active patients to change and modify their lifestyles and thus quality of lives; changing their sport practices to activities that lack pivoting or even quitting altogether. Likewise, sedentary patients are not excluded from unstable knee con-

Table 26.2 Predictive factors help us in choosing treatment

ACL reconstruction outcome—predictive factors	
Primary	Secondary
Instability	BMI
Axis	Previous surgeries
Degenerative Changes (Ahlbäck/ Kellgren– Lawrence)	Acute / chronic

BMI body mass index

sequences, this symptom can occur during daily life activity.

As a result, the decision to indicate a non-operative treatment of older patients is a challenge in itself nowadays. Without a doubt, this decision must be based on a meticulous patient selection that would benefit with one or the other treatment. Many patients are satisfied with the outcomes of non-operative treatments; this must be accounted for along with the patient's decision. Cicotti states that 83% of non-operative patients assessed were satisfied with the outcomes even though they had to modify their lifestyles [15]. Gfoller agrees with Cicotti; in his study, non-operative patients had a high degree of satisfaction but had increased radiologically degenerative changes and reduced physical activity [16]. In conclusion, even if the objective outcomes of different studies and investigations are in favour of ACL reconstruction in patients over 50 years, which is significantly better than non-operative treatments at improving joint functionality, it is not applicable to every patient in this age group with an ACL injury [4, 7–9, 14].

Since this is a special group of patients, with varying functional demands, expectations and overall joint health that must be considered, it is important that we evaluate predictive factors to establish formal and relative contraindications when choosing treatment, while also properly explaining the chances of a successful outcome with either surgical or non-operative treatment. (Table 26.3).

Table 26.3 Contraindications for ACL reconstruction in patients over 50 years old. ACL: anterior cruciate ligament; BMI: body mass index

Formal contraindications	Relative contraindications
Osteoarthritis: Ahlbäck III–IV Kellgren–Lawrence III–IV	Pivot shift negative
Malalignment (Varus >10/Valgus >10)	Sedentary lifestyle
BMI more than 30	Associated injuries *(meniscus, ligament, chondral)*
'Patient doesn't want the surgery option or post ACL reconstruction rehabilitation'	

26.5 Surgical Management of ACL Injury

As previously stated, nowadays, surgery for an ACL injury is a completely valid choice for patients over 50 years old, and is broadly accepted by patients and doctors alike.

What changed in the past 25–30 years that made us rethink our decisions for older ACL patients? Life expectancy increased, quality of life improved, professional athletes have longer career spans and people over 50 years desire to keep on practicing sports. There is medical evidence now that shows good and similar outcomes compared to the younger population, showing the benefit of an ACL reconstruction on a correctly selected patient. It is from this selection that surgical success depends on, which is why it is paramount that we are methodical and make use of diagnosis and treatment algorithms that we are familiar and comfortable with so that we make the best possible decisions [9, 12].

Within this particular group of older adults, any patient showing an ACL injury with low degenerative changes, subjective and objective instability, who hopes to continue their physical activities and is not willing to change their lifestyle, is an ACL reconstruction surgical candidate. Having said that, the main objective with ACL reconstruction is reducing instability and the effects this may have on the knee, not only for active or athletic patients but also for sedentary patients. Instability is key in the therapeutic decision-making and serves as a prognostic indicator [7].For this reason we should make the decision for surgery with the patient, together, explaining all the options, prognosis, chances of success, advantages and disadvantages and, above all, consider the patient's expectations and motivations.

After considering these factors, and if the patient finds quitting their sport practice or changing their lifestyle unacceptable: what would the age limit be to safely indicate an ACL reconstruction? According to a number of studies, where there are practically no significant differences between younger and older populations, age definitely does not constitute a limitation to perform this type of surgery, nor does it condition success. Out of all the variables to consider, age seems to be the least important [9, 12, 14, 17–20].

As we have said before, there are factors specific to the patient and specific to the joint that determine the limitations in the procedure and, in some cases, act as predictive factors for a good, fair or poor postoperative evolution. Patient's expectations, type of job, sports practice and BMI are the predictive factors related to the patient. Surgical history and time elapsed since the injury are points that we must evaluate for the impact they have on joint condition, specifically any history of partial meniscectomies or chronic ACL injuries. Though there is no consensus nor determinations to establish the effect time has on a joint since an ACL injury, there is a prior study indicating there may be no difference between operating within the year of injury or after [6]. Still, we should not ignore the fact that chronic ACL injuries cause significant joint deterioration, multiplying by five times the likelihood of developing osteoarthritis within 10 years. Even if it has not been proven that older patients pose a higher risk of developing osteoarthritis after an ACL injury when compared to younger patients with the same injury, we must consider that this group of patients will show some degree of joint and meniscal deterioration from age itself. ACL injury, combined with a pre-existing joint pathology, would accelerate the risk for an osteoarthritic process [21]. Lastly, and perhaps the leading data for formal contraindications when performing an ACL reconstruction surgery on these patients according our opinion are the degrees of joint degeneration assessed with Ahlbäck classification, Kellgren-Lawrence grading system and the knee joint axis. A valgus or varus misalignment more than 10 degrees are contraindications for surgery in these patients just like an Ahlbäck and Kellgren-Lawrence III and IV [20, 22].

Regarding surgical technique, there are some publications that compare different types of available grafts, fixation methods or single bundle or double bundle techniques and their influence on older patients' outcomes. These investigations do not provide conclusive data, so it could be said that surgical technique has no clinical or functional impact on the final outcome of an ACL reconstruction in these patients [4, 5, 13, 23].

However, we must remember that in older adults the intrinsic biological repair capacity of the native ACL is altered. This could be replicated in the ligamentisation process in any of its three phases but especially in the complete tunnel closure and ossification of the graft inside bone tunnels. Furthermore, there are some articles that alert about the use of Cox-2 inhibitors, vastly utilised as analgesics before or after surgery, could in fact negatively affect cellular response during the early healing phase [24].

Allografts, which are not recommended for young and active patients, seem to be the best option for patients with low functional demand and over 35 years old. However, due to high cost, higher probability of failure and re-ruptures and lack of data and series showing significant benefits, we do not believe they should be our first pick for these patients.

In every day practice, surgical technique and chosen grafts do not differ from our choices in younger patients. We prefer an anatomical transportal single bundle technique with quadruple semitendinosus and gracilis autograft of 8 mm average diameter. As fixation methods we use an extra cortical suspensory fixation for the cortical proximal fixation and an interference peek or titanium distal screw (Figs. 26.6 and 26.7).

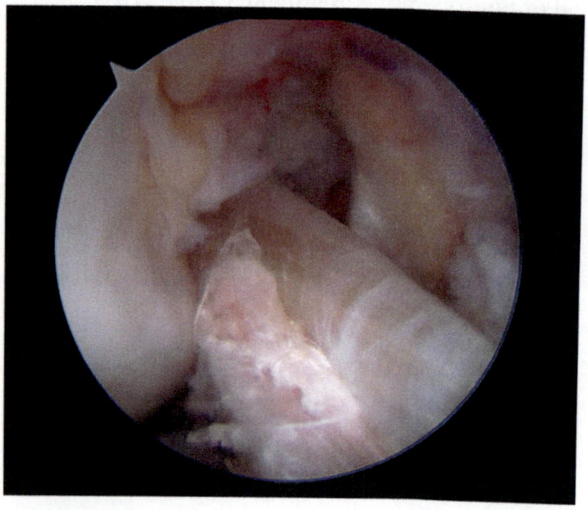

Fig. 26.6 Anterolateral arthroscopic view. ACL reconstruction in patient over 53 years old with hamstrings tendon autograft in right knee. ACL anterior cruciate ligament

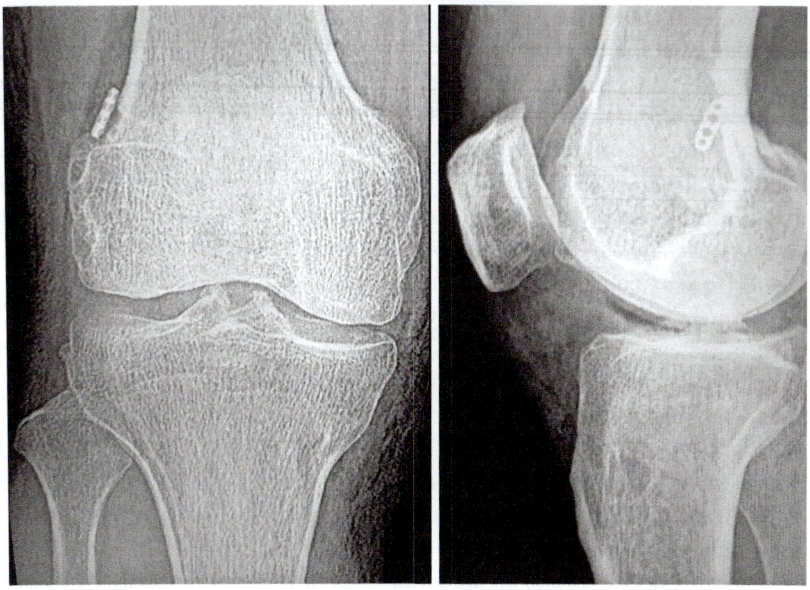

Fig. 26.7 Radiograph after ACL reconstruction in the same patient. ACL anterior cruciate ligament

Despite some studies mentioning that the use of patellar tendon is related to a greater risk of osteoarthritis development, in our experience, the decision to not utilise it is due to the fact that patients over 50 years old already have some degree of patellofemoral joint deterioration; therefore, we prefer not to add any further progression using this graft [25]. Also, some studies mention a higher chance of patellar tendon rupture or patella fracture after an ACL reconstruction [5]. Even if these cases are rare, we would rather minimise risks in this particular group of patients.

Postoperative recovery is no different than what we would do with younger patients. There is no universal protocol for these patients' rehabilitation, nor with younger groups. We would rather have a less aggressive rehabilitation, adapted to each patient's functional demand.

26.6 Why to Perform ACL Reconstruction in Patients Over 50 Years of Age?

26.6.1 Results

First, we must say that not all patients in this age range will benefit from an ACL reconstruction. We have already stated that must be selected considering the formal contraindications related to the degree of osteoarthritis and knee misalignment, as well as other factors that are not a contraindication by themselves but act as predictive factors.

26.6.2 Why Would We Reconstruct an ACL in Older Patients?

Surgery outcomes are comparatively better than the pre-surgery state. Moreover, surgery shows good objective and subjective results, similar to those in younger patients, while non-operative treatment outcomes are more varied and depend on many variables.

The main objective of ACL reconstruction is restoring knee stability. Residual instability and risk associated with it is not resolved by non-operative treatment alone. Even though between 40 and 90% of patients over 50 years have some sort of meniscal or chondral injury, or degenerative changes, these will not have a direct impact on the ACL reconstruction outcome; it will vary its degree of success [12, 22, 23].

These degenerative changes, in addition to non-restored residual instability, accelerate and increase the pre-existing degenerative changes, so stabilising the knee reduces and protects it from degenerative changes.

Lifestyle changes in patients over 50 years old, in direct relation to life expectancy, in the past two or three decades have changed the paradigm of treatment. These patients' expectations are different from past decades', and satisfaction degree assessments are as paramount as the objective results we can find in both surgical and non-operative treatments. This is why proposing a non-operative treatment supposes changing or modifying the activities the patient carried out before the injury. This change does not seem to be the appropriate strategy for these patients, who want to carry on with their lifestyles and functional demand, and quitting their physical activities is not an option. Nowadays, we know that the ACL clinical and functional results in older patients are comparable with the younger population's, and that between 80 and 100% of patients over 50 years old resume their physical activities or sport practices [26–28]. Even if 40–50% of patients do so on the same level, those who do not achieve it will decide to change the activity or sport practice, but will not abandon it [29]. These changes on the activity level do not generally happen due to objective results, but rather to the patients' subjective decisions, which have to be taken into account since it explains why over 95% of surgical patients in this age group are satisfied with the surgery [9].

There is a probability that patients who are not willing to undergo surgery and accept modifying their lifestyles and restricting their activities could have good outcomes, even with the risk that residual instability implies. Over 80% of patients are satisfied with non-operative treatment when accepting to lower their physical activities [30]. Ehlinger compares the outcomes between patients over 50 years old, who underwent surgery and who did not, observing satisfactory results in both groups, though finding differences in the functional scores. The author concluded that patients who do not want surgery should be warned about the risks of getting other joint injuries from knee instability while exercising [7].

In the past, ACL reconstruction outcomes seemed discouraging. Now, there are many studies that assess the outcomes of selected patients excluding those with advanced joint deterioration and misalignment, using Lysholm, Tegner and International Knee Documentation Committee (IKDC) scores, patients' satisfaction degree, radiological studies, and KT1000. In all of them, outcomes improved significantly in relation to pre-surgery scores and when compared to those found in the under 40 years old population, there were no significant differences as shown in the table number 4 (Table 26.4). In a comparative study between patients over 50 years old and under 40 years old, Iorio did not find differences in the Lysholm, IKDC and Tegner scores after a year, nor any osteoarthritic changes observed in radiographs. They did however find better KT1000 results in the over 50 years old age group, which would be explained by the lower muscle elasticity of this group. In this study, even if the patients' satisfaction was shown by the fact that 100% of the over 50 years old age group patients resumed their physical activities, only 56% of them could do so on a pre-injury level [17]. Figueroa found similar results in patients over 50 years old, yet in this case 88% of them recovered a pre-injury activity level and almost 100% were satisfied with the outcomes [22]. Kim compared muscle strength, stability and function in patients over 50 years old and under 40 years old. Even if all parameters were significantly better after an ACL reconstruction in both groups, the objective IKDC showed better results for muscle strength and stability in patients under 40 years old, and did not find significant differences in the other scores and arthrometric tests [18]. Cokyasar observed notable improvement in functional outcomes, IKDC, Tegner and Lysholm scores, as well as improvement in quality of life and psychological state in these patients [8]. Ehlinger found that patients over 50 years old with an ACL reconstruction had less laxity, and their IKDC, Knee and Osteoarthritis Outcome Score (KOOS) and Tegner scores improved. While assessing radiological changes in both study groups found that those who had gone through surgery had less patellofemoral osteoarthritic changes and that those who had not had surgery, had less medial compartment degenerative signs with no differences between groups for

Table 26.4 ACL reconstructions results

Author and follow-up	Patients age	Lysholm	Tegner	IKDC subjective	IKDC objective	KT1000	OA cartilage meniscus	Satisfaction	Return activity
Brandson (2000) F-up >2 years	>40	91	5	*n/i*	100% A-B	*n/i*	*n/i*	*n/i*	*n/i*
Dahm (2008) F-up 6 years	>50	90	4.3	92	*n/i*	*n/i*	*n/i*	*n/i*	*n/i*
Trojani (2009) F-up >2 years	>50	*n/i*	*n/i*	*n/i*	78% A-B	*n/i*	50% meniscal injury	100%	*n/i*
Arbuthnot (2010) F-up >9 years	>55	85,5	3	*n/i*	*n/i*	Better	79% meniscal injury 100% cartilage damage	*n/i*	*n/i*
Ventura (2012) F-up >4 years	>50	90	5	*n/i*	94% A-B	*n/i*	56% meniscal injury 24% cartilage damage	*n/i*	*n/i*
Conteducca (2013) F-up >3 years	>40	93	5.2	*n/i*	89% A-B	Better	*n/i*	*n/i*	*n/i*
Figueroa (2014) F-up >4 years	>50	93.7	5.6	93.7	*n/i*	*n/i*	90% OA 76% meniscal injury 18% cartilage damage	96%	100% return 88% same level
Toanen (2017) F-up >2 years	>60	93.2	*n/i*	83.8	91.66% A-B	*n/i*	100% OA 66% meniscal injury 50% cartilage damage	100%	83% return – 50% same level
Webster (2017) F-up >1 years	>45	*n/i*	*n/i*	79	*n/i*	Better	*n/i*	*n/i*	*n/i*
Wierer (2017) F-up 2 years	>40	94.5	5.5	*n/i*	100% A-B	*n/i*	*n/i*	*n/i*	*n/i*
Iorio (2018) F-up 2 years	>50	94.3	5.4	91.4	97% A-B	Better	*n/i*	100%	100% return 88% same level
Kim D (2018) F-up 1 years	>50	93	*n/i*	75.6	*n/i*	Better	*n/i*	*n/i*	*n/i*

n/i no information, *IDKC* International Knee Documentation Committee, *KOOS* Knee and Osteoarthritis Outcome Score, *OA* osteoarthritis, *F-up* follow-up, age in years the lateral compartment [7]. Tan, comparing the ACL reconstruction outcomes in patients over and under 50 years old, did not find any notable differences in functional scores, arthrometric results or complications either [9].

These and other studies perfectly show and document that, in patients over 50 years old, ACL reconstruction has very good outcomes when patients have been correctly selected, and in all cases they will improve their pre-surgery functional scores and tests as in younger patients [31].

26.6.3 Does ACL Reconstruction Surgery Increase the Risk of Complications in Older Adults?

As we have said, the risk of a post-surgical ACL arthrofibrosis or knee stiffness has nothing to do with age. There are many comparative studies that assess, amongst other things, the complication index not finding significant differences between patients over and under 50 years [9, 31]. On the contrary, Salensky found that younger patients were more prone to complications due to infections, pain and inflammation than patients over 50 years. The latter, however, had more chances of having a post-surgical complication within 90 days due to the risk of having a venous thromboembolic event (VTE), including pulmonary embolism (PE) and deep vein thrombosis (DVT), as well as an acute kidney injury [30].

26.6.4 Is There an Age Limit to Perform This Surgery? What Happens to Patients That Exceed These Study Groups' Limits?

In order to answer we should go back to the beginning of this chapter. Years ago, studies compared groups of people over 40 years old; currently almost every comparative study works with patients over 50 years old. Once again, we must consider if there really is an age limit [32]

Vaishya published a 75-year-old patient's case and Miller the case of an 84-year-old. Both chose reconstruction surgery as treatment for symptomatic ACL instability that restricted them from their daily lives [33, 34]. Our surgical team performed surgery on a 72-year-old female taekwondo black belt who wanted to continue practising and teaching this sport. The outcomes from these three extreme cases were absolutely satisfactory and had great results for the patients.

Toanen assessed ACL reconstruction results in a group of 12 patients over 60 years old (61 years of age on average) with no important osteoarthritis changes after a two-year follow-up.

With the intention of reaching a conclusion about the benefits of performing an ACL surgery, we will take as parameters if these patients will be resuming physical activity, their functional recovery and incidence of osteoarthritis compared to a younger population. As a result we find that all parameters and tests improved significantly with a 100% patients' satisfaction degree, where 83% of them resumed their sports practice, 50% on a pre-injury level and the rest decided to switch to a sport that would not require pivoting. With regard to the osteoarticular state, as is expected for this age group, 100% had some kind of joint deterioration or degenerative meniscal injury that did not affect the final result [29].

As a conclusion to the question posted before, we can say that age is the least important parameter when deciding if we should perform an ACL reconstruction. It does not constitute a limitation in any way, even though we must consider it within the context of the patient treated. An ACL reconstruction in older patients has as good results as in younger patients; the selection of patients that would benefit from an ACL reconstruction is the key to success [19, 29, 35–37].

Ejnar Eriksson, while editor in chief of Knee Surgery, Sports Traumatology, Arthroscopy, wrote in a 2003 editorial about ACL reconstruction in elderly patients: '*...well Dear readers, I am 74 years young. I love down -- hill skiing and take any opportunity to ski. I skate as often as I can and I cannot resist the temptation to play soccer with my grandchildren... If I were to tear my ACL and developed a pivot shift, I definitely would request surgery*' [38]

26.7 Osteoarthritis, Age and ACL Injury

The importance of the ACL in knee biomechanics is well known. Regardless of age, the ACL's function in controlling the knee's stability protects it from developing osteoarthritis and associated cartilage and meniscal lesions even in the presence of previous osteoarthrosis.

According to Johnson, an unstable knee due to an untreated ACL injury has five times more chances of developing osteoarthritis in a 10-year period [21].

Mall observed that adults over 40 years old with an ACL injury have 4.5 times the risk of developing osteoarthritis [20]. However, there is no precise data correlating the time elapsed from injury, reconstructive surgery, osteoarthritic changes and age, in particular in older adults. Either way, we know that restoring rotatory knee stability by reconstructing the ACL in young, active adults diminishes the development of osteoarthritis and the amount of joint replacements in adulthood [6, 23]. Moreover, Salesky found that the chance of a patient over 50 years old with an ACL reconstruction needing a total knee arthroplasty (TKA) within two years is 1.6% [30].

So although older patients do not have an increased risk of developing osteoarthritis due to an ACL injury when compared to younger individuals, it should be taken into account that age, BMI and in some cases surgical history or previous injuries are related to preexisting joint disorders in almost 100% of these patients, specifically over 50 years old [39].

Figueroa observed that more than 90% of the patients in his study, all above 50 years of age, showed some sort of chondral or meniscal injury [22]. Toanen noticed that 66% of the patients over 60 years old showed medial degenerative meniscal injuries and 50% had some degree of cartilage injury, in particular the medial and patellofemoral compartments [29]. Wolfson and Osti, in different studies, found that over 50% of patients over 50 years old with an ACL reconstruction surgery had cartilage injuries grade III and IV (Outerbridge classification), particularly in the patellofemoral and medial compartments, but had no impact on the final outcome [40, 41].

Therefore, even if these degenerative changes act as predictive factors, as stated before, symptoms appear to improve and ACL reconstruction seems to keep osteoarthritis progression under control.

26.8 Non-Operative or Surgical Treatment?

As previously stated, the strategies used when deciding which treatment will benefit an adult patient with an ACL injury vary according to several factors, one of which is age. Age must be viewed within context, amongst other variables, but it is not decisive when choosing between surgical or non-operative treatment. For this particular group of patients, the choice between one road or the other (non-operative or ACL reconstruction) lies in the needs, wishes and motivations of each patient. Unlike what would happen with a 25-year-old patient with an ACL injury, here the patient's decision is as, if not more, important than the symptoms.

Occupation, physical activity, the desire to continue the same activity at the same level, willingness to change their lifestyle and their commitment towards a long recovery are essential data to fully understand the patient's expectations [42, 43].

The knee's osteoarticular state, BMI and the lower limb's axis are crucial predictive factors when deciding between treatments and, without a doubt, necessary to unlock any decision-making is the presence or absence of instability [39, 44].

The patient's acceptance to changing or modifying their activities is key to the success of a non-operative treatment, since we see best results in patients that modify their lifestyles, and if they have a sedentary and low demand lifestyle the results are even better. In general, patients cautioned about the risks of residual instability present very good outcomes, being able to return to their activities almost completely. However, Bogunovic found that 60% of patients with these characteristics would complain about some kind of pain or discomfort during their daily lives [13]. We must take into account that many of these patients, especially ones over 50 years of age,

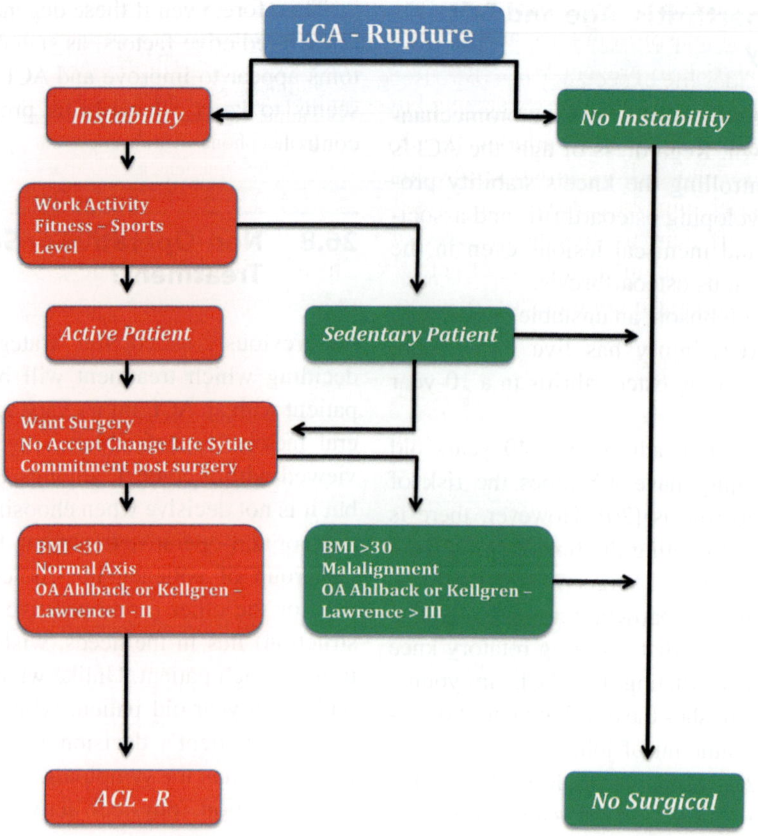

Fig. 26.8 ACL injury management algorithm. *No Instability* = Non operative/ Conservative treatment; *Instability* = *First* we evaluate activity (sports, occupation, etc.). *Active patient* = important to consider: *1.* if he/she wants surgery, *2.* If he/she does not accept to change his lifestyle, *3.* Commitment with post surgery rehabilitation program. *Sedentary patient* = Non operative treatment, except that patient agrees with the 3 points. *Finally* if the BMI is <30, the axis alignment is ok and radiographs changes = Ahlbäck or Kellgren-Lawrence I/II = *ACL Reconstruction*. Otherwise, conservative treatment is suggested. ACL anterior cruciate ligament, BMI body mass index

have chronic ACL injuries that went unnoticed and cartilage alterations or degenerative meniscal injuries that on their own, or combined, can cause pain or discomfort. On the other hand, an ACL chronic injury diagnosis generally presents itself with knee pain with no signs of instability, since it disappears due to joint degeneration many times seen in radiographs imaging as intercondylar stenosis.

As mentioned, the key in deciding for surgical treatment is the presence of instability in a patient that is not willing to modify their lifestyle and wishes to keep on exercising.

'*My knee twists out of place, it won't respond*'. Subjective and objective instability in a knee that is in good condition, no matter the age, is most definitely a candidate for surgical treatment (Fig. 26.8).

26.9 Conclusion

Traditionally, given the low functional demand and unreliable results from surgical treatment in patients aged 40–50 years old, non-operative treatment was the first choice, in addition to changing their lifestyles and activities in order to minimise the effect of residual instability. Over the past two decades life expectancy has increased worldwide, consequently older patients have increasingly been getting ACL injuries due to physical activity. Currently, most patients are not willing to change their lifestyles in any way, arguing that doing so would affect their quality of life hence requiring that we stabilise their knees to be able to continue with their usual lives. For appropriately selected patients, where the result predicting factors for either treatment have been taken into account, ACL reconstruction seems to be the best option. It not only lowers the risks for residual instability and possible progression to osteoarthritis but also has good objective and subjective outcomes, which are no different than in younger patients. Therefore, age as an isolated factor is not conditioning, limiting or a contraindication when deciding for surgical treatment in this group of patients, elderly but young at heart and soul, who feel happy to return to their physical activity or sports practice and who do not accept that their age is a limitation to carry on with their activities. In fact, age is definitely not a problem.

References

1. World Population Prospects 2022: Summary of Results. United Nations. Department of Economical and Social Affairs; 2022. https://doi.org/10.18356/9789210014380.
2. Ortman J, Velkoff V, Hogan H. An aging nation: the older population in the United States, current population reports. U.S. Census Bureau; 2014.
3. Life Expectancy & Probability of Death. Institute for Health Metrics and Evaluation (IHME). University of Washington; 2016.
4. Ode Q, Gonzalez JF, Paihle R, Dejour D, Ollivier M, Panisset JC, Lustig S, French Arthroscopic Society. Influence of operative technique on anterior cruciate ligament reconstruction in patients older than 50 years. Orthop Traumatol Surg Res. 2019;105(8S):S253–8. https://doi.org/10.1016/j.otsr.2019.09.010.
5. Salzler M, Chang J, Richmond J. Management of anterior cruciate ligament injuries in adults aged >40 years. J Am Acad Orthop Surg. 2018;26:553–561. https://doi.org/10.5435/jaaos-d-16-00730.
6. Stone J, Perrone G, Nezwek T, Cui Q, Vlad S, Richmond J, Salzler M. Delayed ACL reconstruction in patients >40 years of age is associated with increased risk of medial meniscal injury at 1 year. Am J Sports Med. 2019;47(1):584–9. https://doi.org/10.1177/0363546518817749.
7. Ehlinger M, Panisset JC, Dejour D, Gonzalez JF, Paihle R, Favreau H, Ollivier M, Lustig S, The Francophone Arthroscopy Society (SFA). Anterior cruciate ligament reconstruction in the over-50s. A prospective comparative study between surgical and functional treatment. Orthop Traumatol Surg Res. 2019;105(8S):S259–65. https://doi.org/10.1016/j.otsr.2019.09.009.
8. Cokyasar B, Altun O, Dasar U. Anterior cruciate ligament reconstruction improves functional scores and quality of life in patients older than 50 years of age. Arthrosc Sports Med Rehabil. 2023;5(6):100806. https://doi.org/10.1016/j.asmr.2023.100806.
9. Tan C, Hsu W, Yu P, Chen C, Kuo L, Chi C, Kim D, Park G. Anterior cruciate ligament reconstruction in patients older than 50 years. A systematic review and meta-analysis. Orthop J Sports Med. 2020;8(4):2325967120915698. https://doi.org/10.1177/2325967120915698.
10. Filbay S, Grindem H. Evidence-based recommendations for the management of anterior cruciate ligament (ACL) rupture. Best Pract Res Clin Rheumatol. 2019;33(1):33–47. https://doi.org/10.1016/j.berh.2019.01.018.
11. Noyes F, Matthews D, Mooar P, Grood E. The symptomatic anterior cruciate-deficient knee. Part II: the results of rehabilitation, activity modification, and counseling on functional disability. J Bone Joint Surg Am. 1983;65(2):163–74. https://doi.org/10.2106/00004623-198365020-00003.
12. Costa G, Grassi A, Perelli S, Agro G, Bozzi F, Lo Presti M, Zaffagnini S. Age over 50 years is not a contraindication for anterior cruciate ligament reconstruction. Knee Surg Sports Traumatol Arthrosc. 2019;27(11):3679–91. https://doi.org/10.1007/s00167-019-05450-1.
13. Bogunovic L, Matava M. Operative and nonoperative treatment options for ACL tears in the adult patient: a conceptual review. Phys Sportsmed. 2013;41(4):33–40. https://doi.org/10.3810/psm.2013.11.2034.
14. Fayard JM, Wein F, Ollivier M, Paihle R, Ehlinger M, Lustig S, Panisset JC, the French Arthroscopy Society. Factors affecting outcome of ACL reconstruction in over-50-year-olds. Orthop Traumatol Surg Res. 2019;105(8S):S247–51. https://doi.org/10.1016/j.otsr.2019.09.011.
15. Ciccotti M, Lombardo S, Nonweiler B, Pink M. Non-operative treatment of ruptures of the anterior cruciate ligament in middle-aged patients. Results after long-term follow-up. J Bone Joint Surg Am. 1994;76(9):1315–21. https://doi.org/10.2106/00004623-199409000-00006.
16. Gföller P, Abermann E, Runer A, Hoser C, Pflüglmayer M, Wierer G, Fink C. Non-operative treatment of ACL injury is associated with opposing subjective and objective outcomes over 20 years

of follow-up. Knee Surg Sports Traumatol Arthrosc. 2019;27(8):2665–71. https://doi.org/10.1007/s00167-018-5296-5.
17. Iorio R, Iannotti F, Ponzo A, Proietti L, Redler A, Conteduca F, Ferretti A. Anterior cruciate ligament reconstruction in patients older than fifty years: a comparison with a younger age group. Int Orthop. 2018;42(5):1043–9. https://doi.org/10.1007/s00264-018-3860-8.
18. Kim D, Park G, Tseng Kuo L, Park W. Patients older than 50 years had similar results of knee strength and anteroposterior stability after ACL reconstruction compared to younger patients. Knee Surg Sports Traumatol Arthrosc. 2019;27(1):230–8. https://doi.org/10.1007/s00167-018-5342-3.
19. Kim K, Kim H, Lee H, Park Y, Kang D, Yoo J, Moon D, Cho S, Hwang S. A comparison of results after anterior cruciate ligament reconstruction in over 40 and under 40 years of age: a meta-analysis. Knee Surg Relat Res. 2018;30(2):95–106. https://doi.org/10.5792/ksrr.17.065.
20. Mall N, Frank R, Saltzman B, Cole B, Bach B Jr. Results after anterior cruciate ligament reconstruction in patients older than 40 years: how do they compare with younger patients? A systematic review and comparison with younger populations. Sports Health. 2016;8(2):177–81. https://doi.org/10.1177/1941738115622138.
21. Johnson V, Roe J, Salmon L, Pinczewski L, Hunter D. Does age influence the risk of incident knee osteoarthritis after a traumatic anterior cruciate ligament injury? Am J Sports Med. 2016;44(9):2399–405. https://doi.org/10.1177/0363546516648318.
22. Figueroa D, Figueroa F, Calvo R, Vaisman A, Espinoza G, Gili F. Anterior cruciate ligament reconstruction in patients over 50 years old. Knee. 2014;21(6):1166–8. https://doi.org/10.1016/j.knee.2014.08.003.
23. Arbuthnot J, Brink R. The role of anterior cruciate ligament reconstruction in the older patients, 55 years or above. Knee Surg Sports Traumatol Arthrosc. 2010;18:73–78. https://doi.org/10.1007/s00167-009-0864-3.
24. Yao S, Fu B, Young P. Graft healing after anterior cruciate ligament reconstruction (ACLR). Asia Pac J Sports Med Arthrosc Rehabil Technol. 2021;25:8–15. https://doi.org/10.1016/j.asmart.2021.03.003.
25. Samuelsson K, Andersson D, Karlsson J. Treatment of anterior cruciate ligament injuries with special reference to graft type and surgical technique: an assessment of randomized controlled trials. Arthroscopy. 2009;25(6):653–85. https://doi.org/10.1016/j.arthro.2009.04.066.
26. Blyth B, Gosal H, Peake W, Bartlett J. Anterior cruciate ligament reconstruction in patients over the age of 50 years: 2- to 8-year follow-up. Knee Surg Sports Traumatol Arthrosc. 2003;11(4):204–11. https://doi.org/10.1007/s00167-003-0368-5.
27. Brown C, McAdams T, Harris A, Maffulli N, Safran M. ACL reconstruction in patients aged 40 years and older: a systematic review and introduction of a new methodology score for ACL studies. Am J Sports Med. 2013;41(9):2181–90. https://doi.org/10.1177/0363546513481947.
28. Dahm D, Wulf C, Dajani K, Dobbs R, Levy B, Stuart M. Reconstruction of the anterior cruciate ligament in patients over 50 years. J Bone Joint Surg Br. 2008;90(11):1446–50. https://doi.org/10.1302/0301-620X.90B11.21210.
29. Toanen C, Demey G, Ntagiopoulos P, Ferrua P, Dejour D. Is there any benefit in anterior cruciate ligament reconstruction in patients older than 60 years? Am J Sports Med. 2017;45(4):832–7. https://doi.org/10.1177/0363546516678723.
30. Salesky M, Oeding J, Zhang A, Ma B, Feeley B, Lansdown D. Patients aged 50 years and older have greater complication rates after anterior cruciate ligament reconstruction: a large database study. Arthrosc Sports Med Rehabil. 2021;3(6):e1827–34. https://doi.org/10.1016/j.asmr.2021.08.010.
31. Desai N, Bjornsson H, Samuelsson K, Karlsson J, Forssblad M. Outcomes after ACL reconstruction with focus on older patients: results from the Swedish National Anterior Cruciate Ligament Register. Knee Surg Sports Traumatol Arthrosc. 2014;22(2):379–86. https://doi.org/10.1007/s00167-013-2803-6.
32. Barber A, Elrod B, McGuire D, Paulos L. Is an anterior cruciate ligament reconstruction outcome age dependent? Arthroscopy. 1996;12(6):720–5. https://doi.org/10.1016/s0749-8063(96)90177-2.
33. Vaishya R, Dhiman R, Vaish A. Anterior cruciate ligament reconstruction in a 75 years old man: a case report with review of literature. Chin J Traumatol. 2014;17(2):121–4. https://doi.org/10.3760/cma.j.issn.1008-1275.2014.02.014.
34. Miller M, Sullivan R. Anterior cruciate ligament reconstruction in an 84-year-old man. Arthroscopy. 2001;17(1):70–2. https://doi.org/10.1053/jars.2001.7802.
35. Trojani C, Sané J, Coste J, Boileau P. Four-strand hamstring tendon autograft for ACL reconstruction in patients aged 50 years or older. Orthop Traumatol Surg Res. 2009;95(1):22–7. https://doi.org/10.1016/j.otsr.2008.05.002.
36. Ventura A, Legnani C, Terzaghi C, Borgo E, M.D. Single- and double-bundle anterior cruciate ligament reconstruction in patients aged over 50 years. Arthroscopy. 2012;28(11):1702–9. https://doi.org/10.1016/j.arthro.2012.04.146.
37. Wierer W, Runer A, Hoser C, Herbst E, Gföller P, Fink C. Acute ACL reconstruction in patients over 40 years of age. Knee Surg Sports Traumatol Arthrosc. 2017;25(5):1528–34. https://doi.org/10.1007/s00167-016-4363-z.
38. Eriksson E. ACL-reconstructions in elderly patients. Knee Surg Sports Traumatol Arthrosc. 2003;11:203. https://doi.org/10.1007/s00167-003-0405-4.
39. Brambilla L, Pulici L, Carimati G, Quaglia A, Prospero E, Bait C, Morenghi E, Portinaro N, Denti M, Volpi P. Prevalence of associated lesions in anterior cruciate ligament reconstruction: correlation with

surgical timing and with patient age, sex, and body mass index. Am J Sports Med. 2015;43(12):2966–73. https://doi.org/10.1177/0363546515608483.
40. Wolfson T, Epstein D, Day D, Joshi B, McGee A, Strauss E, Jazrawi L. Outcomes of anterior cruciate ligament reconstruction in patients older than 50 years of age. Bull Hosp Jt Dis (2013). 2014;72(4):277–83.
41. Osti L, Papalia Del Buono F, Leonardi F, Denaro V, Maffulli N. Surgery for ACL deficiency in patients over 50. Knee Surg Sports Traumatol Arthrosc. 2011;19:412–7. https://doi.org/10.1007/s00167-010-1242-x.
42. Mouton C. CORR Insights1: younger patients and men achieve higher outcome scores than older patients and women after anterior cruciate ligament reconstruction. Clin Orthop Relat Res. 2017;475:2481. https://doi.org/10.1007/s11999-017-5464-9.
43. Webster K, Feller J. Younger patients and men achieve higher outcome scores than older patients and women after anterior cruciate ligament reconstruction. Clin Orthop Relat Res. 2017;475(10):2472–80. https://doi.org/10.1007/s11999-017-5418-2.
44. Legnani C, Terzaghi C, Borgo E, Ventura A. Management of anterior cruciate ligament rupture in patients aged 40 years and older. J Orthop Traumatol. 2011;12(4):177–84. https://doi.org/10.1007/s10195-011-0167-6.

Management of ACL Tears in the Osteoarthritic Knee

27

R. D. Danilkowicz, Alberto Grassi,
Ronald van Heerwaarden, and A. Amendola

27.1 Introduction

Anterior cruciate ligament (ACL) injuries in patients with established degenerative joint disease present a unique set of challenges that go beyond solely the soft tissue structure management. Global care of the knee encompasses an evaluation of the mechanical alignment of the joint and the chondral surfaces and may require an additional intervention concurrently or staged with an ACL reconstruction (ACLR) to ensure optimal outcomes. Despite advances in technique and instrumentation, studies have shown ACLR to fail or have a poor outcome in 11.9% of cases on average, with some studies citing as high as 27%. Many of these poor outcomes can be attributed to the inadequate management of the contributing mechanical and chondral factors. In these specific cases, the addition of osteotomies and/or concomitant chondral procedures has been used to address some of the factors that may lead to long-term failures or dissatisfaction with ACL reconstruction. Typically, osteotomies have been performed to address varus or valgus malalignment that has led to isolated medial or lateral osteoarthritis (OA) by redistributing the joint forces and offloading the diseased compartment; however, the recent re-emphasis on posterior slope of the tibia and its effect on ACL graft forces has introduced additional osteotomy considerations into surgical planning. Chondral injuries due to trauma or malalignment can also be addressed at the time of ACL reconstruction or in subsequent procedures and have been shown to provide improved pain and functionality equal to that of an ACLR alone, despite worse pre-operative metrics [1, 2]. This chapter focuses on the management of ACL tears in osteoarthritic knees, with a focus on additional osseous and chondral considerations.

27.2 Normal Lower Extremity Alignment

Anatomic alignment of the lower extremity traditionally falls within 5 and 7 degrees of valgus angulation in the coronal plane [3]. Outliers outside of the typical angulation can lead to compartmental osteoarthritis based on the direction and degree of coronal malalignment that lead to increased joint forces [4]. Varus deformity from uni-compartmental OA is a common situation in both chronic ACL insufficiency and postoperative ACL reconstruction and presents a potential issue in the long-term health of the reconstructed ACL graft. Studies have found higher forces on

R. D. Danilkowicz · A. Grassi · R. van Heerwaarden
Department of Orthopaedic Surgery, Division of Sports Medicine, Duke University Medical Center, Durham, NC, USA

A. Amendola (✉)
Department of Orthopaedic Surgery, Duke University Medical Center, Durham, NC, USA
e-mail: ned.amendola@duke.edu

M. Karahan et al. (eds.), *Anterior Cruciate Ligament Injuries Throughout the Life Span*,
https://doi.org/10.1007/978-3-031-84501-7_27

the ACL in the setting of varus alignment, particularly in those with a varus thrust, which may help to explain in part the failure rates of up to 25% of ACLR with underlying varus malalignment and the presentation of chronic ACL insufficiency in varus knees [5–7]. The posterior tibial slope is also a critical measurement and typically ranges between 6 and 11 degrees on the medial compartment and between 9 and 11 degrees in the lateral compartment, with the discrepancies attributed to the multiple measurement techniques described in the literature [8–11]. The posterior tibial slope in particular has become a topic of great interest in ACL injury and failed ACL reconstruction, with studies estimating an odds ratio of 4.52 for ACL graft rupture in knees with a posterior slope >12 degrees [12, 13].

27.3 Osteotomies

In patients with malalignment of the knee joint, corrective osteotomies may be undertaken in addition to ACLR that can both re-distribute joint contact forces to reduce the risk of osteoarthritis progression, but also take stress off of the newly placed ACL graft. Additionally, in those with previous ACLR that have failed, a thorough assessment of the alignment is crucial to assess for potential contributions to the failure of the previous surgery. According to the lead author, there are a number of indications of corrective osteotomy in the setting of ACL insufficiency that include but are not limited to:

A. Varus malalignment and uni-compartmental medial OA with chronic anterior laxity.
B. Varus malalignment with thrust and chronic anterior laxity.
C. Failed ACL reconstruction with posterior tibial slope >10–13 ± varus malalignment.

A medial opening-wedge or lateral closing-wedge osteotomy can be utilized to correct coronal plane deformity in an ACL-deficient patient with varus malalignment. Based on anatomic considerations of the proximal tibia, both the medial opening-wedge and lateral closing-wedge will have an effect on the posterior slope. The opening-wedge will create an increased posterior slope, while the closing-wedge reduces slope; however, the extent of each change has been studied in the literature. In a prospective study by Ducat et al. encompassing 321 patients, they found that each of the osteotomies resulted in less than 1 degree of change on average, with 12% of the closing-wedge high tibial osteotomies (HTOs) leading to a decrease of 5 degrees and 7% of the opening-wedge osteotomies resulting in an increase of 5 degrees or more [14]. Additionally, as techniques have advanced there have been methods utilized to reliably reduce the extent of the posterior slope changes with both approaches as studies have identified the importance of hinge and wedge positioning as well as the extent of distraction in relation to the slope produced [15, 16]. Moon et al. found that a posterolateral hinge position led to an increase in posterior tibial slope, while Jo et al. noted a low hinge position led to a greater posterior slope and more occurrences of fracture through the hinge, highlighting the importance of keen intraoperative awareness and planning [15, 16]. A complete osteotomy through the posterior cortex with subsequent posterior distraction is one possible solution for these issues and has been shown to be effective in maintenance of the posterior slope [17, 18]. Regarding wedge placement and position, purposeful placement of distraction devices posteriorly as well as posterior placement of the wedge is also helpful in maintaining the posterior slope after osteotomy.

The medial opening-wedge osteotomy is the preferred method of the senior author, as experience has shown it to provide a more accurate correction. A secondary benefit of the medial approach is the safety of the peroneal nerve and the avoidance of disrupting the tibiofibular joint and posterolateral corner structures of the knee by avoiding an osteotomy of the proximal fibula as is required in the lateral closing-wedge technique [19–21]. Avoiding the fibular osteotomy also forces a hinge at the proximal tibiofibular joint that results in the reduction of the tibial slope, which in turn puts less stress on the ACL. The amount of planned correction can be performed in a number of ways; however, it is the senior authors' preference to measure according to the technique described by Dugdale et al., which involves restoration of the weight-bearing line (center femoral head to center tibiotalar joint) to a pre-selected position along the lateral tibial plateau [22]. The amount of planned correction varies by patient and underlying pathology, with more correction aimed for individuals with advanced OA. In patients with ACL-deficient knees, varus malalignment, and thrust, a goal of neutral leg alignment is reasonable. However, in patients with medial compartment OA along with varus malalignment, the final correction is typically planned for slight valgus to unload the medial compartment, which involves placing the weight-bearing line to pass just lateral to the lateral tibial spine. This correction is slightly less in patients who wish to return to athletic activities, as overcorrection would be a detriment to their functionality. Confirmation of the correction is performed intraoperatively using the wedge base length as planned preoperatively and alignment is checked using an extramedullary alignment rod and fluoroscopy, ensuring that the weight-bearing axis is passing through the center of the knee joint [23]. Caution should be taken in obese patients and those with severe malalignment however, as studies have found supine fluoroscopy alignment measurements with no loading of the joint to be inaccurate [24]. Careful inspection of the intraoperative fluoroscopic films is also important to assess the posterior slope, as mentioned above that the opening wedge can also effect this plane depending on the amount of anterior or posterior distraction at the osteotomy site. In order to correct a biplanar malalignment, the osteotomy hinge should be cut and compressed separately. The medial opening wedge is not ideal for large sagittal corrections however, and if significant posterior slope correction is required, the senior author suggests use of an anterior closing wedge method.

Regarding additional pathology of the knee, the senior author does not perform extensive procedures on the cartilage or meniscus at the same time of the osteotomy and ACLR surgery, instead of choosing to perform them in a staged matter once the osteotomy has healed.

27.4 Outcomes of Alignment Correction in ACLR

Noyes et al. were the first to describe corrective osteotomy high tibial osteotomy (HTO) in the chronically ACL-deficient knee in 1993, reporting on a series of 41 patients with concomitant varus alignment [25]. Despite high overall satisfaction within the cohort, the group recommended that ACLR should be considered if the patient plans to resume athletic activity, or if they choose to forego ACL, they should refrain from athletic activities in order to avoid advanced arthritis [25]. In a follow-up series by the same first author, 41 additional patients with ACL deficiency, varus angulation, and some with additional meniscal or chondral pathology were studied with 34 receiving an ACLR, leaving the authors to conclude that osteotomy should accompany ACLR in knees with complex injury patterns [26]. In 1996, Latterman and Jakob reported on a series of 30 patients with medial compartment OA, varus angulation, and chronic anterior instability of the knee who were treated with either a. isolated HTO, b. HTO with simultaneous ACLR, or c. staged HTO then ACLR 6–12 months later. The decision on treatment was made on patient age, pain level, degree of instability, prior operative history, and amount of activity. Of the initial 30, isolated HTO was performed in 11 patients, with significant improvement in pain and function, with only two patients having a residual pivot shift test [27]. The conclusions from this study were that in older patients >40, isolated HTO is an excellent treatment option; however, in younger patients an HTO can be performed with subsequent ACLR if instability persisted, noting increased complications as the reason to not perform simultaneous procedures [27].

As a result of improved instrumentation and surgical technique, combined HTO and ACL reconstruction is becoming an increasingly common procedure. Biomechanically, combined HTO and ACLR has been shown to substantially decrease the anterior tibial translation, the knee adduction moment of the operative extremity, while slightly increasing it on the contralateral side [28, 29]. Additionally, the same study found a decrease in the knee flexion moment for both the operative and contralateral extremity [28]. These biomechanical changes suggest that postoperatively, the load has been shifted towards the lateral, nondiseased compartment without increasing the total load across the joint [30–32]. Clinically, the operative outcomes have been encouraging. Mehl et al. found no increase in complications using a combined approach in their study comparing HTO alone versus combined HTO and ACLR patients. A 2015 systematic review by Cantin et al. encompassed 321 patients who underwent the combined procedure and found 18% of cases with complications, which ranged in severity from stiffness to revision osteotomy and an almost 10% progression to advanced medial compartment arthritis [33]. A smaller study by Jin at el. looking at 24 patients with the combined procedure in cases of patients with existing medial compartment OA showed improved postoperative functional and activity levels; however, 12.% progressed in their OA [34, 35]. A 2019 meta-analysis on the topic reviewed 18 studies, with complication rates ranging from 0 to 23.5%, included a 6.5% revision osteotomy rate and a 17.5% graft failure rate in studies that reported these outcomes [36].

Outcome studies on the treatment of ACLR and posterior slope correction osteotomies are promising. From a biomechanical standpoint, the linear relationship between posterior slope and graft forces has been established and correction of the slope proven to be protective of the ACL graft [29, 37]. Arum et al. retrospectively analyzed 30 patients with concomitant opening-wedge HTO with ACLR and found that individuals who had a posterior slope correction >5 degrees had significantly better functional scores at follow-up [38]. Similarly, Zaffagnini et al. looked at 32 patients who underwent closing-wedge osteotomy and combined ACLR and found improved function and activity levels postoperatively; however, they also found that residual laxity remained in patients with larger posterior slopes [39]. Outcomes in cases of patients with multiple ACLR failures suggest further benefits to HTO that reduce posterior slope. Reporting on five patients who underwent an anterior based closing-wedge HTO with ALCR, Sonnery-Cottet et al. found that improvement in preoperative slope >13 degrees to an average of 9.2 degrees postoperative in addition to the ACLR also reduced anterior laxity from 10.4 to 2.8 mm and allowed for significant improvements in outcome scores [40]. Similar findings were reported by Dejour et al. who corrected slope from 13.2 to 4.4 degrees on average in multiple revised patients, resulting in no anteroposterior or rotatory laxity. Side-to-side difference was significantly improved with KT-1000 testing from 11.7 to 4.3 mm [41]. More recently, a 2020 study by Song et al. evaluated 18 patients with preoperative posterior slopes >13 degrees and excessive anterior tibial subluxation in extension >10 mm and found that slope reducing HTO with combined ACLR resulted in significantly improved outcome and functional scores and no re-ruptures at a minimum of two years follow-up [42].

27.5 Conclusion

In patients with ACL tears who have underlying malalignment and arthritis, HTO provides a means of both correcting the imbalance and protecting the ACL graft. Based on the degree of coronal and sagittal plane correction needed, there are multiple techniques of osteotomy available to address the specific pathology. In consideration for the choice of an osteotomy, it is important to evaluate not only the correction in the coronal plane, but also the effect or desire to change the posterior slope in the sagittal plane. The medial opening-wedge HTO in combination with ACLR is the preferred choice of the lead author in cases of ACL rupture with underlying medial compartment OA and a posterior slope within normal limits as it provides for the most precise correction. An anterior closing-wedge HTO may be utilized in cases of multiple ACLR failures with a posterior slope >10–13 degrees and normal coronal alignment. In these cases, appropriate patient selection and pre-operative planning are critical as osteotomies can be technically challenging, particularly when combined with ACLR, but can provide excellent patient outcomes when performed correctly.

References

1. Tahami SM, Rad SMD. Outcome of ACL reconstruction and concomitant articular injury treatment. Arch Bone Joint Surg. 2015;3(4):260–3.
2. Pike AN, Bryant T, Ogura T, Minas T. Intermediate-to long-term results of combined anterior cruciate ligament reconstruction and autologous chondrocyte implantation. Orthop J Sports Med. 2017;5(2):2325967117693591. https://doi.org/10.1177/2325967117693591.
3. Hsu RW, Himeno S, Coventry MB, Chao EY. Normal axial alignment of the lower extremity and load-bearing distribution at the knee. Clin Orthop Relat Res. 1990;255:215–27.
4. Sharma L, Song J, Dunlop D, et al. Varus and valgus alignment and incident and progressive knee osteoarthritis. Ann Rheum Dis. 2010;69(11):1940–5. https://doi.org/10.1136/ard.2010.129742.
5. van de Pol GJ, Arnold MP, Verdonschot N, van Kampen A. Varus alignment leads to increased forces in the anterior cruciate ligament. Am J Sports Med. 2009;37(3):481–7. https://doi.org/10.1177/0363546508326715.
6. Noyes FR, Schipplein OD, Andriacchi TP, Saddemi SR, Weise M. The anterior cruciate ligament-deficient knee with varus alignment. An analysis of gait adaptations and dynamic joint loadings. Am J Sports Med. 1992;20(6):707–16. https://doi.org/10.1177/036354659202000612.
7. Noyes FR, Barber-Westin SD. Anterior cruciate ligament revision reconstruction: results using a quadriceps tendon-patellar bone autograft. Am J Sports Med. 2006;34(4):553–64. https://doi.org/10.1177/0363546505281812.
8. Zhang Y, Wang J, Xiao J, et al. Measurement and comparison of tibial posterior slope angle in different methods based on three-dimensional reconstruction. Knee. 2014;21(3):694–8. https://doi.org/10.1016/j.knee.2014.01.008.
9. Dean RS, DePhillipo NN, Chahla J, Larson CM, LaPrade RF. Posterior tibial slope measurements using the anatomic axis are significantly increased compared with those that use the mechanical axis. Arthroscopy. 2021;37(1):243–9. https://doi.org/10.1016/j.arthro.2020.09.006.
10. Liles J, Pereira G, Danilkowicz R, Riboh J, Fletcher A. The reliability of radiographic evaluation of the posterior tibial slope in skeletally immature patients. Orthop J Sports Med. 2020;8(7 suppl 6):2325967120S00469. https://doi.org/10.1177/2325967120S00469.
11. Fletcher AN, Liles JL, Pereira GF, Danilkowicz RM, Amendola A, Riboh JC. The Intrarater and interrater reliability of radiographic evaluation of the posterior tibial slope in pediatric patients. J Pediatr Orthop. 2021;41:e404. https://doi.org/10.1097/bpo.0000000000001792.
12. Napier RJ, Garcia E, Devitt BM, Feller JA, Webster KE. Increased radiographic posterior tibial slope is associated with subsequent injury following revision anterior cruciate ligament reconstruction. Orthop J Sports Med. 2019;7(11):2325967119879373. https://doi.org/10.1177/2325967119879373.
13. Lee CC, Youm YS, Cho SD, et al. Does posterior Tibial slope affect graft rupture following anterior cruciate ligament reconstruction? Arthroscopy. 2018;34(7):2152–5. https://doi.org/10.1016/j.arthro.2018.01.058.
14. Ducat A, Sariali E, Lebel B, et al. Posterior tibial slope changes after opening- and closing-wedge high tibial osteotomy: a comparative prospective multicenter study. Orthop Traumatol Surg Res. 2012;98(1):68–74. https://doi.org/10.1016/j.otsr.2011.08.013.
15. Jo HS, Park JS, Byun JH, et al. The effects of different hinge positions on posterior tibial slope in medial open-wedge high tibial osteotomy. Knee Surg Sports Traumatol Arthrosc. 2018;26(6):1851–8. https://doi.org/10.1007/s00167-017-4526-6.
16. Moon SW, Park SH, Lee BH, et al. The effect of hinge position on posterior tibial slope in medial open-wedge high tibial osteotomy. Arthroscopy. 2015;31(6):1128–33. https://doi.org/10.1016/j.arthro.2015.01.009.
17. Lee YS, Kang JY, Lee MC, Oh WS, Elazab A, Song MK. Effect of the osteotomy length on the change of the posterior tibial slope with a simple distraction of the posterior gap in the uni- and biplanar open-wedge high tibial osteotomy. Arthroscopy. 2016;32(2):263–71. https://doi.org/10.1016/j.arthro.2015.07.018.
18. Westermann RW, DeBerardino T, Amendola A. Minimizing alteration of posterior tibial slope during opening wedge high tibial osteotomy: a protocol with experimental validation in paired cadaveric knees. Iowa Orthop J. 2014;34:16–23.
19. Chun YM, Kim SJ, Kim HS. Evaluation of the mechanical properties of posterolateral structures and supporting posterolateral instability of the knee. J Orthop Res. 2008;26(10):1371–6. https://doi.org/10.1002/jor.20596.
20. Laprade RF, Engebretsen L, Johansen S, Wentorf FA, Kurtenbach C. The effect of a proximal tibial medial opening wedge osteotomy on posterolateral knee instability: a biomechanical study. Am J Sports Med. 2008;36(5):956–60. https://doi.org/10.1177/0363546507312380.
21. Song EK, Seon JK, Park SJ, Jeong MS. The complications of high tibial osteotomy: closing- versus opening-wedge methods. J Bone Joint Surg Br. 2010;92(9):1245–52. https://doi.org/10.1302/0301-620x.92b9.23660.

22. Dugdale TW, Noyes FR, Styer D. Preoperative planning for high tibial osteotomy. The effect of lateral tibiofemoral separation and tibiofemoral length. Clin Orthop Relat Res. 1992;274:248–64.
23. Brinkman JM, Lobenhoffer P, Agneskirchner JD, Staubli AE, Wymenga AB, van Heerwaarden RJ. Osteotomies around the knee: patient selection, stability of fixation and bone healing in high tibial osteotomies. J Bone Joint Surg Br. 2008;90(12):1548–57. https://doi.org/10.1302/0301-620x.90b12.21198.
24. Sabharwal S, Zhao C. Assessment of lower limb alignment: supine fluoroscopy compared with a standing full-length radiograph. J Bone Joint Surg Am. 2008;90(1):43–51. https://doi.org/10.2106/jbjs.F.01514.
25. Noyes FR, Barber SD, Simon R. High tibial osteotomy and ligament reconstruction in varus angulated, anterior cruciate ligament-deficient knees. A two- to seven-year follow-up study. Am J Sports Med. 1993;21(1):2–12. https://doi.org/10.1177/036354659302100102.
26. Noyes FR, Barber-Westin SD, Hewett TE. High tibial osteotomy and ligament reconstruction for varus angulated anterior cruciate ligament-deficient knees. Am J Sports Med. 2000;28(3):282–96. https://doi.org/10.1177/03635465000280030201.
27. Lattermann C, Jakob RP. High tibial osteotomy alone or combined with ligament reconstruction in anterior cruciate ligament-deficient knees. Knee Surg Sports Traumatol Arthrosc. 1996;4(1):32–8. https://doi.org/10.1007/bf01565995.
28. Marriott K, Birmingham TB, Kean CO, Hui C, Jenkyn TR, Giffin JR. Five-year changes in gait biomechanics after concomitant high tibial osteotomy and ACL reconstruction in patients with medial knee osteoarthritis. Am J Sports Med. 2015;43(9):2277–85. https://doi.org/10.1177/0363546515591995.
29. Imhoff FB, Mehl J, Comer BJ, et al. Slope-reducing tibial osteotomy decreases ACL-graft forces and anterior tibial translation under axial load. Knee Surg Sports Traumatol Arthrosc. 2019;27(10):3381–9. https://doi.org/10.1007/s00167-019-05360-2.
30. Hurwitz DE, Sumner DR, Andriacchi TP, Sugar DA. Dynamic knee loads during gait predict proximal tibial bone distribution. J Biomech. 1998;31(5):423–30. https://doi.org/10.1016/s0021-9290(98)00028-1.
31. Kutzner I, Trepczynski A, Heller MO, Bergmann G. Knee adduction moment and medial contact force – facts about their correlation during gait. PLoS One. 2013;8(12):e81036. https://doi.org/10.1371/journal.pone.0081036.
32. Schipplein OD, Andriacchi TP. Interaction between active and passive knee stabilizers during level walking. J Orthop Res. 1991;9(1):113–9. https://doi.org/10.1002/jor.1100090114.
33. Cantin O, Magnussen RA, Corbi F, Servien E, Neyret P, Lustig S. The role of high tibial osteotomy in the treatment of knee laxity: a comprehensive review. Knee Surg Sports Traumatol Arthrosc. 2015;23(10):3026–37. https://doi.org/10.1007/s00167-015-3752-z.
34. Mehl J, Paul J, Feucht MJ, et al. ACL deficiency and varus osteoarthritis: high tibial osteotomy alone or combined with ACL reconstruction? Arch Orthop Trauma Surg. 2017;137(2):233–40. https://doi.org/10.1007/s00402-016-2604-8.
35. Jin C, Song EK, Jin QH, Lee NH, Seon JK. Outcomes of simultaneous high tibial osteotomy and anterior cruciate ligament reconstruction in anterior cruciate ligament deficient knee with osteoarthritis. BMC Musculoskelet Disord. 2018;19(1):228. https://doi.org/10.1186/s12891-018-2161-0.
36. Stride D, Wang J, Horner NS, Alolabi B, Khanna V, Khan M. Indications and outcomes of simultaneous high tibial osteotomy and ACL reconstruction. Knee Surg Sports Traumatol Arthrosc. 2019;27(4):1320–31. https://doi.org/10.1007/s00167-019-05379-5.
37. Bernhardson AS, Aman ZS, Dornan GJ, et al. Tibial slope and its effect on force in anterior cruciate ligament grafts: anterior cruciate ligament force increases linearly as posterior tibial slope increases. Am J Sports Med. 2019;47(2):296–302. https://doi.org/10.1177/0363546518820302.
38. Arun GR, Kumaraswamy V, Rajan D, et al. Long-term follow up of single-stage anterior cruciate ligament reconstruction and high tibial osteotomy and its relation with posterior tibial slope. Arch Orthop Trauma Surg. 2016;136(4):505–11. https://doi.org/10.1007/s00402-015-2385-5.
39. Zaffagnini S, Bonanzinga T, Grassi A, et al. Combined ACL reconstruction and closing-wedge HTO for varus angulated ACL-deficient knees. Knee Surg Sports Traumatol Arthrosc. 2013;21(4):934–41. https://doi.org/10.1007/s00167-013-2400-8.
40. Sonnery-Cottet B, Mogos S, Thaunat M, et al. Proximal tibial anterior closing wedge osteotomy in repeat revision of anterior cruciate ligament reconstruction. Am J Sports Med. 2014;42(8):1873–80. https://doi.org/10.1177/0363546514534938.
41. Dejour D, Saffarini M, Demey G, Baverel L. Tibial slope correction combined with second revision ACL produces good knee stability and prevents graft rupture. Knee Surg Sports Traumatol Arthrosc. 2015;23(10):2846–52. https://doi.org/10.1007/s00167-015-3758-6.
42. Song GY, Ni QK, Zheng T, Zhang ZJ, Feng H, Zhang H. Slope-reducing tibial osteotomy combined with primary anterior cruciate ligament reconstruction produces improved knee stability in patients with steep posterior tibial slope, excessive anterior tibial subluxation in extension, and chronic meniscal posterior horn tears. Am J Sports Med. 2020;48(14):3486–94. https://doi.org/10.1177/0363546520963083.

Part VII

Other Issues

Perioperative Concerns in ACL Reconstruction

28

Ignacio Cardona-Muñoz, Jesus I. Cardona-Medina, and Ariel De la Rosa

28.1 Introduction

Anterior cruciate ligament (ACL) rupture is the most common ligament injury in the knee, with a steady increase in incidence, mainly affecting young and physically active individuals, which increases the risk of added meniscus and/or cartilage damage [1].

ACL reconstruction (ACLR) is the current standard of care for active patients with an ACL rupture. Although the majority of ACLR surgeries successfully restore the mechanical stability of the injured knee, postsurgical outcomes remain widely varied, resulting in less than a half of athletes who undergo ACLR returning to sport within the first year after surgery. It is estimated that 35% cannot recovery their preinjury performance level after ACLR [2, 3].

Surgeons agree that every surgical procedure has its own risk, and ACLR is no exception, and as such, patient risk factors must be been clearly identified in every case [4].

I. Cardona-Muñoz (✉) · J. I. Cardona-Medina · A. De la Rosa
Orthopedic & Sports Medicine Center, Physical Activity and Sports Applied Sciences Institute, Health Sciences Centre, Universidad de Guadalajara, Guadalajara, Mexico

28.2 Our Thoughts of Perioperative Concerns with ACLR—What We Do

I have been involved in the fields of sports medicine and knee surgery for many years, taking care of elite collegiate and recreational athletes, and I have learned that each case represents a different challenge. Every patient has his or her own anatomical and biomechanical characteristics, and own functional expectations with surgery; so, we have to carry out individualized planning and customized execution when performing ACLR [4].

The key point in obtaining successful results in ACLR is making an accurate diagnosis of the injury, and methodologically identifying each of the risk factors involved, without neglecting any, as obvious as it may seem.

28.3 History and Physical Exam

All patients should be evaluated with a strict protocol, even if the diagnosis can be suspected during the consultation. An accurate diagnosis must be made by a proper clinical approach and comprehensive history through physical examination aided by imaging to confirm the diagnosis [5].

When planning the ACLR, it is crucial to obtain a detailed mechanism of injury and other medical findings suggesting additional damage

M. Karahan et al. (eds.), *Anterior Cruciate Ligament Injuries Throughout the Life Span*,
https://doi.org/10.1007/978-3-031-84501-7_28

(i.e., meniscus or chondral tears). Also, we must clearly identify what the patient's expectations are and what level of activity can realistically be postoperatively expected. In other words, identifying whether the patient's expectations are compatible with the patient's knee or if the patient is a "knee abuser" with unrealistic expectations.

Physical examination must be performed methodologically, step by step, including both limbs, paying special attention to limb alignment and walking cycle. Assessing knee stability, muscle strength, patellar tracking, and active and passive range of motion, in addition to looking for any knee extension or flexion deficits, should be performed. It is important to evaluate knee laxity, mainly on women and young patients.

Assessing anterior tibial translation is crucial for clinical diagnosis of an ACL deficient knee. The Lachman and anterior drawer test are the most used to detect abnormal anterior tibial translation. The pivot shift test is considered pathognomonic for anterolateral instability. Studies have shown the high sensitivity, varying from 84 to 98%, with a specificity >98% when the pivot shift is performed under anesthesia [6]. Routinely, all unstable knees are also assessed with Slocum, dial, and Jerk tests [6]. Quantitative assessment of the unstable knee is not always feasible. KT-1000 and KT-2000 are devices widely used to quantify anterior tibial translation. Lately, it is possible to quantify rotatory knee instability with devices based on electromagnetic or global position systems.

28.4 Imaging

Although magnetic resonance imaging (MRI) is considered the "gold standard" to confirm ACL injuries, plain radiograph imaging should not be omitted; it can be utilized to rule out bone fractures, observe the morphology of the lateral femur condyle and tibial slope, especially in knees with deformity of varus or valgus, and avoid missing limb malalignment, given the possibility that another surgery will have to be indicated.

With a conventional MRI, in most cases, we can make an accurate diagnosis, identifying on standard views the type of ACL tear, recognition of associated lesions, for example, meniscal or chondral tears and evaluate other ligaments. So, we can carry out a proper plan of surgery. Some authors recommend special MRI sequence views for recognition of partial ACL tears [7]. Where possible, kinematic MRI, applying instability tests, is possible to predict the grade of pivot shift [8].

28.5 Age and Gender

Certainly, there is not an established age limit to perform an ACLR. In fact, this procedure is most frequently performed in patients younger than 20 years of age [9]. In this group, literature reports that male patients are more likely to sustain a reinjury, approximately 1 in 5, and about 20% will suffer an ACL tear in the contralateral knee [9]. However, risk factors for these reinjuries are controversial and unclear. Published reports differ regarding age, type of graft and fixation system, surgical techniques, postoperative rehabilitation, and criteria to return to sports [9, 15].

Children and adolescents deserve special attention. In recent years, we have seen more cases of ACL injury in early ages due to the practice of extreme sports and high contact physical activities. The timing for ACLR in pediatric patients is a big challenge. Although there is significant evidence that ACLR at this age provides good results [9], in general surgery is not always recommended. Parents must be aware of the potential complication potential and the need of close observation in the postoperative follow-up until skeletal maturity [9].

ACL injuries in female athletes are three times higher than in male athletes [10]. It is important to identify the risk factors that make women more susceptible to having an ACL injury. Most important intrinsic factors are an increased valgus knee and Q angle, an intercondylar notch narrower than males, muscle weakness, greater knee laxity, and an effect of the ovarian hormonal cycle. As such, those particular considerations have to be considered during ACLR in this group of patients [10, 11, 14].

In adults over 40 years of age, the indication of surgery must be reasonably made. Probably, in most cases some degenerative lesions need to be treated at time of ACLR, to decelerate the evolution of osteoarthritis (OA), including additional procedures such as realignment osteotomy, meniscal repair, or allogenic osteochondral transplantation And why not, in patients with isolated symptomatic OA, either medial or lateral compartment, you can consider unicompartmental replacement [12].

28.6 Preoperative

In acute ACL tears, the patient is included in a physical therapy protocol for pain and inflammation management for two to three weeks or until the patient has regained mobility and the inflammation has sufficiently decreased. In most cases, the use of crutches is indicated to reduce weight when stepping or walking to enhance healing of bone, to avoid muscular atrophy and preserve proprioception. These actions reduce the risk of knee stiffness after surgery. When it is feasible, patients are referred to aqua rehabilitation [13].

All patients should be alerted that ACLR requires a long postoperative rehabilitation protocol and also that the results obtained do not always achieve complete joint stability. Patients must understand that the main objective of the surgery is to recover the capability to carry out daily life activities without instability. The return to sports cannot be guaranteed.

28.7 Timing of ACL Surgery

Although there is a lot of evidence in the literature, the optimal time for ACLR currently remains controversial. The debate is, how early the surgery can be performed safely and how long it can be delayed without increasing the risk of developing secondary knee lesions. In acute ACL tears, ACLR should be delayed three to six weeks to eliminate inflammation, but no more than six months, in order to try to prevent concomitant injuries. Unfortunately, in our practice, most of the patients come to our office after having suffered several subluxation episodes [13, 14].

Better results are achieved when patients know their health problem and are involved in decision-making; we ensure the patient clearly understands the natural history of knee instability, highlighting the importance of making a timely decision to reduce the risk of developing OA [15].

28.8 Intraoperative

Any ACLR technique has its own risk factors. The imprecision on performance varies from surgeon to surgeon. Therefore, the surgeon and his surgical team must have the skills and mastery in ACLR. Having participated in the transition on ACLR, from arthrotomies to arthroscopic surgery and having performed different techniques, currently, our choice for ACLR is the arthroscopic anatomical technique. Almost always we reconstruct the ACL using an autograft, either hamstrings or bone patellar tendon bone (BPTB), according to the case.

Tips and pearls for a successful and uncomplicated procedure are as follows:

- Adequate patient position and knee draping.
- Correct selection of arthroscopic portals for obtained adequate visualization.
- Carefully harvesting and preparation of selected graft.
- Wrapping the graft with a gauze soaked in vancomycin, to reduce the risk of infection [17, 18].
- Accurate placement and drilling of femur and tibia tunnels.
- Adequate graft fixation [19].

Tunnel mismatch and deficient graft fixation can result in abnormal knee kinematics which can cause limited range of motion and modify graft tension, which can affect the biological healing process [19].

Tourniquet inflation, when necessary, applied for the shortest possible time, reduces thigh pain, facilitating the prompt mobilization and recovery of the quadricep force, reducing the risk of thrombotic events as well [16].

28.9 Postoperative

To reduce the risk of patients developing knee stiffness in the immediate postoperative period, we use cooling devices to decrease inflammation and pain. All our patients are motivated to begin their rehabilitative program as soon as possible with a customized protocol according to type of injury and specific procedure. We have stopped applying accelerated ACLR protocols. The postoperative follow-up should be very closely monitored, maintaining permanent communication with the therapist. Monitoring patient's recovery progress lets us detect signs of any potential complications, mainly early infection among others. Functional exercises, such as swimming and biking, are allowed in early stages to enhance the rehabilitative process.

Finally, we are also concerned about the issue of return to sport. Return to sports is the main goal after ACLR. The most frequent cause of reinjuries is an early return to pivoting sports before complete graft healing or not having recovered the optimal physical and emotional conditions. It is still controversial on what criteria are used to allow a safe return to sport. Our decision is based on the criteria of time and clinical evaluation. Regarding time, we don't authorize return to sports before nine months, to facilitate the biological healing. In ACLR with BTPB autografts, return to sports can be sooner than using hamstrings tendons or allografts [18–20].

Our clinical evaluation is based on widely accepted recommendations; no swelling, no pain, adequate range of motion, recovery of least 85% of strength quadriceps and hamstrings muscles, and a single one leg hop test [21]. Also, it is important to be sure patients have regained confidence and security for returning to training and competition.

28.10 Summary

In summary, ACLR is a safe and reproducible intervention, but there remains an underlying rate of failure to return to preinjury sporting activity levels. Despite the advances in basic science and the development of more precise surgical techniques, the ACL problem is still not resolved.

References

1. Anderson MJ, Browning WM III, Urband CE, Kluczynski MA, Bisson LJ. The evidence for cute ACL reconstruction without non-operative care Anderson delayed ACL R increases the risk of secondary meniscal and chondral injuries. Rev Orthop J Sports Med. 2021;4(3):2325967116634074. https://doi.org/10.1177/2325967116634074.
2. Dunn KL, Lam KC, Valovich McLeod VTC. Early operative versus delayed or nonoperativetreatment of anterior cruciate ligament injuries in pediatric patients. J Athl Train. 2016;51(5):425–7.
3. Bliss JJ. Anterior cruciate ligament injury, reconstruction, and the optimization of outcome. Indian J Orthop. 2017;51(5):606–13. https://doi.org/10.4103/ortho.IJOrtho_237_17.
4. Figueroa D. Individualized and customized ACL reconstruction. Webinar ISAKS- SLARD. Return to Sports 05/07/2020; 2020.
5. Bergin M, Hofbauer M, Ohashi B, Musahl V. History, physical examination and imaging. In: Anterior cruciate ligament reconstruction, ESSKA 2014. Springer. p. 61–3.
6. Musahl V, Kopf S, Becker R, van der Merwe W, Zaffagnini S, Fu FH, Karlsson J. Rotatory knee laxity test and the pivot shift as tool for ACL treatment algorithm. Knee Surg Sports Traumatol Arthrosc. 2012;20(4):793–800.
7. Steckel H, Vadala G, Davis D, Musahl V, Fu FH. 3-T MR imaging of partial ACL tears: a cadaver study. Knee Surg Sports Traumatol Arthosc. 2007;18(9):1154–63.
8. Okazaky K, Tashiro Y, Izawa T, Matsuda S, Iwamoto Y. Rotatory laxity evaluations of the knee using modified Slocum's test in open magnetic resonance imaging. Knee Sports Traumatol Arthosc. 2012;20(4):679–85.
9. Hewett TE, Di Stasi SL, Myer GD. Current concepts for injury prevention in athletes after anterior cruciate ligament reconstruction. Am J Sports Med. 2013;41(1):216–24. https://doi.org/10.1177/0363546512459638.
10. Snaebjörnsson T, Senorski EH, Sundemo D, Svantesson E, Westin O, Musahl V, Alentorn-Geli EA, Samuelsson K. Adolescents and female patients are at increased risk for contralateral anterior cruciate

ligament reconstruction: a cohort study from the Swedish National Knee Ligament Register based on 17,682 patients. Knee Surg Sports Traumatol Arthrosc. 2017;25:3938–44. https://doi.org/10.1007/s00167-017-4517-7.

11. Lipps DB, Oh YK, Ashton-Miller JA, Wojtys EM. Morphologic characteristics help explain the gender difference in peak anterior cruciate ligament strain during a simulated pivot landing. Am J Sports Med. 2012;40(1):32–40. 12. Prodromos CC, Han Y, Rogowski J, Joyce B, Shi K: A meta-analysis of the incidence of anterior cruciate ligament tears as a function of gender, sport, and a knee injury-reduction regimen. Arthroscopy 2007; 23(12):1320–1325.e6.
12. Beynnon BD, Johnson RJ, Braun S, Sargent M, Bernstein I, Vacek PM. The relationship between the phase of menstrual cycle and anterior cruciate ligament injury: a case-control study of recreational alpine skiers. Am J Sports Med. 2006;34:757–64.
13. Shelbourne KD, Wilcknes JH, Mollabashy A, DeCarlo M. Arthrofibrosis in acute anterior cruciate ligament reconstruction and rehabilitation. Am J Sports Med. 1991;19(4):332–6.
14. Hantes ME, Tsarouhas A. In timing of ACL surgery: any evidence? In: Siebold R, Dejour D, Zaffagnini S, editors. Anterior cruciate ligament reconstruction. A practical surgical guide. Springer ESSKA; 2014. p. 123–6.
15. Smith TO, Postle K, Penny F, McNamara I, Mann CJV. Is reconstruction the best management strategy for anterior cruciate ligament rupture? A systematic review and meta-analysis comparing anterior cruciate ligament reconstruction versus non-operative treatment. Knee. 2014;21(2):462e70.
16. Pérez-Prieto D, Torres-Claramunt R, Gelber PE, Shehata TMA, Pelfort X, Monllau JC. Autograft soaking in vancomycin reduces the risk of infection after anterior cruciate ligament reconstruction. Knee Surg Sports Traumatol Arthrosc. 2016;24(9):2724–8.
17. Crawford MD, Diehl LH, Amendola A. Surgical management and treatment of the anterior cruciate ligament-deficient knee with malalignment. Clin Sports Med. 2017;36(1):119–33.
18. Grindem H, Snyder-Mackler L, Risberg MA, et al. Br J Sports Med. 2016; https://doi.org/10.1136/bjsports-2016-096031.
19. Barber-Westein S, Noyes FR. One in 5 athletes sustain reinjury upon return to high-risk sports after ACL reconstruction: a systematic review in 1239 athletes younger than 20 years. Sports Health. 2020;12(6):587–97.
20. Zhou J, Schiaty ND, Hewelt TE, Bates NA. Analysis of timing of secondary ACL injury in professional athletes does no support the game timing or reason timing as a contributor to injury risk. Int J Sports Phys Ther. 2020;15(2):254.262.
21. Gokeler A, Dingenen B, Hewett TE. Rehabilitation and return to sport testing after anterior cruciate ligament reconstruction: where are we in 2022? Arthrosc Sports Med Rehabil. 2022;4(1):e77–82.

Bleeding and Clotting Control in ACL Reconstruction

29

Bertan Cengiz and Sinan Karaoglu

29.1 Bleeding Control

With the advancement of surgical technique, the number of arthroscopic anterior cruciate ligament (ACL) reconstruction is increasing day by day. Although arthroscopic ACL reconstruction is considered minimally invasive and comparatively safe procedure, it has some potential complications. In a recent study, the overall complication rate was found to be 4.7% for knee arthroscopic procedures; however, arthroscopic ACL reconstruction had a complication rate of 9% [1]. Small et al. reported in a prospective study an overall complication rate of 1.7%. Hemarthrosis was found to be the most common complication with the rate of 1.15%, which accounted for 60.1% of all complications [2]. Also, Andrés-Cano et al. reported that hemarthrosis that required arthrocentesis was one of the most frequent complications with the rate of 4.4% after arthroscopic ACL reconstruction [3].

Hemarthrosis is a major cause of pain after arthroscopic ACL reconstruction and it has potential harmful effects on joint cartilage [4, 5]. It increases susceptibility to infection and limits rehabilitation protocol, which increases morbidity and is associated with poor outcomes and poor patient satisfaction [6, 7]. When single (SB) and double- bundle (DB) ACL reconstructions were compared, it was observed that hemarthrosis and pain were significantly higher in the DB ACL reconstruction group, causing the postoperative rehabilitation to lag during the early period [6].

Clinicians have been trying to develop different methods to prevent the negative effects of these complications due to hemarthrosis for years. First, intra-articular suction drains were used to prevent hemarthrosis. In 1978, Browett found no difference at early postoperative period between pain and formation of hemarthrosis in the patient group that was undergoing open meniscectomy with and without a drain [8]. Nowadays, the necessity of the use of intra-articular drains is questioned. More recently, several studies on arthroscopic ACL reconstructions have reported that there was no difference in the outcomes between the patient groups with or without drains used [9]. Using intra-articular suction drains started to lose its popularity compared with the past, as stated in a recent systematic review study about national surveys in which no clear trends could be delineated about the use of drains among the surveys [10]. Besides all this, drains could be a potential source of retrograde infection [11]. The routine use of intra-articular

B. Cengiz (✉)
Acibadem Kayseri Hospital, Melikgazi/Kayseri, Turkey

S. Karaoglu
Acibadem Kayseri Hospital, Melikgazi/Kayseri, Turkey

Acibadem University, Istanbul, Turkey

M. Karahan et al. (eds.), *Anterior Cruciate Ligament Injuries Throughout the Life Span*,
https://doi.org/10.1007/978-3-031-84501-7_29

drains after arthroscopic ACL surgery was not supported by some studies [9, 12].

One of the controversial topics regarding ACL reconstruction surgery is about using tourniquet during surgery. Depending on the surgeon's preference, ACL reconstruction could be performed either with or without a tourniquet. Many surgeons believe that it improves surgical field visibility, reduces blood loss, and reduces surgical time. In a recent study, Kuo et al. reported that, compared to the nontourniquet group, the tourniquet group had a significant increase with blood loss in the drain, and the operation time was not significantly different between each group. Also, tourniquet usage caused greater decrease in thigh girth in the postoperative period [13]. In another recent study, it was found that tourniquet usage in arthroscopic ACL reconstruction increased pain significantly, and formation of hemarthrosis was significantly lower in the nontourniquet group at two weeks postoperatively. The thigh and calf girth were significantly reduced in the tourniquet group. There was no difference found between the groups in terms of quality of visibility and operative time [14]. Also in a systematic review about tourniquet use in ACL surgery, it was shown that application of tourniquet didn't provide shorter operative time [15]. For these reasons, when evaluated in general, the negative effects of tourniquet usage in arthroscopic ACL reconstruction are more than its positive effects.

Many surgeons generally apply compressive bandages or cold dressings to prevent hemarthrosis after surgery. In their postoperative protocol, Tierney et al. applied cold compressive dressing in the operative suit (Cryocuff) for their outpatient ACL reconstruction group and reported that this method was safe, effective, and easy with better patient satisfaction [16]. However, Dervin reported that, in their study comparing groups that did and did not use cryotherapy and/or compressive dressings in the immediate postoperative period, no significant difference between the groups was detected with respect to length of hospitalization, Hemovac knee drainage, intravenous (i.v.) narcotic requirement, nor pain [17]. In a recent prospective randomized comparative trial, it was stated that postoperative use of continuous cold flow device with elastocompressive bandage after ACL reconstruction lowered the knee volume increase, blood loss, and pain; at the same time, it increased knee range of motion (ROM; [18]). Further studies with longer follow-up period and higher number of patients are required to present beneficial effects of cold dressings clearly.

Tranexamic Acid

Tranexamic acid (TXA) is a synthetic derivate of the lysine amino acid. It binds reversibly to lysine receptors on plasminogen, resulting in the inhibition of plasminogen activation, so that it acts as an antifibrinolytic agent. TXA can stabilize the formed clots and prevent the degradation of fibrin [19]. An additional effect of TXA is directly inhibiting plasmin activity, but to reduce plasmin formation, higher doses are required. TXA is much more potent than other molecules like aminocaproic acid; it also has a 120 min plasma half-life [20]. It has been used in the treatment of hematologic diseases such as von Willebrand disease and hemophilia to reduce blood loss [21, 22]. As stated in the largest trial of antifibrinolytics (CRASH-2 Trial), it was shown that TXA reduced mortality due to bleeding in trauma patients as administered at early period of trauma [23]. A systematic review about the effect of TXA on surgical blood loss showed that the use of TXA was associated with overall reduction in surgical bleeding by about one third [24]. The use of TXA in orthopedic surgery shows promising effects on reducing both blood loss and number of blood transfusions needed postoperatively [25]. Many studies have shown that with the administration of TXA in total hip and knee arthroplasty patients, blood loss was reduced; the amount of blood transfused and number of patients requiring blood transfusion were lower in those patients who were receiving tranexamic acid than in the control group [26–28].

TXA use in arthroscopic ACL reconstruction has increased in popularity recently. A prospective, randomized, double blinded study in which 105 patients who underwent arthroscopic ACL reconstruction were enrolled showed that the usage of i.v. TXA before tourniquet inflation and continuous infusion continued for three hours after completion of the operation reduced postoperative hemarthrosis and decreased the need for a knee aspiration postoperative. It also reduced pain and improved range of motion of the knee without any side effects such as deep venous thrombosis (DVT) and infection [29]. In 2019, Felli et al. evaluated the effects of TXA in ACL reconstruction patients up to three months after surgery. They concluded that administration of TXA reduced hemarthrosis and blood drainage volumes, improved quadriceps strength and ROM and reduced fever episodes during the first two weeks after surgery; however, no differences were found for any outcomes or complications at three months after surgery [30].

Despite the benefits of using TXA in orthopedic surgery that have been shown, the method of TXA administration is still controversial. Many studies have shown that i.v. administration of TXA is significantly effective in reducing formation of hemarthrosis with a minimal decrease in hematocrit levels, as well as improving muscle strength and ROM [26, 28–30]. Two meta-analyses have demonstrated that topical administration of TXA has comparable effectiveness to i.v. administration in reducing blood loss in total knee and hip arthroplasty patients [31, 32]. Other double-blinded, randomized controlled trials with a large number of patients showed that topically applied TXA was effective in reducing blood loss and the need for blood transfusion following total hip arthroplasty (THA) and total knee arthroplasty (TKA), and the authors argued that local administration would be beneficial in avoiding the potential complications of i.v. TXA administration (TRANX-H [33] and TRANX-K [34] studies). A systematic review and meta-analysis concluded that topical administration is superior to i.v. TXA in reducing transfusion rates and decreasing blood loss in TKA and THA [35]. Another meta-analysis of randomized controlled trials evaluated the effectivity of regimens of TXA and found the combined use of TXA, both topical and i.v administration, had better outcomes than topical only or i.v. in TKA and THA patients [36, 37]. When oral and i.v. administration was compared, it was shown that the oral approach had similar clinical outcomes and was less expensive for total joint arthroplasty (TJA) patients [38].

According to previous studies to date, looking from a methodology perspective of administration of TXA, timing, and dosing are also controversial. Some authors recommend only one dose of TXA administration [39, 40], while others adopted multiple doses of TXA administered at different time points before and after surgery [41, 42]. It is stated that higher doses of i.v. or topical TXA give better results by some studies, [43, 44] while different studies did not observe any difference between lower and higher doses of i.v. and topical TXA in TJA [45–47].

Because of the complexity about usage of TXA, a new guideline was published by AAHKS (American Association of Hip and Knee Surgeons) [48]. It was reported that administration of i.v., oral and topical TXA as well as combinations of these are all effective when compared to placebo for reducing the need for transfusion and calculated blood loss in primary TJA. All of these methods of administration demonstrated equivalent efficiency during the perioperative period of primary TJA. No significant difference was found when the efficacy of the dose amount and the number of applications were compared in i.v. and topical application. Timing of administration is important because it is stated that administration of i.v. TXA before the surgical incision significantly reduces blood loss and the need for transfusion than administration after incision [48].

In a recent study, Chiang et al. reported that intra-articular injection of TXA in arthroscopic ACL reconstruction significantly reduced postoperative intra-articular bleeding in the first 24 hours and could be considered as an effective and safe solution to prevent postoperative bleeding and pain [49]. TXA is increasingly being used by local administration due to the lower rates of systemic effects; however, the cytotoxic effects of TXA are still unknown when used intra-articularly. Parker et al. investigated the effects of TXA on human chondrocytes which were harvested from patients undergoing ACL reconstruction, using two-dimensional (2D) and three-dimensional (3D) culture models. They concluded that cytotoxicity was observed in vitro when chondrocytes were exposed to a higher concentration of TXA and the toxicity of TXA was dose-dependent. It was reported that tissue concentration between 10 and 20 mg/mL would be expected to be safe [50].

29.2 Clotting Control (Thromboembolism, Deep Venous Thrombosis (DVT), Pulmonary Embolism (PE))

29.2.1 Incidence

Vascular complications, arterial and/or venous, after arthroscopic ACL reconstruction surgery can cause morbidity and mortality. The arterial complication incidence after ACL surgery is not clearly determined up to date, but the reported incidence is very low, with one retrospective series reporting a rate of 0.3% [51–53]. Pseudoaneurysms were detected as the most frequent arterial complication; however, arterial occlusions were reported in some case reports [51, 54, 55].

Veins are the main source of vascular problems after ACL reconstruction. Deep venous thrombosis (DVT), pulmonary embolism (PE), and post-thrombotic syndrome may be the main clinical presentations of venous thromboembolism (VTE) [51, 56]. After routine arthroscopic interventions except complex procedures like ligament reconstructions, the risk of asymptomatic DVT ranges between 5 and 18%, and symptomatic DVT risk ranges from 1.5 to 2% [57–59]. However, arthroscopic ACL reconstruction is more traumatic than routine arthroscopy of the knee, so that the incidence of the asymptomatic DVT was found to be 17.9% [59]. Struijk-Mulder et al. performed a prospective cohort study to establish the incidence of DVT after arthroscopic ACL reconstruction without thromboprophylaxis as detected by bilateral complete compression ultrasonography. They reported that asymptomatic DVT was detected in 9% of patients, whereas 4% of the patients were symptomatic [60]. Adala et al. reported the incidence of DVT in patients undergoing arthroscopic ACL reconstruction was 1.78% [61]. In a systematic review, after arthroscopic ACL reconstruction without thromboprophylaxis, the incidence of DVT was found to be 9.7% and 2.1% were symptomatic. The PE incidence was found to be 0.1 [53]. In another study, it was stated that the incidence of DVT after ACL reconstruction without thromboprophylaxis was 8.4 and 73% of the patients were asymptomatic; the rate of PE was 0.2% [56]. In a retrospective analysis of 249 ACL reconstruction patients, a unilateral contrast venography examination was performed on the third postoperative day, and a DVT was detected in 24 patients; however, no PEs were detected [51]. These variable rates of DVT and PE depend on the method to determine DVT (ultrasound, venography, MRI, clinic parameters) and patients demographics (tourniquet time, postoperative immobilization, age, risk factors, surgical time).

29.2.2 Risk Factors

Risk factors reported include history of VTE, age (>65 years), obesity, smoking, use of oral contraceptives, and venous insufficiency; operative risk factors were reported to be tourniquet time >1 hour, delay in surgery that causes immobilization; and postoperative risk factors such as knee immobilization [56, 60, 62, 63]. It has been stated that the most important risk factor for the development of DVT after arthroscopic ACL reconstruction is older age [60].

Although tourniquet usage is beneficial for visualization during arthroscopic ACL reconstruction, many studies have shown a significant correlation between VTE and tourniquet time. It is stated that stasis with tourniquet inflation combined with injury to the venous endothelium may result in the formation of a clot [64, 65]. Many studies have shown that when tourniquet time is >90 min, rates of DVT increased [51, 59, 66].

Hirota et al. reported that there was a significant correlation between amount of emboli (this is obtained by monitoring the right atrium with transesophageal echocardiography during the rocedure and calculating the total embolic area n the right atrium with a special formula) and ourniquet time in patients undergoing arthroscopic knee surgeries [65].

29.2.3 Thromboprophylaxis After ACL Reconstruction

It was stated in the American College of Chest Physicians (ACCP) guidelines that for patients undergoing knee arthroscopy who do not have additional thromboembolic risk factors, routine use thromboprophylaxis other than early mobilization is required (Grade 2B level of evidence) [67]. For patients undergoing arthroscopic knee surgery who have additional thromboembolic risk factors or following a complicated procedure, it was recommended to use thromboprophylaxis with low-molecular weight heparin (LMWH) (Grade 1B level of evidence) [67]. Ramos et al. stated in a Cochrane review study that thromboprophylaxis after knee arthroscopy in patients with unknown risk factors for thrombosis was not supported by any strong evidence [57]. Camporese et al. performed a randomized controlled study that compared LMWH and compressive stockings for thromboprophylaxis after knee arthroscopy and they found a significant reduction in symptomatic PE and DVT (symptomatic and asymptomatic) with the use of one week of LMWH after knee arthroscopy [68].

Aspirin is an oral prophylaxis method in which monitoring is not required and is cheaper than other medications. The new guidelines of American Academy of Orthopaedic Surgeons (AAOS) and ACCP have included a full-dose aspirin as an acceptable method for thromboprophylaxis for TJA [56, 69, 70]. Kaye et al. performed a study using aspirin in knee arthroscopy patients with low-risk to prevent VTE, and they stated that the use of aspirin was not warranted in the low-risk population undergoing knee arthroscopy [71]. Some studies have reported that the use of aspirin after immobilization of the lower extremity is insufficient to prevent the risk of VTE [72]. There is an unsignificant effect on VTE protection of aspirin compared to placebo, and LMWH has greater efficacy for preventing VTE as stated in the guidelines of ACCP and American College of Foot and Ankle Surgeons (ACFAS) [58, 73].

In a study by Adala et al., patients undergoing ACL reconstruction who were below the age of 45 years and had no risk factors of DVT were enrolled and thromboprophylaxis was not administered to any patients. Early postoperative routine physical therapy protocol was applied to each patient after surgery. DVT was detected in only 2 out of 112 (1.78%) patients by Doppler venous ultrasonography. The authors recommended thromboprophylaxis in patients with high risks for DVT and older than 45 years old [61]. In another prospective cohort study, it was reported that after arthroscopic ACL reconstruction without thromboprophylaxis, DVT rates were higher than other simple arthroscopic interventions and it was stated that further studies are needed to evaluate the need for thromboprophylaxis after ACL reconstruction in patients with risk factors [60].

Kraus-Schmidt et al. reported the incidence of VTE was 0.4% in a cohort study with a large group of ACL reconstruction patients. Patients with VTE had worse outcomes compared to patients without VTE. The most important risk factor for VTE was age. In conclusion, routine thromboprophylaxis is not recommended after ACL reconstruction, except in patients over 40 years of age [74]. In a recent study that aimed to evaluate the Swedish surgeons strategies for thromboprophylaxis administration following ACL reconstruction, it was reported that thromboprophylaxis was prescribed when the patient had known risk factors for VTE and less than 1/3 patients were being administered thromboprophylaxis [75].

As seen above, recommendations for thromboprophylaxis are still an ongoing debate. It is clearly stated that in the absence of risk factors for VTE, routine thromboprophylaxis is not recommended after ACL reconstruction [76].

29.2.4 Thromboembolism and TXA

The use of antifibrinolytic agents like TXA helps reduce intraoperative blood loss and need for transfusion; however, concerns remain about the risk of thromboembolism (TE) due to the known mechanism of action for TXA as a fibrin clot stabilizer. The CRASH-2 trauma trial showed that TXA had no effect on the rate of TE [23]. Poeran et al. stated that TXA usage in TKR and THR caused no increased risk of vascular occlusive events [77]. In a meta-analysis study performed by Yang et al. in which 15 studies and 877 patients were included, the authors found that TXA use was associated with DVT nor PE after TKA [78].

It was recommended in the Clinical Practice Guideline of American Association of Hip and Knee Surgeons (AAHKS) that use of oral, topical and i.v. TXA in primary TJA patients without VTE history does not increase the risk of VTE compared to placebo. It was stated that this recommendation was supported with "strong" data [48]. Also, for the primary TJA patients with high risk for thromboembolic events such as history of VTE, cerebrovascular accident, transient ischemic attack, vascular stent placement, and myocardial infarction, administration of TXA does not increase the risk of thromboembolic events [48, 79]. They advised to consider the sum of the risks and benefits of the use of TXA in these high-risk patients group. It was also stated in this guideline that there was no evidence to demonstrate TXA increases the risk of arterial thromboembolic events in primary TJA patients [45].

There are very few studies that evaluate the relationship between VTE and TXA administration. In a randomized controlled study by Karaaslan et al., TXA was applied to 53 of 105 patients who underwent arthroscopic ACL reconstruction, and it was reported that no patient developed DVT within three days postoperatively [29]. In another recent randomized controlled study, 80 patients who underwent ACL reconstruction received TXA and followed up for 3 months with no clinical symptoms of DVT recorded [30]. Nugent et al. reported no DVT was seen in the patient group that arthroscopic meniscectomy was performed and TXA was administered i.v. at induction prior to tourniquet inflation [80].

As stated above, it is clearly showed that TXA does not increase VTE in any patients groups that have no risk factors or history of thromboembolic events. However, more further randomized-controlled studies are required with larger groups of patients and longer follow-up.

29.3 Conclusion

Suction drains losing its popularity compared with the past and the routine use of it was not supported by literature.

Tourniquet usage: was seen that negative effects of tourniquet usage in arthroscopic ACL reconstruction are more than its positive effects.

Cold application and compression methods have beneficial effects.

The use of TXA shows promising effects on reducing hemarthrosis and drainage blood volumes improved functional outcomes postoperatively; however, what method of administration, timing and dosing of TXA is still controversial.

Age was stated as the most important risk factor for VTE; at the same time obesity, smoking, use of oral contraceptives, venous insufficiency; tourniquet time >1 hour and immobilization were reported as the other risk factors. Although recommendations for thromboprophylaxis are still an ongoing debate, it is clearly stated that in the absence of risk factors for VTE, there is no current indication for routine thromboprophylaxis after ACL reconstruction.

Various studies have shown that TXA does not increase VTE in any patients groups that have no risk factors or history of thromboembolic events.

References

1. Salzler MJ, Lin A, Miller CD, Herold S, Irrgang JJ, Harner CD. Complications after arthroscopic knee surgery. Am J Sports Med. 2014;42(2):292–6.
2. Small NC. Complications in arthroscopic meniscal surgery. Clin Sports Med. 1990;9(3):609–17.
3. Andres-Cano P, Godino M, Vides M, Guerado E. Postoperative complications of anterior cruciate

ligament reconstruction after ambulatory surgery. Rev Esp Cir Ortop Traumatol. 2015;59(3):157–64.

4. Brown DW, Curry CM, Ruterbories LM, Avery FL, Anson PS. Evaluation of pain after arthroscopically assisted anterior cruciate ligament reconstruction. Am J Sports Med. 1997;25(2):182–6.
5. Hooiveld M, Roosendaal G, Vianen M, van den Berg M, Bijlsma J, Lafeber F. Blood-induced joint damage: long-term effects in vitro and in vivo. J Rheumatol. 2003;30(2):339–44.
6. Bahl V, Goyal A, Jain V, Joshi D, Chaudhary D. Effect of haemarthrosis on the rehabilitation of anterior cruciate ligament reconstruction – single bundle versus double bundle. J Orthop Surg Res. 2013;8:5.
7. Simpson LA, Barrett JP Jr. Factors associated with poor results following arthroscopic subcutaneous lateral retinacular release. Clin Orthop Relat Res. 1984;186:165–71.
8. Browett JP, Gibbs AN, Copeland SA, Deliss LJ. The use of suction drainage in the operation of meniscectomy. J Bone Joint Surg Br. 1978;60-B(4):516–9.
9. McCormack RG, Greenhow RJ, Fogagnolo F, Shrier I. Intra-articular drain versus no drain after arthroscopic anterior cruciate ligament reconstruction: a randomized, prospective clinical trial. Arthroscopy. 2006;22(8):889–93.
10. Grassi A, Carulli C, Innocenti M, Mosca M, Zaffagnini S, Bait C, et al. New trends in anterior cruciate ligament reconstruction: a systematic review of National Surveys of the last 5 years. Joints. 2018;6(3):177–87.
11. Straw R, Colclough K, Geutjens GG. Arthroscopically assisted ACL reconstruction. Is a drain necessary? Knee. 2003;10(3):283–5.
12. Shamim S, Kushwaha SS, Shantanu K, Maurya G, Arshad I. Assessment of an intra-articular drain in arthroscopy-assisted anterior cruciate ligament reconstruction. Cureus. 2021;13(8):e16928. https://doi.org/10.7759/cureus.16928.
13. Kuo LT, Yu PA, Chen CL, Hsu WH, Chi CC. Tourniquet use in arthroscopic anterior cruciate ligament reconstruction: a systematic review and meta-analysis of randomised controlled trials. BMC Musculoskelet Disord. 2017;18(1):358.
14. Reda W, ElGuindy AMF, Zahry G, Faggal MS, Karim MA. Anterior cruciate ligament reconstruction; is a tourniquet necessary? A randomized controlled trial. Knee Surg Sports Traumatol Arthrosc. 2016;24(9):2948–52.
15. Wu H, Chen LX, Li YL, Wu Q, Wu QL, Ning GZ, et al. Tourniquet used in anterior cruciate ligament reconstruction: a system review. Eur J Orthop Surg Traumatol. 2014;24(6):999–1003.
16. Tierney GS, Wright RW, Smith JP, Fischer DA. Anterior cruciate ligament reconstruction as an outpatient procedure. Am J Sports Med. 1995;23(6):755–6.
17. Dervin GF, Taylor DE, Keene GC. Effects of cold and compression dressings on early postoperative outcomes for the arthroscopic anterior cruciate ligament reconstruction patient. J Orthop Sports Phys Ther. 1998;27(6):403–6.
18. Ruffilli A, Buda R, Castagnini F, Di Nicolantonio D, Evangelisti G, Giannini S, et al. Temperature-controlled continuous cold flow device versus traditional icing regimen following anterior cruciate ligament reconstruction: a prospective randomized comparative trial. Arch Orthop Trauma Surg. 2015;135(10):1405–10.
19. Astedt B. Clinical pharmacology of tranexamic acid. Scand J Gastroenterol Suppl. 1987;137:22–5.
20. Hunt BJ. The current place of tranexamic acid in the management of bleeding. Anaesthesia. 2015;70(Suppl 1):50–3. e18
21. Mohri H. High dose of tranexamic acid for treatment of severe menorrhagia in patients with von Willebrand disease. J Thromb Thrombolysis. 2002;14(3):255–7.
22. Nuvvula S, Gaddam KR, Kamatham R. Efficacy of tranexamic acid mouthwash as an alternative for factor replacement in gingival bleeding during dental scaling in cases of hemophilia: a randomized clinical trial. Contemp Clin Dent. 2014;5(1):49–53.
23. Collaborators C-T, Shakur H, Roberts I, Bautista R, Caballero J, Coats T, et al. Effects of tranexamic acid on death, vascular occlusive events, and blood transfusion in trauma patients with significant haemorrhage (CRASH-2): a randomised, placebo-controlled trial. Lancet. 2010;376(9734):23–32.
24. Ker K, Prieto-Merino D, Roberts I. Systematic review, meta-analysis and meta-regression of the effect of tranexamic acid on surgical blood loss. Br J Surg. 2013;100(10):1271–9.
25. Hiippala ST, Strid LJ, Wennerstrand MI, Arvela JV, Niemela HM, Mantyla SK, et al. Tranexamic acid radically decreases blood loss and transfusions associated with total knee arthroplasty. Anesth Analg. 1997;84(4):839–44.
26. Karaaslan F, Karaoglu S, Mermerkaya MU, Baktir A. Reducing blood loss in simultaneous bilateral total knee arthroplasty: combined intravenous-intra-articular tranexamic acid administration. A prospective randomized controlled trial. Knee. 2015;22(2):131–5.
27. Gianakos AL, Hurley ET, Haring RS, Yoon RS, Liporace FA. Reduction of blood loss by tranexamic acid following total hip and knee arthroplasty: a meta-analysis. JBJS Rev. 2018;6(5):e1.
28. El Beheiry H, Lubberdink A, Clements N, Dihllon K, Sharma V. Tranexamic acid administration to older patients undergoing primary total hip arthroplasty conserves hemoglobin and reduces blood loss. Can J Surg. 2018;61(3):177–84.
29. Karaaslan F, Karaoglu S, Yurdakul E. Reducing intra-articular hemarthrosis after arthroscopic anterior cruciate ligament reconstruction by the administration of Intravenous tranexamic acid: a prospective, randomized controlled trial. Am J Sports Med. 2015;43(11):2720–6.

30. Felli L, Revello S, Burastero G, Gatto P, Carletti A, Formica M, et al. Single intravenous administration of tranexamic acid in anterior cruciate ligament reconstruction to reduce postoperative hemarthrosis and increase functional outcomes in the early phase of postoperative rehabilitation: a randomized controlled trial. Arthroscopy. 2019;35(1):149–57.
31. Wang H, Shen B, Zeng Y. Comparison of topical versus intravenous tranexamic acid in primary total knee arthroplasty: a meta-analysis of randomized controlled and prospective cohort trials. Knee. 2014;21(6):987–93.
32. Zhang P, Liang Y, Chen P, Fang Y, He J, Wang J. Intravenous versus topical tranexamic acid in primary total hip replacement: a meta-analysis. Medicine (Baltimore). 2016;95(50):e5573.
33. Alshryda S, Mason J, Sarda P, Nargol A, Cooke N, Ahmad H, et al. Topical (intra-articular) tranexamic acid reduces blood loss and transfusion rates following total hip replacement: a randomized controlled trial (TRANX-H). J Bone Joint Surg Am. 2013;95(21):1969–74.
34. Alshryda S, Mason J, Vaghela M, Sarda P, Nargol A, Maheswaran S, et al. Topical (intra-articular) tranexamic acid reduces blood loss and transfusion rates following total knee replacement: a randomized controlled trial (TRANX-K). J Bone Joint Surg Am. 2013;95(21):1961–8.
35. Alshryda S, Sukeik M, Sarda P, Blenkinsopp J, Haddad FS, Mason JM. A systematic review and meta-analysis of the topical administration of tranexamic acid in total hip and knee replacement. Bone Joint J. 2014;96-B(8):1005–15.
36. Liu X, Liu J, Sun G. A comparison of combined intravenous and topical administration of tranexamic acid with intravenous tranexamic acid alone for blood loss reduction after total hip arthroplasty: a meta-analysis. Int J Surg. 2017;41:34–43.
37. Lin C, Qi Y, Jie L, Li HB, Zhao XC, Qin L, et al. Is combined topical with intravenous tranexamic acid superior than topical, intravenous tranexamic acid alone and control groups for blood loss controlling after total knee arthroplasty: a meta-analysis. Medicine (Baltimore). 2016;95(51):e5344.
38. Ye W, Liu Y, Liu WF, Li XL, Fei Y, Gao X. Comparison of efficacy and safety between oral and intravenous administration of tranexamic acid for primary total knee/hip replacement: a meta-analysis of randomized controlled trial. J Orthop Surg Res. 2020;15(1):21.
39. Yuan X, Wang J, Wang Q, Zhang X. Synergistic effects of intravenous and intra-articular tranexamic acid on reducing hemoglobin loss in revision total knee arthroplasty: a prospective, randomized, controlled study. Transfusion. 2018;58(4):982–8.
40. Wang D, Wang HY, Cao C, Li LL, Meng WK, Pei FX, et al. Tranexamic acid in primary total knee arthroplasty without tourniquet: a randomized, controlled trial of oral versus intravenous versus topical administration. Sci Rep. 2018;8(1):13579.
41. Zhao H, Xiang M, Xia Y, Shi X, Pei FX, Kang P. Efficacy of oral tranexamic acid on blood loss in primary total hip arthroplasty using a direct anterior approach: a prospective randomized controlled trial. Int Orthop. 2018;42(11):2535–42.
42. Cao G, Xie J, Huang Z, Huang Q, Chen G, Lei Y, et al. Efficacy and safety of multiple boluses of oral versus intravenous tranexamic acid at reducing blood loss after primary total knee arthroplasty without a tourniquet: a prospective randomized clinical trial. Thromb Res. 2018;171:68–73.
43. Wong J, Abrishami A, El Beheiry H, Mahomed NN, Roderick Davey J, Gandhi R, et al. Topical application of tranexamic acid reduces postoperative blood loss in total knee arthroplasty: a randomized, controlled trial. J Bone Joint Surg Am. 2010;92(15):2503–13.
44. Maniar RN, Kumar G, Singhi T, Nayak RM, Maniar PR. Most effective regimen of tranexamic acid in knee arthroplasty: a prospective randomized controlled study in 240 patients. Clin Orthop Relat Res. 2012;470(9):2605–12.
45. Sarzaeem MM, Razi M, Kazemian G, Moghaddam ME, Rasi AM, Karimi M. Comparing efficacy of three methods of tranexamic acid administration in reducing hemoglobin drop following total knee arthroplasty. J Arthroplast. 2014;29(8):1521–4.
46. Lin PC, Hsu CH, Huang CC, Chen WS, Wang JW. The blood-saving effect of tranexamic acid in minimally invasive total knee replacement: is an additional preoperative injection effective? J Bone Joint Surg Br. 2012;94(7):932–6.
47. Barrachina B, Lopez-Picado A, Remon M, Fondarella A, Iriarte I, Bastida R, et al. Tranexamic acid compared with placebo for reducing total blood loss in hip replacement surgery: a randomized clinical trial. Anesth Analg. 2016;122(4):986–95.
48. Fillingham YA, Ramkumar DB, Jevsevar DS, Yates AJ, Bini SA, Clarke HD, et al. Tranexamic acid in total joint arthroplasty: the endorsed clinical practice guides of the American Association of Hip and Knee Surgeons, American Society of Regional Anesthesia and Pain Medicine, American Academy of Orthopaedic Surgeons, Hip Society, and Knee Society. Reg Anesth Pain Med. 2019;44(1):7–11.
49. Chiang ER, Chen KH, Wang ST, Ma HL, Chang MC, Liu CL, et al. Intra-articular injection of tranexamic acid reduced postoperative hemarthrosis in arthroscopic anterior cruciate ligament reconstruction: a prospective randomized study. Arthroscopy. 2019;35(7):2127–32.
50. Parker JD, Lim KS, Kieser DC, Woodfield TBF, Hooper GJ. Is tranexamic acid toxic to articular cartilage when administered topically? What is the safe dose? Bone Joint J. 2018;100-B(3):404–12.
51. Janssen RP, Reijman M, Janssen DM, van Mourik JB. Arterial complications, venous thromboembolism and deep venous thrombosis prophylaxis after anterior cruciate ligament reconstruction: a systematic review. World J Orthop. 2016;7(9):604–17.
52. Janssen RP, Scheltinga MR, Sala HA. Pseudoaneurysm of the popliteal artery after anterior cruciate ligament reconstruction with bicortical tibial screw fixation. Arthroscopy. 2004;20(1):E4–6.

53. Janssen RP, Reijman M, Janssen DM. Vascular complications after ACL reconstruction and DVT prophylaxis. In: CC P, editor. The anterior cruciate ligament reconstruction and basic science. Philadelphia: Saunders, An Imprint of Elsevier; 2015.
54. Janssen RP, Sala HA. Embolism of the popliteal artery after anterior cruciate ligament reconstruction: a case report and literature review. Knee Surg Sports Traumatol Arthrosc. 2007;15(12): 1449–51.
55. Panigrahi R, Mahapatra AK, Palo N, Priyadarshi A. Thrombotic popliteal block following simultaneous bi-cruciate single-stage arthroscopic reconstruction in a neglected multiligamentous knee injury in a smoker. BMJ Case Rep. 2015;2015
56. Erickson BJ, Saltzman BM, Campbell KA, Fillingham YA, Harris JD, Gupta AK, et al. Rates of deep venous thrombosis and pulmonary embolus after anterior cruciate ligament reconstruction: a systematic review. Sports Health. 2015;7(3):261–6.
57. Ramos J, Perrotta C, Badariotti G, Berenstein G. Interventions for preventing venous thromboembolism in adults undergoing knee arthroscopy. Cochrane Database Syst Rev. 2008;5(4):CD005259.
58. Falck-Ytter Y, Francis CW, Johanson NA, Curley C, Dahl OE, Schulman S, et al. Prevention of VTE in orthopedic surgery patients: antithrombotic therapy and prevention of thrombosis, 9th ed: American College of Chest Physicians Evidence-Based Clinical Practice Guidelines. Chest. 2012;141(2 Suppl):e278S–325S.
59. Demers C, Marcoux S, Ginsberg JS, Laroche F, Cloutier R, Poulin J. Incidence of venographically proved deep vein thrombosis after knee arthroscopy. Arch Intern Med. 1998;158(1):47–50.
60. Struijk-Mulder MC, Ettema HB, Verheyen CC, Buller HR. Deep vein thrombosis after arthroscopic anterior cruciate ligament reconstruction: a prospective cohort study of 100 patients. Arthroscopy. 2013;29(7):1211–6.
61. Adala R, Anand A, Kodikal G. Deep vein thrombosis and thromboprophylaxis in arthroscopic anterior cruciate ligament reconstruction. Indian J Orthop. 2011;45(5):450–3.
62. Oberstar J, Pereira C. Venous thromboembolic event after elective anterior cruciate ligament reconstruction: a case report. Curr Sports Med Rep. 2018;17(2):48–53.
63. Delis KT, Hunt N, Strachan RK, Nicolaides AN. Incidence, natural history and risk factors of deep vein thrombosis in elective knee arthroscopy. Thromb Haemost. 2001;86(3):817–21.
64. Jarrett PM, Ritchie IK, Albadran L, Glen SK, Bridges AB, Ely M. Do thigh tourniquets contribute to the formation of intra-operative venous emboli? Acta Orthop Belg. 2004;70(3):253–9.
65. Hirota K, Hashimoto H, Kabara S, Tsubo T, Sato Y, Ishihara H, Matsuki A. The relationship between pneumatic tourniquet time and the amount of pulmonary emboli in patients undergoing knee arthroscopic surgeries. Anesth Analg. 2001;93(3):776–80. https://doi.org/10.1097/00000539-200109000-00044.
66. Dong JT, Wang X, Men XQ, Wang XF, Zheng XZ, Gao SJ. Incidence of deep venous thrombosis in Chinese patients undergoing arthroscopic knee surgery for cruciate ligament reconstruction. Knee Surg Sports Traumatol Arthrosc. 2015;23(12):3540–4.
67. Geerts WH, Bergqvist D, Pineo GF, Heit JA, Samama CM, Lassen MR, et al. Prevention of venous thromboembolism: American College of Chest Physicians Evidence-Based Clinical Practice Guidelines (8th Edition). Chest. 2008;133(6 Suppl):381S–453S.
68. Camporese G, Bernardi E, Prandoni P, Noventa F, Verlato F, Simioni P, et al. Low-molecular-weight heparin versus compression stockings for thromboprophylaxis after knee arthroscopy: a randomized trial. Ann Intern Med. 2008;149(2):73–82.
69. An VV, Phan K, Levy YD, Bruce WJ. Aspirin as thromboprophylaxis in hip and knee arthroplasty: a systematic review and meta-analysis. J Arthroplast. 2016;31(11):2608–16.
70. Ogonda L, Hill J, Doran E, Dennison J, Stevenson M, Beverland D. Aspirin for thromboprophylaxis after primary lower limb arthroplasty: early thromboembolic events and 90 day mortality in 11,459 patients. Bone Joint J. 2016;98-B(3):341–8.
71. Kaye ID, Patel DN, Strauss EJ, Alaia MJ, Garofolo G, Martinez A, et al. Prevention of venous thromboembolism after arthroscopic knee surgery in a low-risk population with the use of aspirin. A randomized trial. Bull Hosp Jt Dis. 2015;73(4):243–8.
72. Braithwaite I, Dunbar L, Eathorne A, Weatherall M, Beasley R. Venous thromboembolism rates in patients with lower limb immobilization after Achilles tendon injury are unchanged after the introduction of prophylactic aspirin: audit. J Thromb Haemost. 2016;14(2):331–5.
73. Fleischer AE, Abicht BP, Baker JR, Boffeli TJ, Jupiter DC, Schade VL. American College of Foot and Ankle Surgeons' clinical consensus statement: risk, prevention, and diagnosis of venous thromboembolism disease in foot and ankle surgery and injuries requiring immobilization. J Foot Ankle Surg. 2015;54(3):497–507.
74. Kraus Schmitz J, Lindgren V, Janarv PM, Forssblad M, Stalman A. Deep venous thrombosis and pulmonary embolism after anterior cruciate ligament reconstruction: incidence, outcome, and risk factors. Bone Joint J. 2019;101-B(1):34–40.
75. Ekdahl V, Stalman A, Forssblad M, Samuelsson K, Edman G, Kraus SJ. There is no general use of thromboprophylaxis and prolonged antibiotic prophylaxis in anterior cruciate ligament reconstruction: a nationwide survey of ACL surgeons in Sweden. Knee Surg Sports Traumatol Arthrosc. 2020;
76. Janssen RP, Sala HA. Fatal pulmonary embolism after anterior cruciate ligament reconstruction. Am J Sports Med. 2007;35(6):1000–2.
77. Poeran J, Rasul R, Suzuki S, Danninger T, Mazumdar M, Opperer M, et al. Tranexamic acid use and

postoperative outcomes in patients undergoing total hip or knee arthroplasty in the United States: retrospective analysis of effectiveness and safety. BMJ. 2014;349:g4829.
78. Yang ZG, Chen WP, Wu LD. Effectiveness and safety of tranexamic acid in reducing blood loss in total knee arthroplasty: a meta-analysis. J Bone Joint Surg Am. 2012;94(13):1153–9.
79. Fillingham YA, Ramkumar DB, Jevsevar DS, Yates AJ, Shores P, Mullen K, et al. The safety of tranexamic acid in total joint arthroplasty: a direct meta-analysis. J Arthroplast. 2018;33(10):3070–82 e1.
80. Nugent M, May JH, Parker JD, Kieser DC, Douglas M, Pereira R, et al. Does tranexamic acid reduce knee swelling and improve early function following arthroscopic meniscectomy? A double-blind randomized controlled trial. Orthop J Sports Med. 2019;7(8):2325967119866122.
81. Herzog MM, Marshall SW, Lund JL, Pate V, Mack CD, Spang JT. Trends in incidence of ACL reconstruction and concomitant procedures among commercially insured individuals in the United States, 2002–2014. Sports Health. 2018;10(6):523–31.

30 Prevention and Management of Infection in ACL Reconstruction

Anthony Truong and Christopher J. Vertullo

30.1 Incidence

Post-operative septic arthritis is a rare but potentially devastating complication of primary anterior cruciate ligament reconstruction (ACLR) [1, 2]. The incidence of septic arthritis ranges from 0.14 to 1.8% despite preoperative intravenous (IV) antibiotics [1, 3–6]. Autologous hamstring tendon grafts are at greater risk of infection (1.44%) when compared to bone patellar tendon bone grafts (0.49%), allografts (0.44%) and quadriceps tendon grafts (0.1%) [7–9].

30.2 Pathogen

The most common bacteria involved in septic arthritis following ACLR is *Coagulase Negative Staphylococci* (45%), such as *Staphylococcus epidermidis,* followed by *Staphylococcus aureus* (24%) [7, 10–12]. Other commonly reported pathogens include *Enterobacteriae, Propionibacterium acnes*, *Escherichia coli* and *Pseudomonas aeruginosa* [10, 13, 14]. A small number of devastating fungal infections (*Rhizopus and Candida* albicans) have been reported in the literature [10, 15].

30.3 Clinical Presentation

Patients with post-operative septic arthritis after an ACLR usually present with an increasingly painful knee joint with limited range of motion (ROM), worsening effusion, local erythema and fever usually over 38.5 °C [10]. Over 90% of intra-articular infections post ACLR develop in the acute (<2 weeks) or subacute (2 weeks–2 months) period [8, 10]. Less than 10% of patients developed an infection more than two months following their original surgery [8, 10].

Patients with chronic infections may present with persistent knee pain, swelling and instability which in rare cases have cultured *Propionibacterium acnes* at the time of revision ACLR [16]. Therefore, it is important to consider culturing the index ACL grafts for indolent infection when performing revision reconstruction.

30.4 Diagnosis

There is no gold standard diagnostic tool for diagnosis of deep post-operative joint infection after ACLR. However, prior authors have defined ACLR infection using the following criteria:

A. Truong
School of Medicine, Griffith University, Gold Coast, QLD, Australia

Princess Alexandra Hospital, Brisbane, QLD, Australia
e-mail: Anthony.truong@health.qld.gov.au

C. J. Vertullo (✉)
Center for Musculoskeletal Research Griffith University, Gold Coast, QLD, Australia

Knee Research Australia, Gold Coast, QLD, Australia

M. Karahan et al. (eds.), *Anterior Cruciate Ligament Injuries Throughout the Life Span*,
https://doi.org/10.1007/978-3-031-84501-7_30

A. Positive post-operative culture on aspiration or intra-operative biopsy specimen of a painful knee [17–19].

or

B. Negative aspiration culture with a raised or rising serial C-reactive protein (CRP), an elevated synovial white cell count (WCC) ≥25,000/μL and synovial polymorphonuclear (PMN) cells >90% with a painful effusion, with or without wound drainage, in the absence of a haemarthrosis [17].

or

C. Clinical signs in conjunction with obvious intra-operative findings of infection (purulent secretion, fibrinous deposition, thickening of the synovial membrane, compartment formation; e.g., in patients with previous antibiotic therapy and negative culture results) [18, 19].

A synovial WCC of >25,000 has a likelihood ratio (LR) of 2.9 for septic arthritis compared to a LR of 0.32 when the WCC is <25,000. Likewise, if there is >90% PMN cells in the synovial fluid, the LR is 3.4 compared to 0.34 for when there is <90% PMN cells [20]. The overall sensitivity of synovial fluid cultures and gram stains are 75–95% and 29–50%, respectively, for septic arthritis [20]. It has been reported that a CRP value of 41 (normal <5) was a more sensitive (94.1%) and specific (97.6%) indicator of infection than a erythrocyte sedimentation rate (ESR) value of 32 mm/h (normal <20), which had a sensitivity of 91.2% and specificity of 80.5% [10]. It is recommended to obtain a number of intra-operative samples that include the synovial lining, graft, femoral and tibial tunnels [18, 21]. As extrapolated from the prosthetic joint infection literature, three to five samples have shown to increase the diagnostic yield and facilitate the interpretation of results [21, 22]. Radiological imaging has a limited role in the diagnosis of infection.

30.5 Prevention

Prevention of surgical site infections (SSIs) requires maximising the host to fight infection and minimising the bacterial load. In order to maximise the host, the World Health Organization (WHO) recommendations are to maintain normothermia and have a fraction of inspired oxygen of 80% intra-operatively and two to six hours post-operatively [23]. The panel also suggests intensive peri-operative blood glucose control to prevent SSIs [23]. The World Health Organization and current literature recommend patients cease tobacco smoking at least a minimum of four weeks prior to surgery [24, 25]. Meticulous surgical technique, which involves gentle handling of soft tissue and careful wound closure, should always be performed.

To minimise the patient's bacterial load, pre-operative bacterial decolonisation for carriers of *S. aureus* involves mupirocin 2% ointment with or without a combination of chlorhexidine body wash [23]. Perioperative surgical site skin preparation should be done by clipping hair around the surgical area rather than shaving and using an alcoholic chlorhexidine-based antiseptic solution [23]. Surgical antibiotic prophylaxis, usually a first-generation cephalosporin for those without a penicillin allergy, should be administered intravenously within 120 minutes prior to surgical incision [23].

However, the concentration of IV antibiotics is usually below the minimum inhibitory concentration required to prevent deep joint infections, thus limiting its effectiveness [26]. The 'vancomycin wrap' technique can be used to further eradicate bacterial colonisation of the poorly vascularised graft [14]. First described by Vertullo et al., it involves wrapping the harvested hamstring tendon autograft in a sterile gauze swab saturated with 5 mg/mL vancomycin solution prior to implantation in addition to pre-operative IV antibiotics and the aforementioned recommendations [27]. Numerous cohort studies have reported a joint infection rate of 0% with the 'vancomycin wrap' in primary ACLR [11, 17, 28–31]. This technique has proven to be highly cost-effective in infection prevention [32]. There is emerging evidence that 'vancomycin wrap' technique is effective in reducing infection for bone-patella tendon-bone, quadriceps tendon grafts and in the revision ACLR setting [11, 28, 33]. To date, there have been no randomised controlled trials of the 'vancomycin wrap' technique.

30.6 Management

There is no consensus on an optimal management protocol for the treatment of septic arthritis following an ACLR [10, 18]. However, the current literature recommends management involving urgent arthroscopic debridement for source control combined with four to six weeks of IV antibiotic therapy and two to four weeks of oral antibiotics [3, 10]. The usual antibiotic choice is a first-generation cephalosporin; however, empirical and directed antibiotic therapy should be guided by culture sensitivities and each hospital's local guidelines. Oral antibiotics can be stopped when the CRP normalised twice within an interval of a minimum of 14 days [10]. Post-operatively, weightbearing and range of motion are unrestricted to enable participation in ACL rehabilitation.

A diagnostic and treatment algorithm has been proposed by Phegan et al. in Fig. 30.1 [17].

It is estimated that most patients require 1.5 arthroscopic procedures [3, 10, 34]. Repeated arthroscopic debridements are required in approximately one-third of cases [34]. Despite this, the average success rate of arthroscopic debridement in eradicating infection is 85% [34].

Current literature supports the retention of functional non-infected grafts, which is possible in 80% of cases with early surgical intervention [3, 10, 12, 35, 36]. Indications for the removal of the ACL graft include the presence of one and/or a combination of the below three findings [10, 18]: (1) elongated and non-functional graft during clinical examination and arthroscopic evaluation; (2) severe infection of the graft with thick purulent exudation; and/or (3) persistence of knee infection after two to three debridements.

A small percentage of cases (7.2%) will have persistent bacterial infection [10, 37]. There is no consensus on the optimal surgical algorithm for the treatment of chronic or persistent post-operative ACLR septic arthritis [38]. There are only case series in the literature given the low incidence of infection [37–39]. Treatment will usually consist of a two-stage approach. The first stage aims for infection eradication via an open surgical approach, which consists of an arthrotomy, washout, complete synovectomy, removal of graft and implants and curettage of the femoral and tibial tunnels [10, 37]. Adjacent tunnel osteomyelitis can be a cause for persistent infection [37, 38]. Some have advocated for the use of intra-articular deposition of gentamicin-containing polymethylmethacrylate (PMMA) beads (Septopal, Biomet BioMaterials, Darmstadt, Germany) for 7–10 days to aid in infection eradication [39]. Patients will then have a period of intravenous and oral antibiotic therapy, serial clinical and inflammatory marker monitoring to ensure infection eradication. There may be an intermediatory stage for cases with severe bone tunnel loss that will need bone grafting after infection eradication but before re-implantation of a new graft [37].

The second stage will then involve re-implantation of a new ACL graft using the various graft choices available. Criteria to proceed to the second stage are normal knee motion, no knee joint effusion and normal serum laboratory results [38]. Timing of revision ACLR should be determined on a case-by-case basis and surgeon preference. There is a variation in timing of revision ACLR from 6 weeks after completion of antibiotics through to 12 months or more, in the current literature [4, 10].

Studies have highlighted that infections involving a delayed diagnosis (>7 days), *Staphylococcus aureus*, more than two arthroscopic debridements and allografts, were all associated with a higher rate of graft removal and/or revision ACLR [12, 34, 40].

Serial serum inflammatory markers should be performed during course of treatment. It is not routine to perform repeat knee aspiration or tissue sampling unless there is concern for ongoing infection or prior to a second-stage revision ACLR to ensure infection eradication.

Non-operative management which includes antibiotic therapy with or without multiple irrigations of the joint is not recommended based on the high failure rate to convert to surgical inter-

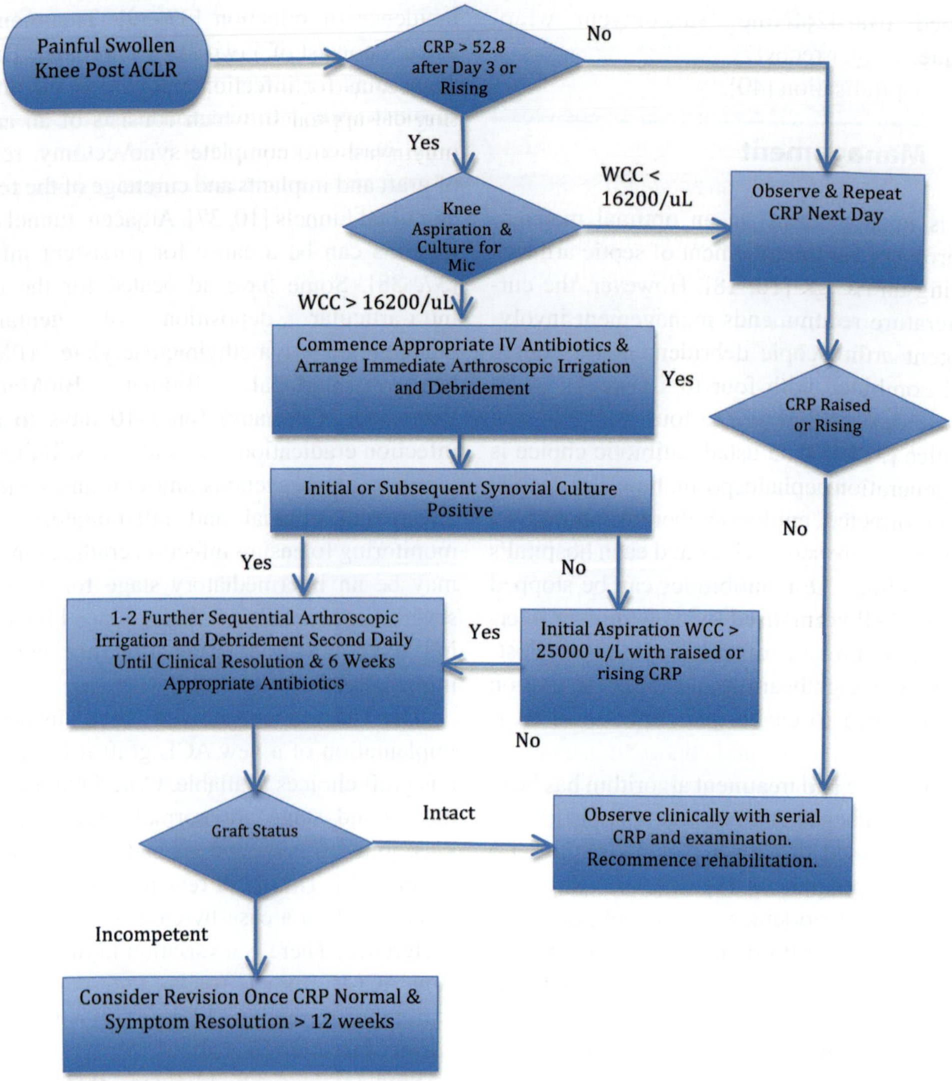

Fig. 30.1 Infected ACLR diagnosis and management flowchart. Abbreviations: ACLR = Anterior cruciate ligament reconstruction. WCC = White cell count. CRP = C reactive protein. IV = Intravenous antibiotics

vention [10, 41, 42]. Non-operative management can lead to longer recovery, articular damage and greater hospitalisation [40].

30.7 Patient Outcomes

Undiagnosed and untreated deep joint infections can lead to ACL graft failure, articular cartilage damage, degenerative arthritis, joint dysfunction and osteomyelitis [7, 10]. Patients can be effectively treated with no evidence of infection reoccurrence, maintain knee stability and achieve near complete ROM [3, 12, 36]. Patient can have flexion and extension deficits of 5.8 and 1.0 degrees respectively, and laxity of 2 mm on KT-100 arthrometry compared to the contralateral knee [36]. The Lysholm scores are not significantly different between patients who are treated for septic arthritis compared to those without septic arthritis, 81 vs. 87 [36, 43]. However, there have been significantly lower Tegner (5.1 vs. 6.5) and International Knee Documentation Committee (IKDC) scores (68.2 vs. 85) in those with infection compared to those without [3, 36, 43]. There is a 67% rate of return to prior level of play in patients with infection after ACLR which is similar to the average rate of 65% [36, 44]. There is limited evidence that effectively treated post-operative septic arthritis following ACLR leads to early degenerative changes on radiographs and magnetic resonance imaging [36, 41, 43].

30.8 Conclusion

Deep joint infection post ACLR is a rare but devastating complication. Hamstring autografts have the highest incidence of post-operative septic arthritis. The most common bacteria involved in septic arthritis following ACLR are *Staphylococcus epidermidis* and *Staphylococcus aureus*. Patients often will present with acute severe knee pain, limited range of motion and inability to weight bear within two weeks of surgery. Patients with chronic infections may have persistent knee pain and swelling. Diagnosis is made usually with a positive tissue and/or synovial fluid sample. Infection prevention includes adherence to the WHO guidelines for the prevention of surgical site infection and use of vancomycin pre-soaking of the graft with intravenous prophylactic antibiotics. Management usually consists of urgent arthroscopic debridement and irrigation, graft retention and a period of directed antibiotic therapy. Persistent infection may require graft removal and a staged revision ACL reconstruction. With early intervention, patients can effectively be treated with no evidence of infection reoccurrence, maintain knee stability and achieve near complete range of motion. Worse outcomes are associated with late presentation and delayed management.

References

1. Judd MAJD, Bottoni LTCC, Kim D, Burke CPTM, Hooker MAJS. Infections following arthroscopic anterior cruciate ligament reconstruction. Arthrosc J Arthrosc Relat Surg. 2006;22:375–84. https://doi.org/10.1016/j.arthro.2005.12.002.
2. Brophy RH, Wright RW, Huston LJ, Nwosu SK, MOON Knee Group, Spindler KP. Factors associated with infection following anterior cruciate ligament reconstruction. J Bone Joint Surg Am. 2015;97:450–4. https://doi.org/10.2106/JBJS.N.00694.
3. Torres-Claramunt R, Pelfort X, Erquicia J, Gil-González S, Gelber PE, Puig L, et al. Knee joint infection after ACL reconstruction: prevalence, management and functional outcomes. Knee Surg Sports Traumatol Arthrosc. 2013;21:2844–9. https://doi.org/10.1007/s00167-012-2264-3.
4. Burks RT, Friederichs MG, Fink B, Luker MG, West HS, Greis PE. Treatment of postoperative anterior cruciate ligament infections with graft removal and early reimplantation. Am J Sports Med. 2003;31:414–8. https://doi.org/10.1177/03635465030310031501.
5. McAllister DR, Parker RD, Cooper AE, Recht MP, Abate J. Outcomes of postoperative septic arthritis after anterior cruciate ligament reconstruction. Am J Sports Med. 1999;27:562–70.
6. Van Tongel A, Stuyck J, Bellemans J, Vandenneucker H. Septic arthritis after arthroscopic anterior cruciate ligament reconstruction: a retrospective analysis of incidence, management and outcome. Am J Sports Med. 2007;35:1059–63. https://doi.org/10.1177/0363546507299443.
7. Gobbi A, Karnatzikos G, Chaurasia S, Abhishek M, Bulgherhoni E, Lane J. Postoperative infection after anterior cruciate ligament reconstruction. Sports

Health Multidiscip Approach. 2016;8:187–9. https://doi.org/10.1177/1941738115618638.
8. Barker JU, Drakos MC, Maak TG, Warren RF, Williams RJ, Allen AA. Effect of graft selection on the incidence of postoperative infection in anterior cruciate ligament reconstruction. Am J Sports Med. 2010;38:281–6. https://doi.org/10.1177/0363546509346414.
9. Özbek EA, Dadoo S, Chang A, Simonian L, Sebastiani R, Herman Z, et al. Rates of septic arthritis after ACL reconstruction: a single-center analysis highlighting quadriceps tendon grafts. Am J Sports Med. 2023;51:1708–14. https://doi.org/10.1177/03635465231165509.
10. Wang C, Lee YHD, Siebold R. Recommendations for the management of septic arthritis after ACL reconstruction. Knee Surg Sports Traumatol Arthrosc Off J ESSKA. 2014;22:2136–44. https://doi.org/10.1007/s00167-013-2648-z.
11. Pérez-Prieto D, Torres-Claramunt R, Gelber PE, Shehata TMA, Pelfort X, Monllau JC. Autograft soaking in vancomycin reduces the risk of infection after anterior cruciate ligament reconstruction. Knee Surg Sports Traumatol Arthrosc. 2014;24:2724. https://doi.org/10.1007/s00167-014-3438-y.
12. Pogorzelski J, Themessl A, Achtnich A, Fritz EM, Wörtler K, Imhoff AB, et al. Septic arthritis after anterior cruciate ligament reconstruction: how important is graft salvage? Am J Sports Med. 2018;46:2376–83. https://doi.org/10.1177/0363546518782433.
13. Jefferies JG, Aithie JMS, Spencer SJ. Vancomycin-soaked wrapping of harvested hamstring tendons during anterior cruciate ligament reconstruction. A review of the 'vancomycin wrap'. Knee. 2019;26:524–9. https://doi.org/10.1016/j.knee.2019.04.008.
14. Pérez-Prieto D, Portillo ME, Torres-Claramunt R, Pelfort X, Hinarejos P, Monllau JC. Contamination occurs during ACL graft harvesting and manipulation, but it can be easily eradicated. Knee Surg Sports Traumatol Arthrosc Off J ESSKA. 2018;26:558–62. https://doi.org/10.1007/s00167-017-4735-z.
15. Muscolo DL, Carbo L, Aponte-Tinao LA, Ayerza MA, Makino A. Massive bone loss from fungal infection after anterior cruciate ligament arthroscopic reconstruction. Clin Orthop Relat Res. 2009;467:2420–5. https://doi.org/10.1007/s11999-009-0714-0.
16. DiBartola AC, Swank KR, Flanigan DC. Anterior cruciate ligament reconstruction complicated by Propionibacterium acnes infection: case series. Phys Sportsmed. 2018;46:273–8. https://doi.org/10.1080/00913847.2018.1463062.
17. Phegan M, Grayson JE, Vertullo CJ. No infections in 1300 anterior cruciate ligament reconstructions with vancomycin pre-soaking of hamstring grafts. Knee Surg Sports Traumatol Arthrosc. 2015;24:2729. https://doi.org/10.1007/s00167-015-3558-z.
18. Mouzopoulos G, Fotopoulos V, Tzurbakis M. Septic knee arthritis following ACL reconstruction: a systematic review. Knee Surg Sports Traumatol Arthrosc Off J ESSKA. 2009;17:1033–42. https://doi.org/10.1007/s00167-009-0793-1.
19. Schuster P, Schulz M, Immendoerfer M, Mayer P, Schlumberger M, Richter J. Septic arthritis after arthroscopic anterior cruciate ligament reconstruction: evaluation of an arthroscopic graft-retaining treatment protocol. Am J Sports Med. 2015;43:3005–12. https://doi.org/10.1177/0363546515603054.
20. Margaretten ME, Kohlwes J, Moore D, Bent S. Does this adult patient have septic arthritis? JAMA. 2007;297:1478–88. https://doi.org/10.1001/jama.297.13.1478.
21. Renz N, Madjarevic T, Ferrari M, Becker R, Ravn C, Vogely C, et al. Recommendations on diagnosis and antimicrobial treatment of infections after anterior cruciate ligament reconstruction (ACL-R) endorsed by ESSKA and EBJIS. J Infect. 2023;86:543–51. https://doi.org/10.1016/j.jinf.2023.03.021.
22. Dudareva M, Barrett L, Figtree M, Scarborough M, Watanabe M, Newnham R, et al. Sonication versus tissue sampling for diagnosis of prosthetic joint and other orthopedic device-related infections. J Clin Microbiol. 2018;56:56. https://doi.org/10.1128/JCM.00688-18.
23. World Health Organization. WHO guidelines to the prevention of SSI. WHO; n.d.
24. Cancienne JM, Gwathmey FW, Miller MD, Werner BC. Tobacco use is associated with increased complications after anterior cruciate ligament reconstruction. Am J Sports Med. 2016;44:99–104. https://doi.org/10.1177/0363546515610505.
25. World Health Organization. Tobacco & postsurgical outcomes. WHO; 2020.
26. Katz LM, Battaglia TC, Patino P, Reichmann W, Hunter DJ, Richmond JC. A retrospective comparison of the incidence of bacterial infection following anterior cruciate ligament reconstruction with autograft versus allograft. Arthrosc J Arthrosc Relat Surg Off Publ Arthrosc Assoc N Am Int Arthrosc Assoc. 2008;24:1330–5. https://doi.org/10.1016/j.arthro.2008.07.015.
27. Vertullo CJ, Quick M, Jones A, Grayson JE. A surgical technique using presoaked vancomycin hamstring grafts to decrease the risk of infection after anterior cruciate ligament reconstruction. Arthrosc J Arthrosc Relat Surg. 2012;28:337–42. https://doi.org/10.1016/j.arthro.2011.08.301.
28. Offerhaus C, Balke M, Hente J, Gehling M, Blendl S, Höher J. Vancomycin pre-soaking of the graft reduces postoperative infection rate without increasing risk of graft failure and arthrofibrosis in ACL reconstruction. Knee Surg Sports Traumatol Arthrosc. 2019;27:3014. https://doi.org/10.1007/s00167-018-5323-6.
29. Figueroa D, Figueroa F, Calvo R, Lopez M, Goñi I. Presoaking of hamstring autografts in vancomycin decreases the occurrence of infection following primary anterior cruciate ligament reconstruction. Orthop J Sports Med. 2019;7:232596711987103. https://doi.org/10.1177/2325967119871038.

30. Schuster P, Schlumberger M, Mayer P, Eichinger M, Geßlein M, Reddemann F, et al. Soaking of the graft in vancomycin dramatically reduces the incidence of postoperative septic arthritis after anterior cruciate ligament reconstruction. Knee Surg Sports Traumatol Arthrosc. 2020;28:2587. https://doi.org/10.1007/s00167-020-05882-0.
31. Baron JE, Shamrock AG, Cates WT, Cates RA, An Q, Wolf BR, et al. Graft preparation with intraoperative vancomycin decreases infection after ACL reconstruction: a review of 1,640 cases. J Bone Joint Surg. 2019;101:2187–93. https://doi.org/10.2106/JBJS.19.00270.
32. Truong AP, Pérez-Prieto D, Byrnes J, Monllau JC, Vertullo CJ. Vancomycin soaking is highly cost-effective in primary ACLR infection prevention: a cost-effectiveness study. Am J Sports Med. 2022;50:922–31. https://doi.org/10.1177/03635465211073338.
33. Schuster P, Schlumberger M, Mayer P, Eichinger M, Geßlein M, Richter J. Soaking of autografts in vancomycin is highly effective in preventing postoperative septic arthritis after revision anterior cruciate ligament reconstruction. Knee Surg Sports Traumatol Arthrosc. 2020;28:1154–8. https://doi.org/10.1007/s00167-019-05820-9.
34. Saper M, Stephenson K, Heisey M. Arthroscopic irrigation and debridement in the treatment of septic arthritis after anterior cruciate ligament reconstruction. Arthrosc J Arthrosc Relat Surg Off Publ Arthrosc Assoc N Am Int Arthrosc Assoc. 2014;30:747–54. https://doi.org/10.1016/j.arthro.2014.02.015.
35. Kusnezov N, Eisenstein ED, Dunn JC, Wey AJ, Peterson DR, Waterman BR. Anterior cruciate ligament graft removal versus retention in the setting of septic arthritis after reconstruction: a systematic review and expected value decision analysis. Arthrosc J Arthrosc Relat Surg. 2018;34:967–75. https://doi.org/10.1016/j.arthro.2017.08.246.
36. Makhni EC, Steinhaus ME, Mehran N, Schulz BS, Ahmad CS. Functional outcome and graft retention in patients with septic arthritis after anterior cruciate ligament reconstruction: a systematic review. Arthrosc J Arthrosc Relat Surg Off Publ Arthrosc Assoc N Am Int Arthrosc Assoc. 2015;31:1392–401. https://doi.org/10.1016/j.arthro.2014.12.026.
37. Zalavras CG, Patzakis MJ, Tibone J, Weisman N, Holtom P. Treatment of persistent infection after anterior cruciate ligament surgery. Clin Orthop. 2005;439:52–5. https://doi.org/10.1097/01.blo.0000181499.49740.e5.
38. Hantes ME, Raoulis VA, Doxariotis N, Drakos A, Karachalios T, Malizos KN. Management of septic arthritis after arthroscopic anterior cruciate ligament reconstruction using a standard surgical protocol. Knee. 2017;24:588–93. https://doi.org/10.1016/j.knee.2017.02.007.
39. Schulz AP, Götze S, Schmidt HGK, Jürgens C, Faschingbauer M. Septic arthritis of the knee after anterior cruciate ligament surgery: stage-adapted treatment regimen. Am J Sports Med. 2007;35:1064–9. https://doi.org/10.1177/0363546507299744.
40. Wang C, Ao Y, Wang J, Hu Y, Cui G, Yu J. Septic arthritis after arthroscopic anterior cruciate ligament reconstruction: a retrospective analysis of incidence, presentation, treatment, and cause. Arthrosc J Arthrosc Relat Surg. 2009;25:243–9. https://doi.org/10.1016/j.arthro.2008.10.002.
41. Monaco E, Maestri B, Labianca L, Speranza A, Vadalà A, Iorio R, et al. Clinical and radiological outcomes of postoperative septic arthritis after anterior cruciate ligament reconstruction. J Orthop Sci Off J Jpn Orthop Assoc. 2010;15:198–203. https://doi.org/10.1007/s00776-009-1447-3.
42. Viola R, Marzano N, Vianello R. An unusual epidemic of Staphylococcus-negative infections involving anterior cruciate ligament reconstruction with salvage of the graft and function. Arthrosc J Arthrosc Relat Surg Off Publ Arthrosc Assoc N Am Int Arthrosc Assoc. 2000;16:173–7.
43. Boström Windhamre H, Mikkelsen C, Forssblad M, Willberg L. Postoperative septic arthritis after anterior cruciate ligament reconstruction: does it affect the outcome? A retrospective controlled study. Arthrosc J Arthrosc Relat Surg Off Publ Arthrosc Assoc N Am Int Arthrosc Assoc. 2014;30:1100–9. https://doi.org/10.1016/j.arthro.2014.03.019.
44. Ardern CL, Taylor NF, Feller JA, Webster KE. Fifty-five per cent return to competitive sport following anterior cruciate ligament reconstruction surgery: an updated systematic review and meta-analysis including aspects of physical functioning and contextual factors. Br J Sports Med. 2014;48:1543–52. https://doi.org/10.1136/bjsports-2013-093398.

GPSR Compliance

The European Union's (EU) General Product Safety Regulation (GPSR) is a set of rules that requires consumer products to be safe and our obligations to ensure this.

If you have any concerns about our products, you can contact us on ProductSafety@springernature.com

In case Publisher is established outside the EU, the EU authorized representative is:

Springer Nature Customer Service Center GmbH
Europaplatz 3
69115 Heidelberg, Germany

Batch number: 10731383

Printed by Printforce, the Netherlands